Phosphor Handbook

Phosphor Handbook

Fundamentals of Luminescence

Edited by

Ru-Shi Liu
Xiao-Jun Wang

Third Edition

CRC Press
Taylor & Francis Group
Boca Raton London New York

CRC Press is an imprint of the
Taylor & Francis Group, an **informa** business

Third edition published 2022
by CRC Press
6000 Broken Sound Parkway NW, Suite 300, Boca Raton, FL 33487-2742

and by CRC Press
2 Park Square, Milton Park, Abingdon, Oxon, OX14 4RN

© 2022 Taylor & Francis Group, LLC

First edition published by CRC Press 1998

CRC Press is an imprint of Taylor & Francis Group, LLC

ISBN: 978-0-367-55512-2 (hbk)
ISBN: 978-1-032-15962-1 (pbk)
ISBN: 978-1-003-09869-0 (ebk)

DOI: 10.1201/9781003098690

Typeset in Times LT Std
by KnowledgeWorks Global Ltd.

In Memoriam the Early Editors of the Handbook.

Shigeo Shionoya
Formerly of The
   University of Tokyo
Tokyo, Japan

Hajime Yamamoto
Formerly of Tokyo
   University of Technology
Tokyo, Japan

William M. Yen
Formerly of The University
   of Georgia
Athens, GA, USA

Contents

Foreword to the Third Edition of the *Phosphor Handbook*

The field of luminescence and phosphors has a long history, starting from early observations of light in the dark from afterglow materials. Centuries of extensive research followed aimed at providing insight into optical phenomena, now resulting in an increasing role of phosphors in our daily lives. Applications of luminescence grow more diverse and include, for example, phosphors in the color displays that our eyes seem to be glued to, energy-efficient LED lighting, data communication, luminescent probes in medical imaging and sensing, gadgets relying on afterglow phosphors and even luminescent lanthanides in our banknotes. It is interesting to note the central role that Asia has played in the discovery and development of new luminescent materials. Early applications involved afterglow paints in China, creating alternative images in the dark. While fundamental luminescence research was carried out in the 20th century at all continents, there has been a remarkably strong role of Japan and China in research, development and discovery of new luminescence processes and phosphors. It is, therefore, not surprising that the first edition of the *Phosphor Handbook* (*Keikotai Handobukku*) was initiated by the Phosphor Research Society in Japan in the 1980s.

The first *Phosphor Handbook* was a great book but with an impact limited to those speaking Japanese. Fortunately, about ten years later, the book was translated into English and edited by two giants in the field of luminescence: Shigeo Shionoya and William Yen. It is this version of the book that I acquired soon after it was released, and it has been a source of information ever since. All aspects of luminescent materials were covered: phosphor synthesis, optical measuring techniques, fundamentals of luminescence processes, operation principles of light emitting devices, light and color perception and of course an almost complete overview of all luminescent materials known, indexed by host material and activator ion. I cannot count how often I consulted this book, to quickly look up the optical properties of an ion-host combination, find a suitable material with specific luminescence characteristics, understand the operation principles of phosphors in various applications and learn about careful measurements and analysis of phosphor properties. The authors, except for one, were Japanese, underpinning the central role of Japan in phosphor research.

As the field of luminescence continued to evolve and expand, it became clear that a second edition of the *Phosphor Handbook* was needed. Sadly Shigeo Shionoya has passed away, and in 2006, William Yen together with Hajime Yamamoto edited the second edition of the *Phosphor Handbook*. The new edition was updated mostly by asking the original authors to adapt the various chapters to include recent developments. The *Phosphor Handbook* continued to play a prominent role in the luminescence community as a source of information on any topic related to phosphors. Almost 15 years later, it was again time to adapt the Handbook to cover important new developments in the rapidly changing phosphor field where new applications and new materials emerge, and also measuring techniques have changed with the introduction of, for example, cheap (pulsed) diode lasers, fiber optics and compact CCD-based spectrometers. Our great colleagues William Yen and Hajime Yamamoto are unfortunately no longer with us and also many of the authors of the various chapters of the first and second edition of the *Phosphor Handbook* have passed away. This made it far from trivial to realize a third edition. We can be extremely grateful that Ru-Shi Liu and Xiao-Jun Wang have taken the initiative to edit and write this third edition of the *Phosphor Handbook*. It is very appropriate that the book is dedicated to the three founders, Shigeo Shionoya, William Yen and Hajime Yamamoto. At the same time, it is appropriate to sincerely thank Ru-Shi Liu and Xiao-Jun Wang for their strong commitment and time invested to organize, write and edit this third edition.

The third edition of the *Phosphor Handbook* is in some aspects different from the two previous editions. The authors are not the same, and it is wonderful to see that so many highly respected colleagues in the field have taken the time to contribute their expertise and knowledge to this third edition. Interestingly, again almost all of the authors of this third edition are Asian (with well over 100 contributing authors, you can count the non-Asian authors on the fingers of one hand). This illustrates the continued strong position of Asia in phosphor research. Just as in the previous editions, all aspects and the broad scope of phosphor research are covered, which makes this Handbook a worthy successor of the previous editions. It will serve as a comprehensive resource describing a wide variety of topics that were also included in the previous editions. It will educate newcomers and help everyone in the field to quickly access all relevant knowledge in the exciting field of phosphor research. In addition to the "classic" topics that continue to be relevant (but sometimes forgotten), many new topics are included, in Theory (e.g., first principle calculations), Materials (e.g., recent developments in quantum dots and upconversion nanocrystals) and Applications (e.g., LED phosphors for NIR sensing and agriculture). All this information no longer fits in a single volume, and this third edition is, therefore, divided into three volumes.

At the time of writing this foreword, I have not read the new edition of the Handbook but did receive an overview of all the chapters and contributing authors. Based on this information, it is clear that the full phosphor community, from students to professors, can benefit from this new comprehensive source of everything you always wanted to know about phosphors – and more. The third edition of the *Phosphor Handbook* will be a classic and continue to promote progress and development of phosphors, in the spirit of the first edition. I look forward to reading it and hope that you as a reader will enjoy exploring this great book and be inspired by it in your research on luminescent materials.

Andries Meijerink
Utrecht, June 2021

Preface to the Third Edition

The last version of the *Phosphor Handbook* was well received by the phosphor research community since its publication in 2007. However, in last 14 years, many notable advances have occurred. The success of the blue LED (Nobel Prize in Physics, 2014) and its phosphor-converted solid illumination greatly advanced the traditional phosphor research. New phosphorescent materials such as quantum dots, nanoparticles and efficient upconversion, quantum cutting phosphors and infrared broadband emission phosphors have been quickly developed to find themselves in ever-broader applications, from phototherapy to bioimaging, optics in agriculture to solar cell coating. These applications have all expanded beyond the traditional use in lighting and display. All of these developments should be included in the popular Handbook, making it necessary to publish a new version that reflects the most recent developments in phosphor research. Unfortunately, all the three well-respected editors of the previous version have passed away. As their former students and colleagues, we, the editors, feel a strong sense of responsibility to carry on the legacy of the Handbook and to update accordingly to continue serving the phosphor community. The aim of the third edition of the Handbook is to continue to provide an initial and comprehensive source of knowledge for researchers interested in synthesis, characterization, properties and applications of phosphor materials.

The third edition of the Handbook consists of three separate volumes. Volume 1 covers the theoretical background and fundamental properties of luminescence as applied to solid-state phosphor materials. New sections include the rapid developments in principal phosphors in nitrides, perovskite and silicon carbide. Volume 2 provides the descriptions of synthesis and optical properties of phosphors used in different applications, including the novel phosphors for some newly developed applications. New sections include Chapters 5 – Smart Phosphors, 6 – Quantum Dots for Display Applications, 7 – Colloidal Quantum Dots and Their Applications, 8 – Lanthanide-Doped Upconversion Nanoparticles for Super-Resolution Imaging, 9 – Upconversion Nanophosphors for Photonic Application, 16 – Single-Crystal Phosphors, 19 – Phosphors-Converting LED for Agriculture, 20 – AC-Driven LED Phosphors and 21 – Phosphors for Solar Cells. Volume 3 addresses the experimental methods for phosphor evaluation and characterization and the contents are widely expanded from the Second Edition, including the theoretical and experimental designs for new phosphors as well as the phosphor analysis through high pressure and synchrotron studies. Almost all the chapters in the third edition, except for some sections in the Fundamentals of Luminescence, have been prepared by the new faces who are actively and productively working in phosphor research and applications.

We commemorate the memory of the three mentors and editors of the previous editions – Professors Shigeo Shionoya, Hajime Yamamoto and William M. Yen. It was their efforts that completed the original Handbook that guided and inspired numerous graduate students and researchers in phosphor studies and applications. We wish to dedicate this new edition to them.

As the editors, we sincerely appreciate all the contributors from across the world who overcame various difficulties through such an unprecedented pandemic year to finish their chapters on time.

We are grateful to Professor Andries Meijerink of Utrecht University for writing the foreword to the Handbook. We also highly appreciate the help from Nora Konopka, Prachi Mishra, and Jennifer Stair of CRC Press/Taylor & Francis Group and perfect editing work done by Garima Poddar of KGL. Finally, we hope that this third edition continues the legacy of the Handbook to serve as a robust reference for current and future researchers in this field.

Co-editors:
Ru-Shi Liu
Taipei, Taiwan

Xiao-Jun Wang
Statesboro, GA, USA
May, 2021

Preface to the Second Edition

We, the editors as well as the contributors, have been gratefully pleased by the reception accorded to the *Phosphor Handbook* by the technical community since its publication in 1998. This has resulted in the decision to reissue an updated version of the Handbook. As we had predicted, the development and the deployment of phosphor materials in an ever increasing range of applications in lighting and display have continued its explosive growth in the past decade. It is our hope that an updated version of the Handbook will continue to serve as the initial and preferred reference source for all those interested in the properties and applications of phosphor materials.

For this new edition, we have asked all the authors we could contact to provide corrections and updates to their original contributions. The majority of them responded, and their revisions have been properly incorporated in the present volume. It is fortunate that the great majority of the material appearing in the first edition, particularly those sections summarizing the fundamentals of luminescence and describing the principal classes of lightemitting solids, maintains its currency and, hence, its utility as a reference source.

Several notable advances have occurred in the past decade, which necessitated their inclusion in the second edition. For example, the wide dissemination of nitride-based LEDs opens the possibility of white light solid-state lighting sources that have economic advantages. New phosphors showing the property of "quantum cutting" have been intensively investigated in the past decade and the properties of nanophosphors have also attracted considerable attention. We have made an effort, in this new edition, to incorporate tutorial reviews in all of these emerging areas of phosphor development.

As noted in the preface of the first edition, the Handbook traces its origin to one first compiled by the Phosphor Research Society (Japan). The society membership supported the idea of translating the contents and provided considerable assistance in bringing the first edition to fruition. We continue to enjoy the cooperation of the Phosphor Research Society and value the advice and counsel of the membership in seeking improvements in this second edition.

We have been, however, permanently saddened by the demise of one of the principals of the society and the driving force behind the Handbook itself. Professor Shigeo Shionoya was a teacher, a mentor and a valued colleague who will be sorely missed. We wish to dedicate this edition to his memory as a small and inadequate expression of our joint appreciation.

We also wish to express our thanks and appreciation of the editorial work carried out flawlessly by Helena Redshaw of Taylor & Francis.

William M. Yen
Athens, GA, USA

Hajime Yamamoto
Tokyo, Japan
December, 2006

Preface to the First Edition

This volume is the English version of a revised edition of the *Phosphor Handbook* (*Keikotai Handobukku*) that was first published in Japanese in December, 1987. The original Handbook was organized and edited under the auspices of the Phosphor Research Society (in Japan) and issued to celebrate the 200th Scientific Meeting of the Society which occurred in April, 1984.

The Phosphor Research Society is an organization of scientists and engineers engaged in the research and development of phosphors in Japan which was established in 1941. For more than half a century, the Society has promoted interaction between those interested in phosphor research and has served as a forum for discussion of the most recent developments. The Society sponsors five annual meetings; in each meeting, four or five papers are presented, reflecting new cutting-edge developments in phosphor research in Japan and elsewhere. A technical digest with extended abstracts of the presentations is distributed during these meetings and serve as a record of the proceedings of these meetings.

This Handbook is designed to serve as a general reference for all those who might have an interest in the properties and/or applications of phosphors. This volume begins with a concise summary of the fundamentals of luminescence and then summarizes the principal classes of phosphors and their light emitting properties. Detailed descriptions of the procedures for synthesis and manufacture of practical phosphors appear in later chapters and in the manner in which these materials are used in technical applications. The majority of the authors of the various chapters are important members of the Phosphor Research Society, and they have all made significant contributions to the advancement of the phosphor field. Many of the contributors have played central roles in the evolution and remarkable development of lighting and display industries of Japan. The contributors to the original Japanese version of the Handbook have provided English translations of their articles; in addition, they have all updated their contributions by including the newest developments in their respective fields. A number of new sections have been added in this volume to reflect the most recent advances in phosphor technology.

As we approach the new millennium and the dawning of a radical new era of display and information exchange, we believe that the need for more efficient and targeted phosphors will continue to increase and that these materials will continue to play a central role in technological developments. We, the co-editors, are pleased to have engaged in this effort. It is our earnest hope that this Handbook becomes a useful tool to all scientists and engineers engaged in research in phosphors and related fields and that the community will use this volume as a daily and routine reference, so that the aims of the Phosphor Research Society in promoting progress and development in phosphors are fully attained.

Co-Editors:
Shigeo Shionoya
Tokyo, Japan

William M. Yen
Athens, GA, USA
May, 1998

About the Author

Ru-Shi Liu

Professor Ru-Shi Liu received his bachelor's degree in chemistry from Soochow University (Taiwan) in 1981. He got his master's degree in nuclear science from the National Tsing Hua University (Taiwan) in 1983. He obtained two PhD degrees in chemistry from National Tsing Hua University in 1990 and from the University of Cambridge in 1992. He joined Materials Research Laboratories at Industrial Technology Research Institute as an Associate Researcher, Research Scientist, Senior Research Scientist and Research Manager from 1983 to 1995. Then he became an Associate Professor at the Department of Chemistry of the National Taiwan University from 1995 to 1999. Then he was promoted to a Professor in 1999. In July 2016, he became the Distinguished Professor.

He got the Excellent Young Person Prize in 1989, Excellent Inventor Award (Argentine Medal) in 1995 and Excellent Young Chemist Award in 1998. He got the 9th Y. Z. Hsu scientific paper award due to the excellent energy-saving research in 2011. He received the Ministry of Science and Technology awards for distinguished research in 2013 and 2018. In 2015, he received the distinguished award for Novel and Synthesis by IUPAC and NMS. In 2017, he got the Chung-Shang Academic paper award. He got "Highly Cited Researchers" by Clarivate Analytics from 2018 to 2021. He got Hou Chin-Tui Award in 2018 due to the excellent research on basic science. He got the 17th Y. Z. Hsu Chair Professor award for the contribution to the excellent research on "Green Science & Technology" in 2019. He then got the 26th TECO award for the contribution to make the combination of the academic and practical application of materials chemistry in 2019. He got the Academic Award of the Ministry of Education and the Academic Achievement Award of the Chemical Society Located in Taipei in 2020.

His research is concerning with materials chemistry. He is the author and co-author of more than 600 publications in international scientific journals. He has also granted more than 200 patents.

Xiao-Jun Wang

Professor Xiao-Jun Wang obtained his BS degree in physics from the Jilin University in 1982, his MS degrees in physics from the Chinese Academy of Sciences in 1985 and the Florida Institute of Technology in 1987 and his PhD in physics from The University of Georgia in 1992 (supervisors: William M. Yen and William M. Dennis). He served as a Research Associate from 1992 to 1993 at the University Laser Center of Oklahoma State University and then received a fellowship from the National Institutes of Health (NIH) as a Postdoctoral Fellow from 1993 to 1995 at the Beckman Laser Institute of the University of California, Irvine. In 1995, he received an NIH training grant and joined Georgia Southern University as Assistant Professor. He was promoted to Full Professor In 2004 and continues to teach there today.

List of Contributors

Pieter Dorenbos
Delft University of Technology
Delft, The Netherlands

Wenna Du
National Center for Nanoscience and Technology
Beijing, China

Sumiaki Ibuki
Formerly of Mitsubishi Electric Corp.
Amagasaki, Japan

Fengyi Jiang
Nanchang University
Nanchang, China

Ming Li
Institute of Semiconductors, CAS
Beijing, China

Song Liang
Institute of Semiconductors, CAS
Beijing, China

Feng Liu
Northeast Normal University
Changchun, China

Weizhen Liu
Northeast Normal University
Changchun, China

Xinfeng Liu
National Center for Nanoscience and Technology
Beijing, China

Yichun Liu
Northeast Normal University
Changchun, China

Hao Long
Xiamen University
Xiamen, China

Dan Lu
Institute of Semiconductors, CAS
Beijing, China

Youming Lu
Shenzhen University
Shenzhen, China

Eiichiro Nakazawa
Formerly of Kogakuin University
Tokyo, Japan

Shigetoshi Nara
Formerly of Hiroshima University
Hiroshima, Japan

Yongqiang Ning
Changchun Institute of Optics, Fine Mechanics
    and Physics, CAS
Changchun, China

Junbiao Peng
South China University of Technology
Guangzhou, China

Yan Peng
Shandong University
Jinan, China

Zhijue Quan
Nanchang University
Nanchang, China

Langping Tu
Changchun Institute of Optics, Fine
    Mechanics and Physics, CAS
Changchun, China

Lijun Wang
Changchun Institute of Optics, Fine
    Mechanics and Physics, CAS
Changchun, China

Xiao-Jun Wang
Georgia Southern University
Statesboro, GA, USA

Xinqiang Wang
Peking University
Beijing, China

Hao Wu
Changchun Institute of Optics,
   Fine Mechanics and
   Physics, CAS
Changchun, China

Haiyang Xu
Northeast Normal University
Changchun, China

Xiangang Xu
Shandong University
Jinan, China

Hajime Yamamoto
Formerly of Tokyo University
   of Technology
Tokyo, Japan

Yugang Zeng
Changchun Institute of Optics,
   Fine Mechanics and
   Physics, CAS
Changchun, China

Baoping Zhang
Xiamen University
Xiamen, China

Cen Zhang
Northeast Normal University
Changchun, China

Hong Zhang
University of Amsterdam
Amsterdam, The Netherlands

Jiahua Zhang
Changchun Institute of Optics, Fine Mechanics
   and Physics, CAS
Changchun, China

Jianli Zhang
Nanchang University
Nanchang, China

Lingjuan Zhao
Institute of Semiconductors, CAS
Beijing, China

1 General Introduction and Physics Background

Feng Liu and Xiao-Jun Wang

CONTENTS

Volume 1 of the handbook serves as a background for Volumes 2 and 3, but it is also an independent volume for readers to gain fundamental knowledge on phosphor luminescence. In this volume, Chapter 2 presents a general background of luminescence; Chapter 3 introduces principal phosphor materials and their optical properties, including II-VI and III-V semiconductors, nitrides, silicon carbide, perovskite, superlattice and quantum structures; Chapter 4 elucidates the different energy transfer processes that often occur in phosphorescence; Chapter 5 discusses the upconversion processes in nanophosphors, another popular phenomena with increasing applications; and Chapter 6 introduces organic and polymer phosphors. Before reading the following chapters in detail, we briefly introduce the luminescence phenomenon and its general physical background using rare-earth or transition metal–doped phosphor as an example.

1.1 WHAT IS PHOSPHOR?

Generally, a phosphor refers to a luminescent material that can convert certain types of incident energy into light emission or luminescence without heating. A wide range of energy sources can excite the phosphors, and the excitation diversity provides a various classification scheme for luminescence phenomena, including photoluminescence, cathodoluminescence, electroluminescence, chemiluminescence and mechanoluminescence. A typical phosphor system consists of a host material and relatively small amounts of luminescent ions, often called activators. The host lattice is often optically inert and constitutes the bulk of the phosphor. An activator is a foreign ion, which may absorb excitation energy and convert it into light emission in the ultraviolet, visible or infrared range. For example, considering the popular phosphor $Y_3Al_5O_{12}:Ce^{3+}$, the host lattice is $Y_3Al_5O_{12}$,

DOI: 10.1201/9781003098690-1

the activator the Ce^{3+} ion. We generally consider the case of a low concentration of activators in host lattice. The concentration of activators is typically on the order of 1 at% and is randomly distributed over the host lattice. Thus, the activators are considered to be far apart from each other. In this sense, when all the activators are identical, we only need to analyze the properties of one representative activator and apply appropriate statistical considerations for all other activators.

In some phosphor systems, it may occur that an activator does not have a significant absorption for a certain excitation energy. In such case, it may be possible to use another dopant ion, which absorbs excitation energy and subsequently transfers the excitation energy to the neighboring activators that yield luminescence. In this case, the absorbing ion is called sensitizer. For the case of activators in high concentrations (e.g., 100%), the adjacent activators can interact each other in phosphors. This illustrates the rather complicated nature of phosphors. A wide variety of activators can give rise to luminescence in appropriate host materials, including rare-earth ions, transition-metal ions, excitons, donor-acceptor pairs and d^{10} or s^2 ions.

1.2 ENERGY LEVELS OF PHOSPHORS

Before introducing the optical spectroscopy of individual activators in phosphors, we first analyze the phosphor using band theory. Instead of having discrete energies as in the case of free atoms, the available energy states form energy bands in solids, in which each atom is subject to a periodic array of potential wells. The solution of the Schrödinger equation for electrons reveals that the allowed energies for the electrons lie only in "allowed zones", each separated by an energy gap.

An important parameter in the band theory is the Fermi level, which is the highest energy occupied electron orbital at absolute zero temperature. For semiconductors and insulators, those energy levels just below the Fermi level are completely full, called valence band. Energy levels just above the Fermi level are completely empty and are called conduction band. Optical absorption may result when the valence electrons are given enough energy to surmount the energy gap E_g to reach the conduction bands. Generally, optical absorption only takes place for frequencies above E_g/h (h is the Planck's constant).

However, whenever lattice defects occur, or if there are impurities within the lattice, there is a breakdown in the periodicity of the crystalline structure and it becomes possible for electrons to possess energies in the energy gap. That is, the existence of structure imperfection in the material or the incorporation of impurities can give rise to lattice distortions, resulting in localized energy levels within the energy gap. The introduced energy levels may be discrete or distributed, depending on the nature of defect and host lattice.

The basic theory used to treat the spectroscopy of phosphor has been developed in the framework of a semiempirical approach, in which the attempt is made to identify the effective interactions operating in the electronic system of the activator that can reproduce the observed spectral results.

For an activator in phosphor, the Hamiltonian can be written

$$H = H_{FI} + H_{CF}, \tag{1.1}$$

where the free-ion Hamiltonian, H_{FI}, describes the activator as if it is isolated from the remainder of the solid. The eigenstates for the H_{FI} are $^{2s+1}L_J$. The crystal-field Hamiltonian, H_{CF}, describes how the activator is affected by the average static environment. For a dopant ion, H_{CF} is its crystal field energy.

1.2.1 EFFECT OF CRYSTALLINE ENVIRONMENT IS NOT REMARKABLE

To discuss the energy levels, a nice example is the $4f^N$ ($N=1$–13) configuration of rare-earth ions in phosphors. For the $4f^N$ configuration, H_{CF} is weak because of the shielding of the $5s^2$ and $5p^6$ outer shells of electrons. So we first consider the free-ion Hamiltonian H_{FI}, and then take the crystal-field Hamiltonian H_{CF} term into account afterwards by perturbation theory.[1,2]

The free-ion Hamiltonian of the $4f^N$ configuration state can be written as

$$H_{FI} = H_0 + H_C + H_{SO} + H_{others}, \qquad (1.2)$$

where H_0 describes the interaction of ion core with an optically active electron. This term gives the energy difference between the ground-state and the $4f^N$ configuration barycenter.

H_C describes the Coulomb interaction between the $4f$ electrons. This interaction splits the $4f^N$ configuration state into a number of states with different energies; each of which is characterized by values of total spin angular momentum S and orbital angular momentum L. H_C has the form:

$$H_C = \sum_{k=2,4,6} F^k \cdot f_k, \qquad (1.3)$$

where F^k is the Slater radial integral and f_k the angular part of the electrostatic interaction.

H_{SO} describes the spin-orbit coupling, which is the interaction of an electron's spin magnetic moment with its orbital magnetic moment. In the scheme of Russell-Saunders approximation, H_{SO} can be written as:

$$H_{SO} = \sum_i \xi(r_i) s_i \cdot l_i, \qquad (1.4)$$

where r_i is the radial coordinate, s_i the spin, l_i the orbital angular momentum of the ith $4f$ electron and $\xi(r_i)$ the spin-orbit coupling parameter. H_{SO} splits up each LS term into a number of states, each of which is characterized by the quantum numbers SLJ (J is the total angular momentum, ranging from $L-S$ to $L+S$). Each of the LSJ states is $(2J+1)$-fold degenerate, and the separation between the SLJ multiplets should be in accordance with the Lande interval rule. In practice, H_{SO} mixes up states of the same J but different LS terms. The intermediate coupling eigenstates can be written in terms of the Russel-Saunders basis states ($|\alpha SLJ\rangle$):

$$|[\alpha'S'L']J\rangle = \sum_{\alpha SL} c_{(\alpha SL)} \cdot |\alpha SLJ\rangle, \qquad (1.5)$$

The coefficient $c_{(\alpha SL)}$ is generally obtained by diagonalizing the $H_C + H_{SO}$ matrix. The label $[\alpha'S'L']$ corresponds to the largest contributing Russell-Saunders state. An energy level schematical diagram may bring us a more intuitive sense, as presented in Figure 1.1. The splitting of the level of the $4f^2$ configuration into LS occurs under the action of H_C and further into SLJ multiplets under the action of H_{SO}.

H_{others} describes the other interactions with higher configurations, including two-electron and three-electron Coulomb correlation contributions, as well as higher order spin-dependent effects. The effect of H_{others} is usually small but important for some accurate calculations.

As a consequence, the free-ion Hamiltonian of the $4f^N$ configuration state may be experimentally fitted. The computed free-ion levels of trivalent rare-earth ions have been summarized by Carnall et al.,[3] as listed in Table 1.1.

Next, the crystal field energy is taken into account. For a free ion, each $4f$ level is $(2J+1)$-fold degenerate. When the rare-earth ion is placed in a lattice, each $4f$ level further splits under the influence of the electric field produced by the environment (i.e., crystal field). That is, the crystal-field Hamiltonian H_{CF} represents the nonspherical part of the interaction with the lattice, causing the splitting of free-ion levels. Some of the crystal-field levels may still degenerate, depending on the

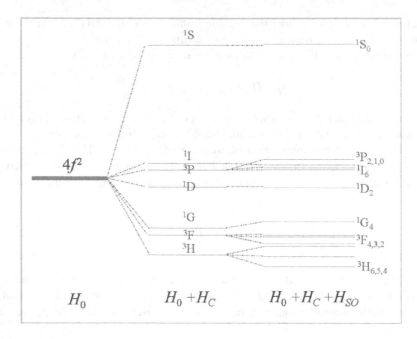

FIGURE 1.1 The Coulomb interaction H_C splits the $4f^2$ configuration into LS terms, while H_{SO} splits the SL terms into SLJ multiplets.

symmetry of the crystal field. For a particular site in lattice, the one-electron H_{CF} is parameterized by a number of parameters, B_q^k. The H_{CF} is expressed as

$$H_{CF} = \sum_{kq} B_q^k \cdot C_q^{(k)}, \tag{1.6}$$

TABLE 1.1
The Computed Free-Ion Levels (Partial SLJ States) of Trivalent Rare-Earth Ions[3]

Pr^{3+}		Nd^{3+}		Pm^{3+}		Sm^{3+}		Eu^{3+}		Gd^{3+}		Tb^{3+}		Dy^{3+}		Ho^{3+}		Er^{3+}		Tm^{3+}	
SLJ	E (cm^{-1})	SLJ	E (cm^{-1})	SLJ	E (cm^{-1})	SLJ	E (cm^{-1})	SLJ	E (cm^{-1})	SLJ	E (cm^{-1})	SLJ	E (cm^{-1})	SLJ	E (cm^{-1})	SLJ	E (cm^{-1})	SLJ	E (cm^{-1})	SLJ	E (cm^{-1})
$^1S_0$	46986	$^2P_{3/2}$	26424	$^3L_9$	25907	$^6P_{7/2}$	26786	$^5L_{10}$	28427	$^6D_{5/2}$	41037	$^5K_7$	41817	$^6P_{5/2}$	27624	$^5F_4$	38515	$^2P_{1/2}$	33319	$^1S_0$	75300
$^3P_2$	22690	$^2D_{5/2}$	24004	$^3D_3$	25895	$^4L_{17/2}$	26762	$^5L_9$	27960	$^6F_{3/2}$	40901	$^5G_6$	41473	$^6P_{3/2}$	27543	$^5I_7$	38339	$^4G_{7/2}$	28224	$^3P_2$	38268
$^1I_6$	21743	$^2P_{1/2}$	23458	$^3P_0$	25811	$^4D_{1/2}$	26446	$^5D_4$	27586	$^6D_{1/2}$	40654	$^5K_5$	41458	$^4M_{19/2}$	26334	$^3L_7$	37900	$^2K_{15/2}$	27922	$^3P_1$	36615
$^3P_1$	21555	$^2K_{15/2}$	21780	$^3L_8$	24907	$^4G_{11/2}$	25829	$^5L_8$	27244	$^6D_{7/2}$	39723	$^5K_8$	40939	$^4K_{17/2}$	25890	$^5P_2$	37794	$^4G_{9/2}$	27637	$^3P_0$	35579
$^3P_0$	20935	$^4G_{11/2}$	21714	$^3G_4$	24840	$^4L_{15/2}$	25667	$^5G_6$	26763	$^6I_{13/2}$	36742	$^5G_6$	40309	$^4F_{7/2}$	25856	$^3H_5$	36720	$^4G_{11/2}$	26631	$^1I_6$	34975
$^1D_2$	17052	$^2D_{3/2}$	21425	$^3H_6$	24702	$^4M_{21/2}$	25434	$^5G_5$	26752	$^6I_{15/2}$	36704	$^5D_2$	39515	$^4I_{13/2}$	25794	$^5L_8$	36314	$^4F_{9/2}$	24756	$^1D_2$	28001
$^1G_4$	10012	$^2K_{13/2}$	19785	$^3P_1$	24216	$^4K_{11/2}$	25201	$^5G_4$	26735	$^6I_{11/2}$	36582	$^5K_9$	39297	$^4M_{21/2}$	25109	$^3F_2$	36294	$^4F_{3/2}$	22712	$^1G_4$	21314
$^3F_4$	7009	$^4G_{9/2}$	19709	$^3L_7$	23772	$^6P_{3/2}$	25064	$^5G_3$	26622	$^6I_{17/2}$	36345	$^5I_5$	38110	$^4G_{11/2}$	23563	$^5D_4$	36009	$^4F_{5/2}$	22376	$^3F_2$	15181
$^3F_3$	6595	$^4G_{7/2}$	19293	$^3D_2$	23140	$^4F_{7/2}$	24995	$^5G_2$	26392	$^6I_{9/2}$	36291	$^5I_4$	37732	$^4H_{15/2}$	22222	$^5P_0$	36008	$^4F_{7/2}$	20715	$^3F_3$	14597
$^3F_2$	5196	$^4G_{5/2}$	17428	$^3G_5$	22475	$^4L_{13/2}$	24676	$^5L_7$	26357	$^6I_{7/2}$	35919	$^5I_6$	37722	$^6F_{1/2}$	13814	$^3G_3$	35203	$^2H_{11/2}$	19337	$^3H_4$	12721
$^3H_6$	4495	$^2H_{11/2}$	16105	$^3G_3$	21935	$^4M_{19/2}$	24138	$^5L_6$	25325	$^6P_{3/2}$	33398	$^5F_1$	37606	$^6F_{3/2}$	13267	$^5G_4$	34812	$^4F_{9/2}$	15455	$^3H_5$	8391
$^3H_5$	2303	$^4F_{9/2}$	14899	$^5G_4$	20554	$^6P_{5/2}$	24132	$^5D_3$	24355	$^6P_{5/2}$	32827	$^5F_2$	37260	$^6F_{5/2}$	12471	$^3L_8$	34156	$^4I_{9/2}$	12597	$^3F_4$	5818
$^3H_4$	191	$^4S_{3/2}$	13691	$^3H_5$	20307	$^4I_{15/2}$	23048	$^5D_2$	21483	$^6P_{7/2}$	32232	$^5I_7$	36713	$^6F_{7/2}$	11070	$^5M_{10}$	34072	$^4I_{11/2}$	10346	$^3H_6$	175
						$^4G_{9/2}$	22873	$^5D_1$	19027	$^8S_{7/2}$	2	$^5F_3$	36674			$^3P_1$	33247				

TABLE 1.1 (Continued)
The Computed Free-Ion Levels (Partial *SLJ* States) of Trivalent Rare-Earth Ions[3]

$^4F_{7/2}$ 13619	$^3K_8$ 19862	$^4M_{17/2}$ 22612	$^5D_0$ 17293	$^5F_4$ 35498	$^6H_{5/2}$ 10273	$^5D_3$ 33063	$^4I_{13/2}$ 6712
$^2H_{9/2}$ 12768	$^5G_3$ 18565	$^4F_{5/2}$ 22301	$^7F_6$ 4907	$^5I_8$ 35255	$^6H_{7/2}$ 9223	$^5G_2$ 30799	$^4I_{15/2}$ 217
$^4F_{5/2}$ 12660	$^3K_7$ 18255	$^4I_{13/2}$ 21644	$^7F_5$ 3849	$^5H_3$ 35060	$^6F_{9/2}$ 9166	$^3F_4$ 29947	
$^4F_{3/2}$ 11660	$^3H_4$ 18075	$^4I_{11/2}$ 21147	$^7F_4$ 2823	$^5F_5$ 35058	$^6F_{11/2}$ 7853	$^3K_6$ 29941	
$^4I_{15/2}$ 6148	$^5G_2$ 18053	$^4M_{15/2}$ 20825	$^7F_3$ 1866	$^5H_4$ 34463	$^6H_{9/2}$ 7806	$^3L_9$ 28895	
$^4I_{13/2}$ 4098	$^3K_6$ 16939	$^4I_{9/2}$ 20660	$^7F_2$ 1026	$^5H_5$ 33891	$^6H_{11/2}$ 5952	$^5G_3$ 28875	
$^4I_{11/2}$ 2114	$^5F_5$ 16223	$^4G_{7/2}$ 20161	$^7F_1$ 372	$^5H_6$ 33015	$^6H_{13/2}$ 3626	$^5F_2$ 28234	
$^4I_{9/2}$ 235	$^5F_4$ 14887	$^4F_{3/2}$ 18982	$^7F_0$ 0	$^5H_7$ 31503	$^6H_{15/2}$ 175	$^3H_6$ 27678	
	$^3S_2$ 14486	$^4G_{5/2}$ 18031		$^5D_0$ 31348		$^5G_5$ 27652	
	$^3F_3$ 13933	$^6F_{11/2}$ 10583		$^5D_1$ 30734		$^3K_7$ 26058	
	$^3F_2$ 13080	$^6F_{9/2}$ 9189		$^5L_6$ 29794		$^5G_4$ 25859	
	$^3F_1$ 12638	$^6F_{7/2}$ 8009		$^5G_2$ 29655		$^5G_5$ 23987	
	$^5I_8$ 6714	$^6F_{5/2}$ 7141		$^5L_7$ 29581		$^5F_1$ 22255	
	$^5I_7$ 4951	$^6F_{3/2}$ 6637		$^5L_8$ 29314		$^5G_6$ 22179	
	$^5I_6$ 3239	$^6H_{15/2}$ 6520		$^5G_3$ 29101		$^3K_8$ 21267	
	$^5I_5$ 1612	$^6F_{1/2}$ 6387		$^5L_9$ 28532		$^5F_2$ 21039	
	$^5I_4$ 120	$^6H_{13/2}$ 5072		$^5G_4$ 28411		$^5F_3$ 20594	
		$^6H_{11/2}$ 3667		$^5D_2$ 28231		$^5F_4$ 18538	
		$^6H_{9/2}$ 2341		$^5G_5$ 27891		$^5S_2$ 18381	
		$^6H_{7/2}$ 1135		$^5L_{10}$ 27095		$^5F_5$ 15456	
		$^6H_{5/2}$ 101		$^5G_6$ 26547		$^5I_4$ 13212	
				$^5D_3$ 26360		$^5I_5$ 11145	
				$^5D_4$ 20568		$^5I_6$ 8578	
				$^7F_0$ 5784		$^5I_7$ 5064	
				$^7F_1$ 5561		$^5I_8$ 9	
				$^7F_2$ 5106			
				$^7F_3$ 4418			
				$^7F_4$ 3439			
				$^7F_5$ 2172			
				$^7F_6$ 124			

where the B_q^k is treated as empirical parameters, and the $C_q^{(k)}$ is spherical tensor operator. For the $4f^N$ configuration, $k=2,4,6$ and the value q depend on the site symmetry of the activator in the lattice.

Figure 1.2 presents a substantial part of the $4f^N$ energy levels of trivalent rare-earth ions in $LaCl_3$. This figure is known as Dieke diagram, in which the $4f$ states are labeled in the form $^{2S+1}L_J$, and the magnitude of crystal-field splitting is indicated by the vertical extent of the bars. The Dieke diagram is applicable to identify the possible transition channels of the trivalent rare-earth ions in almost any phosphor system.

1.2.2 Effect of Crystalline Environment Is Remarkable

Another example for discussing the energy levels is the transition-metal ions from the $3d$ series, involving the configurations d^1 (Ti^{3+}, V^{4+}); d^2 (V^{3+}, Cr^{4+}, Mn^{5+}, Fe^{6+}); d^3 (V^{2+}, Cr^{3+}, Mn^{4+}); d^4 (Mn^{3+}); d^5 (Mn^{2+}, Fe^{3+}); d^7 (Co^{2+}); and d^8 (Ni^{2+}). For the $3d^N$ configuration, the crystal field effect is larger than that of rare-earth ions, since the $3d$ electrons are very sensitive to the crystalline environment. In addition, the spin-orbit couplings of transition metal ions are generally weak, and so the effect of spin-orbit coupling is a smaller than that of crystal field. Therefore, a strong field scheme is generally adopted, in which one first considers how the $3d$ orbitals are affected by the electrostatic field of the surrounding ions. We shall treat the transition metal ions by placing them in a surrounding environment of octahedral symmetry, which splits the fivefold degenerate d orbital into a twofold degenerate e_g state and a threefold degenerate t_{2g} state.

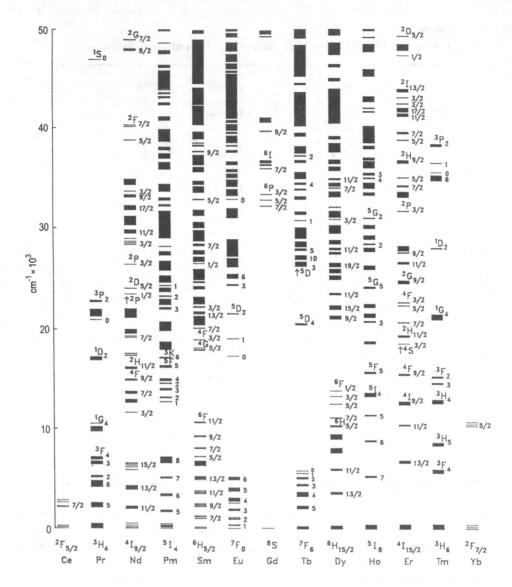

FIGURE 1.2   Energy-level diagram for the trivalent rare-earth ions in LaCl₃.

Starting with these one-electron e_g and t_{2g} orbitals, we subsequently consider the H_C. The new eigenstates are characterized by the values of S and Γ, where Γ is an irreducible representation of the octahedral group to which the eigenstate belongs. For the transition metal ions, spin-orbit interaction is weaker than that of the rare-earth ions; therefore, the H_{SO} has been neglected in the calculation of energy levels in the transition-metal ion-doped phosphors. However, the H_{SO} interaction may split up the $S\Gamma$ terms into multiplets, corresponding to the sharp lines in spectral measurements.

Tanabe and Sugano calculated the $3d^N$ configurations in an octahedral crystal field and summarized the results in the so-called Tanabe-Sugano diagrams.[4] The Coulomb interaction between the $3d$ electrons is characterized by three *Racah* parameters, A, B and C. Parameter A makes equal contribution to the energies of all $3d$ levels. Therefore, we only concern ourselves with B and C. It is expected that the ratio C/B will have almost the same value through the whole series of transition

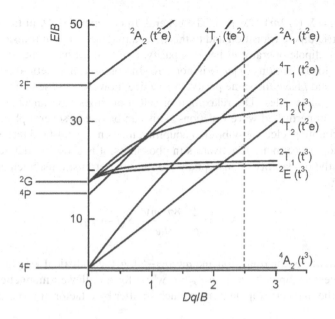

FIGURE 1.3 Energy level of Cr^{3+} ion ($3d^3$) in an octahedral crystal field.

metal ions in any host materials. Hence, for each ion, the energy separations between the various crystal-field levels depend on two parameters: B and the crystal-field energy D_q. The Tanabe-Sugano diagrams show how the level energy E (in unit of B) varies with the D_q. As an example, Figure 1.3 presents the Sugano-Tanabe diagram for Cr^{3+} ion ($3d^3$) in an octahedral crystal field. Free-ion levels are marked as ^{2S+1}L (left-hand side). In phosphors, the free-ion levels split into crystal-filed levels, which are labeled according the strong field configuration in the form $^{2S+1}\Gamma$ (right-hand side). These crystal-field levels may be nondegenerate (Γ=A), twofold degenerate (Γ=E) or threefold degenerate (Γ=T). The energies are measured relative to that of the lowest $S\Gamma$ term, $^4A_2$. The vertical broken line in Figure 1.3 is drawn at the value of D_q/B appropriate to a number of Cr^{3+} ion in ruby. Another example is Mn^{4+} ion ($3d^3$), which share a similar Tanabe-Sugano diagram with Cr^{3+} ion in an octahedral crystal field, since the two ions are isoelectronic. Both ions have the same free-ion levels, but the crystal-field at the quadrivalent Mn^{4+} is larger than that at Cr^{3+}, so that the absorption bands of Mn^{4+} always at shorter wavelengths.

1.3 OPTICAL TRANSITIONS IN PHOSPHORS

The subject of this section is transition intensity of activators in phosphors. A transition intensity from an initial state i to a final state f is proportional to square of the matrix element $\langle f|D|i \rangle$, where D is the general dipole operator, which represents the interaction of the electron with the electromagnetic field. Accordingly, the strength of the transition is given

$$S(i \rightarrow f) = \sum_{i,f} |\langle f|D|i \rangle|^2 , \qquad (1.7)$$

The summation is over the i and f manifolds of states. The transition mechanisms for activator generally relate to electric and magnetic multipoles of the electronic system. In many cases, the electric dipole or magnetic dipole terms are dominant. Higher multipole mechanisms (e.g., electric quadrupole) can also occur but they are negligibly small. The electric dipole moment is $p = \Sigma_i er_i$, where the summation is over all the optically active electrons. Similarly, the magnetic dipole moment is $m = \Sigma_i (e/2m) \cdot (2s_i + l_i)$. The general dipole operator, D, in Eq. (1.7) corresponds to the p for an electric dipole transition and to the m for a magnetic dipole transition.

Since the electric dipole operator er has odd parity, i and f must have opposite parity to ensure the matrix element for transitions $\langle f|er|i \rangle$ is nonzero. In contrast, magnetic dipole transitions are allowed when the i and f have the same parity. In practice, however, the selection rules can seldom be considered as absolute rules. The relaxation of selection rules is connected to wave function admixtures into the unperturbed wave functions. This can be due to several physical mechanisms, such as spin-orbit coupling, electron-vibration coupling, uneven crystal-field terms etc.

In the spectroscopic treatment of activators in phosphors, it is useful to define a dimensionless quantity to express the probability of absorption or emission in transitions. Such quantity is defined as oscillator strength

$$F(i \rightarrow f) = \frac{1}{g_i} \cdot \frac{8\pi^2 m\nu}{3he^2} \cdot S(i \rightarrow f), \qquad (1.8)$$

where ν is the frequency of the transition and $h\nu = |E_f - E_i|$. g_i the statistical weight of the initial state i. For an allowed electric dipole transition $F_{ED} \approx 1$, while for an allowed magnetic dipole transition $F_{MD} \approx 10^{-6}$. That is, the magnetic dipole term is much smaller by a factor of more than five orders of magnitude.

1.3.1 INTRA-CONFIGURATIONAL TRANSITIONS: WEAK INTERACTION WITH THE VIBRATING LATTICE

To simplify the discussion of optical transition, it is better to initially reduce or eliminate the effect of crystalline environment. In this sense, rare earth ion–activated phosphor is a good example, the spectra of which comprise sharp lines arising from purely electronic transitions, and the effect of the environment is felt mainly through their effects on the lifetimes of the transition intensities.

In the case of the rare earths, all states of a $4f^N$ configuration have the same parity, so the electric dipole transitions within the configuration are parity forbidden. Hence, there must be some mixing of the opposite parity $4f^{N-1}5d$ configuration into the $4f^N$ configuration. The odd components of crystal field are present when the rare-earth ion experiences a crystal field that lacks inversion symmetry. These odd components mix a small amount of opposite-parity wave functions (like $5d$) into the $4f$ wave functions. In this way, the intra-configurational $4f \rightarrow 4f$ transitions obtain some intensity.

First, we consider the transitions between free-ion states. The initial state i and final state f refer to the full set of states in the $[\alpha SL]J$ multiplet. In Eq. (1.7), $S_{ED}(i \rightarrow j)$ is the sum of squares of a number of matrix elements. Now, if an odd-parity crystal field term, $H_{CF}(odd)$, mixes states of opposite-parity composition, the matrix element in Eq. (1.7) has the form

$$S_{ED}(i \rightarrow f) = \sum_\beta \frac{\langle f|H_{CF}(odd)|\beta \rangle \langle \beta|D|i \rangle}{E_f - E_\beta} + \text{converse}, \qquad (1.9)$$

where β signifies all the opposite-parity states mixed in by the $H_{CF}(odd)$. We do not have sufficient knowledge about the β states to carry out an exact theoretical calculation of the above matrix element.

Judd and Ofelt introduced a simplification by ascribing the same energy denominators to all β states.[5] They independently derived expressions of $4f \rightarrow 4f$ oscillator. In this way, one can regard $\Sigma_\beta H_{CF}(odd)|\beta\rangle\langle\beta|D$ as an even parity operator. Accordingly, S_{ED} can be written

$$S_{ED}(i \rightarrow f) = \sum_{\lambda=2,4,6} \Omega_\lambda \left|\langle \alpha_f S_f L_f J_f \| U^\lambda \| \alpha_i S_i L_i J_i \rangle\right|^2, \qquad (1.10)$$

where Ω_λ are the Judd-Ofelt intensity parameters, evaluated from experimental data. U^λ is the unit tensor operator of rank λ and reflect the strength and nature of the odd-parity field. The reduced matrix elements $\langle \alpha_f S_f L_f \| U^\lambda \| \alpha_i S_i L_i \rangle$ have been evaluated and are available as a table.[3] Once the Judd-Ofelt parameters have been obtained for a phosphor, these parameters can be used to calculate absorption strengths and radiative decay rates between two $4f$ levels. In addition, selection rules derived from the Judd-Ofelt theory include $|J_f - J_i| = 2, 4, 6$ and $|S_f - S_i| = 0$. For example, $^1D_2 \rightarrow {}^3H_4$ transition of Pr^{3+} in phosphor is much stronger than $^1D_2 \rightarrow {}^3H_5$, and a mixing of singlet and triplet spin wave functions is required.

1.3.2 Intra-configurational Transitions: Strong Interaction with the Vibrating Lattice

In practice, unlike the case of rare earths, many phosphors exhibit luminescence spectra with broad spectral bands. Such broad-band spectra are generally attributed to the effects of the vibrating host lattice. That is, during the luminescence process, the environment of the activator is not static. The surrounding ions vibrate about some average positions, so that the crystal filed varies, affecting the optical transitions of activator. Transition-metal ions in phosphors are characterized by a strong interaction with the surrounding environment. Therefore, we consider here the $3d \rightarrow 3d$ transitions of transition metal ions in phosphors. Optical transitions between $3d^N$ states of transition-metal ions should be parity forbidden but become weakly allowed in phosphors by mixing of odd-parity compositions.

The effect of the surrounding environment on the luminescence of transition-metal ions may be illustrated by a configurational coordinate model, which is successful in explaining the shape of the broadband transitions. In the model, only one representative lattice mode is taken into account for simplicity. Accordingly, a scheme of "breathing mode" is employed, in which the activator is at rest and the surrounding ligands are moving away from the activator and coming back.

The potential energy diagram shown in Figure 1.4 forms the basis of the configurational coordinate model. The potential curve represents the electronic energy of the activator-plus-lattice state (E) as a function of the internuclear separation between it and its nearest neighbors (Q). Figure 1.4 shows a configurational coordinate diagram where E is plotted versus Q. The dependence of a potential energy on Q is parabolic

$$E = \frac{1}{2} k (Q - Q_0)^2,$$
(1.11)

where k is the force constant. The minimum Q_0 of the parabola corresponds to the equilibrium distance in the ground state. The quantum mechanical solution of this problem yields for the energy levels of the oscillator $E_n = (n + 1/2) h\nu$, where $n = 0, 1, 2, \ldots$ and ν is the frequency of the oscillator. The excited state parabola is drawn in such a way that the force constant is weaker than in the ground state. Since the excited state is usually weaker bound than the ground state, this is a representative situation. In practice, for simplicity, we assume that the ground-state parabola and the excited state parabola have the same force constant k (the same shape).

In Figure 1.4, the absorption transition starts from the lowest vibrational level, i.e., $n = 0$. The most probable transition occurs at Q_0 and ends on the edge of the excited state parabola. This transition corresponds to the maximum of the absorption band. If $\Delta Q = 0$, the two parabolas lie exactly above each other, and the band width of the transition vanishes. The larger the value of ΔQ, the broader the absorption band. That is, the width of an absorption band is an indication of how large the difference in ΔQ between the excited state and the ground state. The system can return from the lowest vibrational level of the excited state to the ground state, followed by emission. In the emission process, the activator reaches a high vibrational level of the ground state. Subsequently, relaxation to the lowest vibrational level of the ground state occurs.

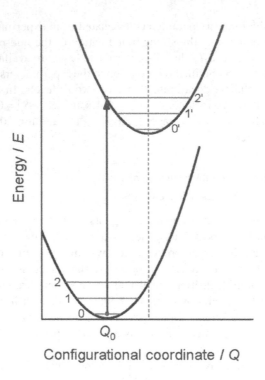

FIGURE 1.4 Configurational coordinate diagram, where the vibrational states $(0,1,2$ and $0',1',2')$ are shown in ground and excited states, respectively.

1.3.3 INTER-CONFIGURATIONAL TRANSITIONS

Besides the $4f$–$4f$ or $3d$–$3d$ intra-configurational transitions, we also consider the inter-configurational transitions, such as $5d$–$4f$ transitions of rare-earth ions in phosphors, to further learn the luminescence performances of phosphors. A $4f^{N-1}5d$ configuration is formed by exciting one of the $4f$ electrons into the $5d$ orbital of a rare-earth ion. Since the interaction between the $5d$ electron and the vibrating lattice is remarkable, the spectra of $5d$–$4f$ transitions always consist of broad spectral bands. In addition, the transition process is allowed and the transition wavelengths are more dependent on the environments than that of $4f$–$4f$ transitions. That is, the energy difference between the $5d$- and $4f$-electrons depends on both the centroid shift and the total crystal field splitting of the $4f5d$ band.

Compared to the free ion of rare earths, the lowest state of the excited $4f^{N-1}5d$ configuration is decreased in energy upon incorporation in a phosphor. This red-shift is due to the nephelauxetic effect, which decreases the barycenter of $4f^{N-1}5d$ configuration. The position of the lowest $4f^{N-1}5d$ energy level also relates to the total splitting of the $4f^{N-1}5d$ state. Such a splitting is determined by the strength of the crystal field. A low covalence of the activator-ligand bond and a weak crystal field may result in a high-energy position of the lowest $4f^{N-1}5d$ level of rare-earth ions in phosphors.

1.3.4 ELECTRON TRANSFER TRANSITIONS

The host materials for phosphors may be insulators or semiconductors. Excitation of electrons from the valence bands to the conduction bands results in the fundamental absorption of the host materials. After the excitation, the host materials may exhibit electron-transfer like emissions, corresponding to recombination luminescence either from various exciton structures (free or bound exciton), or from donor-acceptor recombination. The electronic systems relating to such emission transitions possess spatially extensive wave functions, in contrast to the localized f and d wave functions considered earlier.

Exciton recombination is usually due to bound exciton, i.e., an exciton of which the electron or the hole is trapped at an imperfection in the lattice. The formation of bound exciton may result either from the inter-band excitation or the activator-to-host excitation. The former excitation is of the charge transfer type, since the top of the valence band consists of levels with predominant anion character and the bottom of the conduction band of levels with a considerable amount of cation character. In the latter case, the exciton consists of a state in which the localized orbital of activator is strongly mixed with orbital of the surroundings. Direct transition from the ground state of activator to the conduction band of host is quite unlikely, but an autoionization is much more probable, which consists of a transition from the ground state to one excited state of the activator lying in the conduction band followed by the escape of the electron into the continuum. That is, the autoionization in phosphor can occur if the coupling between the localized excited states and the conduction-band continuum is strong.

There are other types of charge transfer transition in phosphors, including ligand-to-metal, metal-to-ligand and metal-to-metal charge transfers. This kind of transition generally consists of an activator which has a tendency to become oxidized or has a tendency to become reduced. An emission case is the donor-acceptor pair emission, in which an electron trapped at a donor and a hole trapped at an acceptor. The emission wavelength and probability depend on the electron-hole distance in a pair. Moreover, such emissions always consist of broad spectral bands due to strong electron-lattice coupling in phosphors.

Another case on the charge transfer transition relates to the trapping state of phosphors. In phosphors, the existence of structure imperfection or the incorporation of impurities can give rise to a breakdown in the periodicity of the crystalline structure. Such lattice distortion results in a trapping state within the energy gap. Thus, the charge carriers (conduction electrons or valence holes) may be trapped at the trapping levels. On the contrary, transition of the trapped charge carriers to the delocalized conduction or valence bands may occur upon the external stimuli.

1.4 NONRADIATIVE PROCESSES IN PHOSPHORS

When excitation light is incident on a phosphor, some of the excitation energy may be reemitted as light, while the rest may be dissipated into the crystal lattice. That is, during the excitation and emission, the convention of excitation energy generally accompanies with nonradiative processes. This section will discuss the nonradiative ways using the configurational coordination model, which is valuable in helping one to gain a better insight into a number of facts qualitatively, such as multi-phonon relaxation, thermal quenching of luminescence, quenching by electron transfer and energy transfer.

1.4.1 LUMINESCENCE EFFICIENCY

Since nonradiative processes always compete with the radiative processes in phosphors, the efficiency of luminescence will be reduced by the competition. Therefore, before considering the nonradiative processes of phosphors, we would like to give an introduction on the luminescence efficiency, which is defined as the ratio of the number of emitted quanta to the number of absorbed quanta. In practice, the luminescence efficiency is generally expressed in terms of either quantum efficiency (photons/photons) or radiant efficiency (watts/watts).

The effect of nonradiative process on the luminescence dynamics can be simplified illustrated in Figure 1.5. Upon excitation, the activator in a phosphor is raised to an excited state. After the end of the excitation, the time evolution of excited-state population can be described by a rate equation:

$$\frac{dN}{dt} = -\left(W_r + W_{nr}\right)N, \tag{1.12}$$

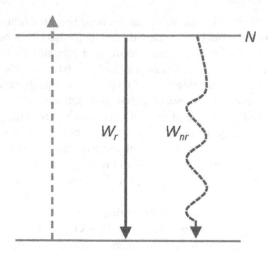

FIGURE 1.5   The simplified energy level scheme for a photoluminescence process of an activator in phosphor.

where N is the electron population of a localized excited state, W_r and W_{nr} the radiative and nonradiative transition probabilities, respectively. The solution of Eq. (1.12) is

$$N(t) = N(0)e^{-(W_r + W_{nr})t}, \tag{1.13}$$

which is also written as

$$N(t) = N(0)e^{-t/\tau}. \tag{1.14}$$

where $N(0)$ is the population at the end of excitation (t=0), τ the lifetime of the excited state. N decreases to $1/e$ (37%) after the time τ [$\tau = 1/(W_r + W_{nr})$]. According to the definitions of W_r and W_{nr}, the quantum efficiency η of the activator is given by

$$\eta = \frac{W_r}{W_r + W_{nr}}. \tag{1.15}$$

Thus, the luminescence lifetime, as well as the quantum efficiency of luminescence, is governed by both radiative and nonradiative processes.

1.4.2   MULTIPHONON RELAXATION

Besides describing the spectral broadening of phosphors caused by the effect of vibrating lattice, configurational coordinate model may also provide insight into the nonradiative processes. In this section, we consider a rare earth–doped phosphor system for illustrating the nonradiative process. Figure 1.6 shows the configurational coordinate diagram, in which the system returns from the lowest vibrational level of the $4f$ excited state to the $4f$ ground state, followed by emission.

The two parabolas in Figure 1.6 are parallel with an energy difference ΔE. When the ΔE is less than about five times the highest vibrational frequency of the host lattice, nonradiative return to the ground state dominates. Such a nonradiative process is called multiphonon relaxation. Note that the process is independent of the concentration of activators in phosphors, since it originates form a spontaneous emission of phonons.

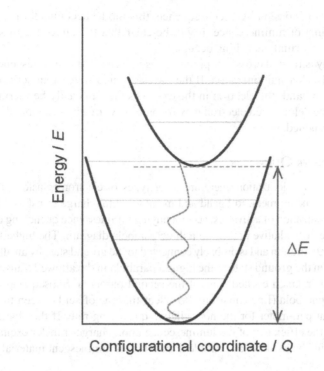

FIGURE 1.6 Configurational coordinate diagram to illustrate the multiphonon relaxation.

1.4.3 THERMAL QUENCHING OF LUMINESCENCE

In the configurational coordinate model, if the excited-state parabola cross the ground-state parabola, the relaxed-excited state may reach the intersection at high temperature (Figure 1.7). By crossing over the intersection of the excited-state parabola, via the intersection of the system, it is possible to return to the ground state in a nonradiative manner. The excitation energy is then lost to the lattice

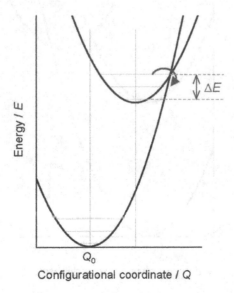

FIGURE 1.7 Configurational coordinate diagram to illustrate the thermal quenching of luminescence.

and does not contribute to emission. As a consequence, this model accounts for the thermally activated crossover quenching of luminescence. It will be clear that the larger the offset between the parabolas, the easier the thermal quenching occurs.

Besides the thermally activated crossover process, in some cases, the luminescence of phosphors may also be quenched by thermal ionization. If the lowest excited state of an activator lies close to the bottom of conduction band, the electron in the excited state can easily be thermally ionized to the conduction band. The delocalized electron may recombine with a hole, so that the luminescence from the activator is quenched.

1.4.4 ELECTRON-TRANSFER QUENCHING

Similar to the case of thermal ionization, there are other types of electron transfer transition in phosphors, including ligand-to-metal, metal-to-ligand and metal-to-metal charge transfers. These transitions generally result in photoionization of activators, accompanying luminescence quenching of the activators. Figure 1.8 shows that the nonradiative process in a three-parabola diagram. The highest parabola originates from a different configuration and is probably connected to the ground state by an allowed transition.

Excitation occurs from the ground state to the highest parabola in the allowed transition. From here, the system relaxes to the relaxed excited state of the second parabola. Subsequently, emission may occur from the second parabola (line emission). It is clear that the offset between the two parabolas (ΔQ) is a very important parameter for the nonradiative quenching rate. If the charge-transfer state moves to lower energy, the efficiency of the luminescence upon charge-transfer excitation decreases. Meanwhile, the ΔQ should be as small as possible if an efficient luminescent material is required.

1.4.5 ENERGY TRANSFER

Nonradiative energy transfer is a well-understood phenomenon and plays an important role in the luminescence of phosphors. The energy transfer processes have been quantitatively studied by

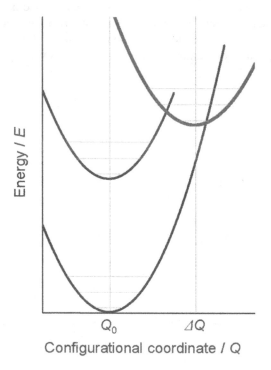

FIGURE 1.8   Configurational coordinate diagram to illustrate the electron transfer quenching.

Forster and Dexter. For convenience, we consider here two centers, i.e., sensitizer S and activator A (S is said to sensitize A). If S is in the excited state and A in the ground state, the relaxed excited state of S may transfer its energy to A. This process is written as

$$S^* + A \rightarrow S + A^*, \qquad (1.16)$$

where the asterisk indicates the excited state.

Nowadays, we have known that the energy transfer from S to A requires a certain interaction between the two centers. The interaction may be either a multipole-multipole interaction or an exchange interaction (wave function overlapping). Moreover, energy transfer can only occur if the excited states of S and A are resonant in energy. Another discussion on the energy transfer is on the distances (R) between S and A. The distance dependence of the transfer rate depends on the type of interaction. For electric multipolar interaction, the distance dependence is given by R^{-n} ($n=6$ and 8 for dipole-dipole interaction and dipole-quadrupole interaction, respectively).[6] For exchange interaction, the distance dependence is exponential, since exchange interaction requires wave function overlap.

If the sensitizer S and activator A are identical ions, the excitation energy in phosphor can be transferred far from the excited site. Such process is called energy migration, in which the excitation energy is transferred in phosphor as exciton or among the excited state of activators. For example, in a phosphor with high activator concentration, the activators may be close enough to each other. In this sense, excitation on one activator may be transferred to another unexcited ion nearby. If this transfer is efficient enough, excitation energy can wander throughout the crystal lattice until it finds a quenching site (impurity or defect) and is then lost to the emission. Consequently, the emission output of the concentrated phosphor is usually lowered. This phenomenon is called concentration quenching. This type of quenching will not occur in phosphors with low activator concentrations because the average distance between the activators may be too large to hamper the migration to reach the quenching sites.

The emphasis of the present handbook will be on the materials. For extensive treatments on luminescence physics and spectroscopy, the readers are referred to the following books:

G. Blasse and B.C. Grabmaier, *Luminescent Materials*, Springer-Verlag, Berlin, 1994.

B. Henderson and G.F. Imbusch, *Optical Spectroscopy of Inorganic Solids*, Oxford Science Publication, Clarendon Press, Oxford, 2006.

B.G. Wybourne, *Spectroscopic Properties of Rare Earths*, John Wiley & Sons, Inc, New York, NY, 1965.

S.W.S. McKeever, *Thermoluminescence of Solids*, Cambridge University Press, Cambridge, 1985.

C.R. Ronda, Luminescence: *From Theory to Applications*, Willy-VCH Verlag GmbH, Weinheim, 2008.

REFERENCES

1. Judd, B.R., *Phys. Rev.*, 127, 750, 1962.
2. Ofelt, G.S., *J. Chem. Phys.*, 37, 511, 1962.
3. Carnall, W.T., Crosswhite, H., and Crosswhite, H.M., *Energy Level Structure and Transition Probabilities of the Trivalent Lanthanides in LaF₃*, Argonne National Lab, Lemont, IL, 1978.
4. Tanabe, Y. and Sugano, S., J. Phys. Soc. Jpn., 9, 753, 1954.
5. Wybourne, B.G., *Spectroscopic Properties of Rare Earths*, John Wiley & Sons, Inc, New York, NY, 1965.
6. Blasse, G. and Grabmaier, B.C., *Luminescent Materials*, Springer-Verlag, Berlin, 1994.

2 Fundamentals of Luminescence

CONTENTS

DOI: 10.1201/9781003098690-2

2.1   ABSORPTION AND EMISSION OF LIGHT

Eiichiro Nakazawa

Most phosphors are composed of a transparent microcrystalline host (or a matrix) and an activator, i.e., a small amount of intentionally added impurity atoms distributed in the host crystal. Therefore, the luminescence processes of a phosphor can be divided into two parts: the processes mainly related to the host and those that occur around and within the activator.

Processes related to optical absorption, reflection, and transmission by the host crystal are discussed, from a macroscopic point of view, in Section 2.1.1. Other host processes (e.g., excitation by electron bombardment and the migration and transfer of the excitation energy in the host) discussed in Section. 2.1.2 deal with phenomena related to the activator atom based on the theory of atomic spectra.

The interaction between the host and the activator is not explicitly discussed in this section; in this sense, the host is treated only as a medium for the activator. The interaction processes such as the transfer of the host excitation energy to the activator will be discussed in detail for each phosphor in Part III.

2.1.1   ABSORPTION AND REFLECTION OF LIGHT IN CRYSTALS

Since a large number of phosphor host materials are transparent and nonmagnetic, their optical properties can be represented by the optical constants or by a complex dielectric constant.

2.1.1.1   Optical Constant and Complex Dielectric Constant

The electric and magnetic fields of a light wave, propagating in a uniform matrix with an angular frequency $\omega\ (= 2\pi\ \nu,\ \nu$: frequency) and velocity $v = \omega/k$ are:

$$E = E_0 \exp[i(\tilde{k} \cdot r - \omega t)] \tag{2.1}$$

$$H = H_0 \exp[i(\tilde{k} \cdot r - \omega t)], \tag{2.2}$$

where r is the position vector and $\tilde{k}$ is the complex wave vector.

E and H in a nonmagnetic dielectric material, with a magnetic permeability that is nearly equal to that in a vacuum ($\mu \approx \mu_0$) and with uniform dielectric constant ε and electric conductivity σ, satisfy the next two equations derived from Maxwell's equations:

$$\nabla^2 E = \sigma\mu_0 \frac{\partial E}{\partial t} + \varepsilon\mu_0 \frac{\partial^2 E}{\partial t^2} \tag{2.3}$$

$$\nabla^2 H = \sigma\mu_0 \frac{\partial H}{\partial t} + \varepsilon\mu_0 \frac{\partial^2 H}{\partial t^2} \tag{2.4}$$

In order that Eqs. (2.1) and (2.2) satisfy Eqs. (2.3) and (2.4), the $\tilde{k}$-vector and its length $\tilde{k}$, which is a complex number, should satisfy the following relation:

$$\tilde{k} \cdot \tilde{k} = \tilde{k}^2 = \left(\varepsilon + \frac{i\sigma}{\omega} \right) \mu_0 \omega^2 = \tilde{\varepsilon}\, \mu_0 \omega^2 \tag{2.5}$$

where $\tilde{\varepsilon}$ is the complex dielectric constant defined by:

$$\tilde{\varepsilon} = \varepsilon' + i\varepsilon'' \equiv \varepsilon + i\frac{\sigma}{\omega} \tag{2.6}$$

Therefore, the refractive index, which is a real number defined as $n \equiv c/v = ck/w$ in a transparent media, is also a complex number:

$$\tilde{n} = n + i\kappa \equiv \frac{c\tilde{k}}{\omega} = \left(\frac{\tilde{\varepsilon}}{\varepsilon_0} \right)^{1/2} \tag{2.7}$$

where c is the velocity of light in vacuum and is equal to $(\varepsilon_0\mu_0)^{-1/2}$ from Eq. (2.5). The last term in Eq. (2.7) is also derived from Eq. (2.5).

The real and imaginary parts of the complex refractive index, i.e., the real refractive index n and the extinction index κ, are called optical constants and are the representative constants of the macroscopic optical properties of the material. The optical constants in a nonmagnetic material are related to each other using Eqs. (2.6) and (2.7),

$$\frac{e'}{\varepsilon_0} = n^2 - \kappa^2 \tag{2.8}$$

$$\frac{\varepsilon''}{\varepsilon_0} = 2n\kappa \tag{2.9}$$

Both of the optical constants, n and κ, are functions of angular frequency ω and, hence, are referred to as dispersion relations. The dispersion relations for a material are obtained by measuring and analyzing the reflection or transmission spectrum of the material over a wide spectral region.

2.1.1.2 Absorption Coefficient

The intensity of the light propagating in a media a distance x from the incident surface having been decreased by the optical absorption is given by Lambert's law.

$$I = I_0 \exp(-\alpha x) \tag{2.10}$$

where I_0 is the incident light intensity minus reflection losses at the surface, and α (cm^{-1}) is the absorption coefficient of the media.

Using Eqs. (2.5) and (2.7), Eq. (2.1) may be rewritten as:

$$E = E_0 \exp(-\omega \kappa x/c) \exp[-i\omega(t+nx/c)] \tag{2.11}$$

and, since the intensity of light is proportional to the square of its electric field strength E, the absorption coefficient may be identified as:

$$\alpha = \frac{2\omega\kappa}{c} \tag{2.12}$$

Therefore, κ is a factor that represents the extinction of light due to the absorption by the media. There are several ways to represent the absorption of light by a medium, as described next.

1. Absorption coefficient, $\alpha(\text{cm}^{-1})$: $I/I_0 = e^{-\alpha x}$
2. Absorption cross section, α/N (cm^2). Here, N is the number of absorption centers per unit volume.
3. Optical density, absorbance, $D = -\log_{10}(I/I_0)$
4. Absorptivity, $(I_0 - I)/I_0 \times 100$, $(\%)$
5. Molar extinction coefficient, $\varepsilon = \alpha\log_{10}e/C$. Here, $C(\text{mol/l})$ is the molar concentration of absorption centers in a solution or gas.

2.1.1.3 Reflectivity and Transmissivity

When a light beam is incident normally on an optically smooth crystal surface, the ratio of the intensities of the reflected light to the incident light, i.e., normal surface reflectivity R_0, can be written in terms of the optical constants, n and κ, by:

$$R_0 = \frac{(n-1)^2 + \kappa^2}{(n+1)^2 + \kappa^2} \tag{2.13}$$

Then, for a sample with an absorption coefficient α and thickness d that is large enough to neglect interference effects, the overall normal reflectivity and transmissivity, i.e., the ratio of the transmitted light to the incident, are, respectively:

$$\bar{R} = R_0\left(1 + \bar{T}\exp(-\alpha d)\right) \tag{2.14}$$

$$\bar{T} = \frac{(1-R_0)^2(1+\kappa^2/n^2)\exp(-\alpha d)}{1 - R_0^2\exp(-2\alpha d)} \simeq \frac{(1-R_0)^2\exp(-\alpha d)}{1 - R_0^2\exp(-2\alpha d)} \tag{2.15}$$

If absorption is zero $(\alpha = 0)$, then:

$$\bar{R} = \frac{(n-1)^2}{(n^2+1)} \tag{2.16}$$

2.1.2 Absorption and Emission of Light by Impurity Atoms

The emission of light from a material originates from two types of mechanisms: thermal emission and luminescence. While all the atoms compose the solid participate in the light emission in the thermal process, in the luminescence process, a very small number of atoms (impurities in most cases or crystal defects) are excited and take part in the emission of light. The impurity atom or defect and its surrounding atoms form a luminescent or an emitting center. In most phosphors, the luminescence center is formed by intentionally incorporated impurity atoms called activators.

This section treats the absorption and emission of light by these impurity atoms or local defects.

2.1.2.1 Classical Harmonic Oscillator Model of Optical Centers

The absorption and emission of light by an atom can be described in the most simplified scheme by a linear harmonic oscillator, as shown in Figure 2.1, composed of a positive charge $(+e)$ fixed at $z = 0$ and an electron bound and oscillating around it along the z-axis. The electric-dipole moment of the oscillator with a characteristic angular frequency ω_0 is given by:

$$M = ez = M_0 \exp(i\omega_0 t) \tag{2.17}$$

and its energy, the sum of the kinetic and potential energies, is $(m_e\, \omega_0^2/2e^2)M_0^2$, where m_e is the mass of the electron. Such a vibrating electric dipole transfers energy to electromagnetic radiation at an average rate of $(\omega_0^4/12\pi\varepsilon_0 c^3)M_0^2$ per second and, therefore, has a total energy decay rate given by:

$$A_0 = \frac{e^2\omega_0^2}{6\pi\varepsilon_0 m_e c^3} \tag{2.18}$$

When the change of the energy of this oscillator is expressed as an exponential function e^{-t/τ_0}, its time constant τ_0 is equal to A_0^{-1}, which is the radiative lifetime of the oscillator, i.e., the time it takes for the oscillator to lose its energy to e^{-1} of the initial energy. From Eq. (2.8), the radiative lifetime of an oscillator with a 600-nm ($\omega_0 = 3 \times 10^{15}$ s^{-1}) wavelength is $\tau_0 \approx 10^{-8}$ s. The intensity of the emission from an electric-dipole oscillator depends on the direction of the propagation, as shown in Figure 2.1.

A more detailed analysis of absorption and emission processes of light by an atom will be discussed using quantum mechanics in the following subsection.

2.1.2.2 Electronic Transition in an Atom

In quantum mechanics, the energies of the electrons localized in an atom or a molecule have discrete values as shown in Figure 2.2. The absorption and emission of light by an atom, therefore, is not a gradual and continuous process as discussed in the abovementioned section using a classical dipole oscillator, but is an instantaneous transition between two discrete energy levels (states), m and n in Figure 2.2, and should be treated statistically.

The energy of the photon absorbed or emitted at the transition $m \leftrightarrow n$ is:

$$\hbar\omega_{mn} = E_m - E_n,\ (E_m > E_n) \tag{2.19}$$

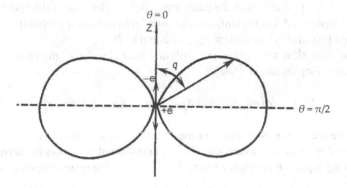

FIGURE 2.1 Electromagnetic radiation from an electric-dipole oscillator. The length of the arrow gives the intensity of the radiation to the direction.

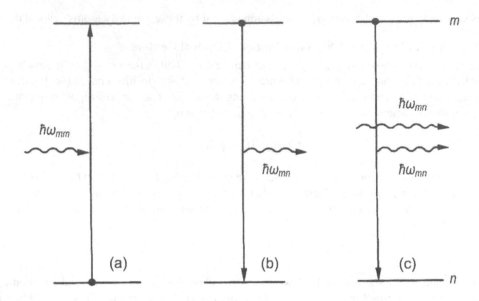

FIGURE 2.2   Absorption (a), spontaneous emission (b), and induced emission (c) of a photon by a two-level system.

where E_n and E_m are the energies of the initial and final states of the transition, respectively, and $\omega_{mn} (=2\pi v_{mn})$ is the angular frequency of light.

There are two possible emission processes, as shown in Figure 2.2; one is called spontaneous emission (b) and the other is stimulated emission (c). The stimulated emission is induced by an incident photon, as is the case with the absorption process (a). Laser action is based on this type of emission process.

The intensity of the absorption and emission of photons can be enumerated by a transition probability per atom per second. The probability for an atom in a radiation field of energy density $\rho(\omega_{mn})$ to absorb a photon, making the transition from n to m, is given by:

$$W_{mn} = B_{n \to m} \rho\left(\omega_{mn}\right) \tag{2.20}$$

where $B_{n \to m}$ is the transition probability or Einstein's B-coefficient of optical absorption and $\rho(\omega)$ is equal to $I(\omega)/c$ in which $I(\omega)$ is the light intensity, i.e., the energy per second per unit area perpendicular to the direction of light.

On the other hand, the probability of the emission of light is the sum of the spontaneous emission probability $A_{m \to n}$ (Einstein's A-coefficient) and the stimulated emission probability $B_{m \to n} \rho(\omega_{mn})$. The stimulated emission probability coefficient $B_{m \to n}$ is equal to $B_{n \to m}$.

The equilibrium of optical absorption and emission between the atoms in the states m and n is expressed by the following equation:

$$N_n B_{n \to m} \rho\left(\omega_{mn}\right) = N_m \left\{ A_{m \to n} + B_{m \to n}\left(\omega_{mn}\right)\rho\left(\omega_{mn}\right) \right\}, \tag{2.21}$$

where N_m and N_n are the numbers of atoms in the states m and n, respectively. Taking into account the Boltzmann distribution of the system and Plank's equation of radiation in thermodynamic equilibrium, the following equation is obtained from Eq. (2.21) for the spontaneous mission probability.

$$A_{m \to n} = \frac{\hbar \omega_{mn}^3}{\pi^2 c^3} B_{m \to n} \tag{2.22}$$

Therefore, the probabilities of optical absorption and the spontaneous and induced emissions between m and n are related to one another.

2.1.2.3 Electric-Dipole Transition Probability

In a quantum mechanical treatment, optical transitions of an atom are induced by perturbing the energy of the system by $\sum_i(-er_i)\cdot E$, in which r_i is the position vector of the electron from the atom center and, therefore, $\sum_i(-er_i)$ is the electric-dipole moment of the atom (see Eq. (2.17)). In this electric-dipole approximation, the transition probability of optical absorption is given by:

$$W_{mn} = \frac{\pi}{3\varepsilon_0 c\hbar^2} I(\omega_{mn})|M_{mn}|^2 \tag{2.23}$$

Here, the dipole moment, M_{mn} is defined by:

$$M_{mn} = \int \Psi_m^* \left(\sum_i er_i\right)\Psi_n d\tau \tag{2.24}$$

where ψ_m and ψ_n are the wavefunctions of the states m and n, respectively. The direction of this dipole moment determines the polarization of the light absorbed or emitted. In Eq. (2.23), however, it is assumed that the optical center is isotropic and then $|(M_{mn})_z|^2 = |M_{mn}|^2/3$ for light polarized in the z-direction.

Equating the right-hand side of Eq. (2.23) to that of Eq. (2.20), the absorption transition probability coefficient $B_{n\rightarrow m}$ and then, from Eq. (2.22), the spontaneous emission probability coefficient $A_{m\rightarrow n}$ can be obtained as follows:

$$B_{n\rightarrow m} = \frac{\pi}{3\varepsilon_0\hbar^2}|M_{mn}|^2$$

$$A_{m\rightarrow n} = \frac{\omega_{mn}^3}{3\pi\varepsilon_0\hbar c^3}|M_{mn}|^2 \tag{2.25}$$

2.1.2.4 Intensity of Light Emission and Absorption

The intensity of light is generally defined as the energy transmitted per second through a unit area perpendicular to the direction of light. The spontaneous emission intensity of an atom is proportional to the energy of the emitted photon, multiplied by the transition probability per second given by Eq. (2.25).

$$I(\omega_{mn}) \propto \hbar\omega_{mn}A_{m\rightarrow n} = \frac{\omega_{mn}^4}{3\pi\varepsilon_0 c^3}|M_{mn}|^2 \tag{2.26}$$

Likewise, the amount of light with intensity $I_0(\omega_{mn})$ to be absorbed by an atom per second is equal to the photon energy ω_{mn} multiplied by the absorption probability coefficient and the energy density I_0/c.

It is more convenient, however, to use a radiative lifetime and absorption cross section to express the ability of an atom to make an optical transition than to use the amount of light energy absorbed or emitted by the transition.

The radiative lifetime τ_{mn} is defined as the inverse of the spontaneous emission probability $A_{m\rightarrow n}$.

$$\tau_{mn}^{-1} = A_{m\rightarrow n} \tag{2.27}$$

If there are several terminal states of the transition and the relaxation is controlled only by spontaneous emission processes, the decay rate of the emitting level is determined by the sum of the transition probabilities to all final states:

$$A_m = \sum_n A_{m \to n}$$ (2.28)

and the number of the excited atoms decreases exponentially, $\propto \exp(-t/\tau)$, with time a constant $\tau = A_m^{-1}$, called the natural lifetime. In general, however, the real lifetime of the excited state m is controlled not only by radiative processes, but also by nonradiative ones (see Section 2.4).

The absorption cross section σ represents the probability of an atom to absorb a photon incident on a unit area. (If there are N absorptive atoms per unit volume, the absorption coefficient α in Eq. (2.10) is equal to σN.) Therefore, since the intensity of the light with a photon per second per unit area is $I_0 = \omega_{mn}$ in Eq. (2.23), the absorption cross section is given by:

$$\sigma_{nm} = \frac{\pi \omega_{mn}}{3 \varepsilon_0 c \hbar} \left| M_{mn} \right|^2$$ (2.29)

2.1.2.5   Oscillator Strength

The oscillator strength of an optical center is often used to represent the strength of light absorption and emission of the center. It is defined by the following equation as a dimensionless quantity.

$$f_{mn} = \frac{2 m_e \omega_{mn}}{\hbar e^2} \left| \left(M_{mn} \right)_z \right|^2 = \frac{2 m_e \omega_{mn}}{3 \hbar e^2} \left| M_{mn} \right|^2$$ (2.30)

The third term of this equation is given by assuming that the transition is isotropic, as it is the case with Eq. (2.24).

The radiative lifetime and absorption cross section are expressed by using the oscillator strength as:

$$\tau_{mn}^{-1} = A_{m \to n} = \frac{e^2 \omega_{mn}^2}{2 \pi \varepsilon_0 m c^3} f_{mn}$$ (2.31)

$$\sigma_{nm} = \frac{\pi e^2}{2 \varepsilon_0 m c} f_{mn}$$ (2.32)

Now one can estimate the oscillator strength of a harmonic oscillator with the electric-dipole moment $\mathbf{M} = -e\mathbf{r}$ in a quantum mechanical manner. The result is that only one electric-dipole transition between the ground state ($n = 0$) and the first excited state ($m = 1$) is allowed, and the oscillator strength of this transition is $f_{10} = 1$. Therefore, the summation of all the oscillator strengths of the transition from the state $n = 0$ is also $\sum_m f_{m0} = 1$ ($m \neq 0$). This relation is true for any one electron system; for N-electron systems, the following f-sum rule should be satisfied, that is:

$$\sum_{m \neq n} f_{mn} = N$$ (2.33)

At the beginning of this section, the emission rate of a linear harmonic oscillator was classically obtained as A_0 in Eq. (2.18). Then, the total transition probability given by Eq. (2.32) with $f = 1$ in a quantum mechanical scheme coincides with the emission rate of the classical linear oscillator A_0, multiplied by a factor of 3, corresponding to the three degrees of freedom of the motion of the electron in the present system.

2.1.2.6 Impurity Atoms in Crystals

Since the electric field acting on an impurity atom or optical center in a crystal is different from that in vacuum due to the effect of the polarization of the surrounding atoms, and the light velocity is reduced to c/n (see Eq. (2.7)), the radiative lifetime and the absorption cross section are changed from those in vacuum. In a cubic crystal, for example, Eqs. (2.31) and (2.32) are changed, by the internal local field, to:

$$\tau_{mn}^{-1} = \frac{n(n^2+2)^2}{9} \cdot \frac{e^2 \omega_{mn}^2}{2\pi\varepsilon_0 mc^3} f_{mn} \tag{2.34}$$

$$\sigma_{nm} = \frac{(n^2+2)^2}{9n} \cdot \frac{\pi e^2}{2\pi\varepsilon_0 mc} f_{nm} \tag{2.35}$$

2.1.2.7 Forbidden Transition

In the case that the electric-dipole moment of a transition M_{nm} of Eq. (2.25) becomes zero, the probability of the electric-dipole (E1) transition in Eqs. (2.25) and (2.26) is also zero. Since the electric-dipole transition generally has the largest transition probability, this situation is usually expressed by the term forbidden transition. Since the electric-dipole moment operator in the integral of Eq. (2.24) is an odd function (odd parity), the electric-dipole moment is zero if the initial and final states of the transition have the same parity; that is, both of the wavefunctions of these states are either an even or odd function, and the transition is said to be parity forbidden. Likewise, since the electric-dipole moment operator in the integral of Eq. (2.24) has no spin operator, transitions between initial and final states with different spin multiplicities are spin forbidden.

In Eq. (2.24) for the dipole moment, the effects of the higher order perturbations are neglected. If the neglected terms are included, the transition moment is written as follows:

$$\left|M_{mn}\right|^2 = \left|(er)_{mn}\right|^2 + \left|\left(\frac{e}{2mc}r \times p\right)_{mn}\right|^2 + \frac{3\pi\omega_{mn}^2}{40c^2}\left|(er \cdot r)_{mn}\right|^2 \tag{2.36}$$

where the first term on the right-hand side is the contribution of the electric dipole (E1) previously given in Eq. (2.24); the second term, in which p denotes the momentum of an electron, is that of magnetic dipole (M1); and the third term is that of an electric quadrupole transition (E2). Provided that $(r)_{mn}$ is about the radius of a hydrogen atom (0.5 Å) and ω_{mn} is 10^{15} rad/s for visible light, radiative lifetimes estimated from Eqs. (2.26) and (2.36) are $\sim 10^{-8}$ s for E1, $\sim 10^{-3}$ s for M1, and $\sim 10^{-1}$ s for E2.

E1 transitions are forbidden (parity forbidden) for f–f and d–d transitions of free rare-earth ions and transition-metal ions because the electron configurations, and hence the parities of the initial and final states, are the same. In crystals, however, the E1 transition is partially allowed by the odd component of the crystal field, and this partially allowed or forced E1 transition has the radiative lifetime of $\sim 10^{-3}$ s. (see Chapter 4).

2.1.2.8 Selection Rule

The selection rule governing whether a dipole transition is allowed between the states m and n is determined by the transition matrix elements $(er)_{mn}$ and $(r \times p)_{mn}$ in Eq. (2.36). However, a group theoretical inspection of the symmetries of the wavefunctions of these states and the operators er and $r \times p$ enables the determination of the selection rules without calculating the matrix elements.

When an atom is free or in a spherical symmetry field, its electronic states are denoted by a set of the quantum numbers S, L, and J in the LS-coupling scheme. Here, S, L, and J denote the quantum number of the spin, orbital, and total angular momentum, respectively, and ΔS, for example, denotes the difference in S between the states m and n. Then, the selection rules for E1 and M1 transitions in the LS-coupling scheme are given by:

$$\Delta S = 0, \Delta L = 0 \text{ or } \pm 1 \tag{2.37}$$

$$\Delta J = 0 \text{ or } \pm 1 \ (J = 0 \to J = 0, \text{ not allowed}) \tag{2.38}$$

If the spin-orbit interaction is too large to use the **LS**-coupling scheme, the **JJ**-coupling scheme might be used, in which many $(\mathbf{S}, \mathbf{L})$ terms are mixed into a **J**-state. In the **JJ**-coupling scheme, therefore, the ΔS and ΔL selection rules in Eqs. (2.37) and (2.38) are less strict, and only the ΔJ selection rule applies.

While the E1 transitions between the states with the same parity are forbidden, as in the case of the f–f transitions of free rare-earth ions, they become partially allowed for ions in crystals due to the effects of crystal fields of odd parity. The selection rule for the partially allowed E1 f–f transition is $|\Delta J| \leq 6$ ($J = 0$–0, 1, 3, 5 are forbidden). M1 transitions are always parity allowed because of the even parity of the magnetic dipole operator $\mathbf{r} \times \mathbf{p}$ in Eq. (2.36).

2.2 ELECTRONIC STATES AND OPTICAL TRANSITION OF SOLID CRYSTALS

Shigetoshi Nara and Sumiaki Ibuki

2.2.1 OUTLINE OF BAND THEORY

First, a brief description of crystal properties is given. As is well known, a crystal consists of a periodic configuration of atoms, which is called a *crystal lattice*. There are many different kinds of crystal lattices and they are classified, in general, according to their symmetries, which specify invariant properties for translational and rotational operations. Figure 2.3 shows a few, typical examples of crystal structures, i.e., a rock-salt (belonging to one of the cubic groups) structure, a zinc-blende (also a cubic group) structure, and a wurtzite (a hexagonal group) structure, respectively.

Second, consider the electronic states in these crystals. In an isolated state, each atom has electrons that exist in discrete electronic energy levels, and the states of these bound electrons are characterized by atomic wavefunctions. Their discrete energy levels, however, will have finite spectral width in the condensed state because of the overlaps between electronic wavefunctions belonging to different atoms. This is because electrons can become itinerant between atoms,

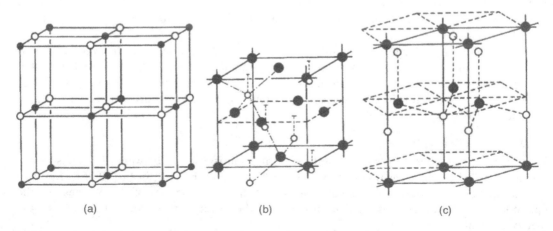

(a)   (b)   (c)

FIGURE 2.3 The configuration of the atoms in three important kinds of crystal structures. (a) rock-salt type, (b) zinc-blende type, and (c) wurtzite type, respectively.

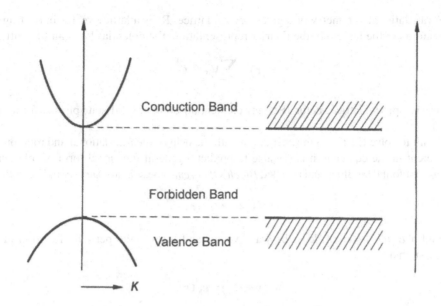

FIGURE 2.4   The typical band dispersion near the minimum bandgap in a semiconductor or an insulator with a direct bandgap in the Brillouin zone.

until finally they fall into delocalized electronic states called *electronic energy bands*, which also obey the symmetries of crystals. In these energy bands, the states with lower energies are occupied by electrons originating from bound electrons of atoms and are called *valence bands*. The energy bands having higher energies are not occupied by electrons and are called *conduction bands*. Usually, in materials having crystal symmetries such as rock-salt, zinc-blende, or wurtzite structures, there is no electronic state between the top of the valence band (the highest state of occupied bands) and the bottom of the conduction band (the lowest state of unoccupied bands); this region is called the *bandgap*. The reason why unoccupied states are called conduction bands is due to the fact that an electron in a conduction band is almost freely mobile if it is excited from a valence band by some method: for example, by absorption of light quanta. In contrast, electrons in valence bands cannot be mobile because of a fundamental property of electrons; as *fermions*, only two electrons (spin up and down) can occupy an electronic state. Thus, it is necessary for electrons in the valence band to have empty states in order for them to move freely when an electric field is applied. After an electron is excited to the conduction band, a hole that remains in the valence band behaves as if it were a mobile particle with a positive charge. This hypothetical particle is called a *positive hole*. The schematic description of these excitations is shown in Figure 2.4. As noted previously, bandgaps are strongly related to the optical properties and the electric conductivity of crystals.

   A method to evaluate these electronic band structures in a quantitative way using quantum mechanics is briefly described. The motion of electrons under the influence of electric fields generated by atoms that take some definite space configuration specified by the symmetry of the crystal lattice can be described by the following Schrödinger equation:

$$-\frac{\hbar^2}{2m}\nabla^2\Psi(\mathbf{r})+V(\mathbf{r})\Psi(\mathbf{r})=E\Psi(\mathbf{r}) \tag{2.39}$$

where $V(\mathbf{r})$ is an effective potential applied to each electron and has the property of:

$$V(\mathbf{r}+\mathbf{R}_n)=V(\mathbf{r}) \tag{2.40}$$

due to the translational symmetry of a given crystal lattice. $\mathbf{R}_n$ is a lattice vector indicating the nth position of atoms in the lattice. In the Fourier representation, the potential $V(\mathbf{r})$ can be written as:

$$V(\mathbf{r}) = \sum_n V_n e^{iG_n \cdot r} \tag{2.41}$$

where $\mathbf{G}_n$ is a reciprocal lattice vector. (See any elementary book of solid-state physics for the definition of $\mathbf{G}_n$.)

It is difficult to solve Eq. (2.39) in general, but with the help of the translational and rotational symmetries inherent in the equation, it is possible to predict a general functional form of solutions. The solution was first found by Bloch and is called *Bloch's theorem*. The solution $\psi(\mathbf{r})$ should be of the form:

$$\Psi(\mathbf{r}) = e^{ik \cdot r} u_k(\mathbf{r}) \tag{2.42}$$

and is called a *Bloch function*. $\mathbf{k}$ is the wave vector and $u_\mathbf{k}(\mathbf{r})$ is the periodic function of lattice translations, such as:

$$u_\mathbf{k}(\mathbf{r} + \mathbf{R}_n) = u_\mathbf{k}(\mathbf{r}) \tag{2.43}$$

As one can see in Eq. (2.40), $u_\mathbf{k}(\mathbf{r})$ can also be expanded in a Fourier series as:

$$u_\mathbf{k}(\mathbf{r}) = \sum_n C_n(\mathbf{k}) e^{iG_n \cdot r} \tag{2.44}$$

where $C_n(\mathbf{k})$ is a Fourier coefficient. The form of the solution represented by Eq. (2.42) shows that the wave vectors $\mathbf{k}$ are well-defined quantum numbers of the electronic states in a given crystal. Putting Eq. (2.44) into Eq. (2.42) and using Eq. (2.41), one can rewrite Eq. (2.39) in the following form:

$$\left\{ \frac{\hbar^2}{2m} (\mathbf{k} + \mathbf{G}_l)^2 - E \right\} C_l + \sum_n C_n V_{l-n} = 0 \tag{2.45}$$

where E eigenvalues determined by:

$$\left| \left\{ \frac{\hbar^2}{2m} (\mathbf{k} + \mathbf{G}_l)^2 - E \right\} \delta_{G_l G_n} + V_{G_l - G_n} \right| = 0 \tag{2.46}$$

Henceforth, the $\mathbf{k}$-dependence of the Fourier components $C_n(\mathbf{k})$ is neglected. These formulas are in the form of infinite-dimensional determinant equations. For finite dimensions by considering amplitudes of V_{Gl-Gn} in a given crystal, one can solve Eq. (2.46) approximately. Then the energy eigenvalues $E(\mathbf{k})$ (energy band) may be obtained as a function of wave vector $\mathbf{k}$ and the Fourier coefficients C_n.

To obtain qualitative interpretation of energy band and properties of a wavefunction, one can start with the 0th order approximation of Eq. (2.46) by taking:

$$C_0 = 1, C_n = 0 \, (n \neq 0) \tag{2.47}$$

in Eq. (2.44) or (2.45); this is equivalent to taking $V_n = 0$ for all n (a vanishing or constant crystal potential model). Then, Eq. (2.46) gives:

$$E = \frac{\hbar^2}{2m} \mathbf{k}^2 = E_0(\mathbf{k}) \tag{2.48}$$

This corresponds to the free electron model.

As the next approximation, consider the case that the nonvanishing components of V_n are only for $n = 0, 1$. Eq. (2.46) becomes:

$$\begin{vmatrix} \dfrac{\hbar^2}{2m}\mathbf{k}^2 - E & V_{G_1} \\[2mm] V_{-G_1} & \dfrac{\hbar^2}{2m}(\mathbf{k}+\mathbf{G}_1)^2 - E \end{vmatrix} = 0 \qquad (2.49)$$

This means that, in $\mathbf{k}$-space, the two free electrons having $E(\mathbf{k})$ and $E(\mathbf{k}+\mathbf{G})$ are in independent states in the absence of the crystal potential even when $\|\mathbf{k}\| = \|\mathbf{k}+\mathbf{G}\|$; this energy degeneracy is lifted under the existence of nonvanishing V_G. In the above case, the eigenvalue equation can be solved easily and the solution gives:

$$E = \frac{1}{2}\{E(\mathbf{k}) + E(\mathbf{k}+\mathbf{G})\} \pm \sqrt{\left\{\frac{E(\mathbf{k}) - E(\mathbf{k}+\mathbf{G})}{2}\right\}^2 + V_G^2} \qquad (2.50)$$

Figure 2.5 shows the global profile of E as a function of $\mathbf{k}$ in one dimension. One can see the existence of energy gap at the wave vector that satisfies:

$$\mathbf{k}^2 = (\mathbf{k}+\mathbf{G}_1)^2 \qquad (2.51)$$

This is called the *Bragg condition*. In the three-dimensional case, the wave vectors that satisfy Eq. (2.51) form closed polyhedrons in $\mathbf{k}$ space and are called the 1st, 2nd, or 3rd, ..., nth Brillouin zone.

As stated so far, the electronic energy band structure is determined by the symmetry and Fourier amplitudes of the crystal potential $V(\mathbf{r})$. Thus, one needs to take a more realistic model of them to

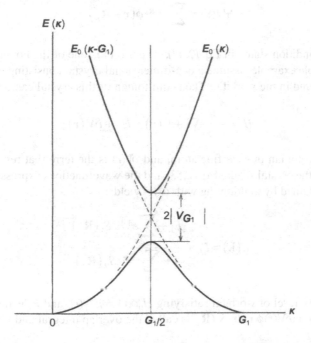

FIGURE 2.5 The emergence of a bandgap resulting from the interference between two plane waves satisfying the Bragg condition, in a one-dimensional model.

get a more accurate description of the electronic properties. There are now many procedures that allow for the calculation of the energy band and to get the wavefunction of electrons in crystals. Two representative methods, the *Pseudopotential method* and the *LCAO method* (Linear Combination of Atomic Orbital method), which are frequently applied to outer-shell valence electrons in semiconductors, are briefly introduced here.

First, consider the pseudopotential method. Eq. (2.46) is the fundamental equation to get band structures of electrons in crystals, but the size of the determinant equation will become very large if one wishes to solve the equation with sufficient accuracy, because, in general, the Fourier components VG_n do not decrease slowly due to the Coulomb potential of each atom. This corresponds to the fact that the wavefunctions of valence electrons are free-electron like (plane-wave like) in the intermediate region between atoms and give rapid oscillations (atomic like) near the ion cores.

Therefore, to avoid this difficulty, one can take an effective potential in which the Coulomb potential is canceled by the rapid oscillations of wavefunctions. The rapid oscillation of wavefunctions originates from the orthogonalization between atomic-like properties of wavefunctions near ion cores. It means that one introduces new wavefunctions and a weak effective potential instead of plane waves and a Coulombic potential to represent the electronic states. This effective potential gives a small number of reciprocal wave vectors (**G**) that can reproduce band structures with a corresponding small number of Fourier components. This potential is called the *pseudopotential*. The pseudopotential method necessarily results in some arbitrariness with respect to the choice of these effective potentials, depending on the selection of effective wavefunctions. It is even possible to parametrize a small number of components in VG_n and to determine them empirically. For example, taking several V_{Gn} values in high symmetry points in the Brillouin zone and, after adjusting them so as to reproduce the bandgaps obtained with experimental measurements, one calculates the band dispersion $E(\mathbf{k})$ over the entire region.

In contrast, the LCAO method approximates the Bloch states of valence electrons by using a linear combination of bound atomic wavefunctions. For example,:

$$\Psi_k(\mathbf{r}) = \sum_n e^{i\mathbf{k}\cdot\mathbf{R}_n} \phi(\mathbf{r} - \mathbf{R}_n) \qquad (2.52)$$

satisfies the Bloch condition stated in Eq. (2.42), where $\phi(\mathbf{r})$ is one of the bound atomic wavefunctions. To show a simple example, assume a one-dimensional crystal consisting of atoms having one electron per atom bound in the s-orbital. The Hamiltonian of this crystal can be written as:

$$H = -\frac{\hbar^2}{2m}\nabla^2 + V(\mathbf{r}) = H_0 + \delta V(\mathbf{r}) \qquad (2.53)$$

where H_0 is the Hamiltonian of each free atom, and $\delta V(\mathbf{r})$ is the term that represents the effect of periodic potential in the crystal. Using Eq. (2.53) and the wavefunctions expressed in Eq. (2.52), the expectation value obtained by multiplying with $\phi^*(\mathbf{r})$ yields:

$$E(\mathbf{k}) = E_0 + \frac{E_1 + \sum_{n\neq 0} e^{i\mathbf{k}\cdot\mathbf{R}_n} S_1(\mathbf{R}_n)}{1 + \sum_{n\neq 0} e^{i\mathbf{k}\cdot\mathbf{R}_n} S_0(\mathbf{R}_n)} \qquad (2.54)$$

where E_0 is the energy level of s-orbital satisfying $H_0\phi(\mathbf{r}) = E_0\phi(\mathbf{r})$, and E_1 is the energy shift of E_0 due to δV given by $\int\phi^*(\mathbf{r})\delta V(\mathbf{r})\phi(\mathbf{r})d\mathbf{r}$. $S_0(\mathbf{R}_n)$ is called the overlap integral and is defined by:

$$S_0(\mathbf{R}_n) = \int \phi^*(\mathbf{r})\phi(\mathbf{r} - \mathbf{R}_n)d\mathbf{r} \qquad (2.55)$$

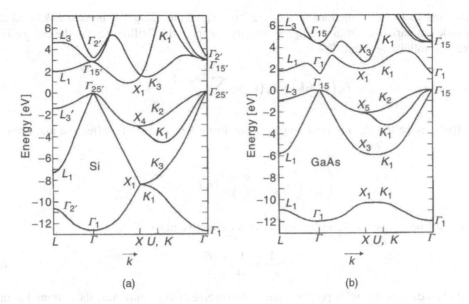

FIGURE 2.6 Calculated band structures of (a) Si and (b) GaAs using a combined pseudopotential and LCAO method. (From Chadi, D.J., Phys. Rev., B16, 3572, 1977. With permission.)

Similarly, $S_1(\mathbf{R}_n)$ is defined as:

$$S_1(\mathbf{R}_n) = \int \phi^*(\mathbf{r})\delta V(\mathbf{r})\phi(\mathbf{r} - \mathbf{R}_n)d\mathbf{r} \tag{2.56}$$

Typically speaking, these quantities are regarded as parameters, and they are fitted so as to best reproduce experimentally observed results. As a matter of fact, other orbitals such as p-, d-orbitals etc. can also be used in LCAO. It is even possible to combine this method with that of pseudopotentials. As an example, Figure 2.6 reveals two band structure calculations due to Chadi;[1] one is for Si and the other is for GaAs.

In Figure 2.6, energy = 0 in the ordinate corresponds to the top of the valence band. In both Si and GaAs, it is located at the Γ point ($\mathbf{k} = (000)$ point). The bottom of the conduction band is also located at the Γ point in GaAs, while in Si it is located near the X point ($\mathbf{k} = (100)$ point).

It is difficult and rare that the energy bands can be calculated accurately all through the Brillouin zone with the use of a small number of parameters determined at high symmetry points. In that sense, it is quite convenient if one has a simple perturbational method to calculate band structures approximately at or near specific points in the Brillouin zone (e.g., the top of the valence band or a conduction band minimum). In particular, such procedures are quite useful when the bands are degenerate at some point in the Brillouin zone of interest.

Now, assume that the Bloch function is known at $\mathbf{k} = \mathbf{k}_0$ and is expressed as $\Psi_{n\mathbf{k}_0}(\mathbf{r})$. Define a new wavefunction as:

$$\eta_{n\mathbf{k}}(\mathbf{r}) = e^{i\mathbf{k}\cdot\mathbf{r}}\Psi_{n\mathbf{k}_0}(\mathbf{r}) \tag{2.57}$$

and expand the Bloch function in terms of $\eta_{n\mathbf{k}}(\mathbf{r})$ as:

$$\zeta_{n\mathbf{k}} = \sum_{n'} C_{n'}\eta_{n'\mathbf{k}}(\mathbf{r}) \tag{2.58}$$

Introducing these wavefunctions into Eq. (2.39) obtains the energy dispersion $E(\mathbf{k}_0 + \mathbf{k})$ in the vicinity of $\mathbf{k}_0$. In particular, near the high symmetry points of the Brillouin zone, the energy dispersion takes the following form:

$$E_n\left(\mathbf{k}_0 + \mathbf{k}\right) = E_n\left(\mathbf{k}_0\right) + \sum_{ij} \frac{\hbar^2}{2}\left(\frac{1}{m^*}\right)_{ij} k_i k_j \tag{2.59}$$

where $(1/m^*)_{ij}$ is called the *effective mass tensor*. From Eq. (2.59), the effective mass tensor is given as:

$$\left(\frac{1}{m^*}\right)_{ij} = \frac{1}{\hbar^2} \cdot \frac{\partial^2 E}{\partial k_i \partial k_j} (i, j = x, y, z) \tag{2.60}$$

For the isotropic case, Eq. (2.60) gives the scalar effective mass m^* as:

$$\frac{1}{m^*} = \frac{1}{\hbar^2} \cdot \frac{d^2 E}{dk^2} \tag{2.61}$$

Eq. (2.61) indicates that m^* is proportional to the inverse of curvature near the extremal points of the dispersion relation, E vs. $\mathbf{k}$. Furthermore, Figure 2.5 illustrates the two typical cases that occur near the bandgap, that is, a positive effective mass at the bottom of the conduction band and a negative effective mass at the top of the valence band, depending on the sign of d^2E/dk^2 at each extremal point. Hence, under an applied electric field $\mathbf{E}$, the specific charge e/m^* of an electron becomes negative, while it becomes positive for a hole. This is the reason why a hole looks like a particle with a positive charge.

In the actual calculation of physical properties, the following quantity is also important:

$$N(E)dE = \frac{1}{3\pi\hbar^2}\left(2m^*\right)^{3/2} E^{1/2} dE \tag{2.62}$$

This is called the *density of states* and represents the number of states between E and $E + dE$. We assume in Eq. (2.62) that space is isotropic and m^* can be used.

The band structures of semiconductors have been intensively investigated experimentally using optical absorption and/or reflection spectra. As shown in Figure 2.7, in many compound semiconductors (most of III–V and II–VI combinations in the periodic table), conduction bands consist mainly of s-orbitals of the cation, and valence bands consist principally of p-orbitals of the anion. Many compound semiconductors have a direct bandgap, which means that the conduction band minimum and the valence band maximum are both at the Γ point ($\mathbf{k} = 0$). It should be noted that the states just near the maximum of the valence band in zinc-blende type semiconductors consist of two orbitals, Γ_8 that is twofold degenerate and Γ_7 without degeneracy; these originate from the *spin-orbit interaction*. It is known that the twofold degeneracy of Γ_8 is lifted in the $\mathbf{k} \neq 0$ region corresponding to a light and a heavy hole, respectively. On the other hand, in wurtzite-type crystals, the valence band top is split by both the spin-orbit interaction and the crystalline field effect; the band maximum then consists of three orbitals: Γ_9, Γ_9, and Γ_7 without degeneracy. In GaP, the conduction band minimum is at the X point ($\mathbf{k} = [100]$), and this compound has an indirect bandgap, as described in the next section.

2.2.2 FUNDAMENTAL ABSORPTION, DIRECT TRANSITION, AND INDIRECT TRANSITION

When solid crystals are irradiated by light, various optical phenomena occur: for example, transmission, reflection, and absorption. In particular, absorption is the annihilation of light (photon) resulting from the creation of an electronic excitation or lattice excitation in crystals. Once electrons

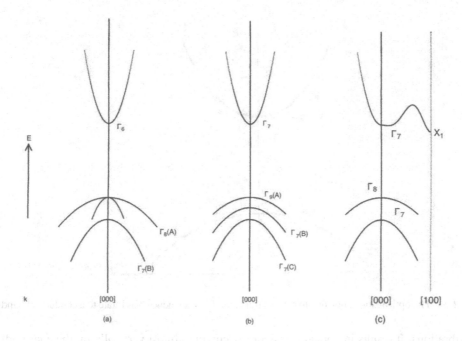

FIGURE 2.7   The typical band dispersion near Γ-point ($\mathbf{k} = 0$) for II-VI or III-V semiconductor compounds. (a) A direct type in zinc-blende structure; (b) a direct type in wurtzite structure; and (c) an indirect type in zinc-blende structure (GaP).

obtain energy from light, the electrons are excited to higher states. In such quantum mechanical phenomena, one can only calculate the probability of excitation. The probability depends on the distribution of microscopic energy levels of electrons in that system. The excited electrons will come back to their initial states after they release the excitation energy in the form of light emission or through lattice vibrations.

Absorption of light by electrons from valence bands to conduction bands results in the fundamental absorption of the crystal. Crystals are transparent when the energy of the incident light is below the energy gaps of crystals; excitation of electrons to the conduction band becomes possible at a light energy equal to, or larger than, the bandgap. The intensity of absorption can be calculated using the absorption coefficient $\alpha(h\nu)$ given by the following formula:

$$\alpha(h\nu) = A \sum p_{if} n_i n_f \tag{2.63}$$

where n_i and n_f are the number density of electronic states in an initial state (occupied by electron) and in a final state (unoccupied by electron), respectively, and p_{if} is the transition probability between them.

In the calculation of Eq. (2.63), quantum mechanics requires that two conditions are satisfied. The first is *energy conservation* and the second is *momentum conservation*. The former means that the energy difference between the initial state and the final state should be equal to the energy of the incident photon, and the latter means that the momentum difference between the two states should be equal to the momentum of the incident light. It is quite important to note that the momentum of light is three or four orders of magnitude smaller than that of the electrons. These conditions can be written as $(\hbar^2/2m^*)k_f^2 = (\hbar^2/2m^*)k_i^2 + h\nu$ (energy conservation), $\hbar k_f = \hbar(k_i + q)$ (momentum conservation), and $\nu = cq$ if one assumes a free-electron-like dispersion for band structure $E(\mathbf{k})$, where k_f and k_i are the final and initial wave vectors, respectively, c is the light velocity, and q is the photon momentum. One can neglect the momentum of absorbed photons compared to those of electrons or

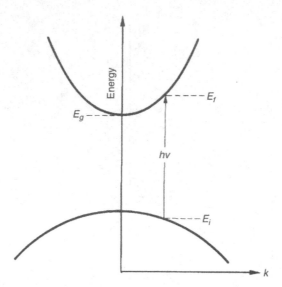

FIGURE 2.8   The optical absorption due to a direct transition from a valence band state to a conduction band state.

lattice vibrations. It results in optical transitions occurring almost vertically on the energy dispersion curve in the Brillouin zone. This rule is called the *momentum selection rule* or **k**-*selection rule*.

As shown in Figure 2.8, consider first the case that the minimum bandgap occurs at the top of valence band and at the bottom of conduction band; in such a case, the electrons of the valence band are excited to the conduction band with the same momentum. This case is called a *direct transition*, and the materials having this type of band structure are called *direct-gap materials*. The absorption coefficient, Eq. (2.63), is written as:

$$\alpha(h\nu) = A^* \left(h\nu - E_g\right)^{1/2} \tag{2.64}$$

with the use of Eqs. (2.63) and (2.64). A^* is a constant related to the effective masses of electrons and holes. Thus, one can experimentally measure the bandgap E_g, because the absorption coefficient increases steeply from the edge of the bandgap. In actual measurements, the absorption increases exponentially because of the existence of impurities near E_g. In some materials, it can occur that the transition at **k** = 0 is forbidden by some selection rule; the transition probability is then proportional to $(h\nu - E_g)$ in the **k** $\neq$ 0 region and the absorption coefficient becomes:

$$\alpha(h\nu) = A' \left(h\nu - E_g\right)^{3/2} \tag{2.65}$$

In contrast to the direct transition, in the case shown in Figure 2.9, both the energy and the momentum of electrons are changed in the process; excitation of this type is called an *indirect transition*. This transition corresponds to cases in which the minimum bandgap occurs between two states with different **k**-values in the Brillouin zone. In this case, conservation of momentum cannot be provided by the *photon*, and the transition necessarily must be associated with the excitation or absorption of *phonons* (lattice vibrations). This leads to a decrease in transition probability due to a higher order stochastic process. The materials having such band structure are called *indirect-gap materials*. An expression for the absorption coefficient accompanied by *phonon absorption* is:

$$\alpha(h\nu) = A \left(h\nu - E_g + E_p\right)^2 \left(\exp\left(\frac{E_p}{k_B T}\right) - 1\right)^{-1} \tag{2.66}$$

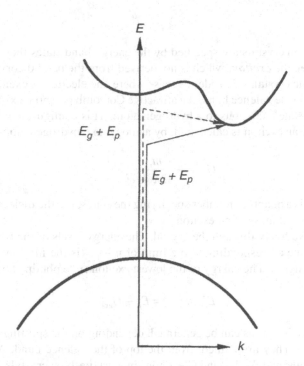

FIGURE 2.9 The optical absorption due to an indirect transition from a valence band state to a conduction band state. The momentum of electron changes due to a simultaneous absorption or emission of a phonon.

while the coefficient accompanied by *phonon emission* is:

$$\alpha(h\nu) = A\left(h\nu - E_g - E_p\right)^2 \left(1 - \exp\left(\frac{-E_p}{k_B T}\right)\right)^{-1} \tag{2.67}$$

where, in both formulas, E_p is the phonon energy.

In closing this section, the light emission process is briefly discussed. The intensity of light emission R can be written as:

$$R = B\sum P_{ul} n_u n_l \tag{2.68}$$

where n_u is the number density of electrons existing in upper energy states and n_l is the number density of *empty states* with lower energy. The large difference from absorption is in the fact that, usually speaking, at a given temperature electrons are found only in the vicinity of conduction band minimum and light emission is observed only from these electrons. Then, Eq. (2.68) can be written as:

$$L = B'\left(h\nu - E_g + E_p\right)^{1/2} \exp\left(-\frac{h\nu - E_g}{k_B T}\right) \tag{2.69}$$

confirming that emission is only observed in the vicinity of E_g. In the case of indirect transitions, light emission occurs from electronic transitions accompanied by phonon emission (cold band); light emission at higher energy corresponding to phonon absorption (hot band) has a relatively small probability since it requires the presence of thermal phonons. Hot-band emission vanishes completely at low temperatures.

2.2.3 EXCITON

Although all electrons in crystals are specified by the energy band states they occupy, a characteristic excited state called the *exciton*, which is not derived from the band theory, exists in almost all semiconductors or ionic crystals. Consider the case where one electron is excited in the conduction band and a hole is left in the valence band. An attractive Coulomb potential exists between them and can result in a bound state analogous to a hydrogen atom. This configuration is called an *exciton*. The binding energy of an exciton is calculated, by analogy, to a hydrogen atom as:

$$G_{ex} = -\frac{m_r^* e^4}{32\pi^2 \hbar^2 \epsilon^2} \cdot \frac{1}{n^2} \tag{2.70}$$

where n (= 1, 2, 3, …) is a quantum number specifying the states, ϵ is the dielectric constant of crystals, and m^* is the reduced mass of an exciton.

An exciton can move freely through the crystal. The energy levels of the free exciton are shown in Figure 2.10. The state corresponding to the limit of $n \to \infty$ is the minimum of the conduction band, as shown in the figure. The energy of the lowest exciton state obtained by putting $n = 1$ is:

$$E_{ex}(n = 1) = E_g - |G_{ex}| \tag{2.71}$$

Two or three kinds of excitons can be generated, depending on the splitting of the valence band, as shown in Figure 2.7. They are named, from the top of the valence band, A- and B-excitons in zinc-blende type crystals; and A-, B-, and C-excitons in wurtzite-type crystals. There are two kinds of A-excitons in zinc-blende materials originating from the existence of a light and heavy hole, as has already been noted.

Excitons create several sharp absorption lines in the energy region just below E_g. Figure 2.11 shows the absorption spectra of excitons in CdS.[2] One can easily recognize the absorption peaks due to A-, B-, and C-excitons with $n = 1$, and the beginning of the interband absorption transition corresponding to $n \to \infty$ (E_g). The order of magnitude of absorption coefficient reaches 10^5 cm^{-1} beyond E_g, as n from the figure. As noted previously, the absorption coefficient in the neighborhood of E_g in a material with indirect transition, like GaP, is three to four orders of magnitude smaller than the case of direct transition.

An exciton in the $n = 1$ state of a direct-gap material can be annihilated by the recombination of the electron-hole pair; this produces a sharp emission line. The emission from the states

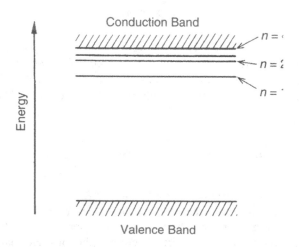

FIGURE 2.10 Energy levels of a free exciton.

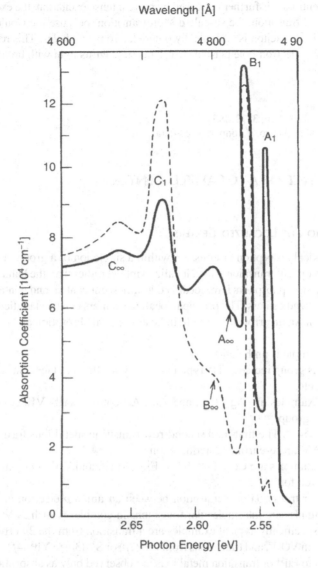

FIGURE 2.11 The exciton absorption spectrum of CdS (at 77K). The solid and broken lines correspond to the cases that the polarization vectors of incident light are parallel and perpendicular to the c-axis of the crystal, respectively. (From Mitsuhashi, H. and Fujishiro, Y., personal communication. With permission.)

corresponding to the larger n states is usually very weak because such states relax rapidly to the $n = 1$ state and emission generally occurs from there.

With intense excitation, excitons of very high concentrations can be produced; excitonic molecules (also called biexcitons) analogous to hydrogen molecules are formed from two single excitons by means of covalent binding. The exciton concentration necessary for the formation of excitonic molecules is usually of the order of magnitude of about 10^{16} cm^{-3}. The energy of the excitonic molecule is given by·

$$E_m = 2E_{ex} - G_m \qquad (2.72)$$

where G_m is the binding energy of the molecule. The ratio of G_m–G_{ex} depends on the ratio of electron effective mass to hole effective mass and lies in the range of 0.03–0.3. An excitonic molecule emits a photon of energy E_{ex}–G_m, leaving a single exciton behind.

If the exciton concentration is further increased by more intense excitation, the exciton system undergoes the insulator-metal transition, the so-called Mott transition, because the Coulomb force between the electron and hole in an exciton is screened by other electrons and holes. This results in the appearance of the high-density electron-hole plasma state. This state emits light with broadband spectra.

REFERENCES

1. Chadi, D.J., *Phys. Rev.*, B16, 3572, 1977.
2. Mitsuhashi, H. and Fujishiro, Y., unpublished data.

2.3 LUMINESCENCE OF A LOCALIZED CENTER

Hajime Yamamoto

2.3.1 CLASSIFICATION OF LOCALIZED CENTERS

When considering optical absorption or emission within a single ion or a group of ions in a solid, it is appropriate to treat an optical transition with a localized model rather than the band model described in Section 2.2. Actually, most phosphors have localized luminescent centers and contain a far larger variety of ions than delocalized centers. The principal localized centers can be classified by their electronic transitions as follows (next, an arrow to the right indicates optical absorption and to the left, emission):

1. $1s \rightleftharpoons 2p$; an example is an F center.
2. $ns^2 \rightleftharpoons nsnp$. This group includes Tl^+-type ions; i.e., Ga^+, In^+, Tl^+, Ge^{2+}, Sn^{2+}, Pb^{2+}, Sb^{3+}, Bi^{3+}, Cu^-, Ag^-, Au^-, etc.
3. $3d^{10} \rightleftharpoons 3d^9 4s$. Examples are Ag^+, Cu^+, and Au^+. Acceptors in IIb–VIb compounds are not included in this group.
4. $3d^n \rightleftharpoons 3d^n$, $4d^n \rightleftharpoons 4d^n$. The first- and second-row transition-metal ions form this group.
5. $4f^n \rightleftharpoons 4f^n$, $5f^n \rightleftharpoons 5f^n$; rare-earth and actinide ions.
6. $4f^n \rightleftharpoons 4f^{n-1}5d$. Examples are Ce^{3+}, Pr^{3+}, Sm^{2+}, Eu^{2+}, Tm^{2+}, and Yb^{2+}. Only absorption transitions are observed for Tb^{3+}.
7. A charge-transfer transition or a transition between an anion p electron and an empty cation orbital. Examples are intramolecular transitions in complexes such as VO_4^{3-}, WO_4^{2-}, and MoO_4^{2-}. More specifically, typical examples are a transition from the $2p$ orbital of O^{2-} to the $3d$ orbital of V^{5+} in VO_4^{3-}, and transitions from O^{2-} ($2p$) or S^{2-} ($3p$) to $Yb^{3+}(4f)$. Transitions from anion p orbitals to Eu^{3+} or transition metal ions are observed only as absorption processes.
8. $\pi \rightleftharpoons \pi^*$ and $n \rightleftharpoons \pi^*$. Organic molecules having π electrons make up this group. The notation n indicates a nonbonding electron of a heteroatom in an organic molecule.

2.3.2 CONFIGURATIONAL COORDINATE MODEL[1-5]

2.3.2.1 Description by a Classical Model

The configurational coordinate model is often used to explain optical properties, particularly the effect of lattice vibrations, of a localized center. In this model, a luminescent ion and the ions at its nearest neighbor sites are selected for simplicity. In most cases, one can regard these ions as an isolated molecule by neglecting the effects of other distant ions. In this way, the huge number of actual vibrational modes of the lattice can be approximated by a small number or a combination of specific normal coordinates. These normal coordinates are called the *configurational coordinates*. The *configurational coordinate model* explains optical properties of a localized center on the basis of potential curves, each of which represents the total energy of the molecule in its ground or excited state as a function of the configurational coordinate (Figure 2.12). Here, the total energy means the sum of the electron energy and ion energy.

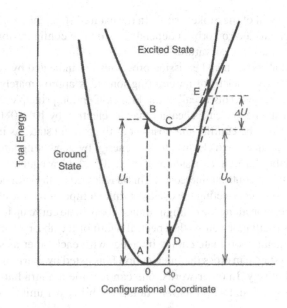

FIGURE 2.12 A schematic illustration of a configurational coordinate model. The two curves are modified by repulsion near the intersection (broken lines). The vertical broken lines A B and C D indicate the absorption and emission of light, respectively.

To understand how the configurational coordinate model is built, one is first reminded of the adiabatic potential of a diatomic molecule, in which the variable on the abscissa is simply the interatomic distance. In contrast, the adiabatic potential of a polyatomic molecule requires a multidimensional space, but it is approximated by a single configurational coordinate in the one-dimensional configurational coordinate model. In this model, the totally symmetric vibrational mode or the "breathing mode" is usually employed. Such a simple model can explain a number of facts qualitatively, such as:

1. Stokes' law; i.e., the fact that the energy of absorption is higher than that of emission in most cases. The energy difference between the two is called the Stokes' shift.
2. The widths of absorption or emission bands and their temperature dependence.
3. Thermal quenching of luminescence. It must be remarked, however, that the one-dimensional model gives only a qualitative explanation of thermal quenching. A quantitatively valid explanation can be obtained only by a multidimensional model.[6]

Following the path of the optical transition illustrated in Figure 2.12, presume that the bonding force between the luminescent ion and a nearest-neighbor ion is expressed by Hooke's law. The deviation from the equilibrium position of the ions is taken as the configurational coordinate denoted as Q. The total energy of the ground state, U_g, and that of the excited state, U_e, are given by the following relations:

$$U_g = K_g \frac{Q^2}{2} \tag{2.73a}$$

$$U_e - K_e \frac{(Q - Q_0)^2}{2} + U_0 \tag{2.73b}$$

where K_g and K_e are the force constants of the chemical bond, Q_0 is the interatomic distance at the equilibrium of the ground state, and U_0 is the total energy at $Q = Q_0$.

The spatial distribution of an electron orbital is different between the ground and excited states, giving rise to a difference in the electron wavefunction overlap with neighboring ions. This difference further induces a change in the equilibrium position and the force constant of the ground and excited states and is the origin of the Stokes' shift. In the excited state, the orbital is more spread out, so that the energy of such an electron orbital depends less on the configuration coordinate; in other words, the potential curve has less curvature.

In Figure 2.12, optical absorption and emission processes are indicated by vertical broken arrows. As this illustration shows, the nucleus of an emitting ion stays approximately at the same position throughout the optical processes. This is called the Franck-Condon principle. This approximation is quite reasonable since an atomic nucleus is heavier than an electron by 10^3–10^5 times. At 0K, the optical absorption proceeds from the equilibrium position of the ground state, as indicated by the arrow A $\rightarrow$ B. The probability for an excited electron to lose energy by generating lattice vibration is 10^{12} to 10^{13} s^{-1}, while the probability for light emission is at most 10^9 s^{-1}. Consequently, state B relaxes to the equilibrium position C before it emits luminescence. This is followed by the emission process C $\rightarrow$ D and the relaxation process D $\rightarrow$ A, completing the cycle. At finite temperature, the electron state oscillates around the equilibrium position along the configurational coordinate curve up to the thermal energy of kT. The amplitude of this oscillation causes the spectral width of the absorption transition.

When two configurational coordinate curves intersect with each other as shown in Figure 2.12, an electron in the excited state can cross the intersection E assisted by thermal energy and can reach the ground state nonradiatively. In other words, one can assume a nonradiative relaxation process with the activation energy ΔU, and with the transition probability per unit time N given by:

$$N = s\exp\frac{-\Delta U}{kT} \tag{2.74}$$

where s is a product of the transition probability between the ground and excited states and a frequency, in which the excited state reaches the intersection E. This quantity s can be treated as a constant, since it is only weakly dependent on temperature. It is called the *frequency factor* and is typically of the order of 10^{13} s^{-1}.

By employing Eq. (2.74) and letting W be the luminescence probability, the luminescence efficiency η can be expressed as:

$$\eta = \frac{W}{W+N} = \left[1 + \frac{s}{W}\exp\frac{-\Delta U}{kT}\right]^{-1} \tag{2.75}$$

If the equilibrium position of the excited state C is located outside the configurational coordinate curve of the ground state, the excited state intersects the ground state in relaxing from B to C, leading to a nonradiative process.

It can be shown by quantum mechanics that the configurational coordinate curves can actually intersect each other only when the two states belong to different irreducible representations. Otherwise, the two curves behave in a repulsive way to each other, giving rise to an energy gap at the expected intersection of the potentials. It is, however, possible for either state to cross over with high probability, because the wavefunctions of the two states are admixed near the intersection. In contrast to the above case, the intersection of two configurational coordinate curves is generally allowed in a multidimensional model.

2.3.2.2 Quantum Mechanical Description

The classical description discussed previously cannot satisfactorily explain observed phenomena, e.g., spectral shapes and nonradiative transition probabilities. It is thus necessary to discuss the configurational coordinate model based on quantum mechanics.

Suppose that the energy state of a localized center involved in luminescence processes is described by a wavefunction Ψ. It is a function of both electronic coordinates $\mathbf{r}$ and nuclear coordinates $\mathbf{R}$ but can be separated into the electronic part and the nuclear part by the *adiabatic approximation*:

$$\Psi_{nk}(\mathbf{r},\mathbf{R}) = \Psi_k(\mathbf{r},\mathbf{R})\chi_{nk}(\mathbf{R}) \tag{2.76}$$

where n and k are the quantum numbers indicating the energy states of the electron and the nucleus, respectively. For the nuclear wavefunction $\chi_{nk}(\mathbf{R})$, the time-independent Schrödinger equation can be written as follows:

$$\left\{ -\sum_{\alpha} \left(\hbar^2/2M_\alpha \right) \Delta \mathbf{R}_\alpha + U_k(\mathbf{R}) \right\} \chi_{nk}(\mathbf{R}) = E_{nk} \chi_{nk}(\mathbf{R}) \tag{2.77}$$

with α being the nuclear number, M_α the mass of the αth nucleus, $\Delta \mathbf{R}_\alpha$ the Laplacian of $\mathbf{R}_\alpha$, and E_{nk} the total energy of the localized center. The energy term $U_k(\mathbf{R})$ is composed of two parts: the energy of the electrons and the energy of the electrostatic interaction between the nuclei around the localized center. Considering Eq. (2.77), one finds that $U_k(\mathbf{R})$ plays the role of the potential energy of the nuclear wavefunction χ_{nk}. (Recall that the electron energy also depends on $\mathbf{R}$.) Thus, $U_k(\mathbf{R})$ is an adiabatic potential and it forms the configurational coordinate curve when one takes the coordinate Q as $\mathbf{R}$. When $U_k(\mathbf{R})$ is expanded in a Taylor series up to second order around the equilibrium position of the ground state, the potentials are expressed by Eq. (2.73). For a harmonic oscillation, the second term is the first nonvanishing term, while the first term is nonzero only when the equilibrium position is displaced from the original position. In the latter case, the first term is related to the Jahn-Teller effect. Sometimes, the fourth term in the expansion may also be present, signaling anharmonic effects. In the following, consider for simplicity only a single coordinate or a two-dimensional model.

Consider a harmonic oscillator in a potential shown by Eq. (2.73). This oscillator gives discrete energy levels inside the configurational coordinate curves, as illustrated in Figure 2.13.

$$E_m = \left(m + 1/2 \right) \hbar \omega \tag{2.78}$$

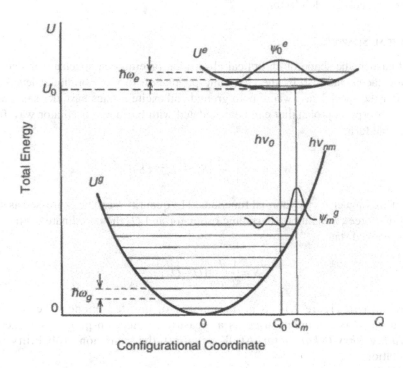

FIGURE 2.13 Discrete energy levels due to lattice vibration, each with the energy of $\hbar \omega$ and the wavefunctions ψ_0^e and ψ_m^g of harmonic oscillators representing the two states. The notation v means the frequency at the emission peak. A luminescent transition can occur at V_{nm}.

where ω is the proper angular frequency of the harmonic oscillator.

The electric-dipole transition probability, W_{nm}, between the two vibrational states n and m is given by:

$$W_{nm} = \left| \iint \Psi_e \chi_{en}^* er \Psi_g \chi_{gm}\, dr dQ \right|^2 = \left| \int \chi_{en}^* \chi_{gm} M_{eg}(Q) dQ \right|^2 \qquad (2.79)$$

Here,

$$M_{eg}(Q) \equiv \int \Psi_e^*(r,Q) er \Psi_g(r,Q) dr \qquad (2.80)$$

When the transition is allowed, M_{eg} can be placed outside the integral, because it depends weakly on Q. This is called the Condon approximation and it makes Eq. (2.79) easier to understand as:

$$W_{nm} = \left| M_{eg}(Q) \right|^2 \cdot \left| \int \chi_{en}^* \chi_{gm}\, dQ \right|^2 \qquad (2.81)$$

The wavefunction of a harmonic oscillator has the shape illustrated in Figure 2.13. For m (or n) = 0, it has a Gaussian shape; while for m (or n) $\neq$ 0, it has maximum amplitude at both ends and oscillates m times with a smaller amplitude between the maxima. As a consequence, the integral $|\int \chi_{en}^* \chi_{gm} dQ|$ takes the largest value along a vertical direction on the configurational coordinate model. This explains the Franck-Condon principle in terms of the shapes of wavefunctions. One can also state that this is the condition for which $W_{nm} \propto |\int \chi_{en}^* \chi_{gm} dQ|^2$ holds. The square of the overlap integral $|\int \chi_{en}^* \chi_{gm} dQ|^2$ is an important quantity that determines the strength of the optical transition and is often called the Franck-Condon factor.

2.3.3 SPECTRAL SHAPES

As described earlier, the shape of an optical absorption or emission spectrum is decided by the Franck-Condon factor and also by the electronic population in the vibrational levels at thermal equilibrium. For the special case where both ground and excited states have the same angular frequency ω, the absorption probability can be calculated with harmonic oscillator wavefunctions in a relatively simple form:

$$W_{nm} = e^{-S} \left[\frac{m!}{n!} \right] S^{n-m} \left[L_m^{n-m}(S) \right]^2 \qquad (2.82)$$

Here $L_\beta^\alpha(z)$ are Laguerre's polynomial functions. The quantity S can be expressed as shown next, with K being the force constant of a harmonic oscillator and Q_0 the coordinate of the equilibrium position of the excited state.

$$S = \frac{1}{2} \frac{K}{\hbar \omega} (Q - Q_0)^2 \qquad (2.83)$$

As can be seen in Figure 2.14, S is the number of emitted phonons accompanying the optical transition. It is commonly used as a measure of electron-phonon interaction and is called the *Huang-Rhys-Pekar factor*. At 0K or $m = 0$, the transition probability is given by the simple relation:

$$W_{n0} = S^n \frac{e^{-S}}{n!} \qquad (2.84)$$

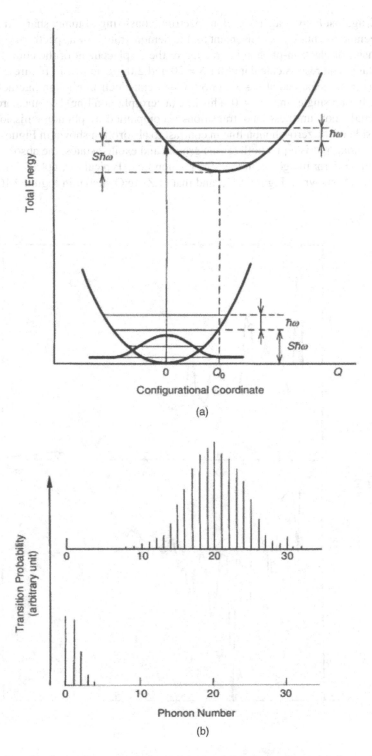

FIGURE 2.14 Part (b) shows the spectral shape calculated for the configurational coordinate model, in which the vibrational frequency is identical in the ground and excited states shown in (a). The upper figure in (b) shows a result for $S = 20$, while the lower figure is for $S = 2.0$. The ordinate shows the number of phonons n accompanying the optical transition. The transition for $n = 0$ is the zero-phonon line.

A plot of W_{n0} against n gives an absorption spectrum consisting of many sharp lines. This result is for a very special case, but it is a convenient tool to demonstrate how a spectrum varies as a function of the intensity of electron-phonon interaction or the displacement of the equilibrium position in the excited state. The results calculated for $S = 20$ and 2.0 are shown in Figure 2.14(a) and (b),[7] respectively. The peak is located at $n \cong S$. For $S \cong 0$ or weak electron-phonon interaction, the spectrum consists only of a single line at $n = 0$. This line (a zero-phonon line) becomes prominent when S is relatively small. For luminescence, transitions accompanied by phonon emission show up on the low-energy side of the zero-phonon line in contrast to absorption shown in Figure 2.14(b). If the energy of the phonon, $\hbar\,\omega$, is equal both for the ground and excited states, the absorption and emission spectra form a mirror image about the zero-phonon line. Typical examples of this case are the spectra of $YPO:_4Ce^{3+}$ shown in Figure 2.15,[8] and that of $ZnTe:O$ shown in Figure 2.16.[9]

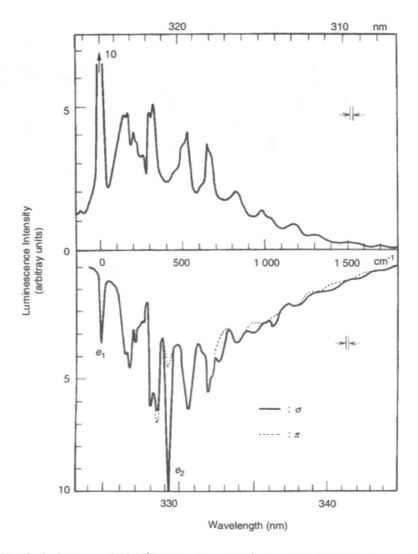

FIGURE 2.15 Optical spectra of $5d\ 4f(^2F_{3/2})$ transition of Ce^{3+} doped in a YPO_4 single crystal. The upper figure is an excitation spectrum, with the lower luminescence spectrum at 4.2K. The two spectra are positioned symmetrically on both sides of the zero-phonon line at 325.0 nm. Vibronic lines are observed for both spectra. The notations π and π indicate that the polarization of luminescence is parallel or perpendicular to the crystal c-axis, respectively.

FIGURE 2.16 Absorption and luminescence spectra of ZnTe:O at 20K. (From Merz, J.L., *Phys. Rev.*, 176, 961, 1968. With permission.)

Examples of other S values are described. For the A emission of KCl:Tl$^+$ having a very broadband width, S for the ground state is found to be 67, while for the corresponding A absorption band, S of the excited state is about 41.[10] Meanwhile, in Al$_2$O$_3$:Cr^{3+} (ruby), $S = 3$ for the narrow $^4A_2 \rightarrow {}^4T_2$ absorption band, and $S \approx 10^{-1}$ for the sharp R lines ($^4A_2 \leftrightarrow {}^4T_2$) were reported.[11] A very small value similar to that of R lines is expected for sharp lines due to $4f^n$ intraconfigurational transitions. The spectra of YPO$_4$:Ce^{3+} in Figure 2.15, which is due to $4f \leftrightarrow 5d$ transition, show $S \approx 1$.[8]

The abovementioned discussion has treated the ideal case of a transition between a pair of vibrational levels (gm) and (en) resulting in a single line. The fact is, however, that each line has a finite width even at 0K as a result of zero-point vibration.

Next, consider a spectral shape at finite temperature T. In this case, many vibrational levels at thermal equilibrium can act as the initial state, each level contributing to the transition with a probability proportional to its population density. The total transition probability is the sum of such weighted probabilities from these vibrational levels. At sufficiently high temperature, one can treat the final state classically and assume the wavefunction of the final state is a δ-function and the population density of the vibrational levels obeys a Boltzmann distribution. By this approximation, the absorption spectrum has a Gaussian shape given by:

$$W(\hbar\omega) = \frac{1}{\sqrt{2\pi}\sigma_a} \exp\left[\frac{-(\hbar\omega - U_1)^2}{2\sigma_a^2}\right] \qquad (2.85)$$

Here,

$$U_1 \equiv U_0 + \frac{Ke}{2Q_0^2} \qquad (2.86)$$

$$\{\sigma_a(T)\}^2 \equiv S_e \frac{(\hbar\omega_e)^3}{\hbar\omega_g} \coth\frac{\hbar\omega_g}{2kT} \qquad (2.87)$$

$$\approx 2S_e \cdot kT \cdot \frac{(\hbar\omega_e)^3}{(\hbar\omega_g)^2} \qquad (2.88)$$

where $\hbar\omega$ is the energy of an absorbed phonon, and S_e denotes S of the excited state. The coefficient on the right-hand side of Eq. (2.85) is a normalization factor defined to give $\int W(\hbar\omega)d\omega = 1$. By defining ω as the spectral width, which satisfies the condition $W(U_1 + w) = W(U_1)/e$, one finds:

$$\omega = \sqrt{2}\sigma_a \qquad (2.89)$$

At sufficiently high temperature, the spectral width w is proportional to $\sqrt{T}$ and the peak height is inversely proportional to $\sqrt{T}$. The relations for the luminescence process are found simply by exchanging the suffixes e and g of the abovementioned equations.

In experiments, a Gaussian shape is most commonly observed. It appears, however, only when certain conditions are satisfied, as is evident from the abovementioned discussion. In fact, more complicated spectral shapes are also observed. A well-known example is the structured band shape of a transition observed for Tl^+-type ions in alkali halides.[6] It has been shown that this shape is induced by the Jahn-Teller effect and can be described by a configurational coordinate model based on six vibrational modes around a Tl^+-type ion. Another example is the asymmetric luminescence band of $Zn_2SiO_4:Mn^{2+}$. To explain this shape, a configurational coordinate model with a small difference between the excited- and ground-state potential minima ($S = 1.2$) has been proposed.[12]

In summarizing the discussion of the spectral shape based on the configurational coordinate model, one can review the experimental results on luminescence bandwidths. In Figure 2.17,[13] the halfwidth of the luminescence band of typical activators in phosphors is plotted against the peak wavelength.[13] The activators are classified by the type of optical transition described in Section 2.3.1. When the $d \rightleftharpoons d$ (Mn^{2+}), $f \rightleftharpoons d$ (Eu^{2+}), and $s^2 \rightleftharpoons sp$ transitions (Sn^{2+}, Pb^{2+}, and Sb^{3+}) are sequentially compared, one finds that the halfwidth increases in the same order. This is apparently because the overlap of the electron wavefunctions between the excited and ground states increases in the abovementioned

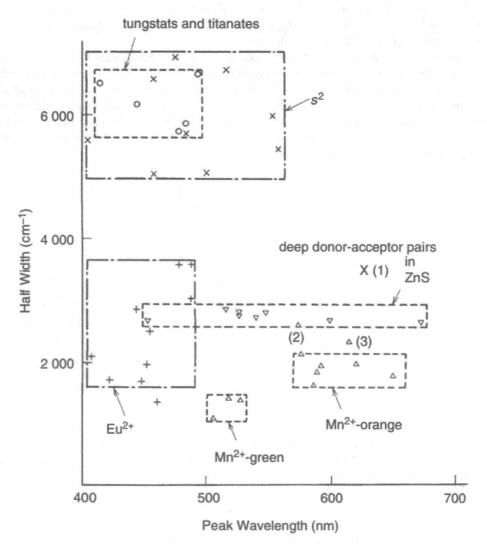

FIGURE 2.17 A plot of peak wavelength and halfwidth of various phosphors. The points (1)–(3) indicate the following materials. The luminescence of (2) and (3) originates from Mn^{2+} principally. (1) $(Sr,Mg)_3(PO_4)_2:Sn^{2+}$; (2) $Sr_5(PO_4)_3F:Sb^{3+}, Mn^{2+}$; (3) $CaSiO_3:Pb^{2+}, Mn^{2+}$. (From Narita, K., Tech. Digest Phosphor Res. Soc. 196th Meeting, 1983 (in Japanese). With permission.)

order. The difference in the wavefunction overlap increases the shift of the equilibrium position of the excited state, Q_0, and consequently Stokes' shift and the halfwidth increase as well.

Weak electron-phonon interactions give line spectra. The line width in this case results from factors other than those involved in the configurational coordinate model. Such factors are briefly reviewed next.

2.3.3.1   Line Broadening by Time-Dependent Perturbation

The most fundamental origin of the line width is the energy fluctuation of the initial and final states of an optical transition caused by the uncertainty principle. With τ being the harmonic mean of the lifetimes of the initial and final states, the spectral line width is given by $\hbar/\tau$ and the spectral shape takes a Lorentzian form:

$$I(v) = \frac{1}{\pi} \cdot \frac{1/v_L}{1 + (v - v_0)^2 / v_L^2} \tag{2.90}$$

where $v_L \equiv (\tau_i^{-1} + \tau_f^{-1})$, v is the frequency of light, v_0 the frequency at the line center, and τ_i and τ_f are the lifetimes of the initial and final states, respectively.

In addition to the spectral width given by Eq. (2.89), there are other kinds of time-dependent perturbation contributing to the width. They are absorption and emission of a photon, which makes "the natural width," and absorption and emission of phonons. The fluorescent lifetime of a transition-metal ion or a rare-earth ion is of the order of 10^{-6} s at the shortest, which corresponds to 10^{-6} cm^{-1} in spectral width. This is much sharper than actually observed widths of about 10 cm^{-1}; the latter arise from other sources, as discussed next.

At high temperatures, a significant contribution to the width is the Raman scattering of phonons. This process does not have any effect on the lifetime but does make a Lorentzian contribution to the width. The spectral width due to the Raman scattering of phonons, ΔE, depends strongly on temperature, as can be seen next:

$$\Delta E = \alpha \left(\frac{T}{T_D}\right)^7 \int_0^{X_0} \frac{x^6 e^x}{(e^x - 1)^2} dx, \; X_0 = \frac{\hbar \omega_a}{KT} \tag{2.91}$$

where T_D is Debye temperature and α is a constant that includes the scattering probability of phonons.

2.3.3.2   Line Broadening by Time-Independent Perturbation

When the crystal field around a fluorescent ion has statistical distribution, it produces a Gaussian spectral shape.

$$I(v) = \frac{1}{\sqrt{2\pi}\sigma} \exp\left\{-\frac{(v - v_0)}{2\sigma^2}\right\} \tag{2.92}$$

with σ being the standard deviation.

Line broadening by an in homogeneously distributed crystal field is called *inhomogeneous broadening*, while the processes described in Section 2.3.3.1 result in *homogeneous broadening*.

2.3.4   NONRADIATIVE TRANSITIONS

The classical theory describes a *nonradiative transition* as a process in which an excited state relaxes to the ground state by crossing over the intersection of the configurational coordinate curve through thermal excitation or other means (refer to Section 2.3.2). It is often observed, however, that the experimentally determined activation energy of a nonradiative process depends upon temperature.

This problem has a quantum mechanical explanation: that is, an optical transition accompanied by absorption or emission of m–n phonons can take place when an nth vibrational level of the excited state and an mth vibrational level of the ground state are located at the same energy. The probability of such a transition is also proportional to a product of the Franck-Condon coefficient and thermal distribution of population in the ground state, giving the required temperature-dependent probability. When the phonon energy is the same both at the ground and excited states, as shown in Figure 2.14, the nonradiative relaxation probability is given by:

$$N_p = N_{eg} \cdot \exp\left\{-S(2\langle n\rangle+1)\right\} \sum_{j=0}^{\infty} \frac{(S\langle n\rangle)^j \left\{S(1+\langle n\rangle)\right\}^{p+j}}{j!(p+j)!} \tag{2.93}$$

where let $p \equiv m - n$, and $\langle n\rangle$ denotes the mean number of the vibrational quanta n at temperature T expressed by $\langle n\rangle = \left\{\exp(\hbar\omega / kT)-1\right\}^{-1}$. The notation N_{eg} implies the overlap integral of the electron wavefunctions.

The temperature dependence of N_p is implicitly included in $\langle n\rangle$. Obviously, Eq. (2.93) does not have a form characterized by a single activation energy. If written in a form such as $N_p \propto \exp(-E_p/kT)$, one obtains:

$$E_p = \left(\langle n\rangle_p - \langle n\rangle\right)\hbar\omega \tag{2.94}$$

where $\langle n\rangle_p\, \hbar\,\omega$ is the mean energy of the excited state subject to the nonradiative process. The energy E_p increases with temperature and one obtains $E_p < \Delta U$ at sufficiently low temperature.

If $S < 1/4$ or if electron-phonon interaction is small enough, Eq. (2.93) can be simplified by leaving only the term for $j = 0$.

$$N_p = N_{eg} \cdot \exp\left\{-S(1+2\langle n\rangle)\right\}\left\{-S(1+\langle n\rangle)\right\}^p \big/ p! \tag{2.95}$$

In a material that shows line spectra, such as rare-earth ions, the dominating nonradiative relaxation process is due to multiphonon emission. If E_{gap} is the energy separation between two levels, the nonradiative relaxation probability between these levels is given by an equation derived by Kiel[15]:

$$N_p = A_K \epsilon^p \left(1 + \langle n\rangle\right)^p \tag{2.96}$$

$$p\hbar\omega = E_{gap} \tag{2.97}$$

where A_K is a rate constant and ϵ is a coupling constant.

Eq. (2.95) can be transformed to the same form as Eq. (2.96) using the conditions $S \approx 0$, $\exp\{-S(1 + 2\langle n\rangle)\} \approx 1$, $S^p/p! \approx \epsilon^p$ and $A_K = N_{eg}$ although Eq. (2.95) was derived independently of the configurational coordinate model. If two configurational coordinate curves have the same curvature and the same equilibrium position, the curves will never cross and there is no relaxation process by thermal activation between the two in the framework of the classical theory. However, thermal quenching of luminescence can be explained for such a case by taking phonon-emission relaxation into account, as predicted by Kiel's equation.

REFERENCES

1. Klick, C.C. and Schulman, J.H., *Solid State Physics*, Vol. 5, Seitz, F. and Turnbull, D., Eds., Academic Press, 1957, pp. 97–116.
2. Curie, D., *Luminescence in Crystals*, Methuen & Co., 1963, pp. 31–68.
3. Maeda, K., *Luminescence*, Maki Shoten, 1963, pp. 6–10 and 37-48 (in Japanese).
4. DiBartolo, B., *Optical Interactions in Solids*, John Wiley & Sons, 1968, pp. 420–427.

5. Kamimura, A., Sugano, S., and Tanabe, Y., *Ligand Field Theory and Its Applications*, First Edition, Shokabo, 1969, pp. 269–321 (in Japanese).
6. Fukuda, A., *Bussei*, 4, 13, 1969 (in Japanese).
7. Keil, T., *Phys. Rev.*, 140, A601, 1965.
8. Nakazawa, E. and Shionoya, S., *J. Phys. Soc. Jpn.*, 36, 504, 1974.
9. Merz, J.L., *Phys. Rev.*, 176, 961, 1968.
10. Williams, F.E., *J. Chem. Phys.*, 19, 457, 1951.
11. Fonger, W.H. and Struck, C.W., *Phys. Rev.*, B111, 3251, 1975.
12. Klick, C.C. and Schulman, J.H., *J. Opt. Soc. Am.*, 42, 910, 1952.
13. Narita, K., *Tech. Digest Phosphor Res. Soc. 196th Meeting*, 1983 (in Japanese).
14. Struck, C.W. and Fonger, W.H., *J. Lumin.*, B111, 3251, 1975.
15. Kiel, A., *Third Int. Conf. Quantum Electronics*, Paris, Grivet, P. and Bloembergen, N., Eds., Columbia University Press, p. 765, 1964.

2.4 TRANSIENT CHARACTERISTICS OF LUMINESCENCE

Eiichiro Nakazawa

This section focuses on transient luminescent phenomena, that is, time-dependent emission processes such as luminescence afterglow (phosphorescence), thermally stimulated emission (thermal glow), photo (infrared)-stimulated emission, and photoquenching. All of these phenomena are related to a quasistable state in a luminescent center or an electron or hole trap.

2.4.1 DECAY OF LUMINESCENCE

Light emission that persists after the cessation of excitation is called *afterglow*. Following the terminology born in the old days, luminescence is divided into fluorescence and phosphorescence according to the duration time of the afterglow. The length of the duration time required to distinguish the two is not clearly defined. In luminescence phenomena in inorganic materials, the afterglow that can be perceived by the human eye, namely that persisting for longer than 0.1 s after cessation of excitation, is usually called phosphorescence. Fluorescence implies light emission during excitation. Therefore, fluorescence is the process in which the emission decay is ruled by the lifetime (<10 ms) of the emitting state of a luminescence center, while the phosphorescence process is ruled by a quasistable state of a center or a trap.

In organic molecules, fluorescence and phosphorescence are distinguished by a quite different definition. The two are distinguished by whether the transition to emit light is allowed or forbidden by spin selection rules. Light emission due to an allowed transition is called *fluorescence*, while that due to a forbidden transition, usually showing a long afterglow, is called *phosphorescence* (see Chapter 13 in Volume 2).

2.4.1.1 Decay of Fluorescence

The decay process of the luminescence intensity $I(t)$ after the termination of excitation at $t = 0$ is generally represented by an exponential function of the elapsed time after the excitation.

$$I(t) = I_0 \exp\left(-t/\tau\right) \tag{2.98}$$

Where τ is the decay time constant of the emission. It should be noted that the emission decay curve of nonlocalized centers, donor-acceptor pairs, for example, is not always represented in the exponential form of Eq. (2.98) (see Chapter 3).

If one denotes the number of excited luminescence centers in a unit volume by n^*, and the radiative and nonradiative transition probabilities by W_R and W_{NR}, respectively, then the rate equation for n^* is:

$$\frac{dn^*}{dt} = -\left(W_R + W_{NR}\right)n^* \tag{2.99}$$

and the solution of this equation is:

$$n^*(t) = n_0^* \exp\left[-(W_R + W_{NR})t\right] \tag{2.100}$$

Here, n^* is the value at $t = 0$, that is, at the endpoint of excitation or, in other words, at the start point of the afterglow.

Therefore, the lifetime of the center, which corresponds to the elapsed time for n^* to be decreased by the factor of e^{-1} of n^*, is $(W_R + W_{NR})^{-1}$. Since the emission intensity is proportional to n^*, the decay time of the afterglow in Eq. (2.98) is equal to the lifetime of the center:

$$\tau = \left(W_R + W_{NR}\right)^{-1} \tag{2.101}$$

and the luminescence efficiency of the center is given by:

$$\eta = \frac{W_R}{W_R + W_{NR}} \tag{2.102}$$

The radiative transition probability W_R of an emitting state is the summation of the spontaneous emission probability $A_{m \to n}$ from the state m to all the final states n (see Eq. 2.29),

$$W_R = \sum_n A_{m \to n} = \sum_n \frac{1}{\tau_{mn}} \tag{2.103}$$

where τ_{mn} is given by Eq. (2.32) or (2.35). The ratio of the transition probability to a particular final state n to W_R, $A_{m \to n}/W_R$, is called the *branching ratio*.

While the nonradiative transition probability W_{NR} is generally ruled by thermal relaxation processes (i.e., the emission of energy into lattice vibrations), it is also increased by the effect of resonant energy transfer between optical centers (see Chapter 4).

The thermal relaxation in a luminescence center can be divided into two types of mechanisms as shown by the two configurational coordinate diagrams (a) and (b) in Figure 2.18. In the first type (a), the center is thermally activated from point A, the point of the lowest energy on the exited state II, to the crossing point C where the electronic states of the excited and ground states are intermixed, and then thermally released from C to B on the ground state I. The energy ε necessary to excite the center from A to C is called the *thermal activation energy*. The probability that the center will make the transition from state II to state I by thermal activation via point C is generally given by:

$$a = s \exp\left(\frac{-\varepsilon}{kT}\right) \tag{2.104}$$

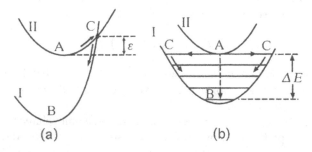

(a)   (b)

FIGURE 2.18   Configurational coordinate models of nonradiative relaxation processes: thermal activation type (a), and multiphonon type (b).

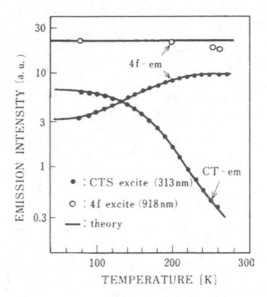

FIGURE 2.19 Temperature dependence of two types of the emission in $Y_2O_2S:Yb^{3+}$, from the CTS (charge-transfer state) and from a 4f-emitting level of Yb3+ ions. (From Nakazawa, E., *J. Lumin.*, 18/19, 272, 1979. With permission.)

Therefore, the nonradiative transition probability by thermal activation is given by:

$$W_{NR} = a_{II \to I} = s \exp\left(\frac{-\varepsilon}{kT}\right) \tag{2.105}$$

where k is the Boltzmann constant and s is the frequency factor. This type of nonradiative transition is strongly dependent on temperature, resulting in thermal quenching, that is, the decrease of emission efficiency and shortening of the emission decay time at high temperature (see Eqs. (2.101) and (2.102)).

An example of the thermal quenching effect is shown in Figure 2.19 for $Y_2O_2S:Yb^{3+}$.[1] The emission from the charge-transfer state (CTS) of Yb^{3+} ions at 530 nm is strongly reduced by thermal quenching at high temperature. The 4f emission under CTS excitation (313 nm), however, is increased at high temperature due to the increased amount of excitation transfer from the CTS. The Figure also shows that, as expected, the emission from the 4f level at 930 nm is not so thermally quenched under direct 4f excitation into the emitting level (918 nm).

The second type of nonradiative transition is a multiphonon process shown in Figure 2.19***. This type is often observed in the relaxation between the 4f excited levels of rare-earth ions, where no cross-point exists between curves I and II in the configuration coordinate diagram because of the similarity of the electronic states. The transition between states I and II occurs at point A, where an energy gap ΔE exists between the states: namely, the transition from the pure electronic state of II to the electron-phonon-coupled state of I with n phonons takes place at A, which is followed by the instantaneous transfer to point C and relaxation to B. The nonradiative transition probability is, therefore, dependent on ΔE or n, the number of phonons necessary to fill the energy gap, since $\Delta E = n\,\omega_p$, where ω_p is the largest phonon energy. The nonradiative multiphonon transition probability is then given by:[2]

$$W_{NR}(\Delta E) = W_{NR}(0) e^{-\alpha \Delta E} \tag{2.106}$$

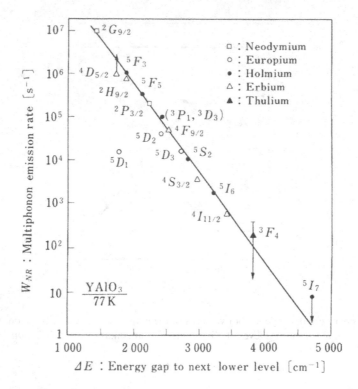

FIGURE 2.20 Energy gap law of nonradiative relaxation due to multiphonon processes. (From Weber, M.J., *Phys. Rev.*, B8, 54, 1993. With permission.)

where α depends on the character of the phonon (lattice vibration). Since the process is mainly due to the spontaneous emission of phonons, the temperature dependence of the probability is small. An experimental result[2] showing the applicability of Eq. (2.106) is shown in Figure 2.20.

2.4.1.2   Quasistable State and Phosphorescence

If one of the excited states of a luminescent center is a quasistable state (i.e., an excited state with very long lifetime), a percentage of the centers will be stabilized in that state during excitation. After excitation has ceased, afterglow is caused by the thermal activation of the state. This situation is illustrated using the configurational coordinate diagram in Figure 2.21, where state III is

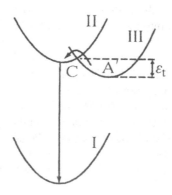

FIGURE 2.21 Configurational coordinate model of the luminescence afterglow (phosphorescence) via a quasistable state.

a quasistable state and state II is an emitting state with a radiative transition probability W_R. The center, once stabilized at A', transfers from state III to state II by thermal activation via point C. The probability of this activation, $a_{III \to II}$, is given by Eq. (2.104). Then, if $W_R \gg a_{III \to II}$, the decay time constant of the emission becomes almost equal to $1/a_{III \to II}$, that is, the lifetime of the quasistable state. The decay curve of the afterglow is represented by an exponential function that is similar to Eq. (2.98), and is strongly temperature dependent. The decay time constant of an emitting center with quasistable states is not usually longer than a second.

2.4.1.3 Traps and Phosphorescence

Excited electrons and holes in the conduction and valence bands of a phosphor can often be captured by impurity centers or crystal defects before they are captured by an emitting center. When the probability for the electron (hole) captured by an impurity or defect center to recombine with a hole (electron) or to be reactivated into the conduction band (valence band) is negligibly small, the center or defect is called a *trap*.

The electrons (holes) captured by traps may cause phosphorescence (i.e., long afterglow) when they are thermally reactivated into the conduction band (valence band) and then radiatively recombined at an emitting center. The decay time of phosphorescence due to traps can be as long as several hours and is often accompanied by photoconductive phenomena.

The decay curve of the afterglow due to traps is not generally represented by a simple exponential function. The form of the curve is dependent on the concentration of the traps and on the electron capture cross sections of the trap and the emitting center. Furthermore, it also depends on the excitation intensity level.

While several kinds of traps usually exist in practical phosphors, only one kind of electron trap is presumed to exist in the simple model shown in Figure 2.22. Let N be the trap concentration, and n_c and n_t the number of electrons per unit volume in the conduction band and trap states, respectively. The number of holes denoted by p is equal to $n_c + n_t$. The rate equation representing the decaying processes of the concentration of electrons and holes after the termination of excitation is:

$$\frac{dn_t}{dt} = -an_t + b(N - n_t)n_c$$
$$\frac{dp}{dt} = -rpn_c$$

(2.107)

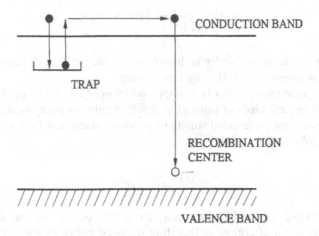

FIGURE 2.22   Luminescence afterglow process via a trap in an energy band scheme.

where a is the probability per second for a trapped electron to be thermally excited into the conduction band and is given by the same form as Eq. (2.45) with the density of states in the conduction band included in s. The probabilities that a free electron in the conduction band will be captured by a trap or to recombine with a hole are given by b and r, respectively. It is supposed that the number of the electrons n_c in the conduction band in the afterglow process is so small that $p \simeq n_t$ and $dp/dt \simeq dn_t/dt$. Then, the abovementioned two equations give:

$$\frac{dn_t}{dt} = \frac{-an_t^2}{n_t + (b/r)(N - n_t)} \tag{2.108}$$

This equation can be solved analytically for two cases: $b \ll r$ and $b \simeq r$.

First, the case of $b \ll r$, which presumes that the electrons once released from traps are not retrapped in the afterglow process. Eq. (2.108) then simplifies to:

$$\frac{dn_t}{dt} = -an_t \tag{2.109}$$

Since the emission intensity is given by $I(t) \propto dp/dt$, and $dp/dt \simeq dn_t/dt$ as mentioned earlier, then

$$I(t) = I_0 \exp(-at) \tag{2.110}$$

This simple exponential decay of afterglow is the same as the one due to the quasistable state mentioned previously and is called a first-order or monomolecular reaction type in the field of chemical reaction kinetics.

In the case $b \simeq r$, which means that the traps and emitting centers have nearly equal capturing cross sections, Eq. (2.108) can be simplified to:

$$\frac{dn_t}{dt} = -\frac{a}{N} n_t^2 \tag{2.111}$$

and then the number of trapped electrons per unit volume is given by:

$$n_t = \frac{n_{t0}}{1 + (N/an_{t0})t} \tag{2.112}$$

Approximating $I(t) \propto dn_t/dt$ as before, the decay curve of the afterglow is obtained as:

$$I(t) = \frac{I_0}{(1 + \gamma t)^2}, (\gamma = N/an_t) \tag{2.113}$$

This form is called the second-order or bimolecular reaction type, where the decay curve is changed by excitation intensity as well as by temperature.

While the treatise mentioned earlier is a simple model presuming a single kind of trap, the real phosphors may have several kinds of traps of different depths. In many real cases, therefore, the afterglow decay curve is not represented simply by a monomolecular or bimolecular reaction curve. It often fits into the following equation.

$$I(t) = \frac{I_0}{(1 + \gamma t)^n} \tag{2.114}$$

where n is around 0.5–2. If $t \gg 1/\gamma$, this decay curve can be approximated by $I(t) \propto t^{-n}$. The decay time constant of an afterglow is therefore denoted either by the $1/e$ decay time or the 10% decay time.

2.4.2 Thermoluminescence

When a phosphor with deep traps is excited for a while at rather low temperatures and then heated, it shows an increased afterglow called thermally stimulated luminescence due to the recombination of electrons thermally reactivated from the deep traps. This emission is also called thermoluminescence, and the temperature dependence of the emission intensity is called the glow curve, which is a good means to measure the depth (i.e., the activation energy of traps). Figure 2.23 shows the glow curves of ZnS:Cu phosphors with various co-activators.[3]

The measurement of a glow curve of a phosphor sample proceeds as follows.

1. The sample is cooled to a low temperature (liquid nitrogen is often used as coolant).
2. The sample is excited by UV light until the traps are filled with electrons or holes.
3. The excitation is terminated, and the temperature of the sample is raised at a constant rate, $dT/dt \equiv \beta$, while the intensity is recorded.
4. The temperature dependence of fluorescence is then measured under a constant UV excitation, which is used to calibrate the effect of temperature quenching on the thermoluminescence intensity.

The glow curve thus obtained is analyzed with the following theory. Assume that (1) a single kind of trap exists; (2) the decay of afterglow is of the first-order type given by Eq. (2.110), and (3) the probability for the trapped electrons to be thermally released into the conduction band is given by Eq. (2.104). Since the retrapping of the released electrons is neglected in the first-order kinetics, the change in the number of trapped electrons is:

$$\frac{dn_t}{dt} = -n_t s \exp(-\varepsilon/kT) \tag{2.115}$$

Integrated from a temperature T_0 to T with the relation $dT/dt = \beta$, this equation gives the number of residual electrons in the traps at T as:

$$n_t(T) = n_{t0} \exp\left(-\int_{t0}^{t} s \exp\left(\frac{-\varepsilon}{kT}\right) \cdot \frac{dT}{\beta}\right) \tag{2.116}$$

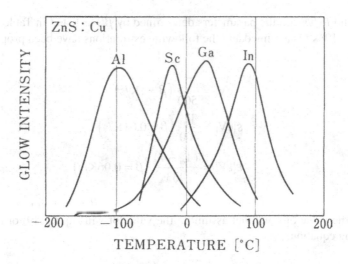

FIGURE 2.23 Glow curves of ZnS:Cu phosphors co-activated with Al, Sc, Ga, and In. (From Hoogenstraaten, W., *Philips Res. Rept.*, 13, 515, 1958. With permission.)

where n_{t0} is the number of the trapped electrons at the initial temperature T_0. Therefore, the emission intensity at T, approximated by $I \propto dn_t/dt$ as mentioned in the previous section, is given by:

$$I(T) \propto n_{t0}s\exp\left(\frac{-\varepsilon}{kT}\right)\exp\left(-\int_{T_0}^{T}s\exp\left(\frac{-\varepsilon}{kT}\right)\cdot\frac{dT}{\beta}\right) \tag{2.117}$$

Based on this equation, the following techniques have been proposed for obtaining the trap depth (activation energy ε) from a glow curve.

1. In the initial rising part of the glow peak on the low-temperature side where the number of trapped electrons is nearly constant, Eq. (2.117) is approximated by

$$I(T) \propto s\exp\frac{-\varepsilon}{kT} \tag{2.118}$$

Then the slope of the Arrhenius plot ($1/T$ vs. $\ln I$) of the curve in this region gives the trap depth ε. In fact, however, it is not easy to determine the depth with this method because of the uncertainty in fixing the initial rising portion.

2. Let the peak position of a glow curve be T_m. Then, the following equation derived from $dI/dT = 0$ should be valid.

$$\frac{\beta\varepsilon}{kT_m^2} = s\exp\left(\frac{-\varepsilon}{kT_m}\right) \tag{2.119}$$

If the frequency factor s is obtained in some manner, ε can be estimated by this relation from the observed value of T_m. Note that the temperature rise rate β should be kept constant throughout the measurement for this analysis. Randall and Wilkins[4] performed a numerical calculation based on this theory and obtained the following equation, which approximates the trap depth ε with 1% error.

$$\varepsilon = \frac{T_m - T_0\left(\beta/s\right)}{K\left(\beta/s\right)} \tag{2.120}$$

Here, $T_0(\beta/s)$ and $K(\beta/s)$ are the parameters determined by β/s as listed in Table 2.1.

For ZnS:Cu, $s = 10^9$ s^{-1} is assumed and the following estimations have been proposed for various values of β.[5]

$$\varepsilon|eV| = \frac{T_m}{500}\ (\beta = 1\text{K/s})$$

$$\varepsilon|eV| \approx \frac{T_m}{400}\ (\beta = 0.01\text{K/s})$$

$$\varepsilon|eV| = \frac{(T_m - 7)}{433}\ (\beta = 0.06\text{K/s})$$

3. If the glow curve is measured with two different rise rates, β_1 and β_2, it is apparent that one can obtain the value of ε without assuming the value of s in Eq. (2.119) or (2.120), using the following equation.

$$\frac{\varepsilon}{k}\left(\frac{1}{T_{m2}} - \frac{1}{T_{m1}}\right) = \ln\left(\frac{\beta_1}{\beta_2}\cdot\frac{T_{m2}^2}{T_{m1}^2}\right) \tag{2.121}$$

TABLE 2.1
Some Parameters (β/s, K(β/s) and $T_n(\beta/s)$)
Related to Thermoluminescence[5]

β/s [K]	K [K/eV]	T_n [K]
10^{-4}	833	35
10^{-5}	725	28
10^{-6}	642	22
10^{-7}	577	17
10^{-8}	524	13
10^{-9}	480	10
10^{-10}	441	7
10^{-11}	408	6
10^{-12}	379	6
10^{-13}	353	5
10^{-14}	331	5
10^{-15}	312	4

Source: From Curie, D., Luminescence in Crystals, John Wiley & Sons, 1963, chap. VI. With permission.

Hoogenstraaten[3] extended this method for many raising rates β_i, and, by plotting the curves ($1/T_{mi}$ vs. $\ln(T_{mi}^2/\beta_i)$), obtained the trap depth from the slope ε/k of the straight line connecting the plotted points as shown in Figure 2.24.[6] A numerical analysis[7] has shown that Hoogenstraaten's method gives the best result among several methods for obtaining trap depths from glow curves.

4. Many methods for obtaining the depth ε from the width of the peak of a glow curve have been proposed.[5] The results are listed next, where the width for a peak is defined in various ways, as shown in Figure 2.25.
$\varepsilon = kT_m^2\sigma$
$\varepsilon = 2kT_m(1.25T_m/\omega - 1)$
$\varepsilon = 1.52\, kT_m^2/\lambda - 3.16\, kT_m$
$\varepsilon = (1+\omega/\lambda)kT_m^2/\sigma$

Note that these methods are usable under certain restricted conditions.[5]

There is a method to obtain trap depth other than the glow-curve method described earlier. It is to use the temperature dependence of the decay time of afterglow, that is, phosphorescence. As mentioned in relation to Eq. (2.120), the decay time constant of the exponential afterglow due to the first-order reaction kinetics is equal to the inverse of the thermal detrapping probability, a^{-1}, and its temperature dependence is given by:

$$\tau_{1/e} = s^{-1}\exp\left(\frac{\varepsilon}{kT}\right) \qquad (2.122)$$

Therefore, the trap depth ε can be obtained from the measurements of the phosphorescence decay time $\tau_{1/e}$ at several different temperatures (T_i). An example is shown in Figure 2.26, in which the depth is obtained from the slope of the straight line connecting the Arrhenius plots of the observed values[8] for τ_i and T_i.

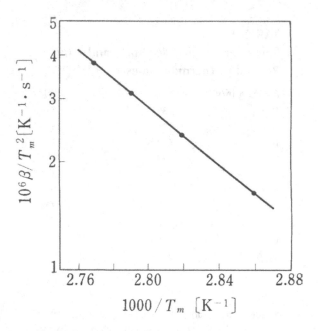

FIGURE 2.24   Hoogenstraaten plot showing the dependence of the peak temperature (T_m) of a glow curve on the temperature-raising rate (β). The slope of this line gives the depth (activation energy) of the trap. (From Avouris, P. and Morgan, T.N., *J. Chem. Phys.*, 74, 4347, 1981. With permission.)

A usable method with which trap depths and relative trap densities are obtained more easily and accurately was recently proposed.[9] In this method, the sample is excited periodically under a slowly varying temperature and the afterglow (phosphorescence) intensity is measured at several delay times (t_d) after the termination of excitation in each cycle.

The temperature dependence of the afterglow intensity at each delay time makes a peak at a certain temperature T_m. From the equation $dI/dT = 0$ and using either Eq. (2.110) or Eqs. (2.113) and (2.114), the following relation is obtained between the peak temperature T_m and the delay time t_d:

$$t_d = s^{-1} \exp\left(\frac{\varepsilon}{kT_m}\right) \tag{2.123}$$

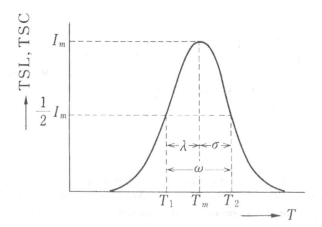

FIGURE 2.25   Predicted shape of a glow curve.

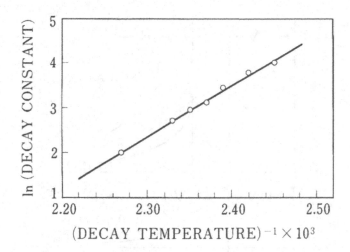

FIGURE 2.26 Temperature dependence of the decay time constant of ZnS:Cu. (From Bube, R.H., *Phys. Rev.*, 80, 655, 1950. With permission.)

Since Eq. (2.123) is similar to Eq. (2.122) (i.e., the decay time method), the method used there for obtaining the trap depth can be applied hereby, substituting t_d for τ, and T_m for T_i. An example of the measured afterglow intensity curves is shown in Figure 2.27.

2.4.3 PHOTOSTIMULATION AND PHOTOQUENCHING

When a phosphor with deep traps is once excited and then irradiated by infrared (IR) or red light during the decay of its phosphorescence, it sometimes shows photostimulation or photoquenching

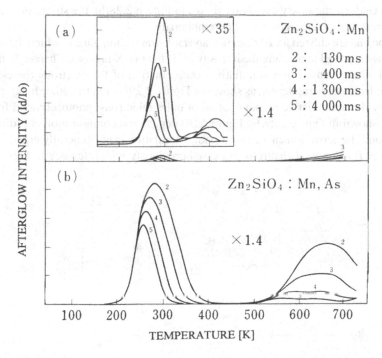

FIGURE 2.27 Temperature dependencies of the afterglow intensity of (a) Zn_2SiO_4:Mn^{2+} and (b) Zn_2SiO_4:Mn^{2+}, As. The delay time (t_d) is 0.13, 0.4, 1.3, and 4.0 s, respectively, for the curves numbered from 2 through 5 in the figures. (From Nakazawa, E., *Jpn. J. Appl. Phys.*, 23(9), L755, 1984. With permission.)

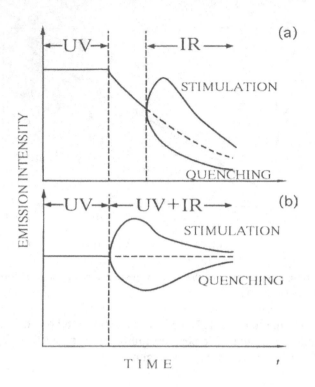

FIGURE 2.28 Photostimulation and photoquenching simulation for the case (b) under a constant excitation and (a) in the afterglow process after the termination of excitation.

of luminescence; that is, an increase or decrease of the emission intensity as schematically shown in Figure 2.28(a). Under a stationary excitation shown in Figure 2.28(b), the stimulation enhances and the quenching reduces the emission intensity temporarily.

These phenomena are utilized for IR detection and radiographic imaging, in which the intensity of the stimulated emission is used to measure the intensity of IR light or X-rays (see Chapter 18 in Volume 2).

Photostimulation is caused by the radiative recombination of the electrons (holes) released by photoactivation from deep trap levels, as shown in Figure 2.29(a). On the other hand, photoquenching is caused by the nonradiative recombination of holes (electrons) photoactivated from luminescent centers as shown in Figure 2.29(b). Figure 2.30 depicts the configuration coordinate model of photostimulation. The activation energy ε_0 of photostimulation is not generally equal to the thermal activation energy ε_t of trapped electrons discussed previously with reference to Figure 2.21. Since

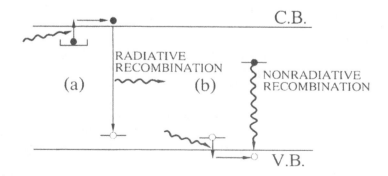

FIGURE 2.29 Photostimulation process (a), and photoquenching process (b) in an energy band scheme.

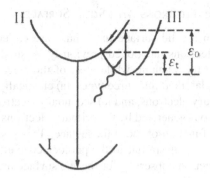

FIGURE 2.30 Photostimulation in configurational coordinate models.

the optical absorption process takes place in a very short time period without changing the configuration of the atoms in the center at that moment, the process is represented by the straight vertical transition in Figure 2.30 from state III (a trap or the quasi-stable state of emitting centers) to state II (the conduction band or emitting centers). On the other hand, the thermal activation needs energy ε_t to overcome at least the lowest barrier between states II and III in Figure 2.30; hence, the activation energy ε_t is generally smaller than ε_0. Photostimulation spectra (i.e., excitation spectra for IR stimulation) can be used to measure the optical activation energy ε_0.

C.B. and V.B. indicate the conduction band and the valence band of the host crystal, respectively.

REFERENCES

1. Nakazawa, E., *J. Lumin.*, 18/19, 272, 1979.
2. Weber, M.J., *Phys. Rev.*, B8, 54, 1973.
3. Hoogenstraaten, W., *Philips Res. Rept.*, 13, 515, 1958.
4. Randall, J.T. and Wilkins, M.H.F., *Proc. Roy. Soc.*, A184, 366, 1945.
5. Curie, D., *Luminescence in Crystals*, John Wiley & Sons, 1963, chap. VI.
6. Avouris, P. and Morgan, T.N., *J. Chem. Phys.*, 74, 4347, 1981.
7. Kivits, P. and Hagebeuk, H.J.L., *J. Lumin.*, 15, 1, 1977.
8. Bube, R.H., *Phys. Rev.*, 80, 655, 1950.
9. Nakazawa, E., *Jpn. J. Appl. Phys.*, 23(9), L755, 1984.
10. Nakazawa, E., *Oyo Buturi*, 55(2), 145, 1986 (in Japanese).

2.5 EXCITATION MECHANISM OF LUMINESCENCE BY CATHODE-RAY AND IONIZING RADIATION

Hajime Yamamoto

2.5.1 INTRODUCTION

Luminescence excited by an electron beam is called *cathodoluminescence* and luminescence excited by energetic particles, i.e., α-ray, β-ray or a neutron beam, or by γ-ray, is called either *radioluminescence* or *scintillation.**

The excitation mechanism of cathodoluminescence and of radioluminescence can be discussed jointly because these two kinds of luminescence have a similar origin. In solids, both the electron beam and the high-energy radiation induce ionization processes, which in turn generate highly energetic electrons. These energetic electrons can be further multiplied in number through collisions, creating "secondary" electrons, which can then migrate in the solid with high kinetic energy, exciting the light-emitting centers. The excitation mechanism primarily relevant to cathodoluminescence is discussed here.

2.5.2 COLLISION OF PRIMARY ELECTRONS WITH SOLID SURFACES

Energetic electrons incident on a solid surface in vacuum are called *primary electrons* and are distinct from the secondary electrons mentioned previously. A small fraction of the electrons is scattered and reflected back to the vacuum, while most of the electrons penetrate into the solid. The reflected electrons can be classified into three types: (a) elastically scattered primary electrons, (b) inelastically scattered primary electrons, and (c) secondary electrons.[1] The secondary electrons mentioned here are those electrons generated by the primary electrons in the solid and are energetic enough to overcome the work function of the solid surface. This phenomenon, i.e., the escape of secondary electrons from the solid, is similar to the photoelectric effect. The relative numbers of the three types of scattered electrons observed for the Ag surface are shown in Figure 2.31.[1,2] As shown in this figure, the inelastically scattered primary electrons are much smaller in number than the other two types.

The ratio of the number of the emitted electrons to the number of the incident electrons is called the *secondary yield* and usually denoted as δ. With this terminology, δ should be defined only in terms of the secondary electrons (c), excluding (a) and (b). However, in most cases, δ is stated for all the scattered electrons—(a), (b), and (c)—for practical reasons.

For an insulator, δ depends on the surface potential relative to the cathode as is schematically shown in Figure 2.32. For $\delta < 1$, the insulator surface is negatively charged; as a consequence, the potential of a phosphor surface is not raised above V_{II} shown in Figure 2.32, even for an accelerating voltage higher than V_{II}. In other words, the surface potential stays at V_{II} and is called the sticking potential. To prevent electrical charging of surface, an aluminizing technique is employed in cathode-ray tubes (CRTs). Negative charging of a phosphor is also a problem for vacuum fluorescent tubes and some of field emission displays, which use low-energy electrons at a voltage below V_I. The aluminizing technique cannot be used in these cases, however, because the low-energy primary electrons, for example, a few ten or hundred eV, cannot go through an aluminum film, even if it is as thin as 100 nm, which is practically the minimum thickness required to provide sufficient electrical conductivity and optical reflectivity. It is, therefore, required to make the phosphor surface electrically conductive.

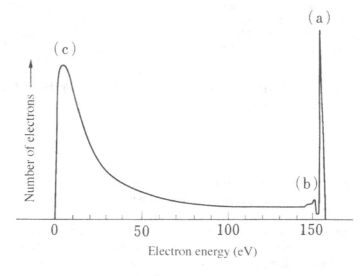

FIGURE 2.31 The energy distribution of electrons emitted from the Ag surface exposed by the primary electrons of 153 eV: (a) electrons emitted by elastic scattering, (b) electrons by inelastic scattering, and (c) secondary electrons. (From Dekker, A.J., *Solid State Physics*, Prentice-Hall, Tokyo, 1960, 418-420. With permission.)

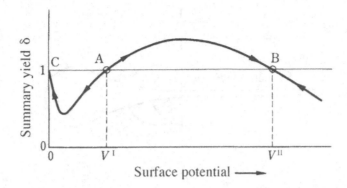

FIGURE 2.32 A schematic illustration of secondary yield as a function of the surface potential of an insulator. The secondary yield δ is unstable at point A, while it is stable at B and C. At these points, the state shifts toward the direction of the given arrows with a change in the potential. Near point C, where the potential is in a region of a few to several tens of volts, the yield approaches 1 because the incident primary electrons are reflected. (From Kazan, B. and Knoll, M., *Electron Image Storage*, Academic, New York, 1968, 22. With permission.)

To evaluate a cathodoluminescence efficiency, one must exclude the scattered primary electrons (a) and (b) in Figure 2.31. The ratio of the electrons (a) and (b) to the number of the incident electrons is called *back-scattering factor*, denoted by η_0. Actually, the electrons (b) are negligible compared with the electrons (a) as shown in Figure 2.31. The value of η_0 depends weakly on the primary electron energy but increases with the atomic number of a solid. η_0 obeys an empirical formula, with the atomic number or the number of electrons per molecule being Z_m^3; that is,:

$$\eta_0 = \left(\frac{1}{6}\right)\ln Z_m - \left(\frac{1}{4}\right) \tag{2.124}$$

The value calculated by this formula agrees well with experimental results obtained for single-crystal samples. For example, the calculated values for ZnS with Z_m of 23 is 0.25 and for YVO_4 with Z_m of 15.7 is 0.21, while the observed values for single crystals of these compounds are 0.25 and 0.20, respectively.[4] In contrast, a smaller value of η_0 is found for a powder layer because some of the reflected electrons are absorbed by the powder through multiple scattering. The observed values of η_0 are 0.14,[4] both for ZnS and YVO_4 in powder form. It has also been reported that η_0 varies by several percent depending on the packing density of a powder layer.[6]

2.5.3 PENETRATION OF PRIMARY ELECTRONS INTO A SOLID

The penetration path of an electron in a solid has been directly observed with an optical microscope by using a fine electron beam of 0.75-μm diameter (Figure 2.33). This experiment shows a narrow channel leading to a nearly spherical region for electron energy higher than 40 keV, while it shows a semispherical luminescent region for lower electron energies.

The former feature is found also for high-energy particle excitation, i.e., the excitation volume is confined to a narrow channel until the energy is dissipated by ionization processes. This result indicates that the scattering cross section of an electron or a particle in a solid is larger for lower electron energy. The energy lost by a charged particle passing through a solid is expressed by Bethe's formula[9]:

$$\frac{dE}{dx} = (2\pi N Z_m e^4 / E)\ln\left(\frac{E}{E_i}\right) \tag{2.125}$$

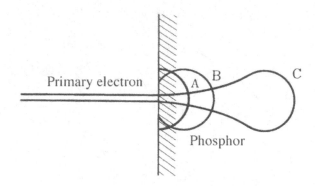

FIGURE 2.33 A schematic illustration of a region excited by an electron beam. This region can be visualized as the luminescent profile of the solid phosphor particle as n through a microscope. The energy of the primary electrons increases in the order of A, B, and C. For A, the energy is several keV and for C, 40 keV or higher. (From Ehrenberg, W. and Franks, J., *Proc. Phys. Soc.*, B66, 1057, 1953; Garlick, G.F.J., *Br. J. Appl. Phys.*, 13, 541, 1962. With permission.)

where E denotes the energy of a primary electron at distance x from the solid surface, N the electron density (cm^{-3}) of the solid, Z_m the mean atomic number of the solid, and E_i the mean ionization potential averaged over all the electrons of the constituent atoms.

Various formulas have been proposed to give the relation between E and x. Among them, the most frequently used is Thomson-Whiddington's formula,[10] which we can derive from Eq. (2.125) simply by putting $\ln(E/E_i) = $ constant.

$$E = E_0(1 - x/R)^{1/2} \tag{2.126}$$

where E_0 is the primary electron energy at the surface and R is a constant called the range, i.e., the penetration depth at $E = 0^*$. It is to be noted that an incremental energy loss, $-dE/dx$, increases with x according to Eq. (2.126).

In a range of $E_0 = 1$–10 keV, the dependence of R on E_0 is given by[11]:

$$R = 250\left(\frac{A}{\rho}\right)\left(E_0/Z_m^{1/2}\right)^n \tag{2.127}$$

where $n = 1.2/(1 - 0.29\log Z_m)$, ρ is the bulk density, A the atomic or molecular weight, Z_m the atomic number per molecule, and E_0 and R are expressed in units of keV and Å, respectively. When $E_0 = 10$ keV, Eq. 2.127 gives $R = 1.5$ μm for ZnS and $R = 0.97$ μm for $CaWO_4$. The experimental values agree well with the calculated values.

When E_0 is decreased at a fixed electron beam current, luminescence vanishes at a certain positive voltage, called the dead voltage. One of the explanations of the dead voltage is that, at shallow R, the primary electron energy is dissipated within a dead layer near the surface, where nonradiative processes dominate as a result of a high concentration of lattice defects.[12] It is also known, however, that the dead voltage decreases with an increase in electrical conductivity, indicating that the dead voltage is affected by electrical charging as well.

2.5.4 IONIZATION PROCESSES

A charged particle, such as an electron, loses its kinetic energy through various modes of electrostatic interaction with constituent atoms when it passes through a solid. Elementary processes leading to energy dissipation can be observed experimentally by the electron energy loss spectroscopy, which measures the energy lost by a primary electron due to inelastic scattering (corresponding to the electrons (b) in Figure 2.31). Main loss processes observed by this method are core-electron excitations and creation of plasmons, which are a collective excitation mode of the valence electrons. Core-electron excitation is observed in the range of 10–50 eV for materials having elements of a large atomic number, i.e., rare-earth compounds or heavy metal oxides such as vanadates or tungstates.[13] The plasmon energy is found in the region of 15–30 eV. Compared with these excitation modes, the contribution of the band-to-band transition is small. As an example exhibiting various modes of excitation, the electron energy loss spectrum of YVO_4 is shown in Figure 2.34.[14]

Plasmons are converted to single-electron excitations in an extremely short period of time, ~10^{-15} s. As a consequence, electrons with energies of 10–50 eV are created every time an energetic primary electron is scattered in a solid as a result of core-electron excitation or plasmon creation. This results in a series of ionization processes in a solid. Most of the electrons generated by the scattering events, or the secondary electrons, are still energetic enough to create other hot carriers by Auger processes. Secondary-electron multiplication can last until the energy of the electron falls below the threshold to create free carriers. All through this electron energy loss process, scattering is accompanied by phonon creation, as schematically shown in Figure 2.35.[15] Secondary-electron multiplication is essentially the same as the photoexcitation process in the vacuum ultraviolet region.

The average energy required to create an electron-hole pair near the band edges, E_{av}, is given by the following empirical formula:[16]

$$E_{av} = 2.67E_g + 0.87 \text{ [eV]} \tag{2.128}$$

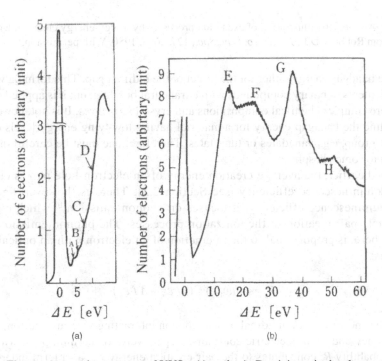

FIGURE 2.34 Electron energy loss spectra of YVO_4: (a) peaks A–D originate in the electronic transitions of the VO_4^{-3} complex; (b) peak E can be assigned to plasmon excitation. Peak G is due to a transition from Y $4p$ orbital to the conduction band, and peak H from V $3p$ to the conduction band. The origin of peak F is not identified. The strong peak at 0 eV indicates the incident electrons with no energy loss. (From Tonomura, A., Endoh, J., Yamamoto, H., and Usami, K., *J. Phys. Soc. Japan*, 45, 1654, 1978. With permission.)

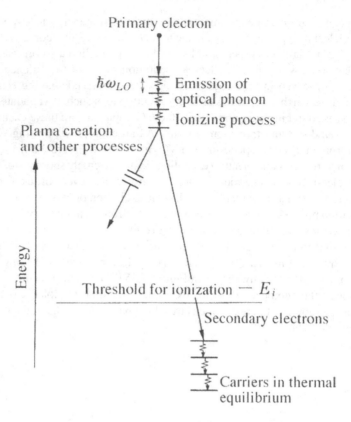

FIGURE 2.35   A schematic illustration of excitation processes by a high-energy electron, which penetrates into a solid. (From Robbins, D.J., *J. Electrochem. Soc.*, 127, 2694, 1980. With permission.)

where E_g is the bandgap energy either for the direct or the indirect gap. This formula was originally obtained for elements or binary compounds with tetrahedral bonding, but it is applied often to phosphors with more complex chemical compositions and crystal structures. It is not, however, straightforward to define the bandgap energy for a material having low-lying energy levels characteristic of a molecular group, e.g., vanadates or tungstates. Therefore, one must be careful in applying the above formula to some phosphors.

As described earlier, the average creation energy of an electron-hole pair is closely related to the cathodoluminescence efficiency (see Section 2.5). There is, however, another way to consider the luminescence efficiency; it focuses on phonon emission,[16] which competes with the electron-hole pair creation in the ionization processes. The phonon emission probability, denoted as R_p here, is proportional to the interaction of an electron with an optical phonon and is expressed as:

$$R_p \propto (\hbar\omega_{LO})^{1/2}(1/\varepsilon_\infty - 1/\varepsilon_0) \qquad (2.129)$$

where ω_{LO} is the energy of a longitudinal optical phonon interacting with an electron, and ε_∞ and ε_0 are high-frequency and static dielectric constants, respectively. When multiplied with the phonon energy, the probability R_p contributes to the pair creation energy E_{av} as a term independent of E_g, e.g., the second term 0.87 eV in Eq. (2.128).

The luminescence excited by energetic particles is radioluminescence.[17] The excitation mechanism of radioluminescence has its own characteristic processes, though it involves ionization processes similar to the cathodoluminescence processes.

For example, the energy of γ-rays can be dissipated by three processes: (1) the Compton effect, (2) the photoelectric effect directly followed by X-ray emission and Auger effect, and (3) the creation of electron-positron pairs. Subsequent to these processes, highly energetic secondary electrons are created, followed by the excitation of luminescence centers, as is the case with cathodoluminescence.

A characteristic energy loss process of neutrons, which has no electric charges but much larger mass than an electron, is due to the recoil of hydrogen atoms. If the neutron energy is large enough, a recoiled hydrogen is ionized and creates secondary electrons. It must be added, however, that hydrogen atoms are not contained intentionally in inorganic phosphors.

2.5.5 ENERGY TRANSFER TO LUMINESCENCE CENTERS

The final products of the secondary-electron multiplication are free electrons and free holes near the band edge, i.e., so-called *thermalized* electrons and holes. They recombine with each other, and a part of the recombination energy may be converted to luminescence light emission.

The process in which either a thermalized electron-hole pair or the energy released by their recombination is transferred to a luminescence center is called *host sensitization* because the luminescence is sensitized by the optical absorption of the host lattice. This process is analogous to the optical excitation near the band edge. Detailed studies were made on the optical excitation of luminescence in IIb–VIb and IIIb–Vb compounds, as described in Sections 3.3.7 and 3.3.8. Luminescence of rare-earth ions and Mn^{2+} ions arises because these ions capture electrons and holes by acting as isoelectronic traps.[18,19] In inorganic compounds having complex ions and organic compounds, the excitation energy is transferred to the luminescence centers through the molecular energy levels.

2.5.6 LUMINESCENCE EFFICIENCY

The cathodoluminescence energy efficiency η, for all the processes described earlier can be expressed by:[20]

$$\eta = \left(1 - \eta_0\right)\eta_x\left(E_{em}/E_g\right)q \tag{2.130}$$

where η_0 is the back-scattering factor given by Eq. (2.124), η_x the mean energy efficiency to create thermalized electrons and holes by the primary electrons or E_g/E_{av}, q the quantum efficiency of the luminescence excited by thermalized electron-hole pairs, and E_{em} the mean energy of the emitted photons. Thus:

$$\eta < \eta_x\left(\frac{E_{em}}{E_g}\right) \tag{2.131}$$

and also $\eta_x < 1/3$ according to Eq. (2.128).

The energy efficiency, luminescence peak wavelength, and color are shown in Table 2.2 for some efficient phosphors. For the commercial phosphors, ZnS:Ag,Cl; ZnS:Cu,Al; $Y_2O_2S:Eu^{3+}$; and $Y_2O_3:Eu^{3+}$, we find $\eta_x \approx 1/3$ from Eq. (2.130) by assuming that $\eta_0 = 0.1$ and $q = 0.9$–1.0. This value of η_x suggests that the energy efficiency is close to the limit predicted by Eq. (2.130) for these phosphors. It is to be emphasized, however, that this estimate does not exclude a possibility for further improvement in the efficiency of these phosphors, for example, by 10 or 20%, since the calculated values are based on a number of approximations and simplifying assumptions. It should also be noted that the bandgap energy is not known accurately for the phosphors given in Table 2.2, except for ZnS, CsI, and CaS. For the other phosphors, the optical absorption edge must be used instead of the bandgap energy, leaving the estimation of η approximate. For CaS, the indirect bandgap, 4.4 eV, gives $\eta_x = 0.21$, while the direct bandgap, 5.3 eV, gives the value exceeding the limit predicted by Eq. (2.130).

TABLE 2.2
Examples of Cathodoluminescence Efficiency[22]

Chemical Composition	WTDS Designation	Energy Efficiency (%)	Peak Wavelength (nm)	Luminescence color
$Zn_2SiO_4:Mn^{2+}$	GJ	8	525	Green
$CaWO_4:Pb$	BJ	3.4	425	Blue
ZnS:Ag,Cl	X	21	450	Blue
ZnS:Cu,Al	X	23, 17	530	Green
$Y_2O_2S:Eu^{3+}$	X	13	626	Red
$Y_2O_3:Eu^{3+}$	RF	8.7	611	Red
$Gd_2O_2S:Tb^{3+}$	GY	15	544	Yellowish green
$CsI:Tl^+$	—	11	—	Green
$CaS:Ce^{3+}$	—	22	—	Yellowish green
$LaOBr:Tb^{3+}$	—	20	544	Yellowish green

Note: The phosphor screen designation by WTDS (Worldwide Phosphor Type Designation System) is presented (Ref. 22).
Source: Many data are collected in Alig, R.C. and Bloom, S., *J. Electrochem. Soc.*, 124, 1136, 1977.

NOTES

* The word originally means flash, as is observed under particle beam excitation.
* Other formulas define the range as the penetration depth at $E = E_0/e$.

REFERENCES

1. Dekker, A.J., *Solid State Physics*, Prentice-Hall, Maruzen, Tokyo, 1960, 418-420.
2. Rudberg, E., *Proc. Roy. Soc. (London)*, A127, 111, 1930.
3. Tomlin, S.G., *Proc. Roy. Soc. (London)*, 82, 465, 1963.
4. Meyer, V.G., *J. Appl. Phys.*, 41, 4059, 1970.
5. Kazan, B. and Knoll, M., *Electron Image Storage*, Academic Press, New York, NY, 1968, 22.
6. Kano, T. and Uchida, Y., *Jpn. J. Appl. Phys.*, 22, 1842, 1983.
7. Ehrenberg, W. and Franks, J., *Proc. Phys. Soc.*, B66, 1057, 1953.
8. Garlick, G.F.J., *Br. J. Appl. Phys.*, 13, 541, 1962.
9. Bethe, H.A., *Ann. Physik*, 13, 541, 1930.
10. Whiddington, R., *Proc. Roy. Soc. (London)*, A89, 554, 1914.
11. Feldman, C., *Phys. Rev.*, 117, 455, 1960.
12. Gergley, Gy., *J. Phys. Chem. Solids*, 17, 112, 1960.
13. Yamamoto, H. and Tonomura, A., *J. Lumin.*, 12/13, 947, 1976.
14. Tonomura, A., Endoh, J., Yamamoto, H., and Usami, K., *J. Phys. Soc. Japan*, 45, 1654, 1978.
15. Robbins, D.J., *J. Electrochem. Soc.*, 127, 2694, 1980.
16. Klein, C.A., *J. Appl. Phys.*, 39, 2029, 1968.
17. Brixner, L.H., *Materials Chemistry and Physics*, 14, 253, 1987; Derenzo, S.E., Moses, W.W., Cahoon, J.L., Perera, R.L.C., and Litton, J.E., *IEEE Trans. Nucl. Sci.*, 37, 203, 1990.
18. Robbins, D.J. and Dean, P.J., *Adv. Phys.*, 27, 499, 1978.
19. Yamamoto, H. and Kano, T., *J. Electrochem. Soc.*, 126, 305, 1979.
20. Garlick, G.F.J., *Cathodo- and Radioluminescence in Luminescence of Inorganic Solids*, Goldberg, P., Ed., Academic Press, New York, NY, 1966, 385–417.
21. Alig, R.C. and Bloom, S., *J. Electrochem. Soc.*, 124, 1136, 1977.
22. Yen, W.M., Shionoya, S. and Yamamota, H., *Phosphor Handbook*, chapter 6, CRC Press, 2nd Ed, Boca Raton, FL, 2007.

2.6 LANTHANIDE LEVEL LOCATIONS AND THEIR IMPACT ON PHOSPHOR PERFORMANCE

Pieter Dorenbos

2.6.1 INTRODUCTION

In the second edition of the *Phosphor Handbook* [1], the topic of lanthanide level locations and their impact on phosphor performance was addressed. It provided the status of our knowledge in the year 2007. Since then our understanding, the accuracy of modeling, and available experimental data on the topic improved considerably. This inspired to revise the original contribution and augment it with the progress of the last 15 years.

The lanthanide ions either in their divalent or their trivalent charge state form a very important class of luminescence activators in phosphors and single crystals [2]. The fast 15- to 60-ns 5d–4f emission of Ce^{3+} in compounds like $LaCl_3$, $LaBr_3$, Lu_2SiO_5, and Gd_2SiO_5 is utilized in scintillators for gamma-ray detection [3]. The same emission has been utilized in cathode-ray tubes and electro-luminescence phosphors. Today, Ce^{3+} in the garnet family of compounds $(Y,Gd)_3(Al,Ga)_5O_{12}$ is the main emitting center in light-emitting diode (LED) lighting systems found in almost every household. The photon cascade emission involving the $4f^2$ levels of Pr^{3+} has been investigated extensively, particularly around the year 2000, in the context of developing high-quantum-efficiency phosphors to be excited by means of the Xe discharge in the vacuum ultraviolet [4]. The narrow $4f^3 \rightarrow 4f^3$ emission lines of Nd^{3+} are used in laser crystals like $Y_3Al_5O_{12}:Nd^{3+}$. Sm^{3+} is utilized as an efficient electron trap and research has been devoted to its information storage properties. For example, $MgS:Ce^{3+};Sm^{3+}$ and $MgS:Eu^{2+};Sm^{3+}$ were studied for applications as optical memory phosphor [5], $Y_2SiO_5:Ce^{3+};Sm^{3+}$ was studied for X-ray imaging phosphor applications [6], and $LiYSiO_4:Ce^{3+};Sm^{3+}$ for thermal neutron imaging phosphor application [7]. The famous $^5D_0 \rightarrow {}^7F_J$ $4f^6$ red line emissions of Eu^{3+} and the blue to red 5d–4f emissions of Eu^{2+} are both used in display and lighting phosphors [2]. The $4f^8$ line emission of Tb^{3+} is often responsible for the green component in tricolor tube lighting [2]. Dy^{3+} plays an important role in the persistent luminescence phosphor $SrAl_2O_4:Eu^{2+};Dy^{3+}$ [8, 9]. Er^{3+} and Tm^{3+} were like Pr^{3+} being investigated for possible photon cascade emission phosphor applications.

Abovementioned summary illustrates the diversity of application involving the luminescence of lanthanide ions. It also illustrates that we can distinguish two types of lanthanide luminescent transitions: (1) transitions between levels of the $4f^n$ configuration and (2) transitions between the $4f^{n-1}$ 5d and the $4f^n$ configurations. In the first approximation, the energy of each $4f^n$ excited state relative to the $4f^n$ ground state can be regarded as invariant with the type of compound. One may then use the Dieke diagram with the extension provided by Wegh *et al.* [10] to identify the many possible luminescence emission and optical absorption lines involving the trivalent lanthanides. The energy of 5d–4f emission, contrary to that of 4f–4f emission, depends very strongly on the type of compound. For example, the wavelength of the 5d–4f emission of Ce^{3+} may range from the ultraviolet in fluorides like $KMgF_3$ to the red in sulfides like Lu_2S_3 [11].

In all phosphor applications, the color of emission and the quantum efficiency of the luminescence process is of crucial importance as is the thermal stability of the emission. These three aspects are related not only to the level locations intrinsic to the lanthanide but also to level location with respect to the host bands. For example, the position of the lowest $4f^{n-1}5d$ state relative to lower lying $4f^n$ states is important for the quenching behavior of 5d–4f emission by multiphonon relaxation. The electron binding energy in the $4f^n$ and $4f^{n-1}5d$ states relative to that in the valence band and conduction band states, which we will call the host-referred binding energies (HRBEs), also affect luminescence quenching and charge carrier trapping phenomena. It was realized soon after the use of lanthanides in phosphors that level location with respect to the host bands is crucial for phosphor performance, and its experimental determination and theoretical understanding has therefore a long history.

In this section, first, a survey is provided on how lanthanide level location affects phosphor performance. Next methods and models to determine level location are presented. After discussing the levels of the free (or gaseous) lanthanide ions, the influence of the host compound on the location of the 5d levels relative to the 4f levels is presented. Next the influence of the host compound on the location of the lowest $4f^n$ state above the top of the valence band is treated. This forms the basis for drawing HRBE schemes with placement of the lanthanide levels in the bandgap and it reflects the state of the art around 2007. Finally, the theory and methods to derive the vacuum referred binding energies (VRBEs) that were developed in 2012 [12] and further refined in 2020 [13, 14] are presented.

2.6.2 Level Locations and Phosphor Performance

In discussing level positions, we need to define the concept of level position. We are all familiar with the Dieke diagram of $4f^n$ level energies of the trivalent lanthanides. In a Dieke diagram, the energy of an excited $4f^n$ level is presented with respect to the energy of the $4f^n$ ground state that is then defined as our zero of energy. Energy refers here to the quantum mechanical energy eigenvalue of the multi-electron state of the lanthanide. Similarly, we can define the energy of an excited $4f^{n-1}$ 5d state, and then locate or place the 5d level with respect to the $4f^n$ levels. Figure 2.36(a) illustrates the location of such 5d state in a free ion lanthanide. When a lanthanide is brought from the gaseous state into a chemical environment (A), there appears a downward shift of the lowest energy 5d level with an amount known as the redshift or depression $D(A)$. Clearly the value of $D(A)$ determines the color of emission and wavelength of absorption of 4f–5d transitions. Figure 2.36(b) illustrates the

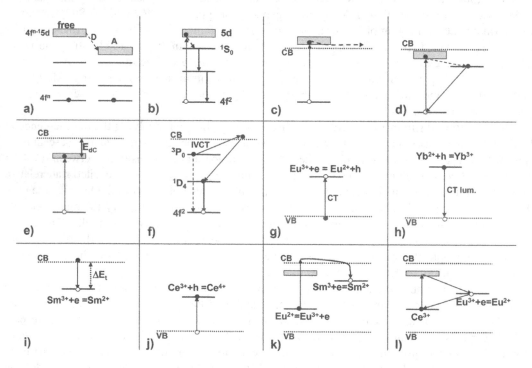

FIGURE 2.36 Illustration of influence of level location on phosphor properties. (a) The redshift D of the $4f^{n-1}5d$ state, (b) photon cascade emission in Pr^{3+}, (c) 5d–4f emission quenching by autoionization, (d) anomalous 5d emission, (e) thermal quenching by ionization, (f) quenching by intervalence charge transfer, (g) valence band charge transfer, (h) charge transfer luminescence, (i) electron trapping by Sm^{3+}, (j) hole trapping by Ce^{3+}, (k) electron transfer from Eu^{2+} to Sm^{3+}, (l) luminescence quenching by lanthanide to lanthanide charge transfer.

importance of lowest energy 4f5d level location relative to $4f^2$ levels in Pr^{3+}. With the 4f5d above the $^1S_0$ level of Pr^{3+}, multiphonon relaxation from the lowest 4f5d state to the lower lying $^1S_0$ level takes place. A cascade emission of two photons may lead theoretically to quantum efficiency larger than 100%. However, with the lowest 5d state below $^1S_0$, broadband 5d–4f emission is observed. Much research has been devoted in the search of Pr^{3+} quantum splitting phosphors for lighting applications [4] and to find efficient 5d–4f emitting Pr^{3+}-doped materials for scintillator applications. Depending on the precise location of the lowest 5d state in Nd^{3+}, Eu^{2+}, Sm^{2+}, and Tm^{2+} also either broadband 5d–4f or narrow line 4f–4f, emission can be observed [15].

Lanthanide level location with respect to, for example, the conduction band bottom is conceptually different than level location, i.e., multi-electron energy eigenvalues, in a Dieke-type diagram. This becomes apparent in Figure 2.36(c), (d), and (e) that illustrates the interplay between the localized $4f^{n-1}5d$ multi-electron state and the delocalized conduction band states. Here level location refers to a binding energy, i.e., the energy needed to remove an electron from a multi-electron state and bring it somewhere else. When the lowest $4f^{n-1}5d$ state is drawn above the bottom of the conduction band as in Figure 2.36(c), it means that the binding energy in that state is smaller than the binding energy of an electron at the bottom of the conduction band. Since nature strives toward lowest energy, ionization will occur spontaneously. An electron is transferred from the $4f^{n-1}5d$ state to the empty conduction band leaving a further oxidized lanthanide in its $4f^{n-1}$ ground state. In such situation 5d–4f, emission is not observed anymore. This is the case in $LaAlO_3:Ce^{3+}$, the rare-earth sesquioxides $Ln_2O_3:Ce^{3+}$ [2], and usually also for Eu^{2+} on trivalent rare-earth sites in oxide compounds [16].

Figure 2.36(d) illustrates the situation with electron binding energy in $4f^{n-1}5d$ a few 0.1 eV below the conduction band bottom. Here the 5d electron again delocalizes but due to Coulomb attraction and stabilization by lattice relaxation remains in the vicinity of the now further oxidized lanthanide. The nature of the state, which is sometimes called an impurity or lanthanide trapped exciton state, is not precisely known. Anyway, the return of the electron to the lanthanide leads to so-called anomalous emission characterized by a very large Stokes shift [17,18]. Finally Figure 2.36(e) shows the situation with the binding energy in the $4f^{n-1}5d$ state well below the conduction band leading to 5d–4f emission. This emission can be quenching by thermally assisted ionization of the 5d electron to conduction band states. The energy E_{dC} between the binding in the $4f^{n-1}5d$ state (d) and the bottom of the conduction band (C) is then proportional with the quenching temperature T_{50} where emission intensity has dropped by 50% [19,20]. In Ref. [20], the relationships between E_{dC}, thermal quenching of the Eu^{2+} 5d–4f emission and the type of host crystals were studied. Knowledge on such relationship is important for developing temperature-stable Eu^{2+}- or Ce^{3+}-doped LED phosphors or temperature-stable Ce^{3+}-doped scintillators. For electroluminescence applications, E_{dC} is an important parameter to discriminate the mechanism of impact ionization against the mechanism of field ionization [21].

Although a Dieke diagram refers to multi-electron energy eigenvalues of a lanthanide, one may unite it with a binding energy diagram as in Figure 2.36(f) that shows a typical situation for Pr^{3+} level locations in a transition metal complex compound like $CaTiO_3$. The $4f^2$ ground-state location with respect to the conduction band bottom equals the minimum energy needed to bring an electron from that ground state to the delocalized conduction band. If Pr^{3+} is first excited to, for example, the $^1D_4$ $4f^2$ level, the minimum energy to ionize the excited Pr^{3+} ion is reduced by an amount that equals the amount of used excitation energy. It implies that we can use the Dieke level energies to locate the excited lanthanide levels in a binding energy scheme. In the situation of Figure 2.36(f), the blue emission from the Pr^{3+} $^3P_0$ level is quenched by intervalence charge transfer (IVCT) [22]. The electron transfers from the $^3P_0$ level to the transition metal Ti^{4+} to form a Pr^{4+}–Ti^{3+} IVCT state. Since the CB bottom is formed by the empty 3d Ti^{4+} orbitals, this is quite similar as the situation in Figure 2.36(d). However, here the electron is transferred back to the Pr^{3+} $^1D_4$ excited state. The position of the $^3P_0$ level relative to the conduction band now controls whether the blue emission from the $^3P_0$ or the red emission from the $^1D_4$ level dominates.

We discussed a few examples where a lanthanide acts as an electron donor to the conduction band. Suppose the situation in Figure 2.36(c) refers to Eu^{2+} where first an electron is excited to the $4f^6 5d$ state and then autoionization takes place. Eu^{3+} is left behind that now can act as an electron acceptor. Clearly, the location of the Eu^{2+} electron donor level must be the same as the location of the Eu^{3+} electron acceptor level and we may equally well denote the level location as a $Eu^{2+/3+}$ donor/acceptor level location. Figure 2.36(g) pertains to an Eu^{3+}-doped compound. Eu^{3+} introduces an electron acceptor state in the forbidden gap. The excitation of an electron from the valence band to that acceptor state creates the ground state of Eu^{2+}. This is a dipole-allowed transition that is used, for example, to sensitize Y_2O_3:Eu^{3+} phosphors to the Hg 254-nm emission for tube lighting [2]. Recombination of the electron with the valence band hole leaves the Eu^{3+} ion in the $^5D_0$ excited $4f^6$ state resulting in red $4f^6 \rightarrow 4f^6$ emission. Figure 2.36(h) shows a similar situation for Yb^{3+}. After electron transfer to Yb^{3+}, the recombination with the hole created in the valence band produces a strongly Stokes shifted charge transfer (CT) luminescence. This type of luminescence gained considerable interest for developing scintillators for neutrino detection [23]. Clearly, for the energy of CT excitation and CT luminescence, the location of the $Yb^{2+/3+}$ D/A level above the valence band top is important.

Figure 2.36(i) shows the trapping of an electron from the conduction band by Sm^{3+} to form the ground state of Sm^{2+}. The location of the $Sm^{2+/3+}$ D/A level below the CB bottom determines the electron trapping depth. Similarly, the location of $Ln^{3+/4+}$ D/A level above the VB top determines the hole trapping depth provided by that Ln^{3+} ion. Figure 2.36(j) illustrates trapping of a hole from the valence band by Ce^{3+}. This hole trapping is an important aspect of the scintillation mechanism in Ce^{3+}-doped scintillators. Similarly Eu^{2+} is an efficient hole trap of importance for the X-ray storage phosphor BaFBr:Eu^{2+}. Phosphor properties become more complicated when we deal with *double lanthanide-doped systems*. Figure 2.36(k) shows the situation for an Eu^{2+} and Sm^{3+} double-doped compound like in SrS and MgS that were studied for optical data storage applications [5,15]. The ultraviolet write pulse excites an electron from Eu^{2+} to the conduction band which is then trapped by Sm^{3+}. Eu^{3+} and Sm^{2+} are created in the process. An infrared read pulse liberates the electron again from Sm^{2+} resulting eventually in Eu^{2+} 5d–4f emission. Similar mechanisms apply for Y_2SiO_5:Ce^{3+};Sm^{3+} and LiYSiO$_4$:Ce^{3+};Sm^{3+} compounds that were developed for X-ray and thermal neutron storage phosphors applications, respectively [6,7]. Double lanthanide doping has been most intensively studied in YPO$_4$ by means of thermoluminescence and optical stimulated emission. Electrons can be juggled back and forth between different lanthanides in a controlled fashion [24]. The mechanism in the persistent luminescence phosphor SrAl$_2$O$_4$:Eu^{2+},Dy^{3+} has been disputed many times. One needs to know the level locations to arrive at a plausible mechanism or to discard implausible ones [8,9,25].

As a last example Figure 2.36(l) shows quenching of emission in, for example, Ce^{3+} and Eu^{3+} co-doped systems. After exciting the Ce^{3+} 4f electron to the lowest 5d state, it can transfer to Eu^{3+} when the $Eu^{2+/3+}$ D/A level is located at lower energy than the binding energy in the lowest Ce^{3+} 5d state. After the transfer, Eu^{2+} and Ce^{4+} are formed. The Eu^{2+} electron can transfer back to Ce^{4+} if the $Ce^{3+/4+}$ D/A level is located below the $Eu^{2+/3+}$ D/A level. The original situation is restored without emission of a photon. Similar quenching routes pertain to Ce^{3+} in Yb-based compounds, and with the appropriate level schemes other "killing" combinations can be found as well.

The abovementioned set of examples shows the importance of energy level locations for the performance of phosphors. That importance was always realized, and around 2005 methods and models were available to routinely determine lanthanide level locations with respect to the host energy bands that are now known as HRBE diagrams. Later in 2012, methods were developed to determine level locations with respect to the vacuum level resulting in VRBE diagrams [12]. In the following sections, the historic developments and current status of lanthanide level positioning are briefly reviewed. For detailed information, the original literature should be consulted.

2.6.3 THE FREE (GASEOUS) LANTHANIDE IONS

The previous section illustrated the importance of lanthanide level locations for phosphor performance. To understand and predict these locations, we first need to understand the properties of the free (gaseous) lanthanide ions. Figure 2.37 shows the data available on the energy E_{fd} needed to excite an electron from the lowest level of the $4f^n 5d^0 6s^m$ configuration ($m=0$, 1, or 2) to the lowest level of the $4f^{n-1} 5d^1 6s^m$ configuration in the gaseous free lanthanide ions or atoms. The data are from Brewer [26] and Martin [27] together with later updates [15]. Data are most complete for the neutral atoms ($m=2$, curve c), the monovalent lanthanides ($m=1$, curve b), and the divalent lanthanides ($m=0$, curve a). A universal curve, curve (a) in Figure 2.37, can be constructed. By shifting this universal curve in energy, the 4f–5d energies as function of n can be reproduced remarkably well irrespective the charge of the lanthanide ion (0, +1, +2, or +3) or the number, m, of electrons in 6s ($m=0$, 1, or 2). This phenomenon is due to the inner shell nature of the 4f orbital. Apparently, the occupation number of electrons in the outer 6s shell has little influence on the universal behavior. The main features of this universal variation have been known for a long time and understood in terms of Jörgensens spin pairing theory for the binding of 4f electrons [28]. The energy is large when the 4f configuration is half ($n=7$) or completely ($n=14$) filled, and the energy is small when it is occupied by one or eight electrons.

Figure 2.38 shows the binding energy (or minus the ionization energy) in the $4f^n$ ground state of the free divalent and free trivalent lanthanide ions with $m=0$. When we add the corresponding energies E_{fd} from Figure 2.37(a) and (e) to curves (b) and (d) in Figure 2.38, we obtain the binding energies in the lowest energy $4f^{n-1}5d$ state, see curves (a) and (c). The stronger binding in the trivalent lanthanide states than in the divalent states is of course due to a stronger Coulomb attraction. The binding energy in the lowest $4f^{n-1}5d$ state appears rather constant and only shows a mild

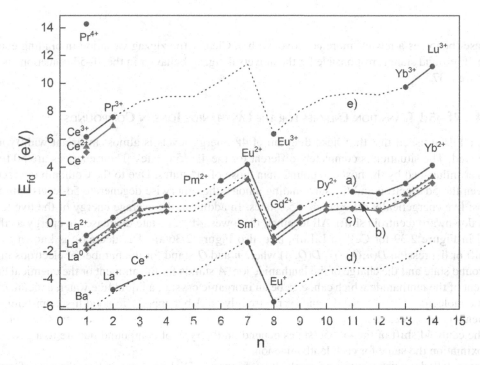

FIGURE 2.37 Experimentally observed energies E_{fd} for the transition between the lowest $4f^n 5d^0 6s^m$ and the lowest $4f^{n-1} 5d^1 6s^m$ states of free (gaseous) lanthanide ions and atoms. A shift of the dashed curve (a) by −0.71, −1.09, −5.42, or +7.00 eV gives curves (b), (c), (d), or (e), respectively.

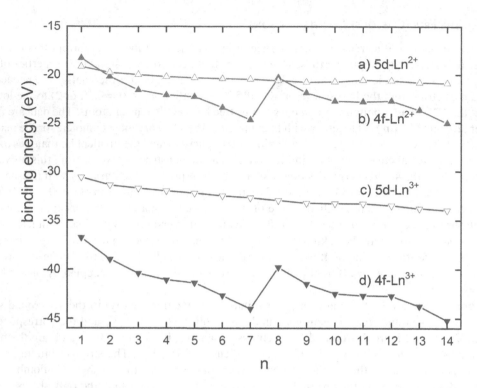

FIGURE 2.38 The binding energy in eV in the lowest $4f^{n-1}5d$ (curves (a) and (c)) state and $4f^n$ ground state (curves (b) and (d)) in the free divalent (curves (a) and (b)) and free trivalent lanthanide ions (curves (c) and (d)).

increase (becomes a few eV more negative) with n. Clearly the zigzag variation in binding energy in the $4f^n$ ground state is responsible for the universal zigzag behavior in the 4f–5d transition energy in Figure 2.37.

2.6.4   4f→5d Transition Energies for the Lanthanide Ions in Compounds

It is well established that the Dieke diagram of $4f^n$ energy levels is almost invariant with type of compound. The situation is completely different for the $4f^{n-1}5d$ states. Their energies are 50 times stronger influenced by the host compound than those of $4f^n$ states. Due to the Coulomb interaction between the 5d-electron and the surrounding anions, the otherwise degenerate 5d-levels split into at most five energetically different $4f^{n-1}5d$ states. In addition, the average energy of the five levels shifts downward (centroid shift). All together, the lowest $4f^{n-1}5d$ state decreases in energy as illustrated in Figure 2.39 for Ce^{3+} in $LiLuF_4$ (see also Figure 2.36(a)). The decrease is known as the redshift or depression $D(n,Q,A) \approx D(Q,A)$ where n and Q stand for the number of electrons in the $4f^n$ ground state and the charge of the lanthanide ion. A stands for the ambient or the chemical environment of the lanthanide which can, e.g., be an inorganic crystal, a liquid like water, a metal, or an organic molecule. The redshift depends very strongly on A but appears to good first approximation independent on n, i.e., the type of lanthanide ion. This implies that both the crystal field splitting and the centroid shift of the $4f^{n-1}5d$ states depend on the type of compound but are to a good first approximation the same for each lanthanide ion.

Figure 2.40 demonstrates this principle. It is an inverted Dieke diagram where the zero of energy is at the lowest $4f^{n-1}5d$ state of the free trivalent lanthanide ion. When the lanthanide ions are in a compound one simply needs to shift the $4f^{n-1}5d$ level energies down by the redshift $D(3+,A)$ to

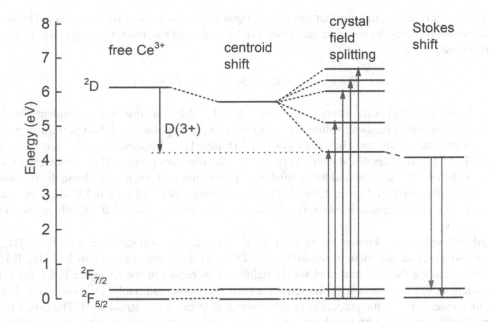

FIGURE 2.39 The effect of the crystal field interaction on the (degenerate) free Ce^{3+} energy states in $LiLuF_4$. The combination of centroid shift and crystal field splitting decreases the lowest 5d state with a total energy D. On the far right, the Stokes shifted 5d–4f emission transitions are shown.

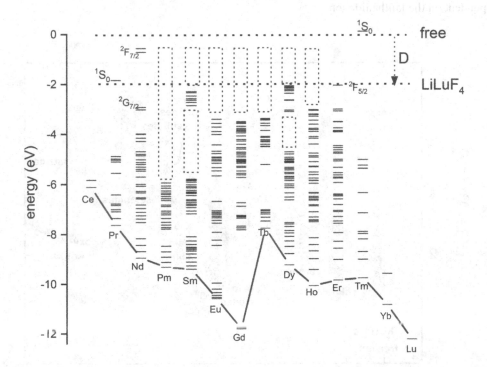

FIGURE 2.40 The *inverted* Dieke diagram where the energies of the lowest $4f^{n-1}5d$ level of the free trivalent lanthanide ions are defined as the zero of energy. A downward shift of the 5d levels with the redshift value $D=1.9$ eV provides the relative position of the lowest 5d level for the trivalent lanthanides in $LiLuF_4$.

find the appropriate diagram for that compound. Figure 2.40 illustrates this for $LiLuF_4$. The lowest 4f–5d transition energy for each lanthanide ion can be read from the diagram. In equation form, this is written as:

$$E_{fd}(n,3+,A) = E_{fd}(n,3+,Afree) - D(3+,A) \qquad (2.132)$$

where $E_{fd}(n,3+,Afree)$ is the energy of the first $4f^n \rightarrow 4f^{n-1}5d$ transition in the trivalent (3+) free lanthanide ion. Initially, these values were derived from extrapolation of data pertaining to lanthanides in compounds and they were compiled in Ref. [29]. Improved values were compiled later in Ref. [14]. Besides 4f–5d energies in $LiLuF_4$, the diagram predicts that the lowest 5d state of Pr^{3+} is below the $^1S_0$ state and broadband 5d–4f emission and not narrow band $^1S_0$ line emission will be observed (cf. Figure 2.36(b)). The lowest energy Nd^{3+} 5d state in $LiLuF_4$ is predicted stable enough against multiphonon relaxation to the $^2G^{7/2}$ level. Indeed, Nd^{3+} 5d–4f emission has been observed.

Redshift values are known for at least 1000 different compounds, see, e.g., Refs. [11,30]. Figure 2.41 gives an overview of redshift values $D(3+,A)$ that were compiled in Ref. [11]. It is by definition zero for the free ions, and for the halides, it increases in the sequence F, Cl, Br, I. For the chalcogenides, an increase in the sequence O, S, Se, and presumably Te is observed. This is directly connected with the properties of the anions that affect the centroid shift. The origin of the centroid shift is related with covalence and polarizability of the anions in the compound [31–33]. The crystal field splitting is related with the shape and size of the first anion coordination polyhedron [33,34]. The small fluorine and small oxygen anions provide the largest values for the crystal field splitting and that is the main reason for the large spread in redshift values for these two types of compounds [33]. Eq. (2.132) equally well applies for the 5d–4f emission because the Stokes shift between absorption and emission is determined by the host lattice and to first approximation also independent on the lanthanide ion.

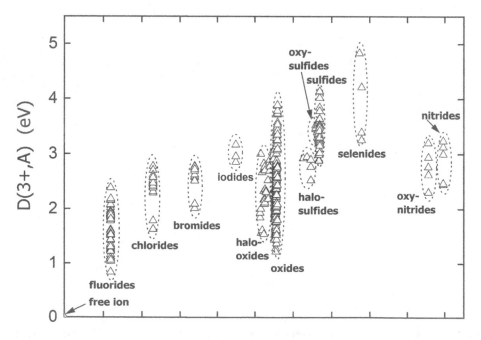

FIGURE 2.41 The redshift $D(3+,A)$ for trivalent lanthanide ions in compounds. The parameter along the horizontal axis groups the data depending on the type of compound.

For the divalent lanthanide ions in compounds, the story is analogous [15,16]. Again one can introduce a redshift $D(2+, A)$ with a similar relationship:

$$E_{fd}(n, 2+, A) = E_{fd}(n, 2+, Afree) - D(2+, A),$$ (2.133)

and construct figures like Figures 2.40 and 2.41.

With all data available on $D(3+, A)$ and $D(2+, A)$, the redshift in divalent lanthanides can be compared with that in the trivalent ones; a roughly linear relationship is found [35].

$$D(2+, A) = 0.64 D(3+, A) - 0.233 eV.$$ (2.134)

Investigations also show a linear relationship between crystal field splitting, centroid shift, and Stokes shift [35]. Combining Eqs (2.132)–(2.134) with the available data on $D(2+, A)$ and $D(3+, A)$, it is now possible to predict 4f–5d energy differences for all 13 divalent and all 13 trivalent lanthanides in about 1000 different compounds; that is more than 25,000 different combinations! Usually, the accuracy is a few 0.1 eV but deviations occur. The work by van Pieterson et al. [36,37] on the trivalent lanthanides in LiYF$_4$, YPO$_4$, and CaF$_2$ shows that the size of crystal field splitting decreases slightly with smaller size of the lanthanide and then the redshift cannot be the same for all lanthanide ions. A study on the size of the 5d-level crystal field splitting in Ce^{3+} and Tb^{3+} also revealed differences in the order of a few tenths of eV in Refs. [14,38]. Smaller lanthanide size implies a longer average distance between the 5d-electron and anion ligands leading to smaller Coulomb interaction and deviations in redshift that correlates with the size of the crystal field interaction.

2.6.5 METHODS TO DETERMINE HOST-REFERRED BINDING ENERGIES

The experimental bases for the results in previous sections are the $4f^n \rightarrow 4f^{n-1}5d$ transition energies that are easily measured by means of luminescence, luminescence excitation, or optical absorption techniques. We deal with a dipole-allowed transition from a localized ground state to a localized excited state involving one and the same lanthanide ion where both states have a well-defined energy. To determine the location of lanthanide levels relative to the host bands or to the vacuum level is not that straightforward, but again one may use information from optical spectroscopy. We will first discuss the HRBEs that reflect the state of knowledge before 2007 followed by the VRBEs reflecting our knowledge of today. Figure 2.36(g) shows the transition of an electron from the top of the valence band (an anion) to Eu^{3+}. The final state is the $4f^7$ ground state of Eu^{2+}. The energy $E^{CT}(6, 3+, A)$ needed for this CT provides then a measure for the difference in binding energy at the valence band top and in the Eu^{2+} $4f^7$ ground state. Wong et al. [39] and Happek et al. [40] assumed that the CT energy provides, what we will call, the Ln$^{2+/3+}$ donor/acceptor (D/A) level location relative to the top of the valence band directly. However, this is not that trivial. The transferred electron and the hole left behind in the VB are still Coulomb attracted to each other and that reduces the transition energy by perhaps as much as 0.5 eV. On the other hand, Eu^{2+} is about 18 pm larger than Eu^{3+}, and the optical transition ends in a configuration of neighboring anions that is not yet in its lowest energy state. It will be followed by lattice relaxation. The binding energy contribution from Coulomb bonding and from lattice relaxation tend to cancel each other and fortuitously the original assumption by Wong et al. and later by Happek et al. in practice appears to work out quite well [41]. The Ln$^{2+/3+}$ and Ln$^{3+/4+}$ D/A level locations can also be obtained from X-ray or ultraviolet photoelectron spectroscopy (XPS or UPS) [42] by determining the energy difference between the valence band and lanthanide XPS peaks.

With the techniques in the preceding paragraph, the $4f^n$ ground-state level locations of lanthanide ions relative to the valence band top can be probed. The level locations relative to the conduction band can be found with other techniques. Various methods rely on the ionization of 5d electrons to conduction band states. The thermal quenching of 5d–4f emission in Ce^{3+} or Eu^{2+} is often due to such ionization process [19,20]. By studying the quenching of 5d–4f luminescence intensity or the shortening of the 5d–4f decay time with temperature, the energy difference E_{dC} between the (lattice relaxed) lowest 5d state and the bottom of the conduction band can be deduced [19]. Such studies were done by, for example, Lizzo et al. [43] for Yb^{2+} in $CaSO_4$ and SrB_4O_7, by Bessière et al. [21] for Ce^{3+} in $CaGa_2S_4$, and by Lyu and Hamilton [19]. Also, the absence of Ce^{3+} emission due to a situation sketched in Figure 2.36(c) or the presence of anomalous emission as in Figure 2.36(d) provides qualitative information on 5d level locations [18]. One may also interpret the absence or presence of vibronic structure in $4f^n \rightarrow 4f^{n-1}5d$ excitation bands as an indication that the $4f^{n-1}5d$ state is located inside the conduction band [36]. One- or two-step photoconductivity provides information on the location of 4f ground states relative to the bottom of the conduction band [44–47]. Another related technique is the microwave conductivity method by Joubert and co-workers that was applied to $Lu_2SiO_5:Ce^{3+}$ [48]. A most sensitive method is thermoluminescence glow peak analysis on lanthanide-doped compounds. Electrons excited across the bandgap can be trapped by a trivalent lanthanide ion as in Figure 2.36(k) in the case of Sm^{3+}. Upon heating the sample linearly with time, the probability of thermal excitation of the trapped electron to the conduction band increases. Once ionized, the electron can recombine with the hole to generate what is called thermoluminescence. The temperature T_m where the TL-glow intensity becomes maximum will be proportional with the electron trapping depth and changes roughly 400 K/eV at a heating rate of 1 K/s. This technique enables to probe energy differences down to few 0.01 eV [13].

2.6.6 Systematic Variation in Lanthanide Level Locations

The previous section provides an impression on the techniques that can and have been used to obtain information on level locations. But often these techniques were applied to a specific lanthanide ion in a specific compound with the aim to understand properties of that combination. Furthermore, each of these techniques provides its own source of unknown systematic error. Those individual studies do not provide us with a broad overview on how level locations change with type of lanthanide ion and type of compound. Such overview is needed to predict phosphor properties and to guide the researcher in the quest for new and better materials. One of the first systematic approaches was by Pedrini et al. by means of photoconductivity measurements to determine the location of the $4f^n$ ground state of different divalent lanthanides in the fluorite compounds CaF_2, SrF_2, and BaF_2 relative to the bottom of the conduction band [46]. They also provided a model to explain the observed variation in $4f^n$ ground state location with n. A systematic approach to determine the level locations of trivalent lanthanides was done by Thiel and co-workers by means of the XPS techniques [49,50]. They studied the trivalent lanthanides in $Y_3Al_5O_{12}$ and determined the $4f^n$ ground-state locations relative to the valence band top of the host crystal. They also combined their findings with the systematic in $4f^n \rightarrow 4f^{n-1}5d$ transition energy found in Ref. [29] to locate the lowest $4f^{n-1}5d$ levels in the bandgap. That location appears relatively constant with the type of lanthanide ion, i.e., with n.

Both the XPS and photoconductivity experiments have drawbacks. The oscillator strength for the transition from the localized $4f^n$ ground state to the delocalized conduction band states is very small and photoconductivity is rarely observed due to such direct transition. Two-step photoconductivity is observed more frequently. After a dipole-allowed excitation to the $4f^{n-1}5d$ state, it is either followed by autoionization, see Figure 2.36(c), or thermally assisted ionization, see Figure 2.36(e). For the XPS experiments, high Ln^{3+} concentrated samples are needed [49,51], and one has to deal with uncertain final state effects to obtain reliable data [52]. At this moment, the amount of information with these two methods is scarce. Although they provided us with valuable ideas and insight on how level locations change with type of lanthanide ion, there is not enough information to obtain detailed insight on how level locations change with type of compound.

Another method to obtain the systematic variation in level location with type of lanthanide is from CT spectroscopy. It appears that the energy of CT to Sm^{3+} is always (at least in oxide

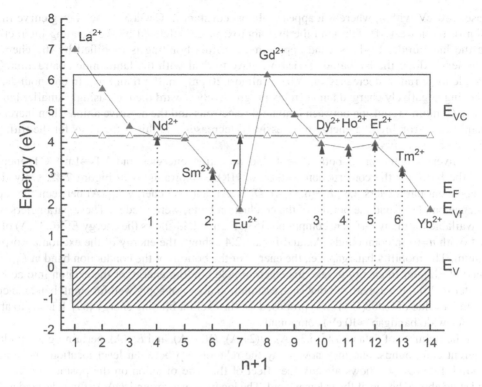

FIGURE 2.42 The location of the $4f^n$ ground states and lowest $4f^{n-1}5d$ states of the divalent lanthanide ions in $CaGa_2S_4$. Arrows 1 through 6 show observed energies of charge transfer to Ln^{3+}. Arrow 7 shows the observed energies for the first $4f^7 \rightarrow 4f^65d$ transitions in Eu^{2+}.

compounds) a fixed energy higher than for the CT to Eu^{3+}. The same applies to Tm^{3+} and Yb^{3+}. This was noticed already in the 1960s [28,53,54] and reconfirmed by later studies [55–57]. An elaborate analysis of data on CT retrieved from literature revealed that the systematic behavior in CT energies holds for all lanthanides in all type of compounds [41].

Figure 2.42 illustrates the state of the art around 2003 to construct diagrams with divalent lanthanide level location in $CaGa_2S_4$. The binding energy at the top of the valence band is defined as zero of energy. The arrows numbered 1 through 6 show the observed energies for CT to trivalent lanthanide ions, and they provide us with the location of the $4f^n$ ground state of the corresponding divalent lanthanides above the VB-top (cf. Figure 2.36(g)). Here the binding energy in the $4f^{n-1}5d$ state was assumed constant with n. Subtracting the E_{fd} values using Eq. (2.133) provides then the typical zigzag shape of $4f^n$ binding energies. Everything is pinned such that this zigzag curve corresponds best with the data from the CT energies. Arrow 7 shows the first $4f^7 \rightarrow 4f^65d$ transition in Eu^{2+}. A constant binding energy in the $4f^{n-1}5d$ state of lanthanides in $CaGa_2S_4$ was also expected for other sulfides but did not apply to the more ionic oxide and fluoride compounds. For those compounds, it appeared that from Eu^{2+} to Yb^{2+} the binding in the $4f^{n-1}5d$ states gradually decrease (become less negative) by about 0.5 eV [18,41]. In other words, the 5d state of Yb^{2+} is found 0.5 eV closer to the bottom of the conduction band than that of Eu^{2+}. This then explains the observation that Yb^{2+} in oxides and fluorides is more susceptible to anomalous emission, as illustrated in Figure 2.36(d), than Eu^{2+} [18].

Comparing Figure 2.38 with Figure 2.42, one may notice that the 4f and 5d binding energy curves for $CaGa_2S_4$ are a more or less tilted version of that for the free divalent lanthanides. The figures show that the bonding in the 4f ground state of free ion Yb^{2+} is stronger than in Eu^{2+} whereas it is about the same in $CaGa_2S_4$. The bonding in the 5d state of the free divalent lanthanides increases few eV with n, whereas it appears almost constant in $CaGa_2S_4$. The zigzag curve in the location of the lowest $4f^n$ state is in the first approximation dictated by the bonding interactions inside the lanthanide. In the second approximation, this bonding is modified by the chemical environment. Along the lanthanide series, we have to deal with the lanthanide contraction. The lanthanide ionic radius decreases and Yb^{2+} is almost 20 pm smaller than La^{2+}. In compounds, the neighboring negatively charged anions relax progressively toward the increasingly smaller lanthanide. The Coulomb repulsion between lanthanide electrons and the negative anions then increases. Lanthanide electron binding energy progressively decreases and this is the basis for the tilting of the binding energy curves.

The universal behavior in both $4f^n \rightarrow 4f^{n-1}5d$ transition energies and $VB \rightarrow Ln^{3+}$ CT energies formed the basis for the construction method of HRBE diagrams as in Figure 2.42. Only three host-dependent parameters, i.e., $E^{CT}(6,3+,A)$, $D(2+,A)$, and the energy $E_{VC}(A)$ between the top of the valence band (V) and the bottom of the conduction band, were needed. These parameters were made available for many different compounds [58]. Figure 2.43 shows the energy $E^{CT}(6,3+,A)$ of CT to Eu^{3+} (with $n=6$) in compounds (A), and Figure 2.44 shows the energy of the exciton absorption maximum. The mobility bandgap, i.e., the energy of the bottom of the conduction band at E_{VC}, was assumed 8% higher in energy to take the electron-hole binding energy of the exciton into account [58]. Later in 2017, it was realized that such assumption does not apply well for small bandgap compound and a value of $0.008(E^{ex})^2$ was proposed for the exciton binding energy [59]. It leads to about $0.08E^{ex}$ for wide bandgap (≈ 10 eV) compounds.

With the richness of data on $E^{CT}(6,3+,A)$, $D(2+,A)$, $D(3+,A)$, and $E^{ex}(A)$, pertaining to hundreds of different compounds, one may now study the relationship between level location and type of compound. Figure 2.43 shows already the effect of the type of anion on the position of the Eu^{2+} ground state above the top of the valence band. The location is at around 8 eV in fluorides and a clear pattern emerges when the type of anion varies. The energy decreases for the halides in the sequence F, Cl, Br, I, and for the chalcogenides in the sequence O, S, Se, and presumably Te. This pattern was

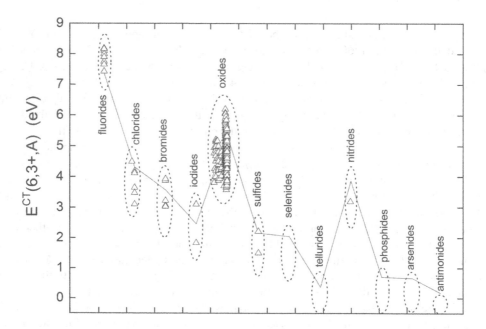

FIGURE 2.43 The energy $E^{CT}(6,3+,A)$ of charge transfer to Eu^{3+} in inorganic crystalline compounds. The parameter along the horizontal axis groups the data depending on the type of compound. The solid curve is given by $E^{CT}=3.72\eta(X)-2.00$ eV where $\eta(X)$ is the Pauling electronegativity of the anion.

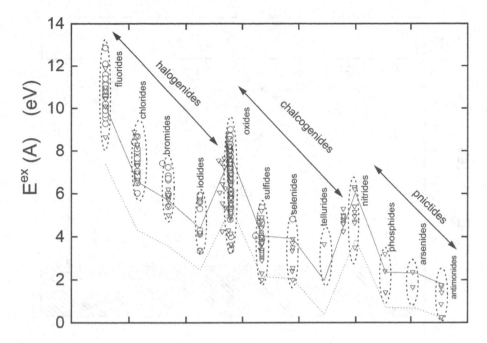

FIGURE 2.44 The energy of host exciton creation E^{ex} of inorganic compounds. The parameter along the horizontal axis groups the data depending on the type of compound. The solid curve is given by $E^{ex}=4.34\eta$ -7.15 eV where η is the Pauling electronegativity of the anion. The dashed curve is the same as in Figure 2.43.

also observed earlier on a much more limited family of compounds and has been interpreted with the Jörgensen model of optical electronegativity [60].

$$E^{CT} = 3.72(\eta(X) - \eta_{opt}(Eu)) \tag{2.135}$$

where $\eta(X)$ is the Pauling electronegativity of the anion X and $\eta_{opt}(Eu)$ is the optical electro-negativity of Eu; a value that must be determined empirically from observed CT energies. With $\eta_{opt}(Eu)=2$, the curve through the data in Figure 2.43 was constructed [58]. The curve reproduces the main trend with type of anion. It also predicts where we can expect the Eu^{2+} ground state in the pnictides; a decrease in the sequence N, O, As, Sb is expected. However, the wide variation of CT energies within, for example, the oxide compounds is not accounted for by the Jörgensen model. Aspects like lanthanide site size, anion coordination number are also important and need to be considered for a refined interpretation of CT data [58].

A similar equation as Eq. (2.135) can be introduced for E^{ex} to illustrate the main trend in the bandgap with type of anion [58]. The bandgap follows the same pattern as the energy of CT with changing type of anion. Interestingly, one may also notice a similar behavior in the values for the redshift in Figure 2.41 with changing type of anion. This shows that the values for the parameters $E^{CT}(6,3+,A)$, $D(2+,A)$, $D(3+,A)$, and $E^{ex}(A)$ used for the construction of diagrams as in Figure 2.42 are strongly correlated.

Analogous to the systematic behavior in the $4f^n$ ground state locations for the divalent lanthanides with changing n, a systematic behavior for the $4f^n$ ground state energies of trivalent lanthanides was proposed in Ref. [41]. One may then use, in principle, the same method as used for the divalent lanthanides to construct level diagrams for trivalent lanthanide ions. For the divalent lanthanide, the "anchor point" of construction is the CT energy to Eu^{3+}, see Figure 2.43. The CT to Ce^{4+} might play the role of such anchor point for the trivalent lanthanide level positions. However, information on CT to Ce^{4+} ion is sparsely available, and not sufficient to routinely construct level diagrams. There

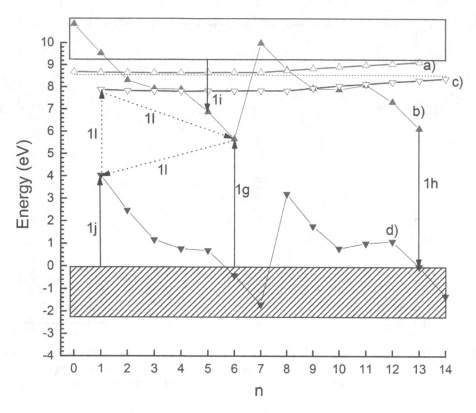

FIGURE 2.45 The location of the 4f^n (curve b and d) and lowest energy 4f^{n-1}5d states (curves (a) and (c)) of the divalent (curves (a) and (b)) and trivalent (curves (c) and (d)) lanthanide ions in YPO$_4$. n and $n+1$ are the number of electrons in the 4f shell of the trivalent and divalent lanthanide ion, respectively. Arrows indicate specific transitions that were also discussed in Figure 2.36. The horizontal dashed line at 8.55 eV is E^{ex}.

was a need for another anchor point, and the energy difference $E_{dC}(1,3+,A)$ between the lowest 5d state of Ce^{3+} and the bottom of the conduction band took that role. Its value (cf. Figure 2.36(d)) was obtained from two-step photoconductivity experiments or from luminescence quenching data.

Figure 2.45 shows, as demonstration for the state of the art around 2007, the level positions of both divalent and trivalent lanthanides in YPO$_4$. With the knowledge of today, the binding energies in the trivalent lanthanide levels need substantial revision. Nevertheless, by comparing with the free ion binding energies in Figure 2.38, several conclusions were made that are still valid today. The binding energy difference of more than 10 eV between the free trivalent and free divalent 4f^{n-1}5d states in Figure 2.38 is drastically reduced to about 0.8 eV in YPO$_4$. Energy differences of 0.5–1.0 eV are commonly observed when constructing diagrams for other compounds. The binding energy difference of 18 eV between the 4f^n ground states of Eu^{2+} and Eu^{3+} is reduced to about 6.1 eV in YPO$_4$.

The full potential of schemes like for YPO$_4$ is demonstrated by comparing Figure 2.45 with the situations sketched in Figure 2.36. Actually, each of the 12 situations in Figure 2.36 can be found in the scheme of YPO$_4$. The arrows marked 1g, 1h, 1i, 1j, 1l show the same type of transitions as in Figure 2.36(g)–(j), (l), respectively. Various other types of transitions, quenching routes, charge trapping depths can be read directly from the diagram. To name a few, (1) the lowest 4f^{n-1}5d states of all the divalent lanthanide ions are between E^{ex} and the bottom of the conduction band. In that situation, the 5d–4f emission is either absent or will have low quenching temperature, see Figure 2.36(c) and (d). (2) The lowest 4f^{n-1}5d states of the trivalent lanthanides are well below E^{ex} and for Ce^{3+}, Pr^{3+}, Nd^{3+}, Er^{3+}, and Tm^{3+} 5d–4f, emission is observed [36,37]. (3) Apart from Eu^{3+}, Gd^{3+}, Yb^{3+}, and Lu^{3+}, all the trivalent lanthanides form valence band hole traps. The trap is deepest for Ce^{3+}

followed by Tb^{3+} and Pr^{3+}. (4) The ground-state energies for the divalent lanthanides are high above the top of the valence band. In practice, this means that the trivalent valence state dominates. (5) The trivalent lanthanides create stable electron traps because the ground states of the corresponding divalent lanthanides are well below the conduction band. (6) The ground states of Sm^{2+}, Eu^{2+}, Tm^{2+}, and Yb^{2+} are below the 5d state of the trivalent lanthanides. This means that Sm^{3+}, Eu^{3+}, Tm^{3+}, and Yb^{3+} can quench the 5d emission of trivalent lanthanide ions.

2.6.7 Vacuum Referred Binding Energy Schemes

The previous section addressed the state of art around 2007 in constructing, what was called, a HRBE scheme for the lanthanides in compounds. This section addresses the progress made since 2007 which developed into the chemical shift model of 2012 [12] and eventually the refined chemical shift model of 2020 [13,14]. The models explain and provide a method to routinely determine the binding energies in all lanthanide $4f^n$ ground states with respect to the vacuum level. In HRBE diagram construction, we already noticed; (1) a more or less compound independent zigzag shape of $4f^n$ binding energy curves, (2) the tilting of the binding energy curves due to the lanthanide contraction and lattice relaxation, (3) a 20 -eV reduction of binding energy difference between divalent and trivalent $4f^n$ curves when in a compound, (4) a difference between sulfide and more ionic oxide and fluoride compounds, (5) the $4f^n$ zigzag shape for the trivalent lanthanides was still lacking a good experimental verification. Each of the above five items is largely resolved with the new models combined with better experimental data.

The main physics behind the chemical shift model is a screening of the positive charge of the lanthanide by the surrounding chemical environment. An Ln^{2+} will be effectively screened by $-2e$ of negative charge and an Ln^{3+} by $-3e$ of negative charge. Screening is most optimal in a metal environment where the free conduction band electrons can approach the lanthanide most closely. Screening is least optimal in a fluoride compound where electrons are strongly bonded in the $2p^6$ fluorine orbitals. Here screening is accomplished by the sheer presence of the anions around the lanthanide at distances given by the lattice structure. The same applies for other anions but then also the polarizability of the anion can contribute to more effective screening. The Coulomb repulsion between the negative screening charge with an electron in the lanthanide causes a reduction in the binding energy of that electron. This reduction is called the chemical shift. Its size depends on the amount of effective screening charge ($-2e$ or $-3e$) and on how close that screening charge can approach the lanthanide. Therefore, chemical shift for Ln^{3+} will be roughly 3/2 times that of Ln^{2+}. This forms already large part of the explanation why the binding energy differences of about 18 eV between the 4f ground states of the free Eu^{2+} and Eu^{3+} are reduced to about 6.1 eV for YPO_4 in Figure 2.45. Because of the lanthanide contraction, the screening charge can approach the lanthanide more closely and chemical shift increases with n. The chemical shift for Yb^{2+} 4f electrons will be larger than that for La^{2+}. This effect can be seen by comparing Figure 2.38 for the free lanthanides with Figure 2.45 for YPO_4.

In Ref. [12], the electrochemical redox potentials for several of the lanthanides in water were translated into binding energies in the $4f^n$ ground states relative to the vacuum level. In addition, XPS and inverse XPS data on the pure lanthanide metals were used to derive VRBEs in the $4f^n$ ground state of the divalent and trivalent lanthanides in metals, i.e., surrounded by free conduction band electrons. Together with the VRBEs for the free ions, three types of chemical environments are available with VRBE data. This appeared sufficient to derive and propose a quantitative model and method to establish the VRBE in divalent and trivalent lanthanide $4f^n$ ground states. The only parameter of the 2012 model was the Coulomb repulsion energy $U(6, A)$ defined as the binding energy difference between the Eu^{2+} and Eu^{3+} ground states in chemical environment A. It is 18.05 eV for free ions and varies from 7.6 eV in highly ionic fluoride compounds down to 5.6 eV in the most polarizable lanthanide metals. In between, it scales with how strong electrons are bonded in the anion that in turn follows the familiar nephelauxetic sequences, i.e., $F > O > Cl > Br > N > I > S > Se$ and within the oxides $SO_4^{2-} > CO_3^{2-} > PO_4^{3-} > H_2O > BO_3^{3-} > SiO_4^{4-} > $ aluminates [61].

In the refined chemical shift model of 2020, a further improvement of locating lanthanide levels was introduced [13,14]. The electron-electron repulsion between the electrons in the 4f orbital creates the different ^{2s+1}L levels in the Dieke diagram, and their level energies can be parameterized in terms of Racah parameters. In the first approximation, the Dieke diagram is the same for all compounds but in the second approximation, there is a compound dependence. Following the nephelauxetic sequences, the e–e repulsion reduces slightly and with that the level spacing and Racah parameters. Since the $4f^n$ ground state in the Dieke diagram is always placed at zero energy, there was no interest in the effect of the chemical environment on the binding energy in that ground state. The refined chemical shift model of 2020 does take that effect into account. It turns out that the binding energy in the $4f^n$ ground states for $n \geq 8$ may increase up to 0.5 eV due to the nephelauxetic effect. The effect for $n \leq 7$ is quite insignificant and hardly to notice experimentally. Altogether, the chemical environment causes (1) a several 10 eV chemical shift of the entire $4f^n$ zigzag binding energy curve. It is compound dependent and derived from the value of $U(6,A)$. (2) A few eV chemical tilt of the zigzag curve due to the lanthanide contraction. Tilting parameters were chosen that are best consistent with experimental data and there appeared no need to make those compound dependent. (3) A few 0.1 eV lowering of the right-hand branch of the zigzag curve due to the nephelauxetic effect. A nephelauxetic parameter β was introduced.

Figure 2.46 shows the VRBE scheme for the lanthanides in YPO_4 constructed with the 2020 refined chemical shift model. There are quite some differences with the earlier version in Figure 2.45.

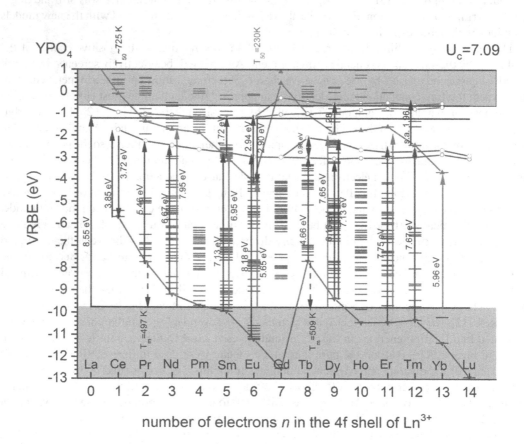

FIGURE 2.46 The VRBE scheme for YPO_4 constructed with $U(6,A) = 7.09$ eV, $\beta(2+)=0.96$, $\beta(3+)=0.94$, $E^{ex}=8.55$ eV, $E^{CT}(6,3+)=5.65$ eV, $D(3+)=2.22$ eV, and $D(2+)=1.26$ eV. The drawn arrows show the experimentally observed transitions. T_{50} refers to the quenching temperature for 5d–4f emission. T_m refers to the temperature at the maximum of the TL-glow peak. Not to overcrowd the figure the arrows and T_m for the release of electrons from divalent lanthanides to the CB are not shown.

Most notable is the VRBE scale instead of the HRBE scale. The divalent lanthanide level locations are still quite similar but here for $n \geq 8$ a distinction is made between the binding energy in the high spin and low spin lowest energy $4f^{n-1}5d$ states. The binding energies in the trivalent lanthanides are different because a too large tilting effect was assumed in the 2007 version. Note that the inverted Dieke diagram illustration of Figure 2.40, representing the eigenenergies of the $4f^n$ states with respect to that of $4f^{n-1}5d$ defined as zero, can be recognized in the VRBE diagram of YPO$_4$ where it represents binding energies with respect to the vacuum level defined as zero.

Figure 2.47 shows the VRBE diagram for CaGa$_2$S$_4$ to be compared with the 2007 version of Figure 2.42. The prior to 2007 assumption that the binding energy in the $4f^{n-1}5d$ state is constant with n still holds reasonably for $n \geq 8$. For YPO$_4$ the binding in $4f^{n-1}$ 5d increases about 0.5 eV in going from Eu^{2+} to Yb^{2+} still consistent with the observation that anomalous 5d–4f emission in oxides and fluorides is more often observed for Yb^{2+} than for Eu^{2+}.

Today, one can routinely construct VRBE schemes with all lanthanide level locations using few experimentally determined parameters. If experimental data are available to located the divalent and trivalent zigzag curves with respect to the host bands, it provides the value for $U(6, A)$. $U(6, A)$ can also be derived if the centroid shift for the Ce^{3+} 5d energies is known [62]. Otherwise, the $U(6, A)$ parameter value can be estimated from the chemical composition of the inorganic compound. The tilting of the $4f^n$ binding energy curve in compounds can be kept compound independent. The nephelauxetic lowering of the right-hand branch is a relatively small effect and a fair estimate can already be made based on the compound composition alone. With this information together with the

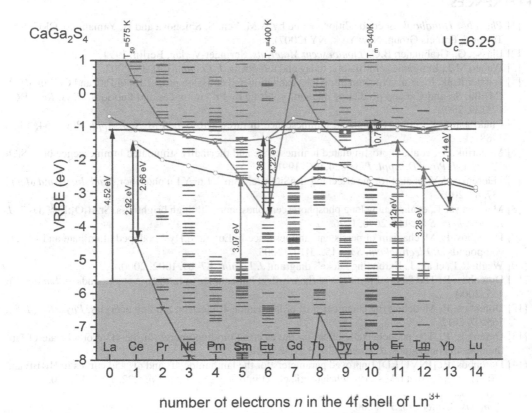

FIGURE 2.47 The VRBE scheme for CaGa$_2$S$_4$ constructed with $U(6, A)$=6.25 eV, $\beta(2+)$=0.906, $\beta(3+)$= 0.906, E^{ex}=4.52 eV, $E^{CT}(6,3+)$=1.9 eV, $D(3+)$=3.25 eV, and $D(2+)$=1.86 eV. The drawn arrows show the experimentally observed transitions. T_{50} refers to the quenching temperature for 5d–4f emission. T_m refers to the temperature at the maximum of the TL-glow peak.

redshift values, the VRBE in all lanthanide $4f^n$ and lowest energy $4f^{n-1}5d$ states are generated. What remains are a CT energy to one of the trivalent lanthanides and the exciton creation energy of the host compound to obtain the VRBE at the VB top and CB bottom. These parameters are available for many hundreds of different compounds.

The schemes still contain systematic errors. Often the bottom of the conduction band is not well defined or known, or level locations may change due to charge compensating defects and lattice relaxation. The bandgap of inorganic compounds tends to lower with increasing temperature with typically 0.05 eV/100K, and this means that the VRBE at the VB-top and/or CB-bottom changes. Such changes or errors are still very important for phosphor performance because a few tenths of eV shift of lanthanide level position with respect to the host bands may change the performance of a phosphor from very good to useless. The level schemes are, however, already very powerful in predicting 4f–5d transition energies, CT transition energies, the electron and hole trapping depths provided by the lanthanides, the preferred lanthanide valence state. The YPO_4 scheme of Figure 2.46 was constructed with 8 different experimental input parameters. This may seem a lot but actually all eight parameters are highly correlated to each other because each one of them is linked to how strong electrons are bonded in the anions and therewith with the nephelauxetic sequences [61]. With future machine learning strategies one might expect that those correlations are better revealed and even correlated to other properties of inorganic compounds like refractive index, band structure, hardness, site symmetry. First attempts into such direction appeared already [63,64].

REFERENCES

[1] *Phosphor Handbook*, Second edition, edited by W.M. Yen, S. Shionoya and H. Yamamoto, CRC Press Taylor & Francis Group, New York, NY (2007).

[2] Blasse, G., Grabmaier, B.C., *Luminescent Materials*, Springer-Verlag, Berlin, 1994.

[3] Weber, M.J., Inorganic scintillators: today and tomorrow, *J. Lumin.*, 100, 35, 2002.

[4] van der Kolk, E., et al., Vacuum ultraviolet excitation and emission properties of Pr^{3+} and Ce^{3+} in MSO_4 (M=Ba, Sr, and Ca) and predicting quantum splitting by Pr^{3+} in oxides and fluorides, *Phys. Rev.*, B64, 195129, 2001.

[5] Chakrabarti, K., et al., Stimulated luminescence in rare-earth-doped MgS, *J. Appl. Phys.*, 64, 1362, 1988.

[6] Meijerink, A., et al., Photostimulated luminescence and thermally stimulated luminescence of Y_2SiO_5 -Ce, Sm, *J. Phys. D: Appl. Phys.*, 24, 997, 1991.

[7] Sidorenko, A.V., et al., Storage effect in $LiLnSiO_4:Ce^{3+},Sm^{3+}$, Ln=Y,Lu phosphor, *Nucl. Instr. and Meth.*, 537, 81, 2005.

[8] Matsuzawa, T., et al., A new long phosphorescent phosphor with high brightness, $SrAl_2O_4:Eu^{2+},Dy^{3+}$, *J. Electrochem. Soc.*, 143, 2670, 1996.

[9] Dorenbos, P., Mechanism of persistent luminescence in Eu^{2+} and Dy^{3+} codoped aluminate and silicate compounds, *J. Electrochem. Soc.*, 152, H107, 2005.

[10] Wegh, R.T., et al., J., Extending Dieke's diagram, *J. Lumin.*, 87–89, 1002, 2000.

[11] Dorenbos, P., The 5d level positions of the trivalent lanthanides in inorganic compounds, *J. Lumin.*, 91, 155, 2000.

[12] Dorenbos, P., Modeling the chemical shift of lanthanide 4f electron binding energies, *Phys. Rev. B* 85, 165107, 2012.

[13] Dorenbos, P., The nephelauxetic effect on the electron binding energy in the $4f^q$ ground state of lanthanides in compounds, *J. Lumin.*, 214, 116536, 2019.

[14] Dorenbos, P., [INVITED] Improved parameters for the lanthanide $4f^q$ and $4f^{q-1}5d$ curves in HRBE and VRBE schemes that takes the nephelauxetic effect into account, *J. Lumin.*, 222, 117164, 2020.

[15] Dorenbos, P., $f \rightarrow d$ transition energies of divalent lanthanides in inorganic compounds, *J. Phys.: Condens. Matter.*, 15, 575, 2003.

[16] Dorenbos, P., Energy of the first $4f^7 \rightarrow 4f^65d$ transition in Eu^{2+} doped compounds, *J. Lumin.*, 104, 239, 2003.

[17] McClure, D.S. and Pedrini, C., Excitons trapped at impurity centers in highly ionic crystals, *Phys. Rev.*, B32, 8465, 1985.

[18] Dorenbos, P., Anomalous luminescence of Eu^{2+} and Yb^{2+} in inorganic compounds, *J. Phys.: Condens. Matter*, 15 2645, 2003.

[19] Lyu, L.-J. and Hamilton, D.S., Radiative and nonradiative relaxation measurements in Ce^{3+} doped crystals, *J. Lumin.*, 48&49, 251, 1991.

[20] Dorenbos, P., Thermal quenching of Eu^{2+} 5d-4f luminescence in inorganic compounds, *J. Phys. Cond. Matter*, 17, 8103, 2005.

[21] Bessière, A., et al., Spectroscopy and lanthanide impurity level locations in $CaGa_2S_4$:Ln (Ln=Ce, Pr, Tb, Er, Sm), *J. Electrochem. Soc.*, 151, H254, 2004.

[22] Boutinaud, P., et al., E, Making red emitting phosphors with Pr^{3+}, *Opt. Mater.*, 28, 9, 2006.

[23] Guerassimova, N., et al., X-ray excited charge transfer luminescence of ytterbium-containing aluminium garnets, *Chem. Phys. Lett.*, 339, 197, 2001.

[24] Dorenbos P., et al., Photon controlled electron juggling between lanthanides in compounds, *J. Lumin.*, 133, 45, 2013.

[25] Joos J.J., et al., Identification of Dy^{3+}/Dy^{2+} as electron trap in persistent phosphors, *Phys. Rev. Lett.*, 125, 033001, 2020.

[26] Brewer, L., *Systematics and the Properties of the Lanthanides*, edited by S.P. Sinha, D. Reidel Publishing Company, Dordrecht, The Netherlands, 1983, 17.

[27] Martin, W.C., Energy differences between two spectroscopic systems in neutral, singly ionized, and doubly ionized lanthanide atoms, *J. Opt. Soc. Am.*, 61, 1682, 1971.

[28] Jörgensen, C.K., Energy transfer spectra of lanthanide complexes, *Mol. Phys.*, 5, 271, 1962.

[29] Dorenbos, P., The $4f^n \leftrightarrow 4f^{n-1}5d$ transitions of the trivalent lanthanides in halogenides and chalcogenides, *J. Lumin.*, 91, 91, 2000.

[30] Dorenbos, P., A review on how lanthanide impurity levels change with chemistry and structure of inorganic compounds, *ECS J. Solid State Sci. Technol.*, 2, R3001, 2013.

[31] Andriessen, J., et al., Ab initio calculation of the contribution from anion dipole polarization and dynamic correlation to 4f-5d excitations of Ce^{3+} in ionic compounds, *Phys. Rev.*, B72, 045129, 2005.

[32] Dorenbos, P., 5d-level energies of Ce^{3+} and the crystalline environment. I. Fluoride compounds, *Phys. Rev.*, B62, 15640, 2000.

[33] Dorenbos, P., 5d-level energies of Ce^{3+} and the crystalline environment. IV. Aluminates and simple oxides, *J. Lumin.*, 99, 283, 2002.

[34] Dorenbos, P., 5d-level energies of Ce^{3+} and the crystalline environment. II. Chloride, bromide, and iodide compounds, *Phys. Rev.*, B62, 15650, 2000.

[35] Dorenbos, P., Relation between Eu^{2+} and Ce^{3+} f→d transition energies in inorganic compounds, *J. Phys.: Condens. Matter.*, 15, 4797, 2003.

[36] van Pieterson, L., et al., $4f^n \rightarrow 4f^{n-1}5d$ transitions of the light lanthanides: Experiment and theory, *Phys. Rev.*, B6, 045113, 2002.

[37] van Pieterson, L., et al., $4f^n \rightarrow 4f^{n-1}5d$ transitions of the heavy lanthanides: Experiment and theory, *Phys. Rev.*, B65, 045114, 2002.

[38] Dorenbos, P., Exchange and crystal field effects on the $4f^{n-1}5d$ levels of Tb^{3+}, *J. Phys.: Condens. Matter*, 15, 6249, 2003.

[39] Wong, W.C., et al., Charge-exchange processes in titanium-doped sapphire crystals. I. Charge-exchange energies and titanium-bound excitons, *Phys. Rev.*, B51, 5682, 1995.

[40] Happek, U., et al., Observation of cross-ionization in $Gd_3Sc_2Al_3O_{12}$:Ce^{3+}, *J. Lumin.*, 94-95, 7, 2001.

[41] Dorenbos, P., Systematic behaviour in trivalent lanthanide charge transfer energies, *J. Phys.: Condens. Matter*, 15, 8417, 2003.

[42] Sato, S., Optical absorption and X-ray photoemission spectra of lanthanum and cerium halides, *J. of the Phys. Soc. of Jpn.*, 41, 913, 1976.

[43] Lizzo, S., et al., Luminescence of divalent ytterbium in alkaline earth sulphates, *J. Lumin.*, 59, 185, 1994.

[44] Jia, D., et al., Location of the ground state of Er^{3+} in doped Y_2O_3 from two-step photoconductivity, *Phys. Rev.*, B65, 235116, 2002.

[45] van der Kolk, E., et al., 5d electron delocalization of Ce^{3+} and Pr^{3+} in Y_2SiO_5 and Lu_2SiO_5, *Phys. Rev.*, B71, 165120, 2005.

[46] Pedrini, C., et al., Photoionization thresholds of rare-earth impurity ions. Eu^{2+}:CaF_2, Ce^{3+};YAG, and Sm^{3+}:CaF_2, *J. Appl. Phys.*, 59, 1196, 1986.

[47] Fuller, R.L. and McClure, D.S., Photoionization yields in the doubly doped SrF_2:Eu,Sm system, *Phys. Rev.*, B43, 27, 1991.

[48] Joubert, M.F., et al., A new microwave resonant technique for studying rare earth photo-ionization thresholds in dielectric crystals under laser irradiation, *Optical Materials*, 24, 137, 2003.

[49] Thiel, C.W., Systematics of 4f electron energies relative to host bands by resonant photo-emission of rare earth ions in aluminum garnets, *Phys. Rev.*, B64, 085107, 2001.

[50] Thiel, C.W., et al., Progress in relating rare-earth ion 4f and 5d energy levels to host bands in optical materials for hole burning, quantum information and phosphors, *J. Modern Opt.*, 49, 2399, 2002.

[51] Pidol, L., et al., Energy levels of lanthanide ions in a $Lu_2Si_2O_7:Ln^{3+}$ host, *Phys. Rev.*, B72, 125110, 2005.

[52] Poole, R.T., et al., Electronic structure of the alkaline-earth fluorides studied by photoelectron spectroscopy, *Phys. Rev.*, B12, 5872, 1975.

[53] Barnes, J.C. and Pincott, H., Electron transfer spectra of some lanthanide(III) complexes, *J. Chem. Soc. (a)*, 842, 1966.

[54] Blasse, G. and Bril, A., Broad band U.V. excitation of Sm^{3+}-activated phosphors, *Phys. Lett.*, 23, 440, 1966.

[55] Krupa, J.C., Optical excitations in lanthanide and actinide compounds, *J. Alloys Compd.*, 225, 1, 1995.

[56] Nakazawa, E., The lowest 4f-to-5d and charge-transfer transitions of rare earth ions in YPO_4 hosts, *J. Lumin.*, 100, 89, 2002.

[57] Krupa, J.C., High-energy optical absorption in f-compounds, *J. Solid State Chem.*, 178, 483, 2005.

[58] Dorenbos, P., The Eu^{3+} charge transfer energy and the relation with the band gap of compounds, *J. Lumin.*, 111, 89, 2004.

[59] Dorenbos, P., Charge transfer bands in optical materials and related defect level location, *Opt. Mater.*, 69, 8, 2017.

[60] Jörgensen, C.K., *Modern Aspects of ligand Field Theory*, North-Holland Publishing Company, Amsterdam, 1971.

[61] Dorenbos, P., Lanthanide 4f-electron binding energies and the nephelauxetic effect in wide band gap compounds, *J. Lumin.*, 136, 122, 2013.

[62] Dorenbos, P., Ce^{3+} 5d-centroid shift and vacuum referred 4f-electron binding energies of all lanthanide impurities in 150 different compounds *J. Lumin.*, 135, 93, 2013.

[63] Pilania G., et al., Physics-informed machine learning for inorganic scintillator discovery, *The J. of Chem. Phys.*, 148, 241729, 2018.

[64] Zhuo Y., et al., Machine learning 5d-level centroid shift of Ce^{3+} inorganic phosphors, *J. Applied Physics*, 128, 013104, 2020.

3 Principal Phosphor Materials and Their Optical Properties

CONTENTS

DOI: 10.1201/9781003098690-3

3.1   WIDE BANDGAP II–VI COMPOUND SEMICONDCTORS

Youming Lu

3.1.1   INTRODUCTION

II–VI compound semiconductors are a compound formed from elements of groups II and VI of the periodic table, such a definition encompasses the oxides, sulfides, selenides and tellurides of beryllium, magnesium, zinc, cadmium and mercury. Among them, wide bandgap II–VI compound semiconductors, represented by ZnSe and ZnO, have the advantages of wide bandgap, large exciton binding energy and energy to form alloys in any proportion. Therefore, they have been considered as important candidate materials for violet, blue and green luminescence, laser as well as photoelectric detection and optical nonlinearity of response in this wavelength range. This section will first focus on the fundamental physical properties, including crystal structure, band structure, crystal growth, optical and electrical properties for wide bandgap II–VI compounds and describe some applications in optoelectronic properties. The interesting exciton behaviors under high-density excitation in this class of II–VI compounds such as ZnSe and CdS are discussed subsequently. Finally, a variety of EL devices such as metal-insulator-semiconductor (MIS) structure, p-n junction and quantum dots (QDs) structure are briefly introduced. The term II–VI compounds used next will be limited to the abovementioned compounds with wide bandgap.

3.1.2   FUNDAMENTAL PROPERTIES

3.1.2.1   Crystal Structure

II–VI compounds usually have two crystal structures, which are zinc blende (ZB) or hexagonal wurtzite (W).[1] ZB structure is similar to the diamond structure, which is constructed from two sets of face-centered cubic lattices that move 1/4 length along the body diagonal. The two sets of lattices are atoms of group II and VI elements, respectively, as shown in Figure 3.1(a). Therefore, the atomic arrangement of the ZB structure is that each atom of group II elements are surrounded by four nearest-neighbor atoms of group VI elements to form a tetrahedron, while each atom of group VI elements is surrounded by four nearest-neighbor atoms of group II to form a tetrahedron (Figure 3.1(b)). II–VI compounds with ZB structure mainly include ZnSe, ZnS, ZnTe, CdTe and so on. W structure consists of two interpenetrating close-packed hexagonal lattices, as illustrated in Figure 3.2, displaced with respect to each other by a distance $3/8c$ along the hexagonal c-axis. BeO, ZnO, ZnS, CdS, CdSe and MgTe have all been observed to take W structure. Note that some II–VI compounds have two crystal structures, for example, ZnS crystallize either in the ZB structure or in the W structure. Usually, the ZB structure corresponds to the low-temperature phase, the phase-transition temperature of the ZB and W structures in ZnS

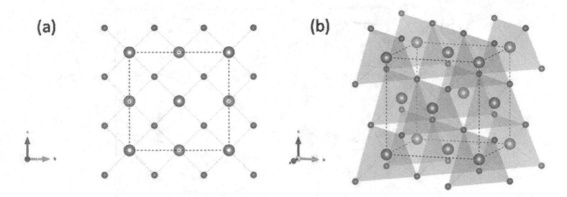

FIGURE 3.1　ZB crystal structure (a) and the polyhedral distribution (b) of II(red)-VI(gray) compounds.

is known to be about 1020°C. Moreover, CdO, MgO, MgS and MgSe take the sodium chloride structure at room temperature (RT), which is a more tightly packed structure. At high pressures, the W and ZB structures of many of II–VI compounds have been observed to revert to the sodium chloride structure.

3.1.2.2　Band Structure

The majority of the knowledge of band structure in the II–VI compounds has been derived from the semiempirical pseudopotential method,[2,3] which have included II–VI compounds with the ZB and wurtzite structures. Figure 3.3 shows the band structure determined by Chelikowsky and Cohen for ZnSe and CdTe, respectively, which take the ZB structure.[2] As can be seen from Figure 3.3, they are all direct-transition-type semiconductors, and both the bottom of the conduction band (CB) and the top of the valence band (VB) are located at the Γ point [k=(0 0 0)] in k-space, and in the CB, there are two minima in upper energy regions at the L [k=(1 1 1)] and the X [k=(1 0 0)] points above the bottom of the CB, respectively. It noted that the VB in the ZB structure is split by the spin-orbit interaction into a higher lying Γ_8 state (in which the orbital

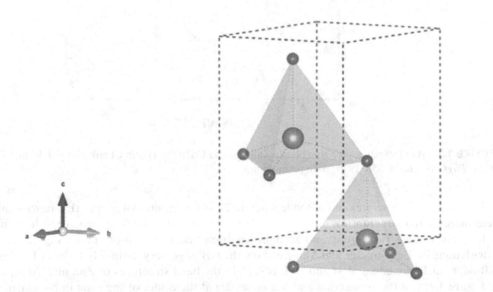

FIGURE 3.2　W crystal structure of II(red)-VI(gray) compounds.

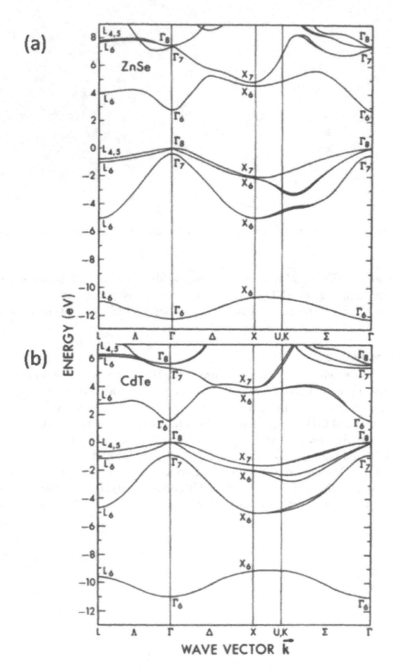

FIGURE 3.3   Band structures for zinc blende ZnSe (a) and CdTe (b). (From Chelikowsky, J. R. and Cohen, M. L., *Phys. Rev. B*, 14, 556, 1976. With permission.)

state is doubly degenerate) and a nondegenerate Γ_7 state. Figure 3.4(a) and (b) shows similarly determined band structure for ZnS and CdSe which take the wurtzite structure.[3] It should be noted that in these two figures, no account has been taken of spin-orbit splitting effects. The calculations by Bergstresser and Marvin show the CB edge very definitely to be at Γ'. Next the VB edges at Γ' are edges at M and H. It is seen in the band structures of ZnS and CdSe, shown in Figure 3.4, that the conduction and VB edges are at the center of the zone in both structures, so that both are direct gap materials.

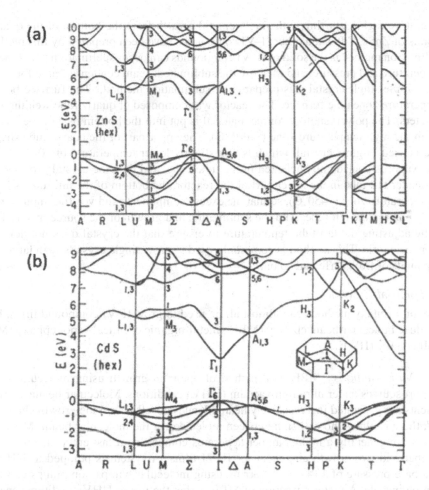

FIGURE 3.4 Band structure of hexagonal ZnS (a) and CdS (b). (From Bergstresser, T. K. and Cohen, M. L., *Phys. Rev.*, 164, 1069, 1967. With permission.)

Thus, for II–VI compounds with either ZB or W structure, because the CB has the character of the s orbital of the cations, while the VB has the character of the p orbital of the anions, this results in that the bottom of the CB and the top of the VB are located at the Γ point in the center of the Brillouin zone, indicating that these compounds are all direct-transition-type semiconductors. If one compares the radiative recombination coefficient of electrons and holes for direct and indirect transitions, the value for the former is four orders of magnitude larger. Therefore, these II–VI compounds have been widely investigated as ideal candidates for the preparation of light-emitting devices in the ultraviolet (UV)-vis wavelength range.

3.1.2.3 Crystal Growth

3.1.2.3.1 Bulk Crystal Growth

The growth methods of II–VI compound bulk crystal include Bridgman method and sublimation method etc. Bridgman method is also called high-pressure melting method, which is due to large vapor pressure and dissociation degree of II–VI compounds at melting point, and the growth of crystals from the melt must be carried out in a high-pressure single-crystal furnace. Taking the growth CdTe as an example,[4] the Cd and Te elements were first sealed in a quartz tube under vacuum and placed in a vertical high-pressure furnace for heating to ~800°C to synthesize the CdTe. The prepared CdTe was heated to 1150°C to melt, and then the reaction tube was slowly dropped to a low-temperature region, starting with single-crystal growth from the bottom of the quartz tube. Single crystals of ZnSe, CdSe and other II–VI compounds have been prepared by this method.

The sublimation method is the solid of II–VI compounds can occur sublimation phenomenon at a certain temperature and pressure, and then use the sublimated steam condensation to form crystals. For example, ZnSe single crystal was prepared by sublimation method.[5] The furnace body was a double-temperature zone tube furnace. The reactor was composed of quartz tube welding with different diameters. The polycrystalline source material is put into the bottom part of the coarse tube and placed in the high-temperature zone (~1005°C). The upper part of the coarse tube shaped like a tube cone is used to grow crystal, which is at a slightly lower temperature of 1000°C. A sublimated ZnSe vapor is cooled and nucleated in the tube cone, and then the crystal grows gradually. There is a slender tail tube in the lower part of the reactor, the bottom of the tail tube is located in the lower temperature zone (~450°C), so that the excess components and volatile impurities can be preferentially entered the tail tube. And the sublimation and growth of the source materials can be controlled by adjusting the tail tube temperature to ensure that the crystal does not deviate from the stoichiometric ratio. This method is commonly used to grow single crystals with large sizes for II–VI compounds of CdS, ZnTe and so on.

3.1.2.3.2 Epitaxial Growth

III–V compound epitaxy methods can almost all be used to grow II–VI compound films. Here we briefly introduce molecular-beam epitaxy (MBE), metal-organic chemical vapor epitaxy (MOVPE) and hot-wall epitaxy (HWE).

3.1.2.3.2.1 MBE Methods
MBE is a method of epitaxial growth using molecular or atomic beam transport sources under ultrahigh vacuum (UHV) conditions. Molecular beams of II and VI group elements are generated in a resistively heated Knudsen cell within the growth chamber.

In UHV, there is almost no collision between molecules in the molecular beam. Modern II–VI MBE systems are generally a multichamber apparatus comprising a fast entry chamber, a preparation chamber and two growth chambers for II–VI films. A system is pumped to UHV conditions with a base pressure of 10^{-11}–10^{-10} Torr by using molecular pumps, ion pumps and titanium sublimation pumps etc. A major attraction of MBE is that the use of UHV conditions enables the incorporation of high-vacuum-based surface analytical and diagnostic techniques. Reflection high-energy electron diffraction (RHEED) is commonly employed to examine the substrate and the actual epitaxial film during growth. A (quadrupole) mass spectrometer is essential for monitoring the gas composition in the MBE growth chamber. In the growth chamber with UHV, there is almost no collision between molecules in the molecular beam. Therefore, the purity of the material can be greatly improved, the quality of crystallography is good and the surface is bright and flat. MBE has the characteristics of low growth temperature and slow growth rate, so it is an atomic-level growth technique, which is beneficial to grow the multilayer heterostructures.[6] Since MBE does not need to grow under thermal equilibrium conditions, it is a kinetic process, so the self-compensation effect can be suppressed when preparing II–VI group of compounds, which makes the thin-film material have better electrical properties. MBE, as an important ultrathin layer growth technology, has been widely used for epitaxial growth of II–VI compounds, such as ZnTe, ZnSe, ZnS thin films, and their superlattices (SLs), quantum well (QW) structure materials prepared by traditional MBE methods and ZnO-based films, heterostructures and their devices prepared by laser-assisted MBE technology.[7]

3.1.2.3.2.2 MOVPE Methods
MOVPE method, which has the characteristics of high growth rate, low growth temperature, easy doping and favorable growth of multilayer structure, is one of the commonly used methods for epitaxial growth of II–VI compounds. It uses metal-organic (MO) of II groups and hydride of VI groups as source materials for crystal growth. The thin films and their heterostructures of II–VI compound semiconductors are epitaxially grown on the substrates by

TABLE 3.1

The Growth Conditions for the II–VI Compound Semiconductor Films by MOVPE Method

Compounds	Substrate	Reactants	Growth Temperature (°C)
ZnS	Al_2O_3, $MgAl_2O_4$, BeO	$(C_2H_6)_2Zn+H_2S$	750
ZnSe	Al_2O_3, $MgAl_2O_4$, BeO, GaAs	$(C_2H_6)_2Zn+H_2Se$	725–750
ZnTe	Al_2O_3	$(C_2H_6)_2Zn+(CH_3)_2Te$	500
CdS	Al_2O_3, GaAs, CdS	$(CH_3)_2Cd+H_2Se$	475–560 (AP)
CdSe	Al_2O_3, Glass	$(CH_3)_2Cd+H_2S$	600
CdTe	Al_2O_3, $MgAl_2O_4$, BeO	$(CH_3)_2Cd+(CH_3)_2Te$	300–350 (LP)
HgTe	CdTe	$(CH_3)_2Te+Hg+H_2$	500

thermal decomposition. Table 3.1 gives the conditions for the growth of II–VI compound films using the MOVPE method. The MO is of II group generally use alkyl compounds such as dimethylzinc (DMZn), diethylcadmium (DECd) etc.

Most of these MOs are liquids with high vapor pressures, which are stored in temperature-controlled containers. Transport of the MOs to the growth zone is achieved by bubbling H_2 through the liquid sources. The organic sources of II groups fed into the reactor were mixed with hydride (e.g., H_2S, H_2Se) of the VI groups, a thermal decomposition reaction occurs on the surface of the heated substrate and then the film of II–VI compounds is grown. The substrates are usually heated by radio frequency (RF). Because all the source materials of the MOVPE are passed into the reactor in gaseous state, the parameters such as composition, conductive type, carrier concentration and thickness of the epitaxial layers can be controlled by accurately controlling various gas flow rates. Thus, the distribution of components and impurities in the films is uniform and steep, which is conducive to the growth of nanoscale thin layers and multilayer structures. Besides, the MOVPE equipment is easy to control and easy to grow in multi-chip and large sizes, and the lower cost is beneficial to batch production. For example, the MOFPE growth of ZnSe involves pyrolysis of a vapor-phase mixture of H_2Se or dimethylselenide (DMSe) and the most commonly DMZn or diethylzinc (DEZn). Free Zn atoms and Se molecules are formed and these species recombine on the hot substrate surface in an irreversible reaction to form ZnSe.[8,9] Growth is carried out in a cold-wall reactor in flowing H_2 at atmospheric or low pressure. The substrate is heated to temperatures of 300–400°C, typically by RF heating.

3.1.2.3.2.3 HWE Methods HWE is a vapor-phase epitaxial growth technique in which the source material is heated and evaporated in a vacuum ($\sim10^{-4}$ Pa). In a closed system, the temperature of the growth chamber wall is kept close to the temperature of the source material, and compound gas molecules or their component atoms are evaporated onto the substrate to deposit, nucleate and grow up, thus forming epitaxial films. This method has the advantages of simple equipment, low cost and saving source materials and so on. Because of the hot wall, the epitaxial growth is carried out in close proximity to the source material temperature, and it is the epitaxial growth under the condition close to the thermodynamic equilibrium condition. When the CdS film is grown, for example, the substrate temperature is 450°C, and the source temperature is only 25°C higher than the substrate temperature. Therefore, the epitaxial layer grown by this method contains low impurities and defects. Moreover, a variety of II–VI group heterostructure materials, such as ZnTe/ZnSe,[10] can be prepared by using two hot-wall source tubes.

3.1.2.4   Optical Properties

The reflection coefficient R and absorption coefficient α of the II–VI compounds are closely related to the photon energy $h\nu$. When the photon energy is larger than the bandgap ($h\nu>E_g$), the magnitude

TABLE 3.2

Optical Energy Gapes of II–VI Compounds at Low Temperatures

Compounds	E_g (eV)	dE_g/dT (10^{-4} eV/K)
ZnO	3.44 W	−9.5
ZnS	3.91 W/3.84 ZB	−8.5/−4.6
CdS	2.58 W	−5.2
HgS	2.10 ZB	
ZnSe	2.80 W/2.83 ZB	/−8
CdSe	1.84 W	−4.6
HgSe	−0.1 ZB	
ZnTe	2.39 ZB	−5.0
CdTe	1.60 ZB	−2.3
HgTe	−0.1 ZB	

W, wurtzite; ZB, zinc blende structure.

order of α usually reaches 10^4–10^5 cm^{-1}, which is related to its characteristic of direct bandgap. When $h\nu < E_g$, α drops sharply ($h\nu < 0.1$ eV, $\alpha \sim 1$ cm^{-1}). The appearance color of these compound crystals under visible light irradiation is related to the position of their intrinsic absorption edges. For example, the absorption edges of ZnO and ZnS are in the UV region, so the two crystals are transparent and colorless to the visible light; the absorption edges of CdSe and CdTe are in the infrared region, and their crystals are gray; other II–VI compounds with absorption edges in the visible region decrease with the E_g, and the color of the crystals changes from yellow (ZnSe) to dark red (ZnTe). The low-temperature values of the energy gap or absorption edge energy as determined from transmission and reflectivity measurements are listed in Table 3.2.[1]

Because II–VI compounds are semiconductor materials with direct transitions, which have the advantages of wide bandgap and high luminescence efficiency. In particular, the large exciton binding energy makes II–VI compounds still have abundant exciton effect at high temperature, it is considered as an ideal candidate to achieve laser emission. No matter electroluminescence (EL) or photoluminescence (PL), there is a great deal of work to report exciton emission in II–VI compounds. For example, strong sharp emission lines observed near the band edge in ZnS, ZnSe, CdS materials are attributed to exciton recombination. Moreover, laser emission is realized for all Zn, Cd compounds, and the emission with photon energy close to E_g is generated by exciton recombination. The exciton recombination will be highlighted in later chapters. Moreover, the luminescence of II–VI compounds is closely related to defects and impurities in the samples. The band-edge emission can be converted to the recombination related to donor, or acceptor, or donor-acceptor pair (DAP) by incorporating the acceptor impurities of certain group-I and group-V elements, and the donor impurities of group-III and group-VII elements. Especially, the desired luminescence properties can be obtained by doping transition group elemental impurities. For example, a wide green emission band due to DAP recombination is observed in the Cu- and Al-doped ZnS, and a red luminescence band can be obtained in Mn-doped ZnS. Most of the cathode-ray luminescence studies were conducted on powder and film samples. The electron-hole pairs can be excited by high-energy electron bombardment, and the resulting emission spectrum is similar to the PL result obtained at $h\nu > E_g$. The laser emission has been successfully achieved by electron-beam excitation for all sulfur compounds of Zn and Cd, in which the light emission spectrum with photon energy close to E_g consists of very narrow lines arising from exciton recombination.

3.1.2.5 Electrical Properties

II–VI compound crystals are prone to produce defects than III–V compound crystals, in which point defects can cause deviation of their composition stoichiometry and cause changes in the type of conduction. II–VI compounds are represented by MX, the point defects of which include vacancy (V_M, V_X), interstitial atoms (M_i, X_i), antistructural defects (M_X, X_M) and foreign impurity (F). These point defects will ionize under certain valence <X atom valence. Considering ZnSe as an example, the group-III elements serve as donor dopants, e.g., Al, Ga or In on Zn site,[13] as do the group-VII elements, e.g., F, Cl or I on Se site.[8] For doping of p-ZnSe, suitable acceptor materials are the group-I elements, e.g., Li, Na or Cu on Zn site,[13] or the group-V elements, e.g., N, P or As on Se site.[14] Additionally, it was reported that oxygen, which is an isoelectronic impurity in ZnSe, can act as an acceptor. However, also the defect V_{Se} may serve as donor, while V_{Zn} acts as an acceptor.

The presence of point defects in II–VI compound crystals can cause a deviation of the atom stoichiometry. When the concentration of X_i or V_M is high in MX crystals, the amount of X will be larger than M, the stoichiometry of the material deviates positively. The negative deviation of stoichiometric ratio occurs when the amount of component M is excess, that is, M_i, V_X concentration is large. Positive and negative deviations result in different conductive types of semiconductor materials, namely n or p. It can be seen that doping impurity and deviating from stoichiometric ratio will lead to the change of conductive type of the semiconductor, and this kind of semiconductor that shows different conductive types due to the change of composition is amphoteric semiconductor. However, only CdTe is amphoteric material in II–VI compounds. For II–VI compounds, it is noted that ZnS, CdS, ZnSe, CdSe can only present the n-type conductance and ZnTe can only present p-type conductance in the case of undoped, and it is difficult for these materials to obtain the inverse type conductance by the usual doping method. Further studies have found that this is due to the existence of charge-different impurities and lattice defects in the crystal to occur compensation effect. The doped impurities are compensated by the defect center of the opposite charge type formed by the impurity incorporation, and this phenomenon is called self-compensating effect.[12]

Self-compensation is widespread in II–VI compounds. The classical self-compensation theory holds that the wider the bandgap is, the higher the vacancy formation energy is and the larger the ionic bond composition is, the stronger the self-compensation degree becomes. Therefore, the self-compensation effect has become an obstacle to the application of II–VI compounds. Especially for short-wavelength materials, such as ZnO, ZnS, obtaining low resistance p-type materials is still an urgent problem to be solved. The impurity compensation is caused by high-temperature thermal equilibrium required for material preparation, so the self-compensation can be restricted by the use of nonequilibrium doping. The p-type ZnSe with a high conductivity was successfully prepared by N plasma-assisted doping in 1990,[11] which led to the realization of blue light-emitting diodes (LEDs) and lasers, and thus the first application of short-wavelength semiconductor light-emitting devices was pioneered.

3.1.2.6 Applications[15]

II–VI compounds are important optical and optoelectronic materials. Because of the direct energy gaps in the whole visible range, it can be used to fabricate planar color displays, lasers, luminescent devices, optical waveguides, optical modulators and optical bistability devices.

1. Light-emitting devices: ZnSe-based materials can be prepared into green-blue light LED and laser diode (LD) is an important candidate for the realization of all-color solid-state light sources. Recently, the research of ZnO-based UV luminescence and laser devices has attracted wide attention.

2. Thin-film electroluminescent display: The phosphors, such as ZnS:Ag (blue) and ZnS:Cu, have been widely used in cathode-ray tubes. Especially ZnS AC film EL can be made into a planar light source of various areas and shapes, which has the advantages of high light efficiency and low power consumption.

3. Light/radiation detectors: CdTe is an important material for the preparation of detectors of high energy radiation and high energy particle, which has important applications in medicine, nuclear safety, quality control of automobile manufacturing etc.
4. Photodetector: CdS, ZnSe and other materials are used to fabricate photoconductive detectors because of their high photoresponse in the visible range. ZnO-based materials have become the research hotspot of UV photodetectors in recent years.
5. Solar cells: The theoretical conversion efficiency of CdS/CdTe solar cells is 30%. At present, the solar cell based on CdTe/CdS thin films is progressing. The record efficiency of this solar cell is 22.1% and commercial module record efficiency is above 18.6%.

Because of the diversity of properties, II–VI compound semiconductors have many potential applications, such as nonlinear optics, second harmonic generation, acousto-optic effects etc. The preparation of novel devices using the best matching of various materials is a current development trend. For example, optoelectronic integrated systems made of II–VI compounds and GaAs or Si have a good application prospect in the fields of optical processing, optical storage and calculation.

3.1.3 EXCITON LUMINESCENCE

3.1.3.1 Exciton and Its Binding Energy

Excitons are a meta-excited state in semiconductors, that is, the quantization unit of bound electron-hole waves. The excitons, as an entity, can move in semiconductors, known as free excitons; the excitons can also be bound by impurities and defect states, known as bound excitons. At high excitation density, exciton interaction will occur, which may form exciton molecules; under high excitation density and strong coupling conditions, a large number of excitons can be further condensed into electron-hole droplets.

For direct bandgap semiconductors, the exciton energy can be converted into light energy by emitting photons, which is called exciton radiative recombination. There are two types of free exciton and bound exciton in excitonic recombination. The free exciton can move freely in the crystal and has a certain lifetime, while the free exciton disappears by emitting photons to form radiative recombination transition. The photon energy is $h\nu = E_g - E_x$, where the E_g is the bandgap and the E_x is the free exciton binding energy. Since excitons are electron-hole pairs bound by Coulomb action, in which electron is negative and hole is positive, excitons can be regarded as electrons orbiting holes and can be solved by using the hydrogen-like model, where the energy required for ground state-level electrons to be excited to the continuous state level is defined as the exciton binding energy. It can be expressed as follows:

$$E_b = E \frac{\mu e^4}{2\hbar^2 \varepsilon^2 n^2} = \frac{\mu}{m_e \varepsilon_r^2} \times 13.6 \, \text{eV} \qquad (3.1)$$

where μ is the equivalent mass, and the m_e and m_h are the effective masses of electrons and holes, respectively. ε_r is the relative dielectric constant. Because the excitons exist in many excited states, the emission spectra caused by free exciton recombination can consist of a series of narrow bands. At the same time, because the free exciton can move freely in the crystal and the exciton kinetic energy increases with the increase of temperature, the emission band widens with the increase of temperature, and the intensity decreases and the peak position moves to the low energy side.

For bound exciton radiative recombination, because the exciton is confined near the impurity center and cannot move, the radiation line of bound exciton is much narrower than that of free exciton and the width of spectral line does not change with temperature. Under the same conditions, the bound exciton radiation line is located on the long wave side of the free exciton radiation line. The energy of emitting photons is $h\nu = E_g - E_x - E_b$, where E_b is the binding energy of bound excitons.

Excitonic stability depends on temperature, electric field, carrier concentration and other factors. If the kT value of the sample is close to or greater than the exciton binding energy, the exciton will decompose due to the intense thermal motion. So the clear exciton luminescence can be observed only at low temperature in many semiconductor materials.[5,16] When the temperature increases, the exciton line will widen and the exciton luminescence intensity will decrease, so that quenching occurs. In addition, because electrons and holes move in the opposite direction under the action of electric field, the exciton effect in the semiconductor will be weakened or even failed due to electric field. While when the carrier concentration in the sample is large, excitons may also decompose due to the shielding effect of the free charge on the Coulombic field. To sum up, in a wide gap of II–VI compound semiconductor materials, as well as in of the semiconductor SL and QW structures to be discussed in detail in the next section, the exciton binding energy is generally larger. Even at RT, the exciton binding energy is much larger than kT. The strong exciton effect leads to the obvious exciton luminescence observed in the emission spectra, and even the laser devices with exciton recombination effect have been successfully realized.[17,64]

3.1.3.2 Recombination Luminescence of Free and Bound Excitons

II–VI compounds, represented by ZnSe and ZnO, have the advantages of wide bandgap, direct band transition and large exciton binding energy. Their main application is to realize purple, blue and green luminescence and laser devices by using energy band engineering and exciton effect. Therefore, it is necessary to study the exciton-related luminescence properties. This section mainly takes ZnSe and ZnO thin films as examples to introduce the research work on exciton recombination luminescence in recent years.

Since the 1980s, high-quality ZnSe films have been successfully prepared by MBE[6] and metal-organic chemical vapor deposition[9] methods. One of the crucial questions concerning the growth of ZnSe layers is the choice of a suitable substrate material. The quite small lattice mismatch of only 0.27% at RT when ZnSe is grown on a GaAs substrate is attractive for making use of ZnSe/GaAs structures and QWs. However, there are compressive and tensile strains in different directions of the films due to lattice mismatch between heterostructures. Much of the research works focused on the effect of the strain properties on thin-film crystalline quality, so exciton luminescence properties of ZnSe films grown on different substrates have been widely reported.[18–21]

Because of the quite small lattice mismatch (~0.27%), ZnSe thin films grown on GaAs substrate are expected to be under nearly uniform compressional strain in the plane of the heterojunction, with an equivalent uniaxial tensile strain in the perpendicular dirction.[19,20] Such a tetragonal distortion reduces the type of symmetry of the ZnSe lattice from T_d to D_{2d}. The degenerate valance band, Γ_8 band, splits intro the Γ_7 and the Γ_6 bands. For the transverse electric (TE)-polarized light (i.e. the polarization parallel to the heterojunction plane), both the transitions of the Γ_7 valance band $\rightarrow \Gamma_6$ CB and the Γ_6 valance band $\rightarrow \Gamma_6$ CB are allowed. Figure 3.5 shows the TE-polarized PL spectrum and the reflectivity spectrum of ZnSe thin films grown GaAs substrates for the TE-polarized light.[19] From the reflectivity spectrum, it can be seen clearly that two dips at 440.33 nm (2.8156 eV) (E_x^U) and 442.86 nm (2.7995 eV) (E_x^L) are associated with the Γ_6–Γ_6 exciton and the Γ_7–Γ_6 exciton, respectively. The oscillatory structures above 443.7 nm (below 2.794 eV) are due to interference effects. Correspondingly, there appear two free exciton emission lines at 442.3 nm (2.8031 eV) (E_x^U) and at 442.85 nm (2.7996 eV) (E_x^L) in the PL spectrum. As shown in Figure 3.6, the bound exciton emissions at neutral donors, $I_x(D^0,X)$ and $I_2(D^0,X)$, are clearly observed. An important conclusion is that the origin of the I_x line is different from the I_2 line, where, I_x line is considered to be associated with the native donor, while the I_2-bound exciton line is generally believed to be associated with the Ga donor diffused from the GaAs substrate.

Figure 3.6 shows a comparison of the low-temperature luminescence of such ZnSe/ZnSe layers (curve b) with heteroepitaxial ZnSe/GaAs layers (curve a) and bulk ZnSe (curve c).[18] Free exciton luminescence lines are clearly present in both ZnSe layers but are missing in the bulk. This indicates the generally better quality and purity of the films, compared to bulk ZnSe, where efficient

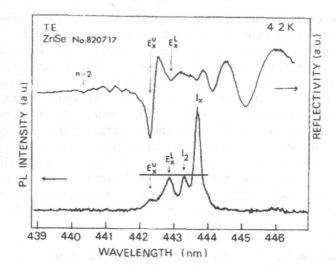

FIGURE 3.5 Reflectivity and PL spectra of MBE ZnSe grown at 400°C. TE-polarized light (the polarization is parallel to the heterojunction plane) was incident on the sample for the reflectivity measurement, while TE-polarized emission was detected in the PL measurements, in which the 3250 Å line from a He–Cd laser was used. (From Yao, T., *J. Cryst. Growth*, 72, 31, 1985. With permission.)

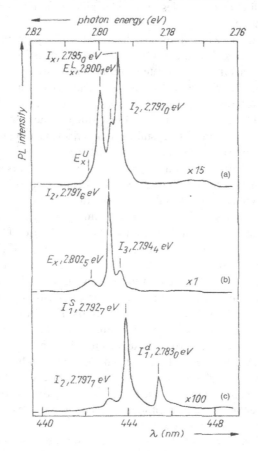

FIGURE 3.6 Photoluminescence at 4.2K in the excitonic region of nominally undoped ZnSe layers grown on (a) GaAs (0 0 1), (b) ZnSe (0 0 1) wafers and (c) of bulk ZnSe. (From Menda, K., Takayasu, I., Minato, T. and Kawashima, M., *J. Cryst. Growth*, 86, 342, 1988. With permission.)

radiative recombination of free excitons is mostly prevented by their fast capture at impurities. Spectrum b of a homoepitaxial film exhibits one free exciton peak E_x (2.802 eV) only, found at the same energy position as in bulk ZnSe, for the latter being determined from other experiments using very pure samples.[9] This underscores that strain does not exist at the ZnSe/ZnSe interface. The spectrum a of a ZnSe/GaAs sample shows two free exciton features, E_x^U and E_x^L, which are characteristic of the photoluminescence of heteroepitaxially grown layers, as a direct consequence of the stress present in the film.

It can be seen that the exciton luminescence of ZnSe films can only be obtained at low temperature due to the relatively low exciton binding energy, so later ZnSe-based light-emitting devices are all selected as active layers by using the QW and SL structures, which will be described in detail in the next chapter. As a comparison, II–VI groups of ZnO films exhibit abundant exciton effects at RT due to their exciton binding energy of 60 meV, which leads to the obvious optical propertices of excitons, which is more conducive to the realization of laser emission devices at RT. In 1997, the optical propertices of excitons at RT were first reported in ZnO thin films grown by laser MBE (L-MBE).[22] Figure 3.7(a) shows the absorption spectra of ZnO films at 70, 165 and 295K, respectively. There is an obvious 1s-A(B) free exciton absorption peak at 70K, with a photon energy 3.38 eV. As the temperature rises, this absorption peak gradually widens and shifts toward lower energy. At RT, the absorption peak of this free exciton can still be clearly observed, and the peak energy is located at 3.32 eV. When the sample is excited at 325 nm lines of the He–Cd laser, a strong

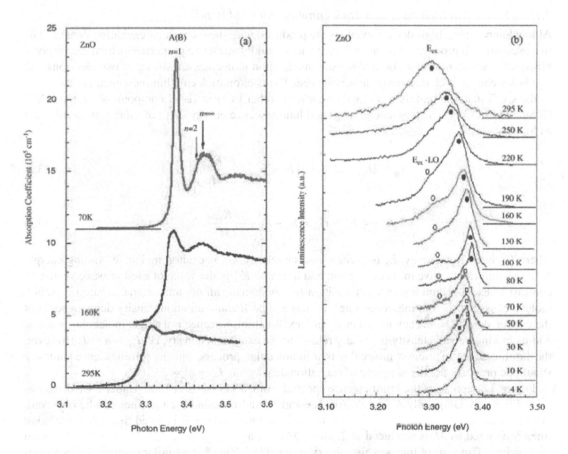

FIGURE 3.7 Absorption spectra (a) and emission spectra (b) of the ZnO films at different temperatures. (From Zu, P., Tang, Z. K., Wong, G. K. L., Kawasaki, M., Ohtomo, A., Koinuma, H. and Segawa, Y., *Solid State Commun.*, 103, 459, 1997; Tang, Z. K., Kawasaki, M., Ohtomo, A., Koinuma, H. and Segawa, Y., *J. Cryst. Growth*, 287, 169, 2006. With permission.)

band-edge emission can be observed. Figure 3.7(b) shows the photoluminescence spectra of the samples at different temperatures. At temperatures below 80K, the emission spectra are dominated by ZnO-bound exciton peaks at positions indicated by open and solid squares, which was attributed to excitons bounded at donor centers (open square) and acceptor centers (solid square), respectively. The intensities of the bound exciton emission peaks decrease as the temperature is increased. These peaks gradually merge into the free-exciton emission band which appears at their higher energy side (indicated by solid circles). The free exciton emission band dominates the luminescence spectra as the temperature is raised above 70K. A weak one LO-phonon replica (indicated by the open circle) of the free exciton band can also be seen in the spectra. At RT, only free exciton emission band is present. This is the first observation of clear RT free-exciton emission from ZnO, bulk single crystals as well as thin films. Subsequently, the authors reported that the UV-stimulated emission of ZnO films at RT excited by 355-nm line of YAG:Nd laser with high excitation intensity (see Figure 3.11).[23,29]

3.1.3.3 Exciton Behavior under High-Density Excitation

Under extreme conditions of high excitation density and low temperature, the process of interaction between different excited states in semiconductors can be observed, including exciton molecule (M), exciton-exciton scattering and electron-hole plasma (EHP) and so on. The interactions between these excited states are found in the study of luminescence phenomena.

3.1.3.3.1 Exciton Molecules and Their Luminescence (M Band)

At low temperature, high-density excitons are produced with the increase of excitation density. The two excitons will produce covalent exchange with mutually attractive properties, which can be combined into exciton molecules or double excitons. Exciton molecules consisting of two electrons and two holes can form characteristic luminescence. The exciton molecule luminescence is first found in the CuCl material and then is found one after another in most II–VI compounds, such as CdS, CdSe and ZnO.[24–26] The total energy (E_m) and luminescence energy (hv_m) of exciton molecules are as follows:

$$E_m = 2E_{ex} - G_m = \left[2\left(E_g - E_b\right) + \frac{\hbar^2 K_{ex}^2}{2M} \right] - G_m \qquad (3.2)$$

$$hv_m = E_{ex} - G_m = E_g - E_b + \frac{\hbar^2 K_{ex}^2}{2M} - G_m \qquad (3.3)$$

where E_{ex} is exciton energy, E_b is exciton binding energy, G_m is exciton molecule binding energy, M is the sum of effective mass of electron and hole and K_{ex} is the value of exciton wave vector. In contrast to exciton luminescence, in addition to low energy, an obvious characteristic of exciton molecules is that the luminescence intensity has a superlinear excitation density dependence. Let the exciton density be n, excitation density is J, exciton luminescence is a single molecular process, and its luminescence intensity is linearly related to the excitation density, i.e. $I_{ex} \propto n \propto J$. However, the luminescence of exciton molecules is a bimolecular process, and its luminescence intensity should be proportional to the square of excitation density, i.e. $I_m \propto n^2 \propto J^2$.

Figure 3.8 represents the luminescence spectral[24] of CdS crystals at 1.8K. Under a weak excitation of mercury lamp, only A free exciton line and I_1- and I_2-bound exciton lines can be observed. When the sample is excited by N_2 laser with high excitation power, it is noted that a new emission line, designated as M, is produced at 2.546–2.544 eV, an energy position a little below I_2-bound exciton line. This type of line was also observed in CdSe,[25] ZnO[26] at a similar position, i.e. at a position a little below I_2 line. M line has been observed in all the crystals studied independently of the observed intensities of I_1- and I_2-bound exciton lines, indicating that M line is of intrinsic nature. The intensity I of M line grows superlinearly against excitation power J with a relation of $I_m \propto J^{1.7}$.

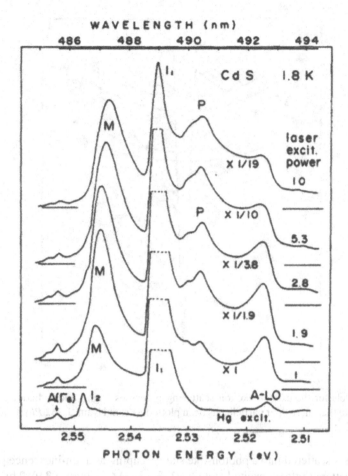

FIGURE 3.8 Luminescence spectra of a CdS crystal at 1.8K under 337.1-nm laser light excitation. The highest excitation level corresponds to about 40 kW/cm^2. The bottom shows the spectrum under mercury lamp (365.0 nm) excitation. (From Shionoya, S., Saito, H., Hanamura, E. and Akimoto, O., *Sol. Stat. Commun.*, 12, 223, 1973. With permission.)

The fact that M line is of intrinsic nature and shows a superlinear dependence on excitation power makes one confirm that this line should be due to excitonic molecules. With further increase of excitation power, the intensity of M line tends to be saturated, and another new line at about 2.53 eV, labeled P band, becomes dominant. The origin of the P band observed here is attributed to exciton-exciton scattering, which will be discussed next.

3.1.3.3.2   Exciton-Exciton Scattering and Its Luminescence (P Band)

When the excitation density reaches high enough, there will be a strong interaction between excitons and excitons, resulting in radiate luminescence by exciton-exciton threshold. It is considered to be one of the main reasons for the stimulated emission in many semiconductor materials. As shown in Figure 3.8, although M line grows superlinearly with increasing excitation power, it can be clearly observed that the P line caused by exciton-exciton scattering appears, even in the region of lower excitation powers. Further increasing excitation power, M line tends to be saturated. At the same time, P line starts to increase very rapidly and outstrips M line in intensity, exhibiting higher optical gains.

Figure 3.9 is a simple diagram of the exciton-exciton collision process responsible for the P-band luminescence. It is an inelastic scattering process between two free excitons in their ground state

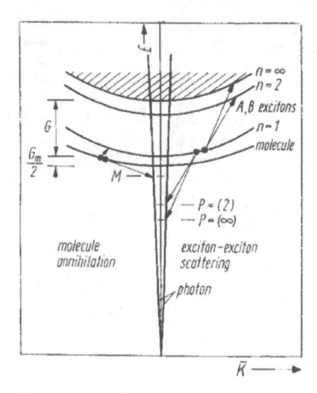

FIGURE 3.9 Models for the exciton-exciton scattering processes (right) and radiative decay of excitonic molecules (left). Note that energy per particle has been plotted. (From Hvam, J. M., *Phys. Stat. Sol. B*, 63, 511, 1974. With permission.)

(n=1). One exciton is scattered into a photon-like state escaping as a luminescence photon while the other exciton is scattered into an excited state, n=2, 3, ..., ∞ (see Figure 3.9). This process can be described by the following equations:

$$\text{Exciton}\,(A) + \text{exciton}\,(B) \rightarrow \text{exciton}^* + \text{photon} \tag{3.4}$$

As shown in Figure 3.9, here the wave vector is determined. If the momentum ($\hbar\vec{q}$) of the photon is ignored, according to the conservation of energy and momentum, the emitted photon energy ($\hbar\omega$) in this process are expressed as:

$$\hbar\omega = E_{ex} - E_b\left(1 - \frac{1}{n^2}\right) - \frac{\hbar^2 K_1^i K_2^i}{M},\ n = 2,3,\ldots, \tag{3.5}$$

where E_{ex}, E_b and M are exciton ground-state energy (at K=0), binding energy and mass, respectively. n is the quantum number of the exciton in the excited state, K_1^i and K_2^i are wave vectors of the initial excitons. Neglecting the kinetic energy term ($T\approx 0$), Equation (3.6) predicts a series of lines with the approximate peak energies

$$E_{P(n)} = E_{ex} - E_b\left(1 - \frac{1}{n^2}\right),\ n = 2,3,\ldots, \tag{3.6}$$

Figure 3.10 shows the photoluminescence spectra of ZnO at 1.9K, excited by the 337.1-nm line of an N_2 laser with the highest excitation power corresponds to about 500 kW/cm². At low excitation levels, the bound exciton lines I_6 and I_9 are dominant.[27] At 1.9K the I_6 line at 3.361 eV was

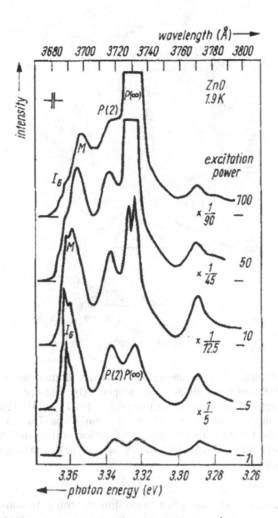

FIGURE 3.10 Photoluminescence spectra in ZnO at 1.9K mid-3371 Å laser excitation. The numbers to the right represent relative excitation powers and the highest excitation power corresponds to about 600 kW/cm². Note the scale factors on the different spectra. The assignment of the peaks is discussed in Hvam, J. M., *Phys. Stat. Sol. B*, 63, 511, 1974. With permission.

the strongest in most of the samples. Furthermore, it is seen the exciton-exciton interaction lines (P-band) with the two pronounced peaks P (2) at 3.337 eV and P (∞) at 3.324 eV.[28] With increasing excitation the P-band luminescence increases superlinerly with laser power. Particularly, the P (∞)-line is enhanced, probably due to stimulation effects that also produce a line narrowing resolving the P (∞)-line in two lines with peaks at 3.326 and 3.322 eV, i.e. with a separation of 4 meV. This splitting of the P (∞)-line is close to the separation between the A and B exciton series in ZnO, revealing the formation that it is due to the splitting between A and B excitons.

The features of *M* lines that appear at intermediate excitation power are the same as those observed in CdS.[24] Therefore, *M* line observed here is identified to be due to the recombination of exciton molecules. It noted that the peak intensity of the *M* line increases more than linearly with exciton density. The fact that the increase with excitation is less than for the *P* lines (quadratic in the exciton density) is explained by an increased broadening and by dissociation of the excitonic molecules with increasing effective exciton temperature. The radiative decay of excitonic molecules is also illustrated in Figure 3.9. One exciton is radiatively annihilated via a photon-like state and the other exciton is left as a free exciton in state *n*=1 (A or B).

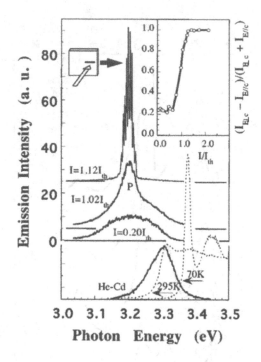

FIGURE 3.11 The lower trace shows the absorption spectrum (dotted curve) and photoluminescence spectrum, measured at 70K and room temperature, respectively. The upper traces show spontaneous and stimulated emission spectra under pumping intensities of 0.2 I_{th}, 1.02 I_{th} and 1.12 I_{th} provided by the frequency-tripled and mode-locked Nd:YAG laser ~355 nm, 15 ps. The inset at right side shows the degree of polarization of the UV emissions plotted as a function of the pumping intensity. (From Tang, Z. K., Wong, G. K. L., Yu, P., Kawasaki, M., Ohtomo, A., Koinuma, H. and Segawa, Y., *Appl. Phys. Lett.*, 72, 3270, 1998. With permission.)

Generally, the luminescence (P band) due to exciton-exciton collisions can only be observed at low temperatures. However, because the exciton binding energy of ZnO is as high as 60 meV, this makes ZnO have very abundant exciton effect at RT, which leads to the realization of RT UV laser emission caused by exciton-exciton scattering in ZnO microcrystalline films.[29] Figure 3.11 (the upper traces) show spontaneous and stimulated emission spectra under pumping intensities (I) of 0.2 I_{th}, 1.02 I_{th} and 1.12 I_{th} (I_{th}=40 kW/cm²) provided by the frequency-tripled and mode-locked Nd:YAG laser (355 nm, 15 ps). The inset of Figure 3.11 at right side shows the degree of polarization of the UV emissions plotted as a function of the pumping intensity. At low pumping intensities, a broad emission band is observed at 3.2 eV that is lower than free-exciton emission (E_{ex}) by about 110 meV. The intensity of this emission band increases quadratically when the pumping intensity is below the I_{th}, in contrast to the linear increase in the free exciton emission intensity. When the ZnO film is pumped at an intensity just above the threshold, I=1.02 I_{th}, a narrow emission band P emerges directly from the broad spontaneous spectrum. The P band grows superlinearly with pumping intensity. The authors attribute the RT UV-stimulated emission in the ZnO microcrystallites to the radiative recombination process of the exciton-exciton collision as discussed in Ref. [29] in detail.

3.1.3.3.3 Electron-Hole Droplets and Their Luminescence
At low temperatures and high excitation densities, not only exciton molecules can be formed, but also more complex systems may be formed—EHP. With the increase of excitation density,

the electron-hole gas is dense and condenses into electron-hole droplets. Because the particles are dense to a certain extent, they will produce covalent exchange and strong correlation effects with mutually attractive properties, forming EHP. EHP is different from exciton and exciton molecules. The characteristic peak of its luminescence appears in the low energy direction of the exciton peak, which can be understood as that the EHP has a low energy gap E_g^*. The transition from an exciton gas to the EHP is a nonequilibrium, first-order-phase transition in a driven system. Furthermore, this transition involves many-particle renormalization effects of the band structure. If the chemical potential μ of the electron-hole pair system is raised by sufficiently strong pumping above the reduced gap E_g^*, one has population inversion between conduction and VB and in direct gap materials considerable optical amplification (gain) which is the basis of most LDs.

An EHP has been observed under sufficiently strong pumping in all groups of semiconductors. The EHP can form a bound state also in direct gap materials with respect to a gas of free excitons. One of the important contributions in this respect is the more polar binding of the III–V and especially of the II–VI compounds, resulting in a coupling between optical phonons and plasmas, which stabilizes the EHP.[30,31]

Figures 3.12 and 3.13 show the luminescence spectra under high excitation density with an N_2 laser in CdS epitaxial layers grown by HWE or by laser ablation,[32] ZnSe and ZnTe bulk crystals.[33] For these samples, the PL spectra at low temperature and weak excitation show free and bound exciton luminescence, and only weak DAP recombination. This is quite different from the high excitation spectra shown in Figures 3.12 and 3.13, where a new broadband formation lies on the low energy side of the exciton peak. As the pump intensity increases, the structure becomes smoothed out and the full-width-at-half maximum (FWHM) becomes broader. The maximum of this band shifts to lower energies with increasing intensities. This band cannot be attributed to any of the excitonic high excitation luminescence process.[31] So the emission features that appear at high excitation indicate the formation of an EHP.

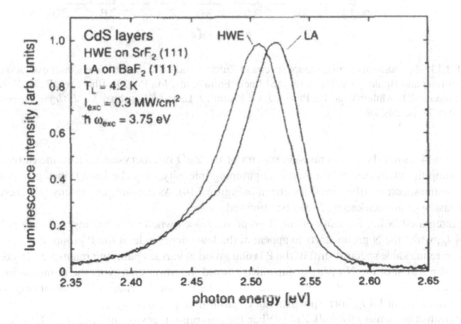

FIGURE 3.12 The luminescence of an electron-hole plasma in CdS layers grown by hot-wall epitaxy (HWE) and by laser ablation (LA). (From Klingshirn, C., *J. Cryst. Growth*, 117, 753, 1992. With permission.)

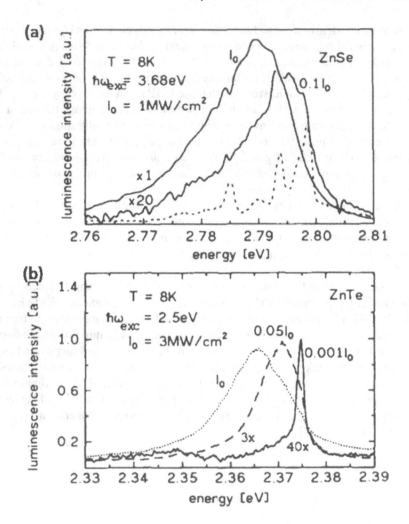

FIGURE 3.13 Spontaneous luminescence spectra of ZnSe (a) and ZnTe (b) at different laser excitation intensities (full line) and Hg lamp excitation (dashed line). (From Kunz, M., Pier, T., Bhargava, R. N., Reznitsky, A., Kozlovskii, V. I., Müller-Vogt, G., Pfister, J. C., Pautrat, J. L. and Klingshirn, C., *J. Cryst. Growth*, 101, 734, 1990. With permission.)

Figure 3.14 shows the laser emission spectra of the ZnO microcrystallite film measured under higher pumping intensities.[29] At a moderate pumping intensity, only the lasing lines grouped by P are seen in the spectrum (the lowest spectrum in Figure 3.14). As the pumping intensity is increased, another lasing group marked by N can be observed.

The integrated lasing intensity of the P group decreases when the pumping intensity is higher than 1.4 I_{th}, while the N group starts to appear at the low energy side of the P group. It is noted that there is no remarkable spectral shift of the P lasing group at various pumping intensities. In contrast, the central position of the N lasing group shifts toward the lower energy as the pumping intensity is increased. Obviously, the P-group and the N-group emission lines are of different origins. The pumping intensity at 1.4 I_{th} corresponds to an electron-hole pair density of 10^{18} cm^{-3}, which is near the Mott transition density for bulk ZnO. When the pumping intensity changed from 1.44 I_{th} to 2.73 I_{th}, the N band downshifted from 3.175 to 3.145 eV by 30 meV, which is in good agreement with the redshift of the EHP observed in bulk ZnO.[32] Hence, the N group lasing can be reasonably attributed to the EHP radiative recombination.

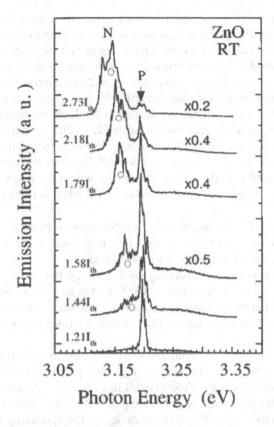

FIGURE 3.14 Lasing spectra of the ZnO microcrystallite film pumped using the frequency-tripled output of a mode-locked Nd:YAG laser at various pumping intensities. (From Tang, Z. K., Wong, G. K. L., Yu, P., Kawasaki, M., Ohtomo, A., Koinuma, H. and Segawa, Y., *Appl. Phys. Lett.*, 72, 3270, 1998. With permission.)

3.1.4 ELECTROLUMINESCENCE

EL refers to the luminescence of semiconductor materials after adding voltage, including high electric field luminescence and low electric field junction luminescence. The former luminescent materials are powders and films, and the latter is generally crystal material. As early as the 1950s and 1960s, a lot of researches have been carried out on the field luminescent of II–VI compounds represented by ZnS powder, and it has been widely used in plane luminescent light source and plane display device. Here, the research progress on low electric field junction light-emitting devices (LEDs) of II–VI compounds is mainly introduced.

3.1.4.1 Electroluminescence of MIS Structure

Because the wide gap II–VI compounds have a direct bandgap at RT, they are able to emit photons by direct recombination of free holes and electrons when the minority carriers are injected efficiently. Hence, II–VI compounds are considered to be important semiconductor materials for preparing short-wavelength electroluminescent devices. However, it is very difficult to prepare p-n junctions in II–VI compounds due to the self-compensation effect. It has been found that Schottky diodes consisting of a metal contact on semiconductors can achieve the injection luminescence from minority carrier when biased in the forward direction if there is a layer of insulating or semi-insulating material between the metal and the semiconductor. That is so-called MIS structure. As an alternative to a p-n junction, the MIS structure for minority carrier injection, such as ZnSe,[34,35] CdS[36] and ZnTe[37] etc., was reported by several investigators in early works.

Allen et al.[38] reported electroluminescent results under reverse bias ZnSe Schottky diodes in 1972. ZnSe containing aluminum donors was used in their investigations A Schottky barrier was formed by evaporating a gold contact on ZnSe. Ohmic contacts were made by alloying indium pellets. Au–ZnSe Schottky diodes were constructed by evaporating a gold dot on ZnSe surface, and Ohmic contacts were made by alloying indium pellets in ZnSe opposite surface. There is a good understanding of the luminescence for ideal Schottky diodes under reverse bias, electrons tunnel from the metal into the semiconductor, are accelerated in the depletion field and then impact-ionize luminescence centers. Subsequently, they reported that EL of ZnSe Schottky diodes was achieved by injecting minority carrier (hole) under forward bias, in which the luminescence mechanism of Schottky diode under forward bias is discussed in detail.[34] The minority carrier injection is usually negligible in an ideal Schottky diode because the barrier height for holes, namely the difference between the metal Fermi energy and the semiconductor VB edge, is large compared with thermal energies.

It is found that if an oxide layer is produced by chemical process between metal and semiconductor in making Schottky diode, some new features will be introduced in the behavior of diode, which is beneficial to minority carrier injection under forward bias.[34]

On the one hand, inclusion of an insulating layer between the metal and the semiconductor allows the metal Fermi level to move relative to the VB edge in such a way that holes from states near and above the metal Fermi level can be injected into the semiconductor.[39] On the other hand, the insulating layer of the Schottky diode is thin enough for tunneling to cause minority carrier injection.[35,40]

Thus, the insulating layer between metal and semiconductor in Schottky diode is beneficial to improve carrier injection efficiency. The insulator layer was prepared by using oxidation of ZnSe, such as oxidation was made by dipping ZnSe crystals in a 30–35% solution of H_2O_2 at RT.[41] Others have constructed ZnSe MIS structures by evaporating SiO_2,[35] ZnS[40] films as insulator layer.

Yamaguchi et al.[35] prepared a SiO_2 film on ZnSe by RF sputtering as an insulating layer and constructed an Au–ZnS–ZnSe diode, achieving RT blue EL from a ZnSe MIS structure. Figure 3.15 shows typical emission spectra of ZnSe MIS diode under forward bias at 300 and 90K. At RT, the MIS diode exhibits only a blue emission with a peak wavelength at 4650 Å (2.666 eV) and half-width of 130 Å. The light originating from the ZnSe surface near the *I–S* interface is observed under the semitransparent Au electrode. Comparing the EL with the photoluminescence,

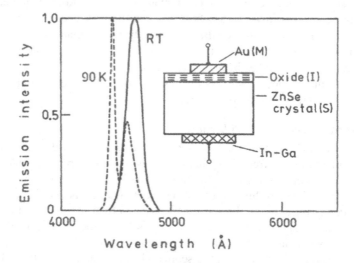

FIGURE 3.15 Typical ZnSe MIS diode emission spectra under forward bias at 300 and 90K. (From Yamaguchi, M. and Yamamoto, A., *Jpn. J. Appl. Phys.*, 16, 77, 1977. With permission.)

this near-bandgap emission in Figure 3.15 is considered to be due to the radiative recombination of bound excitons. At 90K, ZnSe MIS diode exhibits two strong peaks, that is, exciton emission (E_x) at high energy side and edge emission originating from a free-to-bound (FB) transition at low energy side, where FB recombination process is attributed to the transition of a free electron to the bound states of some acceptors. The authors suggested that the EL mechanism of ZnSe MIS diode under forward bias arises from the fact that the hole in the metal can be injected into the ZnSe through the effectively thinner part of the insulator layer by tunneling and recombines with another electron through recombination center.

Fan and Woods reported the EL of forward-biased MIS diodes prepared on high-purity single-crystal CdS[36] and ZnSe.[40] The origin of EL emission band at RT was investigated by variable temperature spectra. Figure 3.16 shows the EL spectra of forward-biased CdS MIS diodes over the temperature range 50–290K. An insulator layer has been fabricated by electron-beam deposition a ZnSe film on CdS, and the MIS structure was completed with an evaporated layer of gold. At 50K, the emission consisted exclusively of the A-1LO band at 4830 Å (2.5149 eV) and the A-2LO band at 5013 Å (2.4733 eV) which are associated with the $\Gamma 9 \rightarrow \Gamma 7$, free (A) exciton recombination with the emission of one and two LO phonons, respectively. It is observed clearly that the A-1LO replica was thermally quenched in EL as the temperature was raised, and at the same time, the relative intensity of the A-2LO band increased. In contrast with the 1LO replica, the A-2LO band was clearly resolved right up to RT where its maximum lay at 5288 Å. It is noted that a new band appeared in the spectrum run at low temperature, which is called the E_S band. As the temperature increases, the E_S band gradually increases in intensity until it became the broadband at 5135 Å at RT. The authors demonstrated that the band at 5135 Å is associated with free exciton recombination following scattering from free electrons in the CB.

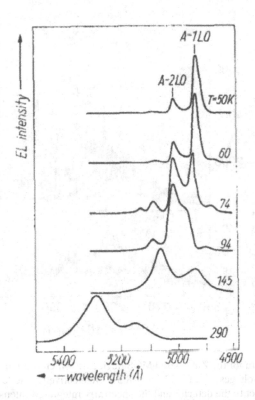

FIGURE 3.16 EL emission spectra of a forward-biased CdS MIS device. (From Fan, X. W. and Woods, *J. Phys. Stat. Sol. A*, 70, 325, 1982. With permission.)

Random lasing (RL) is a phenomenon that has been discovered in some disordered media and extensive attention has been paid to this field in recent years. In particular, the realization of random lasers on ZnO polycrystalline films or low-dimensional structures has been extensively reported,[42–45] fully reflecting the unique properties of ZnO materials. To avoid the use of p-type ZnO, MIS structure are usually used to fabricate electrically pumped ZnO random lasers.[42] An electrically pumped random laser was realized in a simple $Au/SiO_x/ZnO$ metal-oxide-semiconductor (MOS) structure.[42] The device consists of an $n\pm Si$ substrate, which acts as an electron injection layer, deposited with a layer of ZnO polycrystalline thin film. A very thin layer of amorphous SiO_y ($y<2.0$) is unintentionally formed between the $n+Si$ substrate and the ZnO thin film during the deposition process. Finally, an Au layer is deposited on the SiO_x dielectric layer to form the MIS structure. Figure 3.17 shows RT EL spectra of the ZnO-based MIS device applied with different forward bias voltages. A broadband from ZnO near-bandedge emission (NBE) is observed under a forward bias of 3.5 V. As the forward bias voltage increases to 6.3 V, the EL spectrum becomes narrower. When the forward bias increases above 7 V and above, several sharp peaks emerge in the emission spectra. The larger the forward bias, the more the sharp peaks appear in the EL spectra. The inset of Figure 3.17 shows the spectrally integrated emission intensity as a function of the current, indicating the threshold current equal to ~68 mA. Hence, this clearly shows that the ZnO MIS device exhibits coherent RL

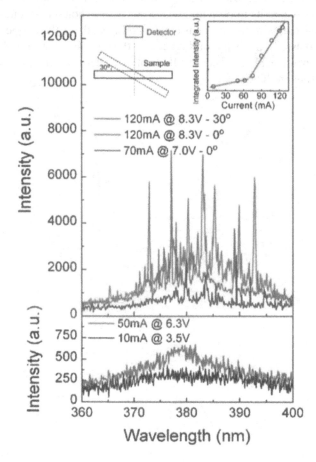

FIGURE 3.17 RT EL spectra for the ZnO-based MOS device, in which ZnO film is ~300-nm thick, applied with different forward bias voltages: 3.5, 6.3, 7.5 and 8.3 V, respectively. The left and right insets show the sample placement with respect to the detector and the spectrally integrated intensity as a function of the current, respectively. (From Ma, X., Chen, P., Li, D., Zhang, Y. and Yang, D., *Appl. Phys. Lett.*, 91, 251109, 2007. With permission.)

action when the Au electrode is applied with a sufficiently high positive voltage. In this context, in the region adjacent to SiO_y/ZnO interface, stimulated emission from ZnO occurs due to population inversion and, moreover, light is scattered by ZnO grains and SiO_y films. Therefore, RL proceeds due to optical gain achieved by the stimulated emission and multiple scattering. Nevertheless, one drawback of the ZnO-based MIS devices is their high threshold current. This is because (1) the impact ionization process is not an effective method to supply a hole concentration to the ZnO random media and (2) there is a SiO_y layer unintentionally formed at the interface between n+Si and ZnO random medium so that high-energy electrons are required to tunnel through the SiO_y blocking layer. The use of n+Si substrate to supply electrons to the ZnO random medium is also not effective.

To further reduce the threshold current of ZnO-based MIS random lasers, Zn_2TiO_4 nanoparticles (NPs) (size <20 nm) were incorporated into the polycrystalline ZnO thin film.[43] This is because the large difference in refractive index between Zn_2TiO_4 and ZnO materials can enhance the scattering strength to sustain RL action.

In 2010, Zhu et al.[44] achieved low-threshold electrically pumped random lasers in ZnO nanocrystallite films grown on glass substrates in a simple way. To realize low-threshold electrically pumped RL, the MIS structure can be modified as a metal-semiconductor-insulator-semiconductor (MSIS) structure such as the Au/i-ZnO/MgO/ZnO configuration where i-ZnO/MgO formed the insulator. In the Au/i-ZnO/MgO/ZnO structure, i-ZnO is the part of the insulator that generates holes through the impact ionization process. As the threshold ionization energy is proportional to the bandgap energy, it is easier to generate holes from i-ZnO than from either SiO_x or MgO. The RT EL spectra of the Au/i-ZnO/MgO/ZnO structure are shown in Figure 3.20. When the injection current is 5.0 mA, a broad EL emission with two broad emissions centered at about 400 and 530 nm can be observed. The emission at around 530 nm arises from the deep-level emission of the i-ZnO layer, while the one at around 400 nm corresponds to the NBE of the ZnO underneath the MgO layer. Some sharp lines appear when the injection current is increased to 6.0 mA and they become more dominant when the injection current is further increased to 9.0 mA. The FWHM of these sharp peaks is less than 0.8 nm. The integrated intensity of the emission at around 400 nm vs. the injection current is shown in the inset of Figure 3.18, from which a threshold current of about 6.5 mA can be derived.

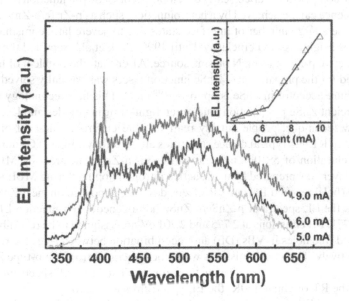

FIGURE 3.18 EL spectrum of the Au/i-ZnO/MgO/ZnO structure under different injection currents. The inset shows the integrated intensity of the emission at around 400 nm from this structure as a function of injection current. (From Zhu, H., Shan, C.-X., Zhang, J.-Y., Zhang, Z.-Z., Li, B.-H., Zhao, D.-X., Yao, B., Shen, D.-Z., Fan, X.-W., Tang, Z.-K., Hou, X. and Choy, K.-L., *Adv. Mater.*, 22, 1877, 2010. With permission.)

Note that the threshold is over one order of magnitude smaller than the value reported in a similar Au/SiO$_x$/ZnO MOS structure (68 mA)[42] and is significantly smaller than the corresponding value obtained in the Au/MgO/ZnO structure (43 mA) mentioned in their work, which confirms that the i-ZnO layer indeed increases the generation efficiency of holes, thus lower threshold lasing has been obtained, as expected.

In 2020, Ma et al.[45] studied EL in Er-doped ZnO MIS diodes prepared on silicon substrate by RF magnetron sputtering. A type of MIS structured LED, where a semitransparent Au film, a polymethyl methacrylate film and an Er-doped ZnO (ZnO:Er) film on a silicon substrate act as the "*M*", "*I*" and "*S*" components, respectively, has been prepared. It has been demonstrated that the ZnO:Er-based MIS structured LED exhibits the evolution from RL to Er^{3+}-related EL with increasing forward bias voltage. Under relatively lower forward bias voltages, the concentrations of holes and electrons in the region near the PMMA/ZnO interface are considerably high, resulting in the stimulated emission. Overall, the EL intensity increases with the applied forward bias voltage. The FWHM of any one sharp peak is less than 3 Å. Therefore, it can be believed that RL also occurs in the present ZnO:Er-based MIS device using PMMA as the insulating layer. In this context, during certain multiple light scattering processes, the photons can pick up optical gain larger than optical loss, thus enabling the RL to occur in the region near the PMMA/ZnO interface.

3.1.4.2 Electroluminescence of p-n Junction

For some time now, there has been a great deal of attention in wide bandgap semiconductors, in particular there is strong commercial desire to produce efficient and lasting blue light-emitting diodes (LEDs) and short-wavelength LDs. Worldwide research efforts were initially focused on ZnSe and its alloys,[47–49] but then, GaN-based technologies have progressed more rapidly.[46] These developments have culminated in the demonstration of RT operating green-blue as well as blue LD structure.[50–56] Since 1997 the UV-stimulated emission of ZnO films at RT has been reported,[29] which has set off the research upsurge of ZnO-based p-n junction UV light-emitting diodes and lasers.[64–72]

3.1.4.2.1 ZnSe p-n Junction LEDs and Lasers

Because of the self-compensation effect, early development of ZnSe p-n junction is difficult, although injection luminescence can be achieved by a heterojunction, such as p-ZnTe/n-ZnSe. However, these devices would cause a large number of interface states due to severe lattice mismatch, resulting in the lower luminous brightness and efficiency. Until 1990, Park et al.[11] reported the doping technology of high conductive p-ZnSe using N plasma source. After that, the problem of high carrier concentration required for the preparation of ZnSe junction lasers was basically solved. The maximum concentration of pure acceptor in ZnSe is up to 1×10^{19} cm^{-3}. This new technology lays the foundation to realize efficient ZnSe p-n junction blue-green light-emitting diodes and laser devices.

Using this new technique, people quickly realized blue electroluminescent of ZnSe homogeneous p-n junction at RT.[47,48] Typical device structures are as follows: First, a Cl-doped n-ZnSe layer with carrier concentration of 5×10^{18} cm^{-3} was grown on n-ZnSe substrate by MBE, and then an N-doped p-ZnSe layer was prepared by using active nitrogen doping during MBE growth. The carrier concentration (hole) reached 10^{16} cm^{-3}. Pt electrodes were deposited on the p-ZnSe by sputtering. Figure 3.19 shows the EL spectra of p-ZnSe/n-ZnSe homogeneous p-n junction LEDs at low temperature and RT. 77K EL emissions at 2.775 and 2.701 eV are assigned to be recombination between donor electrons and free holes in VBs (DF) and recombination between free electrons and acceptor holes (FA), respectively. These EL emissions would be generated from the p-type ZnSe region. FA emission decreased gradually with increasing temperature. The 110K EL spectrum is dominated by DF emission. At the RT of about 300K, the EL spectrum shows strong DF emission. Other deep-level emissions were well suppressed, so the color of the light was pure blue to the unaided eye.

Meanwhile, 3M research group[49] first reported the ZnCdSe/ZnSe QW structured blue-green semiconductor laser, realizing a laser with a wavelength of 490 nm operating in pulse mode at 77K. The origin of blue-green light at low temperature is attributed to exciton emission in ZnCdSe/ZnSe

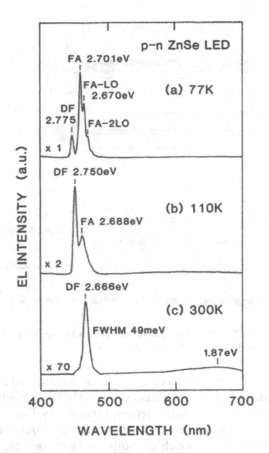

FIGURE 3.19 Electroluminescence spectra from the ZnSe p-n junction LED. (From Ohkawa, K., Ueno, A. and Mitsuyu, T., *Jpn. J. Appl. Phys.*, 30, 3873, 1991. With permission.)

QWs. The QWs are used as activation layer due to their remarkable characteristics (the properties of SL QWs will be introduced in the next chapter), resulting in that the first blue-green semiconductor LD in the world is realized. This work is considered to be an important milestone in the history of semiconductor laser development. Figure 3.20 shows the structure of the first blue-green semiconductor laser. Besides, N plasma doping, the core of the laser is a single QW (SQW) composed of n-ZnSe/ZnCdSe/p-ZnSe. The structure of this device contains ZnCdSe, ZnSe and ZnS0.07Se0.93 structure, which is ZnCdSe potential well layer, ZnSe the barrier layer on the two sides of well layer is used as the optical waveguide region, and then the ZnS0.07Se0.93 is cladding layer. Figure 3.21 shows optical spectra that are characteristic of this device, part (a), part (b) is the spectrum below and above the threshold and part (c) is the magnification of the laser spectrum (b). As it can be clearly seen from Figure 3.24 that when the injection current of the laser exceeds the threshold, the emission spectrum consists of many longitudinal modes, the peak spacing is 0.03 nm, the central wavelength is 490 nm. When the working temperature is up to 200K, the laser can also be observed. However, a serious problem of the laser is sample heating and threshold operating voltage up to 20 V, which is due to the failure to solve the p-ZnSe ohmic contact.

To solve this problem, scientists from the Brown-Pudu University Joint Research Group designed a laser structure using p-GaAs as a substrate and successfully obtained a blue-green laser with pulse operation at RT from (Zn,Cd)Se/ZnSe QW structure.[50] Figure 3.22(a) is a schematic illustration of the heterostructures that have been fabricated into LDs, including doping levels and layer thicknesses. Figure 3.22(b) shows the stimulated emission spectra of the laser at different

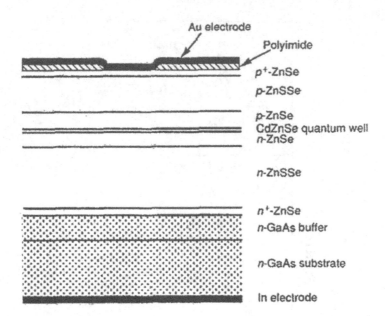

FIGURE 3.20   A cross section of a blue-green laser diode. (From Haase, M. A., Qiu, J., DePuydt, J. M. and Cheng, H., *Appl. Phys. Lett.*, 59, 1272, 1991. With permission.)

temperatures. It can be seen that the stimulated emission can be observed at RT. The mechanism of laser production is attributed to the exciton emission related to the heavy hole of $n=1$. At 77K, the devices have lasted for many hours, including repeated thermal cycling, without serious degradation. Subsequently, Sony group[51] adopt ZnS/ZnSSe/ZnS multi-QW structure and ZnMgSSe/ZnSSe/ZnMgSSe/GaAs multi-QW structure, which can shift the laser wavelength to pure blue or even UV region. By improving laser structures, coupled with a greater understanding of the associated

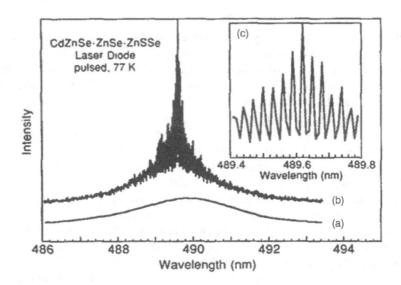

FIGURE 3.21   The optical spectra for a blue-green laser diode: (a) below threshold, (b) above threshold and (c) an expanded view of the lasing spectrum, taken with 0.01-nm steps. The device is 1020-μm long. Intensity scales for these three graphs are in arbitrary units and are not the same. (From Haase, M. A., Qiu, J., DePuydt, J. M. and Cheng, H., *Appl. Phys. Lett.*, 59, 1272, 1991. With permission.)

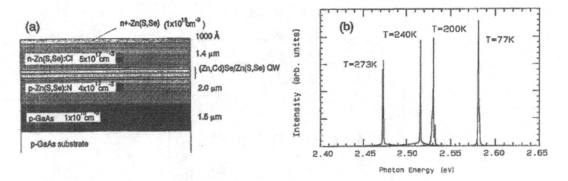

FIGURE 3.22   (a) Schematic illustration of the heterostructures for laser diodes fabricated on p-GaAs substrate. (b) Emission spectra from T=77 to 273K for a (Zn,Cd)Se/Zn(S,Se) QW laser diode in the n=1 HH exciton region. (From Jeon, H., Ding, J., Nurmikko, A. V., Xie, W., Grillo, D. C., Kobayashi, M., Gunshor, R. L., Hua, G. C. and Otsuka, N., *Appl. Phys. Lett.*, 60, 2045, 1992. With permission.)

materials problems, the laser has realized from pulse operation at RT to continuous operation at RT. Following this encouraging starts all groups switched to work on magnesium-containing lasers. Operation of a pulsed laser at temperatures as high as 394K was achieved by the Philips group.[52]

However, it is found that near or at RT conditions, the devices are frequently much more shorten lived, apparently chiefly due to thermally induced deterioration. Therefore, most of the researches have concentrated on improving the lasing lifetime. It is worth noting that the laser lifetimes increase by approximately two orders of magnitude every year and soon produced devices that operate for over an hour at RT.[53] Nowadays, although blue-green ZnCdSe/MgZnSSe-based LDs achieve to operate under the continuous-wave (CW) condition, the lifetime was limited to 400 hours.[54] This short device lifetime is considered due to higher ionicity of chemical bonds, resulting in lower formation energy of various defects. To overcome the weakness in crystal hardness of II–VI material, alloying with beryllium has been proposed.[55] ZnSe-based alloy containing beryllium has a much higher degree of covalency than other II–VI compounds. Thus the beryllium-containing material system is expected to overcome a problem of limited lifetime due to weakness inherent to II–VI materials.[55,56] On the other hand, II–VI light-emitting devices can be grown on InP substrates while one preserves the lattice match of active layers, so that a long lifetime is equally expected.[57]

In 2002, the research group from Sophia University of Japan reported that yellow-green (560 nm) II–VI LDs on InP substrates were successfully operated under the pulsed current injection at 77K.[58] The device structure is shown in Figure 3.23. The BeZnSeTe active layer was lattice-matched to InP substrates. A separate confinement heterostructure (SCH) was formed by employing MgSe/BeZnTe:N SLs as p-cladding layers and MgSe/ZnCdSe:Cl SL as n-cladding layers. The gain-guiding LDs with Au electrodes are fabricated, the stripe width which is 12 mm, by standard photolithographic processing. As an n-contact, Au/Ge was used. The above structural improvements were effective for suppressing defect generation and diffusion during the operation, and for lengthening the lifetimes, and the degradation factors could be solved for the BeZnSeTe devices developed in their studies. The threshold current density was about 2.5 kA/cm².

In 2006, this group studied the aging characteristics of BeZnSeTe yellow light-emitting diodes (LEDs) fabricated on InP substrates by MBE under direct current injections at RT. It was shown that the decay speed of the light output during the aging was slower than that of conventional ZnCdSe/MgZnSSe LEDs. A long lifetime more than 5000 hours and a half lifetime of 5180 hours were obtained at a current density of 130 A/cm².[59] In 2013, they also proposed BeZnTe/ZnSeTe SL quasi-quaternaries (SLQQs) as an active layer material replaced with BeZnSeTe quaternaries for yellow/green light-emitting devices. Applying the SLQQ for the active layer, LEDs were fabricated on n-type InP substrates. Single-peak yellow EL at 584 nm was observed with the FWHM of

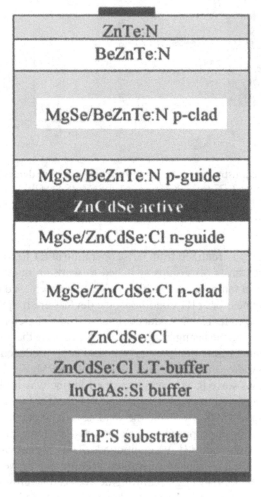

FIGURE 3.23 Device structure of ZnCdSe/BeZnTe LDs. (From Che, S.-B., Nomura, I., Kikuchi, A. and Kishino, K., *Appl. Phys. Lett.*, 81, 972, 2002. With permission.)

105 meV, as shown in Figure 3.24. These results show that BeZnTe/ZnSeTe SLQQs are promising active layer materials of wide-range visible (including yellow and green regions) light-emitting devices.[60]

Although LDs based on beryllium-containing II–VI material have long been reported,[58,59] CW lasing at RT were not reported until near 2010.[61–63] The Joint Research Group of AIST and Hitachi[61] have demonstrated CW operation of BeZnCdSe QW LDs at RT in the green to yellow spectrum range. The LDs structures were grown by MBE. A schematic representation of the LD structures is shown in Figure 3.25(a). The LD structure had a SCH, which consisted of a BeZnCdSe single QW sandwiched between BeZnSe optical guiding layers. The n-type BeMgZnSe:Cl was used for cladding layers of Be containing ZnSe-based LDs. However, as the p-type cladding layer, a BeMgZnSe/ZnSe:N short-period SL (SPSL) was used to obtain a higher p-type carrier concentration. A BeTe buffer layer was used to reduce extended defects that nucleated at the II–VI/GaAs interface. A nitrogen-doped BeTe/ZnSe pseudograding supperlattice was used as an ohmic contact to the p-type cladding layer. To confine injected current, a stripe mesa with width of 2–20 μm is formed by wet

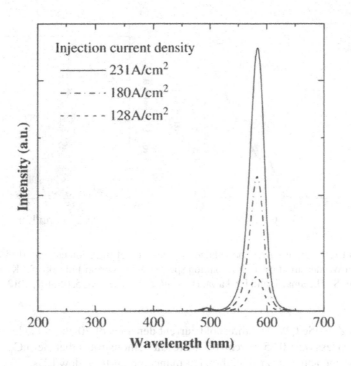

FIGURE 3.24 Electroluminescence spectra of the SLQQ LED under pulsed current injections at RT. (From Kobayashi, T., Nomura, I., Murakami, K. and Kishino, K., *J. Cryst. Growth*, 378, 263, 2013. With permission.)

etching a ZnSe/BeTe ohmic layer and a ZnSe/BeTe GSL contact layer. Then, area other than stripe is protected by an insulator (SiO₂). As a p-electrode, Ti/Pt/Au is deposited on the upper side, while AuGe/Au is deposited on back side for n-electrode. LDs with three different Cd alloy compositions of 32, 39 and 42% in BeZnCdSe active layer, corresponding lattice mismatch of 2.1, 2.6 and 2.8%, respectively, were fabricated by MBE.[58] CW lasing spectra in these LDs at 25°C are shown in Figure 3.25(b). CW lasing in the green-to-yellow spectral region (543–570 nm) at RT was successfully obtained. The green-to-yellow lasing color was tuned by simply varying the Cd content of

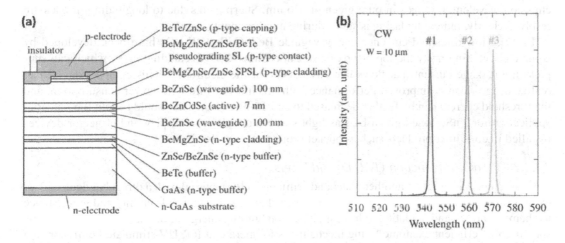

FIGURE 3.25 (a) Schematic structure of BeZnCdSe QW laser diode (b) CW lasing spectra in these LDs at 25°C. (From Kasai, J.-i., Akimoto, R., Hasama, T., Ishikawa, H., Fujisaki, S., Tanaka, S. and Tsuji, S., *Appl. Phys. Express*, 4, 082102, 2011. With permission.)

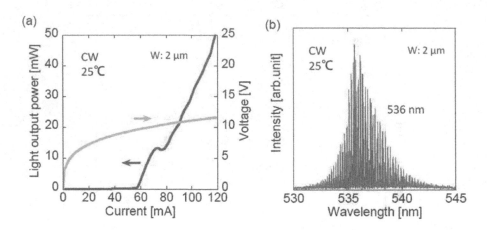

FIGURE 3.26 (a) Light output power and voltage as a function of injection current. (b) Lasing spectrum at output power of 1 mW, measured by high-resolution spectrometer. (From Fujisaki, S., Kasai, J.-i., Akimoto, R., Tanaka, S., Tsuji, S., Hasama, T. and Ishikawa, H., *Appl. Phys. Express*, 5, 062101, 2012. With permission.)

the quaternary BeZnCdSe QW. The threshold current densities of 20-μm-wide lasers were found to be sufficiently low (less than 0.85 kA/cm^2). This result demonstrates that BeZnCdSe is a promising material for use as the active layer in high-performance green-to-yellow LDs.

Subsequently, the carrier confinement in the active layer was improved by increasing the effective bandgap energy in SPSL from 2.81 to 2.86 eV. A high-power lasing over 50 mW was achieved, indicating that the BeZnCdSe SQW LD performance was improved.[62] Figure 3.26(a) shows the light output power and voltage as a function of injected current in LD with a stripe width of 2 μm. The measurement was done at 25°C and DC current injection condition. Threshold current density and threshold voltage for lasing are 55 mA and 9.8 V, respectively. The output power is as high as 50 mW at the injection current of 118 mA and voltage of 11.5 V, where the reflectivity of both facets in the LD chip is 90%. The maximum power reaches 70 mW, when a reflectivity in a front facet is decreased to 70%. The output power in this LD is much improved, compared with our previous LDs with the maximum output power of 1.5 mW.[61] Figure 3.26(b) shows emission spectrum of LD that is operated at output power of 1 mW, measured by high-resolution spectrometer at 25°C. The peak emission wavelength ranges in pure green of 536 nm. Sharp peaks due to longitudinal modes are resolved clearly, indicating lasing is stable during measurement.

In 2015, low-threshold current ridge-waveguide BeZnCdSe QW LDs have been developed by completely etching away the top p-type BeMgZnSe/ZnSe:N SPSL cladding layer, which can suppress the leakage current that flows laterally outside of the electrode.[63] This compact device can realize a significantly improved performance with much lower threshold power consumption, and the threshold current can be further decreased to sub-10-mA level. This would benefit the potential application for ZnSe-based green LDs as light sources in full-color display and projector devices installed in consumer products such as pocket projectors.

3.1.4.2.2 ZnO p-n Junction UV LEDs and Lasers

ZnO, as a metallic oxide, is another wideband semiconductor material with a direct bandgap of 3.37 eV at RT. It is superior over nitrides and selenides in chemical and thermal stability and in resistance to chemical attack and oxidation. It has a high exciton binding energy of 60 meV, which in principle should allow efficient excitonic lasing mechanisms to operate at RT. UV-stimulated emission and lasing have been extensively reported in ZnO microcrystallite thin films[29] and its QW structure.[64,65] To realize these device applications, an important issue is to fabricate both high-quality p-type and n-type ZnO films. However, like most wide bandgap semiconductors, ZnO has the "asymmetric doping" limitation, i.e., it can be an easily doped high-quality n-type, but it is difficult to dope p-type. This is due to the existent of some problems such as its self-compensating effect, deep acceptor level and low solubility of the acceptor dopant.

Nowadays, the bottleneck of ZnO homojunction luminescence and laser devices is mainly focused on how to obtain stable and low resistance p-type materials. Although some progress has been made in p-type doping of ZnO with the continuous efforts of researchers, the fabricated devices are far from reaching the application level. Nevertheless, the outstanding characteristics and advantages of ZnO materials make many researchers devote themselves to the research of ZnO p-n junction LEDs and lasers.

Early in this century, some research groups have prepared ZnO homogeneous p-n junction devices by P diffusion and achieved low-temperature EL. The luminescence wavelength is located in the UV-white region.[66] The Kawasaki research group of Northeast University of Japan reported the electroluminescent of ZnO homogeneous p-i-n junction devices prepared by plasma-assisted MBE (P-MBE) at RT in 2004.[67] First, the ScAlMgO$_4$ (SCAM) substrate almost exactly matched with the ZnO lattice was used for the ZnO film growth. The excellent quality ZnO film was obtained because the problem of epitaxial mismatch was well solved. Second, a repeatable p-ZnO film was obtained, in which the N was incorporated at 450°C to increase the N doping concentration, and then the substrate temperature was rapidly increased to 900°C to improve the crystal quality. The ZnO homogeneous p-i-n junction LED is constructed. Figure 3.27 shows the structure diagram and I-V characteristic curve of the p-i-n junction LED. As can be seen, this device shows good rectifier characteristics and threshold voltage is about 7 V. Figure 3.28 shows the electroluminescent spectrum of the device. It can be seen from the figure that the electroluminescent spectrum consists of two luminescence peaks, which are located around 570 and 440 nm, respectively. The luminescence peak of 440 nm is attributed to the radiative recombination of DAPs in the p-ZnO. Although exciton emission is not obtained, this result increases researchers' confidence in ZnO UV emission devices. As a milestone, this work has accelerated the research process of ZnO junction devices.

In 2006, the Lu and Shen team from the Chinese Academy of Sciences reported that p-ZnO films were prepared on c-Al$_2$O$_3$ substrates by using NO as source materials of N doping and O.[68] ZnO p-n homojunction was fabricated by combining undoped n-type ZnO. The EL spectra of ZnO homojunction were obtained at the temperatures of 11–200K, as shown in Figure 3.29. At low temperature, the main emission peak is located at about 410 nm and another weaker emission peak is

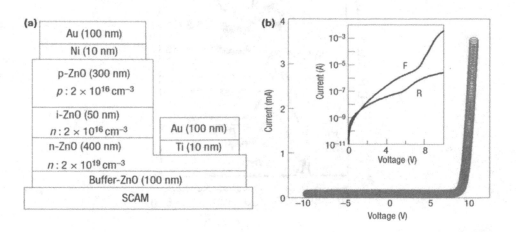

FIGURE 3.27 The structure diagram and I-V characteristic curve of the p-i-n junction LED. (a) The structure of a typical p–i–n junction LED. (b) I-V characteristics of a p–i–n junction. The inset has logarithmic scale in current with F and R denoting forward and reverse bias conditions, respectively. (From Tsukazaki, A., Ohtomo, A., Onuma, T., Ohtani, M., Makino, T., Sumiya, M., Ohtani, K., Chichibu, S. F., Fuke, S., Segawa, Y., Ohno, H., Koinuma, H. and Kawasaki, M., *Nat. Mater.*, 4, 42, 2005. With permission.)

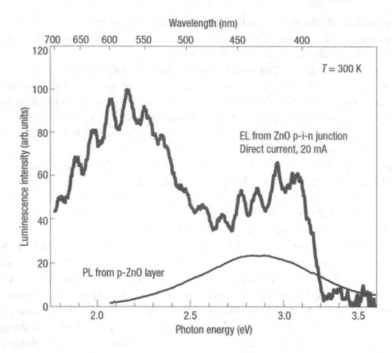

FIGURE 3.28 The electroluminescent spectrum of the p-i-n junction LED device. (From Tsukazaki, A., Ohtomo, A., Onuma, T., Ohtani, M., Makino, T., Sumiya, M., Ohtani, K., Chichibu, S. F., Fuke, S., Segawa, Y., Ohno, H., Koinuma, H. and Kawasaki, M., *Nat. Mater.*, 4, 42, 2005. With permission.)

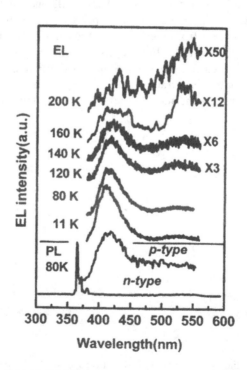

FIGURE 3.29 EL spectra of the ZnO LED measured at temperatures of 11–200K. The lower panel of the figure shows the PL spectra at 80K. of the n-type and p-type ZnO films grown on *a*-plane Al₂O₃ substrates, respectively. (From Jiao, S. J., Zhang, Z. Z., Lu, Y. M., Shen, D. Z., Yao, B., Zhang, J. Y., Li, B. H., Zhao, D. X., Fan, X. W. and Tang, Z. K., *Appl. Phys. Lett.*, 88, 031911, 2006. With permission.)

at 520 nm. When the temperature exceeds 200K, the emission peak at 410 nm gradually quenched, while the luminescence peak at 520 nm red-shifted to 540 nm and dominated. Subsequently, this research group successfully realized the RT EL of ZnO LED prepared on sapphire substrate by improving the doping technology of the N,[69] this work has attracted attention due to the lower cost of constructing devices on cheap sapphire substrates.

S. J. Park team from Korea[70] demonstrated the operation of a UV light-emitting ZnO homo-junction LED by growing P-doped p-type ZnO on Ga-doped n-type ZnO. The ZnO LED emitted 380-nm UV light at RT and showed clear rectification with a threshold voltage of 3.2 V. It can be clearly seen that in addition to the UV emission at 380 nm, there is a strong luminescence band in the visible region. In particular, after inserting an MgZnO barrier layer in the junction, the inten-sity of NBE was further increased and the deep level emission was greatly suppressed by using $Mg_{0.1}Zn_{0.9}O$ layers as energy barrier layers to confine the carriers to the high-quality n-type ZnO.

Meanwhile, the preparation of p-type ZnO using As as dopant has also attracted the attention of many researchers. Moxtronic company's Y. R. Ryu et al.[64] reported the successful preparation of UV LDs based on ZnO/BeZnO films. The devices have p-n heterojunction structures with a multiple-QW (MQW) active layer sandwiched between guide-confinement layers. The MQW active layer comprises undoped ZnO and BeZnO, while the two guide-confinement layers were As-doped p-type ZnO/BeZnO and Ga-doped n-type BeZnO/ZnO films, respectively. Figure 3.30 shows the electroluminescent spec-tra and electrically pumped lasing spectra of ZnO homojunction LEDs. A strong free exciton emission can be observed at a 10-mA injection current. The exciton binding energy in the MQW region is excep-tionally large (263 meV). The intense P band emission due to exciton-exciton scattering is observed as the injection current gradually increases to 60 mA. The emission intensity of the P band far exceeds the emission of free excitons and shows a superlinear increase. At high pulse excitation density, a mode of obviously stimulated emission occurs near the P_n band. This indicates that excitonic UV lasing is achieved by current injection at RT in ZnO/BeZnO MQW LED.

Liu's team from the University of California reported that based on p-type Sb-doped ZnO n-type Ga-doped ZnO thin films, ZnO p-n homojunction light-emitting diodes were fabricated on Si sub-strate by MBE.[71] Low resistivity Au NiO and Au Ti contacts were formed on top of p-type and n-type ZnO layers, respectively. The light-emitting diodes yielded strong NBEs in temperature-dependent and injection current-dependent EL measurements. At RT, although the luminescence peak from the UV part can be seen, the main luminescence is still from the defect-related lumi-nescence. Subsequently, they also reported electrically pumped ZnO QW diode lasers.[65] Sb-doped

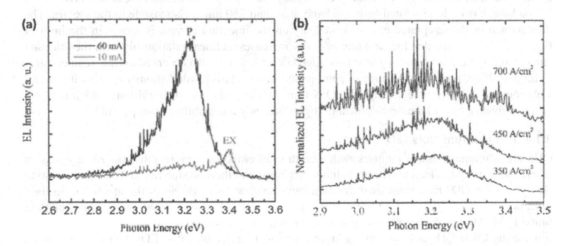

FIGURE 3.30 Electroluminescent (a) and lasing spectra (b) at RT of ZnO-based homojunction LEDs pre-pared using As doping. (From Ryu, Y. R., Lubguban, J. A., Lee, T. S., White, H. W., Jeong, T. S., Youn, C. J. and Kim, B. J., *Appl. Phys. Lett.*, 90, 131115, 2007. With permission.)

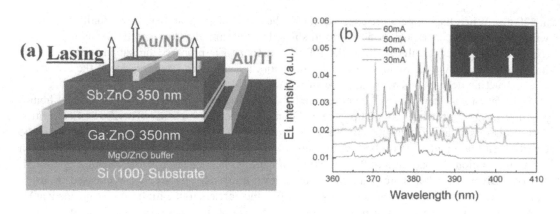

FIGURE 3.31 (a) Structure diagram of Sb-doped ZnO homojunction laser diodes and (b) electroluminescent spectra of laser diodes at different injection currents. (From Chu, S., Olmedo, M., Yang, Z., Kong, J. and Liu, J., *Appl. Phys. Lett.*, 93, 181106, 2008. With permission.)

p-type ZnO/Ga-doped n-type ZnO with an MgZnO/ZnO/MgZnO QW embedded in the junction was grown on Si substrate by MBE. The diodes emit lasing at RT with a very low-threshold injection current density of 10 A/cm². Figure 3.31 shows the structure diagram of the Sb-doped ZnO-based homojunction LDs (a) and the electroluminescent spectra at different electric injection currents (b). Under forward bias, lasing mode peaks can be seen when the injection current is 30 mA. When the injection current increases to 60 mA, a large number of sharp lasing mode peaks appear, and the intensity of luminescence increases. The lasing mechanism is exciton-related recombination and the feedback is provided by close-loop scattering from closely packed nanocolumnar ZnO grains formed on Si.

Recently, Tang's team from Sun Yat-sen University of China[72] reported the EL of ZnO homojunction LED, in which Be assisted p-type doping of ZnO:N. A p-i-n junction LED was fabricated by P-MBE with a combination of an undoped i-ZnO layer and a p-BeZnO:N layer on the n-ZnO:Ga film. While the Ni/Au and In act as ohmic contact electrodes of the p-type ZnO and n-type ZnO, respectively the corresponding p-i-n junction exhibits excellent diode characteristics. The EL spectra of the device and the optical images of the device with different drive current are given in Figure 3.33, respectively. A NBE at around 390 nm dominates in the spectra that can be interpreted as the DAP transition relate to the acceptors in the p-BeZnO:N film. Besides the strong NBE band, a broad weak deep level-related emission band at around 550 nm is also visible in the spectra. The dependence of the integrated emission intensity on the injection current is shown in the inset of Figure 3.32. It reveals that the emission intensity increases as linear relationship with the injection current at RT. Noticeable, strong near bandedge UV emission is also observed even at temperatures as high as 400K under the injection of continuous current. This result demonstrates that it is a feasible route to enhance hole doping in ZnO:N films by the assistance of Beryllium, which is also a critical advance toward the development of high-efficiency and stabilized p-type ZnO.

3.1.4.3 Quantum Dots LED

QDs are a unique class of emitters with size-tunable emission wavelengths, saturated emission colors, near-unity luminance efficiency, inherent photo- and thermal stability and excellent solution processability. QDs have been used as downconverters for backlighting in liquid-crystal displays to improve color gamut, leading to the booming of quantum dot televisions in consumer market. Since 1994, Alivisatos group[73] has reported QD electroluminescent devices (QD-LEDs) for the first time, using CdSe QDs as light-emtting layer materials to assemble QD-LEDs devices. Since then, researchers have carried out related research on QD-LEDs devices, including red QD-LEDs,[74,75] green QD-LEDs[76,77] and blue QD-LEDs.[78,79] The efficiency and lifetime of QD-LEDs have been improved significantly in the past decade, as detailed in Table 3.3.[80]

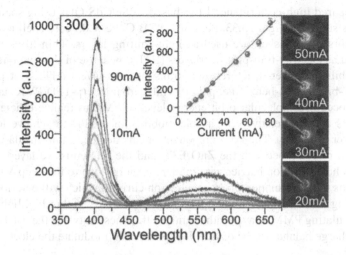

FIGURE 3.32 The EL spectra and the optical images of the ZnO LED under different injection currents at 300K, the insets show the dependence of the integrated near-bandedge emission intensity on the injection current. (From Chen, A., Zhu, H., Wu, Y., Chen, M., Zhu, Y., Gui, X. and Tang, Z., *Adv. Funct. Mater.*, 26, 3696, 2016. With permission.)

TABLE 3.3
Recent Progress of High-Performance QLEDs

QDs Structure	Ligands	EL Peak (nm)	Device Structure	Turn-on Voltage (V) @1 cd/m²	Peak EQE (%)	Lifetime (h) T_{50} @100 cd/m² (n=1.5)[a]
CdSe/CdS	Decylamine and TOPO	620	ITO/ZnO/QDs/NPB/LG101/Al	1.7	18.5	4000
CdSe/CdS	Dodecanethiol	640	ITO/PEDOT:PSS/p-TPD/PVK/ QDs/PMMA/ZnO/Ag	1.7	20.5	100,000
CdSe/CdS/ ZnS	Oleic acid	600	ITO/ZnO/Al₂O₃/QDs/Al₂O₃/CBP/ MoO₃/Al	2.0	~11	110,000
CdSe/CdS/ ZnS	Oleic acid	628	ITO/ZnO/QDs/CBP/MoO₃/Al	2.0	7.3	7000
CdSe@ ZnSe/ZnS	Oleic acid and TOPO	625	ITO/PEDOT:PSS/TFB/QDs/ZnO/Al	1.8	12.0	300,000[b]
CdSe@ ZnSe/ZnS	Oleic acid and TOPO	537	ITO/PEDOT:PSS/TFB/QDs/ZnO/Al	2.3	14.5	90,000[b]
CdSe@ ZnS/ZnS	Mercaptopropionic acid	516	ITO/PEDOT:PSS/PVK/QDs/ZnO/Al	6.0	12.6	–
ZnCdS/ZnS	Octanethiol	443	ITO/PEDOT/TFB/QDs/ZnO/Al	2.8	12.2	–
ZnCdS/ZnS	Oleic acid	452	ITO/PEDOT:PSS/PVK/QDs/ZnO/Al	4.0	7.1	–
CdSe@ ZnSe/ZnS	Oleic acid and TOPO	455	ITO/PEDOT:PSS/PVK/QDs/ZnO/Al	4.0	10.7	1000
InP@ZnSeS	Oleic acid	518	ITO/ZnO/QDs/CBP/MoO₃/Al	2.3	3.46	–
CuInS/ZnS	Oleic acid	580	ITO/PEDOT:PSS/TFB/QDs/ZnO/Al	3	7.3	35
CsPbBr₃	Oleic acid and oleylamine	512	ITO/PEDOT:PSS/p-TPD/QDs/ TPBI/LiF/Al	3.4	6.27	–

Source: From Dai, X., Deng, Y., Peng, X. and Jin, Y., *Adv. Mater.*, 29, 1607022, 2017. With permission.

[a] To compare the device lifetime reported by different groups, the authors assumed a value of 1.5 for the acceleration factor.

Dai et al.[75] prepared high-performance LED-based CdSe/CdS QDs using solution method, the device structure is shown in Figure 3.33. Phase-pure ZB CdSe–CdS core-shell nanocrystals (NCs) with ten monolayers of CdS shell are used as light-emtting layers. Thin films of colloidal ZnO NCs are employed as electron-transport interlayers (ETLs) because of their unique combination of high electron mobility. Bilayer-structured hole-transport interlayers (HTLs) of poly-(N,N′-bis(4-butylphenyl)-N,N′-bis(phenyl)-benzidine)/poly (9-vinlycarbazole) (p-TPD/PVK) take advantage of the deep highest occupied molecular-orbit energy level of PVK to realize efficient hole injection into the QD layers and the relatively high hole mobility of p-TPD to achieve low turn-on voltage and high power efficiency. The key component of this device, a thin insulating PMMA layer (Figure 3.33(d)), is inserted between the ZnO ETL and the QD emissive layer (EML). For their devices, there is a hole injection barrier with moderate energy due to the deep VB energy level of the QDs. This leads to the formation of excess electron current, which will deteriorate rapidly QDs based light-emitting diode (QLED) under operational conditions. Author highlights that the incorporation of the insulating PMMA layer with a suitable thickness between the ZnO ETL and the QD EML optimizes charge balance in the device. It is possible to modulate the electron injection from

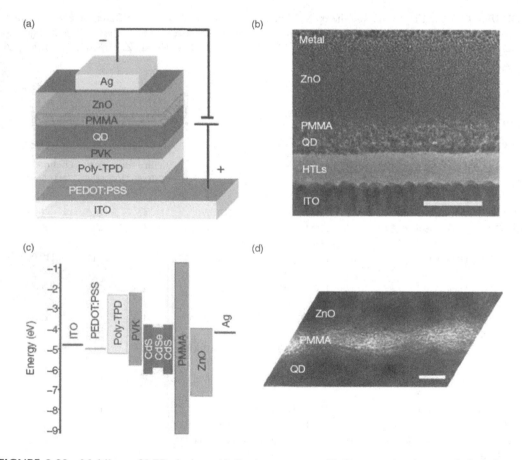

FIGURE 3.33 Multilayer QLED devices. (a) Device structure. (b) Cross-sectional transmission electron microscopy image showing the multiple layers of material with distinct contrast. Scale bar, 100 nm. The PMMA layer is evident only when the cross-sectional sample is sufficiently thin (d) because the neighboring quantum dot layer and the ZnO layer can obstruct the imaging of the PMMA layer. HTL, hole-transport interlayer. (c) Flat-band energy-level diagram. (d) High-magnification transmission electron microscopy image of an extremely thin cross-sectional sample revealing the presence of the PMMA layer between the ZnO layer and the quantum dot layer. Scale bar, 5 nm. (From Dai, X., Zhang, Z., Jin, Y., Niu, Y., Cao, H., Liang, X., Chen, L., Wang, J. and Peng, X., *Nature*, 515, 96, 2014. With permission.)

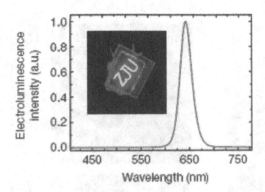

FIGURE 3.34 Electroluminescence spectrum at an applied voltage of 3 V and, inset, a photograph of a device with the Zhejiang University logo. (From Dai, X., Zhang, Z., Jin, Y., Niu, Y., Cao, H., Liang, X., Chen, L., Wang, J. and Peng, X., *Nature*, 515, 96, 2014. With permission.)

the ZnO ETLs to the QD layers and eliminate excess electron currents in the QLEDs by adjusting the thickness of the inserted PMMA layers. Figure 3.34 shows the normalized EL spectrum of the QLED. The symmetric emission peak at 640 nm has a narrow fullwidth at half-maximum of 28 nm. It exhibits color-saturated deep-red emission, subbandgap turn-on at 1.7 V, high external quantum efficiencies (EQEs) of up to 20.5%, low-efficiency roll-off (up to 15.1% of the EQE at 100 mA/cm²) and a long operational lifetime of more than 100,000 hours at 100 cd/m², making this device the best-performing solution-processed red LED, of the time.

Lee et al.[77] reported efficient green QD electroluminescent devices based on CdSe@ZnS/ZnS core-shell structure. Green CdSe@ZnS QDs of 9.5-nm size with a composition gradient shell were first prepared by a single-step synthetic approach, and then 12.7-nm CdSe@ZnS/ZnS QDs were obtained through the overcoating of an additional 1.6-nm thick ZnS shell. Two QDs of CdSe@ZnS and CdSe@ZnS/ZnS were incorporated into the solution-processed hybrid QD-based LED structure, where the QD EML was sandwiched by poly(9-vinylcarbazole) and ZnO NPs as hole and electron-transport layers, respectively. Figure 3.35(a) shows a schematic illustration of all-solution-processed multilayered green QLED, the electrode fabrication was completed by thermal evaporation of Al. Relatively uniform surface morphologies of spin-deposited, compactly packed QD and ZnO NP layers are also seen from scanning electron microscopic (SEM) images of Figure 3.35(b) and (c), respectively. Figure 3.35(d) presents normalized EL spectrum from CdSe@ZnS/ZnS QLED at a driving voltage of 9 V (corresponding to 322 mA/cm²). It can be clearly observed a green emission band with the peak wavelength of 516 nm and bandwidth of 21 nm. The authors found that the presence of an additional ZnS shell makes a profound impact on device performances such as luminance and efficiencies. Compared to CdSe@ZnS QD-based devices, the efficiencies of CdSe@ZnS/ZnS QD-based devices are overwhelmingly higher, specifically showing unprecedented values of peak current efficiency (CE) of 46.4 cd/A and EQE of 12.6%. The authors believe that such excellent results are attributable to a unique structure in CdSe@ZnS/ZnS QDs with a relatively thick ZnS outer shell as well as a well-positioned intermediate alloyed shell, enabling the effective suppression of non-radiative energy transfer between closely packed EML QDs and Auger recombination at charged QDs.

Shen et al.[78] reported high-efficiency blue-violet QD-based light-emitting diodes (QD-LEDs) by using high quantum yield ZnCdS/ZnS-graded core-shell QDs with proper surface ligands. As schematically shown in Figure 3.36(a), the QD-LED devices reported here have a typical multilayer structure consisting of ITO/PEDOT:PSS (30 nm)/TFB (40 nm)/QDs (45 nm)/ZnO (30 nm)/Al (100 nm). Except for Al that is deposited by thermal vacuum deposition, all the other layers are sequentially spin-coated on glass substrates with a prepatterned ITO transparent anode. It is found to cause a twofold increase in the electron mobility within the QD film by replacing the oleic acid ligands on

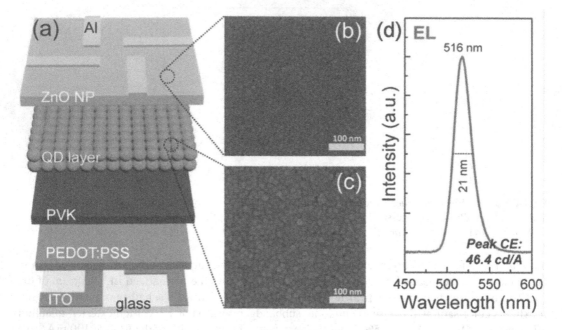

FIGURE 3.35 (a) Schematic illustration of all-solution-processed, multilayered green QLED consisting of ITO/PEDOT: PSS/PVK/CdSe@ZnS or CdSe@ZnS/ZnS QDs/ZnO NPs/Al. Surface SEM images of uniformly, compactly packed (b) ETL of ZnO NPs and (c) EML of CdSe@ZnS/ZnS QDs. (d) Normalized EL spectra from CdSe@ZnS/ZnS QLED at a driving voltage of 9 V. (From Lee, K.-H., Lee, J.-H., Kang, H.-D., Park, B., Kwon, Y., Ko, H., Lee, C., Lee, J. and Yang, H., *ACS Nano*, 8, 4893, 2014. With permission.)

the as-synthesized QDs with shorter 1-octanethiol ligands. Such a ligand exchange also results in an even greater increase in hole injection into the QD layer, thus improving the overall charge balance in the LEDs and yielding a 70% increase in quantum efficiency. Figure 3.36(b) shows normalized PL spectra (solid lines) and EL spectra (dashed lines) of three blue-violet QD-LEDs with different emission wavelengths. By selectively choosing the core reaction time and the core size, QDs with PL emission peaks at 427 and 453 nm are successfully obtained. It is obvious that the position of

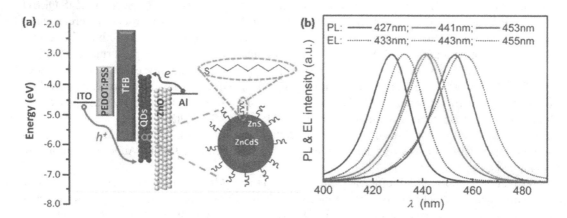

FIGURE 3.36 (a) Schematic illustration of energy levels for the multilayer QD-LEDs. (b) Normalized PL spectra (solid lines) and EL spectra (dashed lines) of three blue-violet QD-LEDs with different emission wavelengths. (From Shen, H., Cao, W., Shewmon, N. T., Yang, C., Li, L. S. and Xue, J., *Nano Lett.*, 15, 1211, 2015. With permission.)

EL peak is slightly red-shifted compared to the PL peak for all the devices, but FWHM of the EL is approximately the same as that of the PL. For QD-LEDs devices with an EL peak at 443 nm, a maximum luminance of 7600 cd/m² and a maximum EQE of 10.3±0.9% (with the highest at 12.2%) are obtained by using 1-octanethiol-capped QDs. Similar quantum efficiencies are also obtained for other blue/violet QD-LEDs with peak emission at 455 and 433 nm.

Based on high EQE (>10%) and the low turn-on voltage (~2.6 V), these blue-violet ZnCdS/ZnS QD-LEDs show great promise for use in next-generation full-color displays.

Nowadays, the EQEs of all the primary red (R), green (G) and blue (B) QLEDs have exceeded 15%, which are comparable to those of state-of-the-art organic LEDs (OLEDs), rendering QLED technology great commercial potential in high-quality lighting and display products. Despite the fast advances in monochrome QLEDs, the performance for white QLEDs (WQLEDs) with great potential for illumination and backlight applications to date has fallen short of that of state-of-the-art white OLEDs. To date, most were fabricated by mixing different-emission-wavelength QDs such as red, green and blue QDs together as a single white EML. The highest EQE of WQLEDs is only 10% due to inefficient exciton formation and non-radiative. As a result, many attempts have been made to prepare efficient WQLEDs, in which a tandem structure by stacking several emitting units is considered to be an effective method to fabricate efficient white LEDs, which has been used in the practical applications of lighting and displays.

Recently, a highly efficient WQLEDs with serially stacked red/green/blue units was prepared by an all solution process.[79] Figure 3.37(a) gives a schematic diagram of WQLED adopting an inverted device structure, including a multilayer architecture of ITO/ZnO NPs/PEIE/QDs/PVK/PEDOT:PSS/Al, where PEIE-modified ZnO NPs serve as ETL, the QDs as emitters, PVK as HTL and PEDOT:PSS as hole injection layer (HIL). ITO and Al are used as the cathode and anode for the devices, respectively. Figure 3.37(b) shows the band structure of the device. It can be seen clearly, compared with electron injection into QDs through ZnO electronic transport layer (ETL), there is a relatively large energy barrier for hole injection between the highest occupied molecular orbital (HOMO) of PVK and the VB maximum of QDs, which leads to the presence of excess electrons in QDs EML. The authors chosen to deposit the PEIE surface modifier with a wide bandgap of 6.2 V on ZnO ETL can not only function as an insulator exhibiting a large electron injection barrier but also upshift the CB minimum (CBM) of ZnO, thus blocking excess electrons into QDs EML and improving the charge balance in QLEDs. The normalized EL spectra of the red, green and blue QLEDs are presented in Figure 3.37(c). The pure QDs emissions at different peak wavelengths of 624, 534 and 456 nm, corresponding to the narrow FWHM of 32, 27 and 21 nm, respectively, indicate efficient recombination of electrons and holes in the red, green and blue QDs layers.

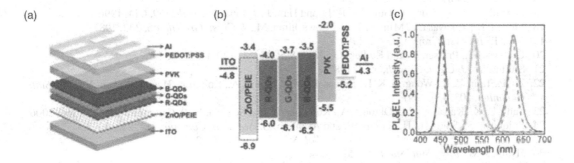

FIGURE 3.37 (a) Device architecture and (b) energy band diagram. The dashed line in (b) represents the energy level of ZnO. (c) Normalized photoluminescence (PL) and EL spectra of blue, green and red QDs. The solid line and dashed line represent EL and PL emission, respectively. (From Cao, F., Zhao, D., Shen, P., Wu, J., Wang, H., Wu, Q., Wang, F. and Yang, X., *Adv. Opt. Mater.*, 6, 1800652, 2018. With permission.)

It is found from the measurement that the maximum luminance values of 68,100 cd/m² for red emission, 46,430 cd/m² for green and 1708 cd/m² for blue are achieved at the driving voltages of 10.5, 8.5 and 8.5 V, respectively. All the turn-on voltages for these QLEDs are at around 3.7–4.0 V. The peak CEs of 30.2 cd/A for red QLED and 91.2 cd/A for green QLED correspond to the EQE of 20.7 and 21.2%, respectively. At the time, this efficiency value achieved efficiency breakthroughs in red and green QLEDs. Meanwhile, the maximum CEs of 2.7 cd/A and EQE of 8.4% for our blue QLED are close to the best-reported CE and EQE of blue QLEDs emitting at 450–460 nm.

After optimizing single red, green and blue QLEDs, tandem WQLEDs consisting of three R/G/B emitting units were prepared, in which all functional layers are deposited by the solution process except for Al electrode. There, the PVK/PEDOT:PSS/ZnO/PEIE layer is designed and used as the interconnecting layer (ICL) for the tandem WQLEDs. The EL spectra of tandem WQLEDs with the ICL made of 50% isopropanol mixed PEDOT:PSS show R/G/B three emission wavelengths peaking at 456, 534 and 624 nm at a current density of 1.33 mA/cm², which are identical with those of single QLEDs. By adjusting the thickness of different color QDs, the components of the corresponding colors can be changed and better performance of the devices can be obtained, where the peak CE and the EQE of WQLED are 79.9 cd/A and 28.0%, respectively. This is the highest efficiency value reported in the WQLED at that time.

REFERENCES

1. Ray, B., *II–VI Compounds*, Pergamon: Amsterdam, pp. 3–5, 1969.
2. Chelikowsky, J. R. and Cohen, M. L., *Phys. Rev. B*, 14, 556, 1976.
3. Bergstresser, T. K. and Cohen, M. L., *Phys. Rev.*, 164, 1069, 1967.
4. Yamada, S., *J. Phys. Soc. Jpn.*, 15, 1940, 1960.
5. Huang, X. M. and Igaki, K., *J. Cryst. Growth*, 78, 24, 1986.
6. Gunshor, R. L. and Kolodziejski, L. A., *IEEE J. Quantum Electron*, 24, 1744, 1988.
7. Segawa, Y., Ohtomo, A., Kawasaki, M., Koinuma, H., Tang, Z. K., Yu, P. and Wong, G. K. L., *Phys. Stat. Sol. B*, 202, 669, 1997.
8. Blanconnier, P., Hogrel, J. F., Jean-Louis, A. M. and Sermage, B., *J. Appl. Phys.*, 52, 6895, 1981.
9. Yodo, T., Koyama, T. and Yamashita, K., *J. Cryst. Growth*, 86, 273, 1988.
10. Yang, H., Ishida, A., Fujiyasu, H. and Kuwabara, H., *J. Appl. Phys.*, 65, 2838, 1989.
11. Park, R. M., Troffer, M. B., Rouleau, C. M., DePuydt, J. M. and Haase, M. A., *Appl. Phys. Lett.*, 57, 2127, 1990.
12. Mandel, G., Morehead, F. F. and Wagner, P. R., *Phys. Rev.*, 136, 826, 1964.
13. Merz, J. L., Nassau, K. and Shiever, J. W., *Phys. Rev. B*, 8, 1444, 1973.
14. Kosai, K., Fitzpatrick, B. J., Grimmeiss, H. G., Bhargava, R. N. and Neumark, G. F., *Appl. Phys. Lett.*, 35, 194, 1979.
15. Strauss, A. J., *II–VI and IV–VI Semiconductors*, Oxford: Pergamon, pp. 427–437, 1992.
16. Klick, C. C., *Phys. Rev.*, 89, 274, 1953.
17. Nurmikko, A. V., Jeon, H., Gunshor, R. L. and Han, J., *J. Cryst. Growth*, 159, 644, 1996.
18. Menda, K., Takayasu, I., Minato, T. and Kawashima, M., *J. Cryst. Growth*, 86, 342, 1988.
19. Yao, T., *J. Cryst. Growth*, 72, 31, 1985.
20. Gutowski, J., Presser, N. and Kudlek, G., *Phys. Stat. Sol. A*, 120, 11, 1990.
21. Stutius, W., *J. Cryst. Growth*, 59, 1, 1982.
22. Zu, P., Tang, Z. K., Wong, G. K. L., Kawasaki, M., Ohtomo, A., Koinuma, H. and Segawa, Y., *Solid State Commun.*, 103, 459, 1997.
23. Tang, Z. K., Kawasaki, M., Ohtomo, A., Koinuma, H. and Segawa, Y., *J. Cryst. Growth*, 287, 169, 2006.
24. Shionoya, S., Saito, H., Hanamura, E. and Akimoto, O., *Sol. Stat. Commun.*, 12, 223, 1973.
25. Saito, H., Shionoya, S. and Hanamura, E., *Sol. Stat. Commun.*, 12, 227, 1973.
26. Hvam, J. M., *Phys. Stat. Sol. B*, 63, 511, 1974.
27. Reynolds, D. C., Litton, C. W. and Collins, T. C., *Phys. Rev.*, 140, A1726, 1965.
28. Hvam, J. M., *Solid State Commun.*, 12, 95, 1973.
29. Tang, Z. K., Wong, G. K. L., Yu, P., Kawasaki, M., Ohtomo, A., Koinuma, H. and Segawa, Y., *Appl. Phys. Lett.*, 72, 3270, 1998.
30. Klingshirn, C. and Haug, H., *Phys. Rept.*, 70, 315, 1981.

31. Leheny, R. F. and Shah, J., *Phys. Rev. Lett.*, 37, 871, 1976.
32. Klingshirn, C., *J. Cryst. Growth*, 117, 753, 1992.
33. Kunz, M., Pier, T., Bhargava, R. N., Reznitsky, A., Kozlovskii, V. I., Müller-Vogt, G., Pfister, J. C., Pautrat, J. L. and Klingshirn, C., *J. Cryst. Growth*, 101, 734, 1990.
34. Livingstone, A. W., Turvey, K. and Allen, J. W., *Solid State Electron.*, 16, 351, 1973.
35. Yamaguchi, M. and Yamamoto, A., *Jpn. J. Appl. Phys.*, 16, 77, 1977.
36. Fan, X. W. and Woods, J., *Phys. Stat. Sol. A*, 70, 325, 1982.
37. Donnelly, J. P., Foyt, A. G., Lindley, W. T. and Iseler, G. W., *Solid State Electron.*, 13, 755, 1970.
38. Allen, J. W., Livingstone, A. W. and Turvey, K., *Solid State Electron.*, 15, 1363, 1972.
39. Card, H. C. and Smith, B. L., *J. Appl. Phys.*, 42, 5863, 1971.
40. Fan, X. W. and Woods, J., *IEEE Trans. Electron Devices*, 28, 428, 1981.
41. Yamaguchi, M., Yamamoto, A. and Kondo, M., *J. Appl. Phys.*, 48, 196, 1977.
42. Ma, X., Chen, P., Li, D., Zhang, Y. and Yang, D., *Appl. Phys. Lett.*, 91, 251109, 2007.
43. Li, Y., Ma, X., Xu, M., Xiang, L. and Yang, D., *Opt. Express*, 19, 8662, 2011.
44. Zhu, H., Shan, C.-X., Zhang, J.-Y., Zhang, Z.-Z., Li, B.-H., Zhao, D.-X., Yao, B., Shen, D.-Z., Fan, X.-W., Tang, Z.-K., Hou, X. and Choy, K.-L., *Adv. Mater.*, 22, 1877, 2010.
45. Chen, J., Wang, Z., Cao, J., Yang, D. and Ma, X., *J. Appl. Phys.*, 127, 055705, 2020.
46. Nakamura, S., *Proc. Int. Symp. on Blue Laser and Light Emitting Diodes*, Tokyo: Ohmsha, pp. 119, 1996.
47. Haase, M. A., Cheng, H., DePuydt, J. M. and Potts, J. E., *J. Appl. Phys.*, 67, 448, 1990.
48. Ohkawa, K., Ueno, A. and Mitsuyu, T., *Jpn. J. Appl. Phys.*, 30, 3873, 1991.
49. Haase, M. A., Qiu, J., DePuydt, J. M. and Cheng, H., *Appl. Phys. Lett.*, 59, 1272, 1991.
50. Jeon, H., Ding, J., Nurmikko, A. V., Xie, W., Grillo, D. C., Kobayashi, M., Gunshor, R. L., Hua, G. C. and Otsuka, N., *Appl. Phys. Lett.*, 60, 2045, 1992.
51. Okuyama, H., Nakano, K., Miyajima, T. and Akimoto, K., *J. Cryst. Growth*, 117, 139, 1992.
52. Gaines, J. M., Drenten, R. R., Haberern, K. W., Marshall, T., Mensz, P. and Petruzzello, J., *Appl. Phys. Lett.*, 62, 2462, 1993.
53. Ishibashi, L., *Proceedings of the Seventh International Conference on II-VI Compounds and Devices*, Elsevier: Edinburgh, 1995.
54. Kato, E., Noguchi, H., Nagai, M., Okuyama, H., Kijima, S. and Ishibashi, A., *Electron. Lett.*, 34, 282, 1998.
55. Vèrié, C., *J. Cryst. Growth*, 184–185, 1061, 1998.
56. Landwehr, G., Waag, A., Fischer, F., Lugauer, H. J. and Schüll, K., *Physica E*, 3, 158, 1998.
57. Faschinger, W. and Nürnberger, J., *Appl. Phys. Lett.*, 77, 187, 2000.
58. Che, S.-B., Nomura, I., Kikuchi, A. and Kishino, K., *Appl. Phys. Lett.*, 81, 972, 2002.
59. Nomura, I., Nakai, Y., Hayami, K., Saitoh, T. and Kishino, K., *Phys. Stat. Sol. B*, 243, 924, 2006.
60. Kobayashi, T., Nomura, I., Murakami, K. and Kishino, K., *J. Cryst. Growth*, 378, 263, 2013.
61. Kasai, J.-i., Akimoto, R., Hasama, T., Ishikawa, H., Fujisaki, S., Tanaka, S. and Tsuji, S., *Appl. Phys. Express*, 4, 082102, 2011.
62. Fujisaki, S., Kasai, J.-i., Akimoto, R., Tanaka, S., Tsuji, S., Hasama, T. and Ishikawa, H., *Appl. Phys. Express*, 5, 062101, 2012.
63. Feng, J. and Akimoto, R., *Appl. Phys. Lett.*, 107, 161101, 2015.
64. Ryu, Y. R., Lubguban, J. A., Lee, T. S., White, H. W., Jeong, T. S., Youn, C. J. and Kim, B. J., *Appl. Phys. Lett.*, 90, 131115, 2007.
65. Chu, S., Olmedo, M., Yang, Z., Kong, J. and Liu, J., *Appl. Phys. Lett.*, 93, 181106, 2008.
66. Aoki, T., Hatanaka, Y. and Look, D. C., *Appl. Phys. Lett.*, 76, 3257, 2000.
67. Tsukazaki, A., Ohtomo, A., Onuma, T., Ohtani, M., Makino, T., Sumiya, M., Ohtani, K., Chichibu, S. F., Fuke, S., Segawa, Y., Ohno, H., Koinuma, H. and Kawasaki, M., *Nat. Mater.*, 4, 42, 2005.
68. Jiao, S. J., Zhang, Z. Z., Lu, Y. M., Shen, D. Z., Yao, B., Zhang, J. Y., Li, B. H., Zhao, D. X., Fan, X. W. and Tang, Z. K., *Appl. Phys. Lett.*, 88, 031911, 2006.
69. Wei, Z. P., Lu, Y. M., Shen, D. Z., Zhang, Z. Z., Yao, B., Li, B. H., Zhang, J. Y., Zhao, D. X., Fan, X. W. and Tang, Z. K., *Appl. Phys. Lett.*, 90, 042113, 2007.
70. Lim, J. H., Kang, C. K., Kim, K. K., Park, I. K., Hwang, D. K. and Park, S. J., *Adv. Mater.*, 18, 2720, 2006.
71. Chu, S., Lim, J. H., Mandalapu, L. J., Yang, Z., Li, L. and Liu, J. L., *Appl. Phys. Lett.*, 92, 152103, 2008.
72. Chen, A., Zhu, H., Wu, Y., Chen, M., Zhu, Y., Gui, X. and Tang, Z., *Adv. Funct. Mater.*, 26, 3696, 2016.
73. Colvin, V. L., Schlamp, M. C. and Alivisatos, A. P., *Nature*, 370, 354, 1994.

74. Mashford, B. S., Stevenson, M., Popovic, Z., Hamilton, C., Zhou, Z., Breen, C., Steckel, J., Bulovic, V., Bawendi, M., Coe-Sullivan, S. and Kazlas, P. T., *Nat. Photon.*, 7, 407, 2013.
75. Dai, X., Zhang, Z., Jin, Y., Niu, Y., Cao, H., Liang, X., Chen, L., Wang, J. and Peng, X., *Nature*, 515, 96, 2014.
76. Lim, J., Park, M., Bae, W. K., Lee, D., Lee, S., Lee, C. and Char, K., *ACS Nano*, 7, 9019, 2013.
77. Lee, K.-H., Lee, J.-H., Kang, H.-D., Park, B., Kwon, Y., Ko, H., Lee, C., Lee, J. and Yang, H., *ACS Nano*, 8, 4893, 2014.
78. Shen, H., Cao, W., Shewmon, N. T., Yang, C., Li, L. S. and Xue, J., *Nano Lett.*, 15, 1211, 2015.
79. Cao, F., Zhao, D., Shen, P., Wu, J., Wang, H., Wu, Q., Wang, F. and Yang, X., *Adv. Opt. Mater.*, 6, 1800652, 2018.
80. Dai, X., Deng, Y., Peng, X. and Jin, Y., *Adv. Mater.*, 29, 1607022, 2017.

3.2 II–VI GROUP WIDE BANDGAP SEMICONDUCTOR SUPERLATTICES

Youming Lu

The development of modern film growth technology allows to prepare a new type of artificial material, namely quantum well (QW) and superlattice (SL). This is a revolution in the field of semiconductor science, which has aroused great interest. The concept of SL was first introduced by the Esaik and Tsu of American IBM Corporation in 1969,[1] the AlGaAs–GaAs SL structure[2] was prepared by molecular-beam epitaxy (MBE) for the first time. Since then, various types of semiconductor SLs have emerged. However, most works are focused on the growth and optoelectronic properties of III–V semiconductor SLs. Osbourn[3] put forward the concept of stained-layer SL (SLS) in 1982, and Fujiyasu et al.[4] prepared the ZnSe–ZnS/GaAs SLS by hot-wall method (HWE) in 1984. From then on, it has opened a new page in the important field of II–VI wide bandgap semiconductor SL and has been continuously developed and paid attention to. This section focuses on the growth, luminescence and optical nonlinearity of II–VI wide bandgap semiconductor SLs dominated by ZnSe, ZnS and ZnTe.

3.2.1 OVERVIEW OF THE DEVELOPMENT FOR SEMICONDUCTOR SUPERLATTICES

In 1970, the experimental results of GaAs/GaAsP SLs were first reported by Esaki and Tsu,[5] and the used samples were grown by Blakeslee and Aliotta[6] through chemical vapor deposition (CVD). The existence of SLs was confirmed by X-ray analysis, but no expected electronic properties were obtained. In fact, the semiconductor SL structure is composed of two different semiconductors stacked alternately in layers and has a quantum confine effect in only one direction. In this structure, the electron (hole) system changes from the three-dimensional (3D) structure of the bulk material to the two-dimensional (2D) structure. Brillouin zone is split into a series of minizones after the SL shown in Figure 3.38 is formed. Accordingly, the electronic structure of the semiconductor

FIGURE 3.38 Heterojunction superlattice structure.

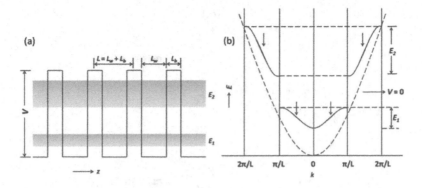

FIGURE 3.39 Schematic diagram of subband formation (a) and dispersion relations (b) in superlattices.

is deformed to form multiple subbands, generates subbandgap in the $E-k$ dispersion relation (refer to Figure 3.39). Furthermore, the behavior of electrons moving toward SL potential in the ground-state subband is analyzed, and a characteristic of generating negative differential conductance is proposed theoretically.

In 1972, the current-voltage characteristics of GaAs/GaAlAs SLs fabricated with MBE were measured by Esaki et al.,[7] and the negative resistance characteristics found therein are well explained by the quantum effect mentioned previously. Since the 1980s, with the rapid development of film growth technology by MBE and metalorganic CVD (MOCVD) methods, the control of the thickness of monoatomic layers has become possible.[8] As a result, researchers are able to design and fabricate new semiconductor materials. SLs and QWs under the condition of interdisciplinary mutual promotion in the three fields of physics, materials and devices, so far has been greatly developed. In 1978, it was first proposed that a new type of SL consists of ultrathin n- and p-type doped layers of the same semiconducting material.[9] This kind of SL, which does not contain any interface and is crystallographically only slightly perturbed by a relatively small amount of dopants, is called doped SL. After that, the doping SLs were deeply studied and the corresponding results were obtained.[10] CdTe/CdMnTe, ZnSe/ZnMnSe dilute-magnetic SLs have become another research hotspot, and their characteristic magneto-optical properties have been reported.[11]

3.2.2 Energy Band Engineering of Semiconductor Superlattice

3.2.2.1 Semiconductor Band Structure and Superlattice Type

The band structure of semiconductor shows that the conduction band (CB) is composed of s orbit and the valence band (VB) is composed of p orbit. If the spin-orbit interaction is ignored, the top of the VB corresponds to the triple degeneracy of the three p orbitals. The spin-orbit splitting band (SO) is split by considering the spin-orbit interaction, Kane[12] proposed an eight-band model with double degeneracy at the top of the VB. That is, it has two Γ_6 CBs ($J=1/2$), four Γ_8 VBs (the corresponding state of $J=3/2$) and two Γ_6 VBs (SO of $J=1/2$). There are two kinds of mass holes, which are heavy holes (HHs) and light holes (LHs), respectively. Figure 3.40 shows Kane's eight-band model of cubic crystal structure. Δ_0 is the spin-orbit splitting band constant. The degeneracy of HH band and LH band is removed in the QW structure due to the quantum size effect in the z direction, and the energy level is further divided into two bands corresponding to $J = \pm3/2$ bands (HH) and $J = \pm1/2$ bands (LH) of the z direction.

When the two materials of the SL structure are bonded, the barrier shown in Figure 3.41 will be generated because of the different electron affinity and bandgap. If the energy at the bottom of the CB of one semiconductor is lower than that of another semiconductor and the energy at the top of the VB is higher than that of another semiconductor, such an SL system is called a I-type SL. As an

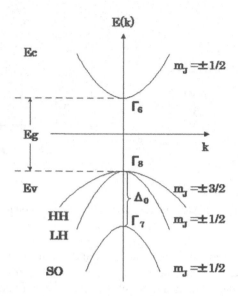

FIGURE 3.40   Energy band diagram of quantum well structure.

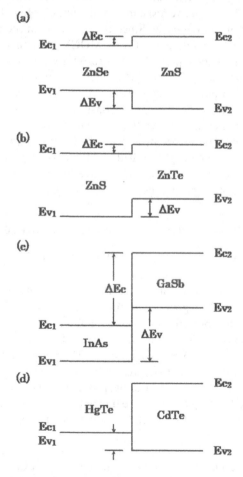

FIGURE 3.41   Different types of superlattice structure (a): type I structure, (b): type-II structure, (c): type-IIA structure (d): type-III structure.

example, the SLs consisting of ZnS and ZnSe, or ZnSe and ZnCdSe are equivalent to this case, as shown in Figure 3.41(a). Compared with this, if the energies of the CB bottom and the VB top of one semiconductor are lower than that of the another semiconductor, the SL system is called the II-type SL, such as SLs shown in Figure 3.41(b) consisting of ZnS and ZnTe or ZnSe and ZnTe. For the II type, when the energy at the bottom of the CB of one semiconductor is lower than the energy at the top of the VB of another semiconductor, the phase transition from semiconductor to semimetal can be seen. This is called the SL of II type A. The example is an SL consisting of InAs and GaSb of III–V group compounds, as shown in Figure 3.41(c). Figure 3.41(d) shows the III-type SL structure, which can only be seen in semimetallic Hg systems. For example, HgTe/CdTe is a good example.

The exciton effect in the I-type and II-type structures of SL is significantly different. Because of the quantum confinement effect, the Coulomb interaction between electrons of CB and holes of VB in the I structure is strong, and the bound state of excitons increases significantly. This leads to:

1. Increase of binding energy (the binding energy of 2D excitons is four times that of 3D excitons[13]).
2. Increase in vibration intensity.
3. Reduction of wave function distribution.

For II-type structure, the electron and hole wells are separated in real space. The overlap integral of the electron and hole wave function is several orders of magnitude smaller than that of the I-type structure, which leads to the weak luminescence associated with the exciton. Especially in this system, exciton recombination is an indirect transition in real space, resulting in a long-lifetime feature.[14] On the other hand, it is expected that there are large differences in 2D electronic states in I and II SLs systems. For the I-type structure, the selection rule that allows direct transitions of the electrons and holes in the square potential well under the infinite depth barrier model is $n_e = n_h$. In addition, the transition probability is 0, that is, only the transition between states with the same quantum number is allowed. Actually, the energy barrier of the I-type SL structure is limited, and the transition between the energy levels of different quantum numbers is also observed,[15] which is approximately established as a selective rule. However, in II SL structure, the selection rule of $\Delta n = 0$ for the optical transition of electrons and holes separated in real space is destroyed and the subband transition of $\Delta n \neq 0$ can be seen.[16]

3.2.2.2 Energy Band Discontinuity of Heterostructures

In a heterostructure interface of two semiconductors 1 and 2, the finite energy difference from the CB bottom and VB top of substance 1 to the CB bottom and VB top of substance 2 is called band discontinuity (ΔE_c) and (ΔE_v). Though several attempts have been made to calculate the ΔE_c and ΔE_v,[17-20] so far the theory of giving sufficiently accurate values has not been completed. The commonly used methods are Anderson electron affinity rules and Harrison tight-binding theory.

3.2.2.2.1 Anderson Electron Affinity Rules[17]

At the bottom of the CB produced by the electron affinity when the two materials are bonded, the band discontinuity ΔE_c is generated as shown in Figure 3.42.

$$\Delta E_c = \chi^{(1)} - \chi^{(2)} \tag{3.7}$$

$$\Delta E_v = E_g^{(1)} - E_g^{(2)} - \chi^{(1)} + \chi^{(2)} \tag{3.8}$$

where $E_g^{(1)}, E_g^{(2)}$ and $\chi^{(1)}, \chi^{(2)}$ are the bandgaps and electron affinity for two semiconductors 1 and 2, respectively. Table 3.4 shows the bandgaps[21] and the electron affinities[17] for II–VI group of semiconductors.

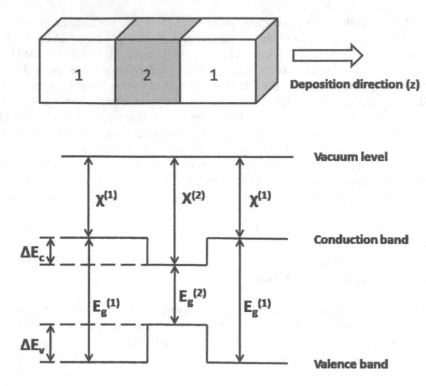

FIGURE 3.42 Relationship between band discontinuity and electron affinity.

3.2.2.2.2 Harrison-Based Tight-Binding Theory[18]

According to Harrison's theory of atomic orbital linear combination (AOLC), the energy positions of each material energy band can be determined by the common criteria based on the energy levels that make up the elements. That is, the energy at the top of the semiconductor VB is given by the following formula:

$$E_{v,\max} = \frac{E_p^c + E_p^a}{2} \sqrt{\left(\frac{E_p^c - E_p^a}{2}\right)^2 + V_{xx}^2} \tag{3.9}$$

TABLE 3.4
Bandgap Values, Electron Affinity and $E_{v,\max}$ for II–VI Group Semiconductors

Material	E_g (eV) (4.2K)	dE_g/dT (10^{-4} eV/K)	χ (eV)	$-Ev_{,\max}$ (eV)
ZnS	3.91 W	−8.5		
	3.84 ZB	−4.6	3.9	11.40
ZnSe	2.83 ZB	−8	4.09	10.58
ZnTe	2.39 ZB	−5.0	3.53	9.5
CdS	2.58 W	−5.2	4.5	11.12
CdSe	1.84 W	−4.6	4.58	10.35
CdTe	1.60 ZB	−2.3	4.28	9.02

Note: W: hexagonal wurtzite structure, ZB: zinc-blende structure.

Source: The data are acquired from Anderson, R. L., *Solid State Electron.*, 5, 341, 1962; Harrison, W. A., *J. Vac. Sci. Technol.*, 14, 1016, 1977; Ray, B., *II–VI Compounds*. Pergamon: Amsterdam, pp. 54, 1969.

$$V_{xx} = \frac{2.16\hbar^2}{m_e d^2} \tag{3.10}$$

where E_p^c and E_p^a are p state energies on the metallic atom (cation) and the nonmetallic atom (anion), respectively. V_{xx} is an appropriate interatomic matrix element between atomic p states on adjacent atoms. The m_e is the mass of electrons and d is the distance between atoms. The VB discontinuity of the heterojunction is given by the difference of $E_{v,max}$ between the various substances (refer to Table 3.4).

$$\Delta E_v = E_{v,max}^{(1)} - E_{v,max}^{(2)} \tag{3.11}$$

$$\Delta E_c = E_g^{(1)} - E_g^{(2)} - E_{v,max}^{(1)} + E_{v,max}^{(2)} \tag{3.12}$$

3.2.2.3 Strain Calculation Due to Lattice Mismatch

When multilayer structures are formed between materials with different lattice constants, it is usually difficult to grow high-quality crystals due to dislocation caused by lattice mismatch. However, it is found that when the film thickness is less than a certain value, the lattice mismatch is moderated by the strain of the lattice, and the dislocations are suppressed to occur. The film thickness is called critical film thickness. In 1973, Matthews et al.[22] pointed out that SL structures can be fabricated when the film thickness is thicker than the critical film thickness. If the lattice constants and layer thickness of the two materials are set to a_1, a_2 and d_1, d_2, the lattice constants ($a_{//}$) of the strain SLs parallel to the interface will be given by the following formula:

$$a_{//} = \frac{a_1 G_1 d_1 + a_2 G_2 d_2}{G_1 d_1 + G_2 d_2} \tag{3.13}$$

$$G_i = 2\left[C_{11}^i + C_{12}^i - \frac{2(C_{12}^i)^2}{C_{11}^i} \right] \quad i = 1, 2 \tag{3.14}$$

where G_1 and G_2 are the rigid coefficients of the material, respectively. $C_{11}^i, C_{12}^i (i=1, 2)$ is the elastic constant of two kinds of semiconductor materials (refer to Table 3.5 [23,24]). According to Ref. [23], the relationship between the lattice constants parallel and perpendicular to the interface of the strain SL is given by Equation (3.15), respectively.

$$a_{//} - a_0^i = -\left(\frac{C_{11}^i}{2C_{12}^i} \right)(a_\perp^i - ai_0) \tag{3.15}$$

TABLE 3.5

Lattice Constants, Static Dielectric Constant and Modulus of Elasticity for II–VI Group Compounds

Material	A_{ZB} (Å)	α (10^{-6} K^{-1})	ε_∞	C_{11} (10^{10} N/m²)	C_{12} (10^{10} N/m²)
ZnS	5.409	3.814	5.1	10.46	6.53
ZnSe	5.668	3.996	5.4	8.10	4.88
ZnTe	6.089	4.270	5.3	7.13	4.07
CdS	5.818	4.315	5.3	9.07	5.81
CdSe	6.052	4.299	5.8	7.41	4.52
CdTe	6.480	4.57	7.3	5.35	3.68

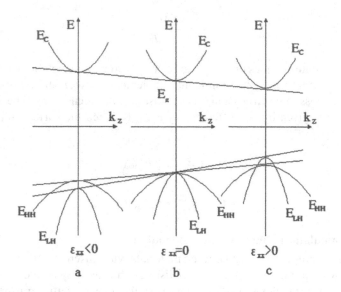

FIGURE 3.43 Bandgap variations due to strain.

where $a_0^i(i=1,2)$ is the lattice constant of two semiconductor bulk materials. The strains $\varepsilon_{xx}^i, \varepsilon_{yy}^i$ and $\varepsilon_{zz}^i(i=1,2)$ in parallel (xy) or vertical (z) directions to the (1 0 0) interface are given as follows:

$$\varepsilon_{xx}^i = \varepsilon_{yy}^i = \frac{a_{//} - a_0^i}{a_0^i} \tag{3.16}$$

$$\varepsilon_{zz}^i = \frac{a_\perp^i - a_0^i}{a_0^i} \tag{3.17}$$

According to Equation (3.15):

$$\varepsilon_{xx}^i = -\left(\frac{C_{11}^i}{2C_{12}^i}\right)\varepsilon_{zz}^i \tag{3.18}$$

Furthermore, because the strain is applied to each material, not only the lattice constant and the electronic state are affected, but also the bandgap of the material changes, which leads to the energy splitting of the heavy and LHs at the top of the VB ($k=0$). Figure 3.43 represents the strain-induced energy band shift.

As a method to estimate the band shift caused by strain, two formulas are listed next:

1. Asai and Oe formulas[25]

$$\Delta E_g(HH) = \left[2a\left(\frac{C_{11} - C_{12}}{C_{11}}\right) - b\left(\frac{C_{11} + 2C_{12}}{C_{11}}\right)\right]\varepsilon_{xx} \tag{3.19}$$

$$\Delta E_g(LH) = \left[2a\left(\frac{C_{11} - C_{12}}{C_{11}}\right) + b\left(\frac{C_{11} + 2C_{12}}{C_{11}}\right)\right]\varepsilon_{xx} \tag{3.20}$$

$$\Delta(E_g + \Delta_0) = 2a\left(\frac{C_{11} - C_{12}}{C_{11}}\right)\varepsilon_{xx} \tag{3.21}$$

where $\Delta E_g(HH)$, $\Delta E_g(LH)$ and $\Delta(E_g+\Delta_0)$ are strain-induced energy difference between the conduction and VBs associated with HH, LH and spin-orbit splitting, respectively. E_g represents the bandgap of bulk materials. ε_{xx} is a strain component parallel to the (1 0 0) plane, C_{11} and C_{12} are elastic coefficients. a is the hydrostatic deformation potential, b is the shear deformation potential, and Δ_0 is the spin-orbit splitting of the VBs at $k=0$.

2. Van de wall formulas[26]

$$\Delta E_{v,av} = a_v \left(\varepsilon_{xx} + \varepsilon_{yy} + \varepsilon_{zz} \right) \tag{3.22}$$

$$\Delta E_c = a_c \left(\varepsilon_{xx} + \varepsilon_{yy} + \varepsilon_{zz} \right) \tag{3.23}$$

$$\Delta E_{v1} = \frac{1}{3}\Delta_0 - \frac{1}{2}\delta E_{001} \tag{3.24}$$

$$\Delta E_{v2} = -\frac{1}{6}\Delta_0 + \frac{1}{4}\delta E_{001} + \frac{1}{2}\left[\Delta_0^2 + \Delta_0\delta E_{001} + \frac{9}{4}\left(\delta E_{001}\right)^2 \right]^{\frac{1}{2}} \tag{3.25}$$

$$\Delta E_{v3} = -\frac{1}{6}\Delta_0 + \frac{1}{4}\delta E_{001} - \frac{1}{2}\left[\Delta_0^2 + \Delta_0\delta E_{001} + \frac{9}{4}\left(\delta E_{001}\right)^2 \right]^{\frac{1}{2}} \tag{3.26}$$

$$\delta E_{001} = 2b\left(\varepsilon_{zz} - \varepsilon_{xx} \right) \tag{3.27}$$

Here, the relevant band edges are the (nondegenerate) conduction-band minimum at Γ, E_c, and the VBs at Γ (degenerate in the absence of strain and spin-orbit splitting), which the heavy, LH bands and the spin-orbit split-off band will be labeled by E_{v1}, E_{v2} and E_{v3}, respectively. The average of these VBs will be referred to as $E_{v,av}$. $\Delta E_{v,av}$ is the relative variation of the mean energy of the VB, $\Delta E_i (i=c, v1, v2, v3)$ is the relative energy change of CB (c), HH band ($v1$), LH band ($v2$) and spin-orbit splitting band ($v3$). Table 3.6 shows the strain parameter values for II–VI group semiconductors.

3.2.2.4   Calculation of Quantum Energy Levels and Subbands Due to Quantum Size Effects in Superlattices

When the thickness of the semiconductor is reduced to close to the average free path of the carrier, there will be the quantum size effect that cannot be seen on the bulk material. Figure 3.44 shows the potential distribution and the wave function in the heterostructure. In the case of Figure 3.44(b), the

---

TABLE 3.6

Parameters for the Band Structure of II–VI Group of Semiconductors

Material	Δ_0 (eV)	a_v (eV)	a_c (eV)	a (eV)	b (eV)	$E_{v,av}$ (eV)
ZnS	0.07	2.31	−4.09	−6.40	−1.25	−9 15
ZnSe	0.43	1.65	−4.17	−5.82	−1.2	−8.37
ZnTe	0.91	0.79	−5.83	−6.62	−1.26	−7.17
CdTe	0.93	0.55	−3.96	−4.52	−1.10	−7.07

Source: From Van de Walle, C. G., Shahzad, K. and Olego, D. J., *J. Vac. Sci. Technol.*, 6, 1350, 1988. With Permission.

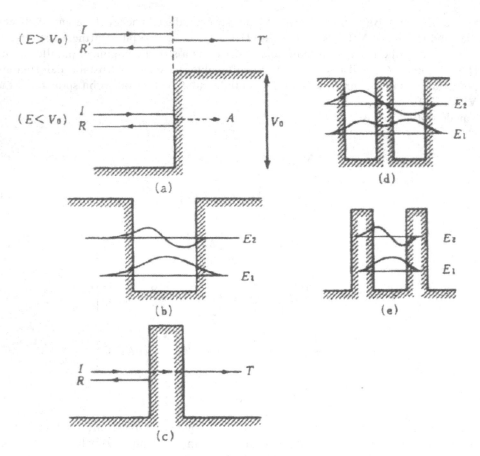

FIGURE 3.44 Potential distribution and wave function in heterostructures. Single heterojunction (a), quantum well (b), tunnel barrier (c), double quantum well (d) and double tunnel barrier (e).

formation of quantum level can be seen when the free degree of electrons (holes) is limited in one direction. The energy eigenvalues (E_n) in this direction (z) are given by:

$$E_n = \left(\frac{\hbar^2}{2m^*} \right) \left(\frac{\pi n}{L_z} \right)^2, \quad n = 1, 2, \ldots \tag{3.28}$$

where m^* is the effective mass of the carrier, L_z is the width of the QW. Furthermore, considering the in-plane energy of the potential well, the energy eigenvalue of the carrier is given in the following form:

$$E\left(k_x, k_y, n\right) = \left(\frac{\hbar^2}{2m^*} \right)\left(k_x^2 + k_y^2\right) + E_n \tag{3.29}$$

where k_x and k_y are the x and y components of the wavenumber, respectively. The electron state density is:

$$g(z) = \frac{m^*}{\left(\pi \hbar L_z\right)} \tag{3.30}$$

The density of states in the QW structure has a stepped distribution, as is shown in Figure 3.45. This results in a significant increase in the density of states near the minimum energy. The excellent

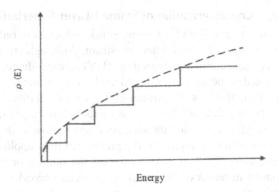

FIGURE 3.45 The state density of the superlattice structure. The dashed line represents the state density of the bulk crystal.

properties of QW lasers are usually caused by this phenomenon.[27,28] In Figure 3.44(d), when the QW approaches, the decay of the electron wave function within the barrier becomes incomplete, resulting in coupling between the quantum levels within the adjacent QW (tunneling effect). As a result, degenerate quantum energy-level splitting. When the number of coupled QWs is large enough, the split quantum energy levels are basically continuous, forming a small band. In this way, a semiconductor SL is formed by a periodic structure formed by alternating stacking electron confining layers (well layers) and tunneling layers (barrier layers), and when there is no tunneling effect, it is called multi-QW structure.

In SLs, electrons propagate along the vertical direction of the film, so their wavelength λ_e and wave vector $k_z = (2\pi/\lambda_e)$ can be selected for various values. In particular, when k_z equals to an integer multiple of (π/L), the reflection of the electron wave is generated due to satisfying the Bragg condition, and the bandgap is generated in the dispersion relation (refer to Figure 3.39). However, because the period $L = L_w + L_b$ is larger than the lattice constant, it is shown to be located in a position with smaller wave vector. Dispersion relation of SL can be obtained by Kronig-Penney equation of periodic square potential well.[29] For the case of $E_z < V_0$, the following formula is used:

$$\frac{\beta^2 - \alpha^2}{2\alpha\beta} \sin(\alpha L_w) \sinh(\beta L_b) + \cos(\alpha L_w) \cosh(\beta L_b) = \cos(k_z L) \tag{3.31}$$

$$\alpha = \left(\frac{2m_w E}{\hbar^2}\right)^{\frac{1}{2}} \tag{3.32}$$

$$\beta = \left[\frac{2m_b(V_0 - E)}{\hbar^2}\right]^{\frac{1}{2}} \tag{3.33}$$

where $L_w(L_b)$ and $m_w(m_b)$ are layer thickness and effective mass of the potential well (the potential barrier), respectively. When the left side of Equation (1.33) is used as a function of $E_z(k_z)$, its value will produce subbands in the range of ±1, in addition to which the subbandgap will be generated.

3.2.3 Optical Properties of II–VI Compound Superlattices

Wide gap II–VI group semiconductors, represented by ZnSe, exhibit room temperature (RT) exciton effects when they are used to form SLs due to their direct bandgap and large exciton binding energy. It can be used to realize short-wavelength RT laser devices, which has attracted wide interest. The related research mainly focuses on the following aspects.

3.2.3.1 Preparation and Characterization of Stained-Layer Superlattices

Since the lattice mismatch between II and VI group semiconductors is rather large, the epilayers in their QWs and SLs are exposed to a considerable strain. Although strained-layer SLs with up to 13% mismatch have been successfully prepared on II–VI semiconductor systems,[30,31] the misfit dislocations that are generated at the interfaces due to the lattice mismatch, which makes it difficult to fabricate II–VI group semiconductor SL structures with good quality.[32] In the majority of the semiconductors devices, the interfaces of the heterostructures have to be the sharpest possible to minimize effects due to roughness and fluctuations. As a result, how to design and prepare II–VI group SL and QW structures with high quality is the premise of their applications.

Photoluminescence is commonly used to characterize the quality of SL QWs. The interface properties can be understood in terms of the excitonic linewidth broadening which is influenced by either the interface roughness or by well width fluctuation. Taguchi et al.[33] studied photoluminescence and absorption spectra of the ZnSe/ZnS and ZnSe/ZnSSe SLSs to characterize the interface and to clarify the energy separation of the degenerate hole bands due to a compressive strain in the ZnSe QW. The thickness of the ZnSe/ZnS interfaces is found to be equal to at least three ZnSe monolayers by the excitonic linewidth broadening. Maia Jr. et al.[34] studied the excitonic properties of ZnSe/ZnS$_x$Se$_{1-x}$ QW by taking into account spin-orbit splitting, strain and interfacial effects, where the authors analyzed three systems of strained QWs and SLs: the first and second ones considering a pure pseudomorphic growth where all the strain is in the ZnS$_x$Se$_{1-x}$ layer; and the third for the free-standing case, when the strain is shared between the well and the barrier layer. Figure 3.46 shows the ZnSe/ZnS$_{0.18}$Se$_{0.82}$ QW band offset for pseudopotential growth and free growth. The existence of thick interfaces (5 Å) can blue shift considerably (13 meV) the exciton energy in ZnSe/ZnS$_{0.18}$Se$_{0.82}$ QWs. In this case, the interface-related enhancement of the carrier's and confined exciton energy in ZnSe/ZnS$_{0.18}$Se$_{0.82}$ QWs and SLs can reach several meV. The interface-related broadening of the excitonic energy in ZnSe/ZnS$_{0.18}$Se$_{0.82}$ QWs can be of the order of 30 meV for interfaces 10 Å thick and a well 30 Å wide.

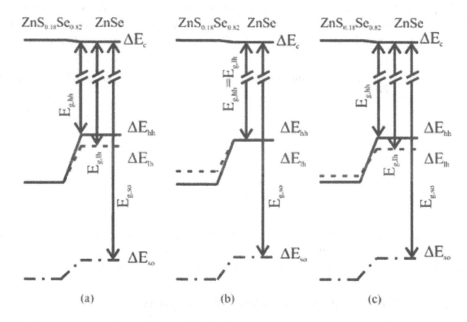

FIGURE 3.46 The band offsets of a ZnSe/ZnS$_{0.18}$Se$_{0.82}$ structure for pseudomorphic growth with: (a) compressed ZnSe and unstrained ZnS$_{0.18}$Se$_{0.82}$, (b) tensile ZnS$_{0.18}$Se$_{0.82}$ and unstrained ZnSe (ZnSe buffer), and (c) for the free-standing case with tensile ZnS$_{0.18}$Se$_{0.82}$ and compressed ZnSe. (From Maia, F. F., Freire, J. A. K., Freire, V. N., Farias, G. A. and da Silva, E. F., *Appl. Surf. Sci.*, 237, 261, 2004. With permission.)

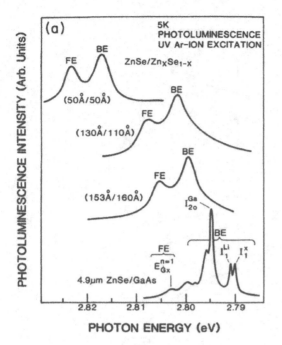
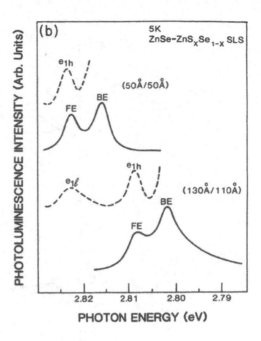

FIGURE 3.47 (a) Low-temperature photoluminescence spectra of three ZnSe–ZnS$_x$Se$_{1-x}$ strained-layer superlattices with well widths of 50, 130 and 153 Å. The spectrum of a 4.9-μm ZnSe/GaAs is given in the lower right corner of part (a). (b) Photoluminescence (solid) and excitation (dashed) spectra at 5K for two of the strained-layer superlattices of ZnSe–ZnS$_x$Se$_{1-x}$ with well widths of 130 and 50 Å. (From Shahzad, K., Olego, D. J. and Van de Walle, C. G., *Phys. Rev. B*, 38, 1417, 1988. With permission.)

Shahzad et al.[35] carried out photoluminescence and excitation experiments to study effects of the strain and carrier confinement in ZnSe–ZnS$_x$Se$_{1-x}$ SLS grown by MBE on GaAs substrates. The emission spectra of the QWs with good interfaces demonstrate the contribution due to the free exciton recombination peak (FE) at higher energy and a bound exciton emission peak (BE), shown in Figure 3.47(a). Note a very clear shift of the exciton peaks to higher energies as the 11 width decreases. The free exciton intensity relative to bound excitons has increased tremendously even at very low temperatures. The energy separation between the two peaks remains constant at about 6 meV as the SL period is changed. Figure 3.47(b) shows photoluminescence (solid) and excitation (dashed) spectra at 5K for two of the ZnSe–ZnS$_x$Se$_{1-x}$ SLSs with well widths of 130 and 50 Å. For the excitation case, the detector was set at 2.801 eV for the former case and at 2.816 eV for the latter case. e_{1h} and e_{1l} indicate the transition from $n=1$ electron to heavy- and light-hole subbands, respectively. For the sample of 130Å/130Å, the ground-state free exciton (e_{1h}) and the first excited state (e_{1l}) can be detected. The energy positions of e_{1h} and e_{1l} agree very well with those calculated based on the band offsets and a strain splitting between heavy- and light-hole bands of 8 meV. It is found that the free exciton peaks on PL are consistent with excitation measurements that show them to be due to ground-state electron to heavy-hole subband-related transitions. Comparing the free exciton energy position in the cases of the SLs with that of the thick epilayer, it can be seen that there is a clear shift of the excitons to higher energies in the case of SLs (see Figure 3.47). This blue shift is caused by a combination of strain effects due to the lattice mismatch discussed earlier and also due to the quantum confinement of the electrons and holes.

The ZnSe–ZnTe SLS is widely studied as possible a tunable light emitter by variation of the constituent material layer thicknesses. The successful growth of these SLs has been achieved by a variety of methods such as MBE,[36] HWE[37] and shows strong photoluminescence properties. ZnSe–ZnTe SLS belongs to type-II SL system, it is necessary to determine the value of VB offset

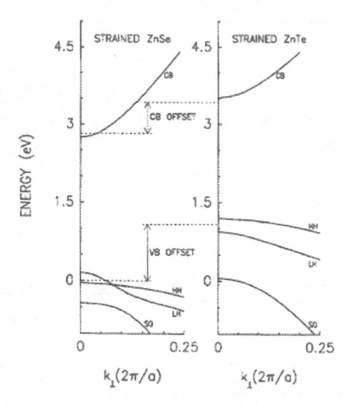

FIGURE 3.48 Band structure of bulk ZnSe and ZnTe, including strain appropriate to a 15×15 Å free-standing superlattice. Zero of energy is taken to be the valence-band edge of ZnSe before strain is introduced. The band structure shown is along the growth direction [1 0 0]. (From Rajakarunanayake, Y., Miles, R. H., Wu, G. Y. and McGill, T. C., *J. Vac. Sci. Technol.*, 6, 1354, 1988. With permission.)

between ZnSe and ZnTe. Rajakarunanayake et al.[38] reported the first study of the band structure of ZnSe–ZnTe SLS. Based on the second-order k.p theory, the effects of the strain and spin-orbit splitting on the SL band structure are calculated. The dependence of the SL bandgap on the valence-band offset is studied. Assuming that the photoluminescence of the SL is due to the isoelectronic Te bound excitons in the ZnSe, the best value of VB offset is 0.975 eV by fitting the experimental photoluminescence data with k.p theory. If the photoluminescence is assumed to be due to band-to-band transitions, the VB offset is 1.196+0.134 eV. Figure 3.48 shows the relative arrangement of the energy bands for ZnSe–ZnTe SLSs along the [1 0 0] growth direction, in which Harrison's valence-band offset $E_V^{ZnTe} - E_V^{ZnSe} = 1.080$ eV is used. Dashed lines indicate the positions of the band edges of the unstrained bulks. Since strain moves the band edges, the effective valence-band offset in the SL is different from the unstrained offset and depends on the particular strain distribution in the structure.

The ZnS–ZnTe SLS makes a type-II SL system consisting of electron wells in the ZnS layers and hole wells in the ZnTe layers. Because the lattice mismatch of ZnS–ZnTe system is about 13%, it is very difficult to fabricate the SL and QW structures of such a high mismatch system. Most of the previous works were focused on the fabrication and confirmation of SL structure.[39,40]

Only a few works were reported on the photoluminescence (PL) properties.[30] Lu et al.[31] studied the origin of the newly observed PL spectrum in the ZnS–ZnTe SLS, in which a multiband structure with equal energy spacing has been observed for the first time, as shown in Figure 3.49. This multiband structure can be explained by electron-hole recombinations from electrons at levels of $n=0, 1, 2, \ldots$ in n-ZnS to the holes at the $n=0$ level in p-ZnTe, in which electrons and holes are confined in nearly parabolic potential wells formed by the modulation of space charges at the p-n

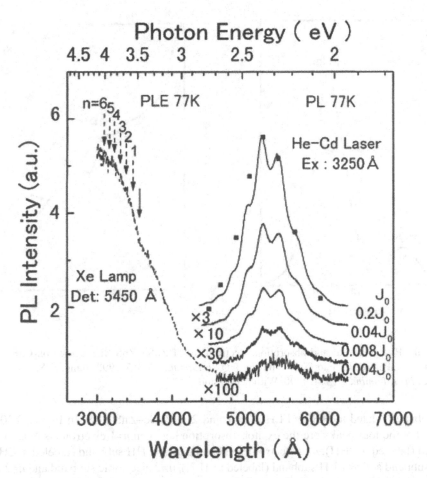

FIGURE 3.49 PLE spectrum (broken line) and emission spectra (solid lines) with different excitation intensities of the ZnS–ZnTe SLS at 77K. The solid line arrow indicates a step structure in the PLE spectrum. The broken line arrows show the expected positions of transitions from different harmonic oscillator levels. Solid squares (■) represent the calculated luminescence intensities based on the harmonic oscillator model. (For details, see Lu, Y. M., Kato, A., Matsumoto-Aoki, T., Sakamoto, Y. and Iida, S., *J. Cryst. Growth*, 214–215, 245, 2000. With permission.)

heterojunctions. By taking into account the Boltzmann distribution of the electrons among the levels $n=0, 1, 2, \ldots$ in wells, the observed spectral shape can be fitted. Thermal quenching of the luminescence in high-temperature region shows the hole release from the level in ZnTe with an activation energy of 0.2 eV. This is believed to be the first paper demonstrating the electron and hole confinements in one-dimensional parabola-like potential wells induced by space charges associated with p-n heterojunction in II–VI compounds.

3.2.3.2 Excitonic Properties of Superlattice and Quantum Well

The exciton binding energy can be improved by using SL and QW structure, thus achieving exciton luminescence at high temperature and even RT. Fu et al.,[41] for the first time observed the luminescence of exciton molecules in the ZnSe–ZnMnSe SLS at 10K, with a maximum exciton binding energy of up to 40 meV. Fan et al.[42] observed free exciton luminescence associated with $n=1, 2$ HH subband in ZnSe–ZnS SLS at 77K, and at RT, there is only the emission of $n=1$ exciton. This is because the $n=2$ HH exciton with higher energy level is thermally dissociated when the temperature increase, as shown in Figure 3.50(a). Subsequently, they first reported RT excitonic absorption from

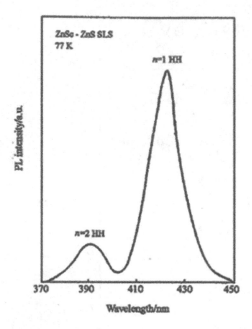
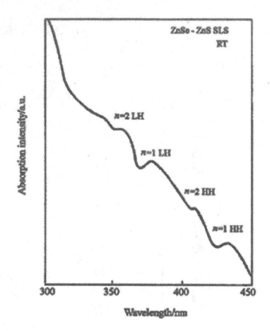

FIGURE 3.50   PL spectra (a) and absorption spectrum (b) of ZnSe–ZnS SLS grown on CaF$_2$ substrate. (From Guan, Z. P., Fan, G. H., Fan, X. W. and Yu, J. Q., *J. Lumin.*, 53, 355, 1992; Jiang, F. Y., Fan, X. W. and Fan, G. H., *Chin. J. Lumin.*, 11, 174, 1990. With permission.)

the two subbands related to HH and LH of $n=1$ and 2 in ZnSe–ZnS SLS. In Figure 3.50(b), it is considered that the four peaks are the exciton absorption between $n=1$ electronic subband and $n=1$ HH subband (labeled n=1HH), $n=2$ electronic subband and $n=2$ HH subband (labeled n=2HH), $n=1$ electronic subband and $n=1$ LH subband (labeled n=1LH), $n=2$ electronic subband and $n=2$ LH subband (labeled n=2LH), respeetively.[43]

Figure 3.51(a) shows the absorption spectra of a series of ZnSe–ZnS SLS with different well widths.[44] The quantized $n_z=1$ HH and LH excitons are the main features, while the excited, $n\geq2$, states become discernible in samples with wider wells. Because of the quantum size effect, the exciton absorption peak shifts to the high energy side with decreasing well width. Figure 3.51(b) also gives the reflection, the luminescence and the photoluminescence excitation spectrum for one sample.[45] It is obvious that the structures are much broader than for the III–V compounds-based MQW in Ref. [46].

Yamada et al.[13] studied that excitonic properties of the ZnSe–ZnS SLSs fabricated on (1 0 0) ZnS substrates by a low-pressure MOCVD (LP-MOCVD) method. Under high excitation conditions by an N$_2$ pulsed laser, the peak observed in PL spectra will be ascribed to the intrinsic recombination of excitons occurring between the $n=1$ elctrons and the $n=1$ heavy-holes in the QW. With increasing excitation density, it was found that a dominant band moves toward higher photon energy. The dissociation mechanisms of excitons were analyzed by measuring the temperature dependence of the linewidth and emission intensity obtained under high excitation. Figure 3.52 shows the temperature dependence of the emission intensity of the excitonic peak obtained under the weak and high excitation conditions. The thermal quenching of the emission intensity, under weak excitation conditions, starts from 50K and shows an activation energy of about 48 meV. However, in the case of the high excitation, the thermal quenching starts from about 100K and the emission intensity is maintained up to 300K. The activation energy of this quenching is estimated to be 71.3 meV that is approximately four times larger than that of the exciton binding energy in ZnSe bulk (about 19 meV).[47]

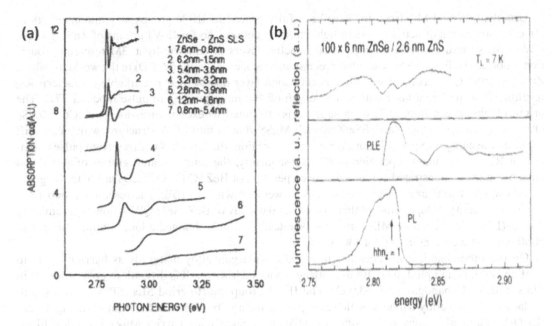

FIGURE 3.51 Absorption spectra (a) of ZnSe–ZnS strained-layer superlattices. Reflection, PL and PL excitation spectra (b) at low temperature and excitation. (From O'Donnell, K. P., Parbrook, P. J., Yang, F., Chen, X., Irvine, D. J., Trager-Cowan, C., Henderson, B., Wright, P. J. and Cockayne, B., *J. Cryst. Growth*, 117, 497, 1992; Sack, W., Oberhauser, D., O'Donnell, K. P., Parbrook, P. J., Wright, P. J., Cockayne, B. and Klingshirn, C., *J. Lumin.*, 53, 409, 1992. With permission.)

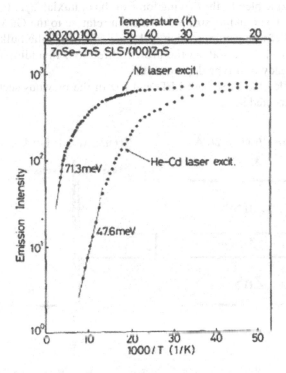

FIGURE 3.52 Temperature dependence of the excitonic emission intensity in a ZnSe–ZnS SLS (L=~20 A and L_b=30 A, 150 periods) on a (1 0 0)-oriented ZnS substrate under high and weak excitation conditions. (From Yamada, Y. and Taguchi, T., *J. Cryst. Growth*, 101, 661, 1990. With permission.)

Usually, QW and SL structures have been fully proved in III–V groups of semiconductors as effective structures of active layers in light-emitting devices, so for II–VI groups of ZnSe groups, QW or SL structures are also widely used as active layers of emitting light. 3M research group[48] first reported the first ZnSe-based blue-green semiconductor laser diode (LD) in the world, in which ZnCdSe/ZnSe QW structure was used as activation layer due to their remarkable characteristics, resulting in to realize a laser with a wavelength of 490 nm operating in pulse mode at 77K. The origin of blue-green light at low temperature is attributed to exciton emission in ZnCdSe/ZnSe QWs. Sony group[49] adopt ZnMgSSe/ZnSSe/ZnMgSSe/GaAs multi-QW structure, which can shift the laser wavelength to pure blue or even ultraviolet region, the laser has realized from pulse operation at RT to continuous operation at RT. Subsequently, the joint research groups of AIST and Hitachi[50] have demonstrated continuous-wave operation of BeZnCdSe QW LDs at RT in the green to yellow spectrum range. The LDs structures were grown by MBE. The research group from Sophia University of Japan studied that carried out the study of BeZnSeTe QW yellow light-emitting diodes (LEDs) prepared by MBE on InP substrates,[51,52] which obtained a long lifetime more than 5000 hours at a current density of 130 A/cm².

On the other hand, it is necessary to prepare wide gap alloy materials as barrier layers to confine carriers and photons in light-emitting devices. However, it is difficult to grow alloys thin films with wide bandgap of E_g>3.0 eV. The II–VI group short-period SLs (SPSLs) are used to replace wide-gap alloy barriers, which are applied to ZnSe base II–VI blue light-emitting diodes (LEDs). Wu et al.[53] grew II–VI SPSL by MBE to replace wide barrier MgZnSSe alloy films, preparing blue light-emitting diodes. The device consists of a ZnSe active layer, SPSL MgZnSSe cladding layers and a ZnSeTe digital graded contact. The SPSL structure is made of alternating layers of MgSe, ZnSe, ZnS and ZnSe, as shown in Figure 3.53. The RT electroluminescence (EL) spectrum shows only a single peak at 460 nm with full width at half maximum of 13 nm. Advantages of the SPSL structure are as follows: First, all the constituent layers of the SPSL block are structurally zinc blende, the driving force of the epitaxial layer toward twinning can be greatly reduced. SPSL layer and its strain distribution relative to the GaAs substrate are shown in the figure, strong alternating compression and tensile strain in the bulk can be used to block the propagation of thread defects along the growth direction. In addition, for ZnSe layers can improve the low efficiency of p-type doping ability.

Since we have detailed the ZnSe base LED and LD in the previous section, see Section 1 for details, it will not be repeated here.

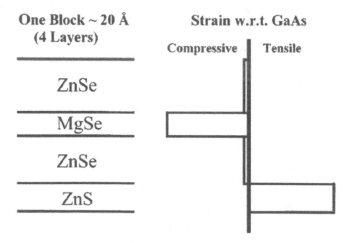

FIGURE 3.53 Schematic of SPSL structure and the layer strains with respect to the GaAs substrate. (From Wu, B. J., Kuo, L. H., DePuydt, J. M., Haugen, G. M., Haase, M. A. and Salamanca-Riba, L., *Appl. Phys. Lett.*, 68, 379, 1996. With permission.)

3.2.3.3 Optical Properties of Asymmetric Double Quantum Wells

Carrier recombination in semiconductor heterostructures such as QWs has been a field of active research for the last few years because this process is not only of basic physical interest but also of fundamental importance for optoelectronic devices. Many kinds of optoelectronic devices, including LD and light-emitting diodes (LEDs), are based on this mechanism. The studies ZnSe on the basis LD show that improving the structure is one of the important steps to increase the quantum efficiency and reduce the threshold current density. Therefore, some have proposed the use of a special QW structure—asymmetric double QW (ADQW)—to try a new structure for the LD's active layers.

ADQW structure consisting of two wells of different widths coupled by a thin barrier, shown in Figure 3.54, is more interesting because it is modulated by carrier tunneling that is one of the main characteristics of this structure.[54,55]

Figure 3.54 shows the band structure of the ADQW that includes two different width QWs, a wide well (WW) and a narrow well (NW), coupled by a thin barrier. As a result of the different widths of the two QWs, the $n=1$ electronic energy levels, E_{1we} corresponding to the WW and E_{1ne} corresponding to the NW, are different. Then electrons excited in the NW can tunnel through the thin barrier to the WW, and so can holes. It has been shown that when the energy difference ΔE_{1e} between E_{1we} and E_{1ne} is larger than an LO phonon energy (E_{LO}), the electron tunneling can be assisted by LO phonons, which is a fast process, and the tunneling time (T_t) may be smaller than the lifetime of the electrons in the NW (T_{ne}).[56–58] Therefore, most of the electrons in the NW tunnel to the WW before recombination with holes in the NW.[59,60] However, because of the small discontinuity of the VB (for the ZnCdSe/ZnSe ADQW) and the large effective mass of the HHs, it is difficult to design a structure with the energy difference (ΔE_{1h}) between E_{1wh} and E_{1nh} larger than E_{LO}, where E_{1wh} and E_{1nh} are $n=1$ HH energy levels in the WW and NW, respectively. As a result, the hole tunneling from the NW to the WW cannot be assisted by LO phonons. According to Krol,[61] however, nonresonant delocalization of hole wave functions combined with alloy scattering can provide an efficient mechanism for fast hole transfer from the NW to the WW at finite in-plane momenta, and the holes tunnel to the WW before reaching the lowest subband in the NW. As both kinds of carriers have a fast tunneling process, the quantity of carriers in the WW of the ADQW should be more than that in a normal QW, and the efficiency of irradiative recombination of carriers in the WW of the ADQW may be higher than that of a normal QW, which is the main topic of our study in this article. Although in II–VI compound semiconductors the carriers exist as excitons due to the large Coulomb interactions between electrons and holes, it does not prevent carriers from tunneling.[62,63] Fan's group studied in detailed the exciton tunneling process, and exciton formation, relaxation and

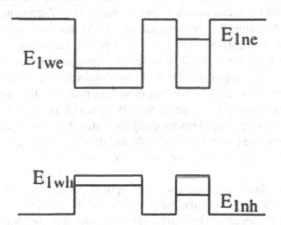

FIGURE 3.54 Band structure of the asymmetric double quantum well. (From Yu, G., Fan, X., Zhang, J., Yang, B., Zhao, X. and Shen, D., *Solid State Commun.*, 110, 127, 1999. With permission.)

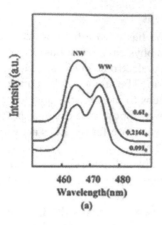 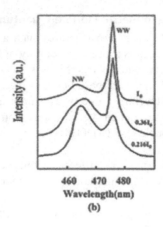 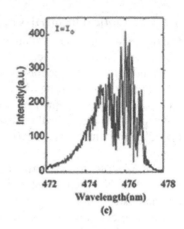

FIGURE 3.55 Emission spectra of the sample of 5 nm/3 nm/3 nm structure at 77K (I_0=1 MW/cm²) (a) surface emission, (b) edge emission, (c) oscillating modes of the lasing shown in (b). (From Yu, G., Fan, X., Zhang, J., Yang, B., Zhao, X. and Shen, D., *Solid State Commun.*, 110, 127, 1999. With permission.)

recombination process modulated by the tunneling in ZnCdSe/ZnSe ADQW.[64,65] The laser emission due to the tunneling was observed in the ZnCdSe/ZnSe ADQW structure, and it was found that the tunneling can increase the carrier density of the QWs and decrease the threshold of the stimulated emission.[65]

Figure 3.55(a) shows the PL spectra from top surface emission of the ZnCdSe/ZnSe ADQW (5 nm/3 nm/3 nm) structure and Figure 3.55(b) shows the emission from the cleaved edge of this sample. It can be seen that the emission intensities from different wells in Figure 3.55(a) and (b) change differently with the excitation intensity (I_{ex}). Under the condition of surface emission, the emission intensity from the NW (I_n) changes faster than that from the WW (I_w) with the excitation, and the emission from the NW dominates the spectrum at a high excitation intensity. In contrast, under the condition of edge emission, the lasing arises from the WW and dominates the spectra with increasing I_{ex}. The lasing feature is proven by oscillating modes of this emission shown in Figure 3.55(c).

Under the condition of edge emission, because of the formation of a Fabry-Perot (FP) cavity, the lasing arises from the WW at a certain I_{ex}, which changes the lifetime of carriers in the WW. The relationship between the lifetimes of the stimulated emission and the spontaneous emission is expressed as: T_{st}, T_{sp}=nL, where T_{st}, T_{sp} are the lifetimes of the stimulated and spontaneous emission, respectively, and nL is the photon quantity in the L mode. From the equation, it is easy to find T_{st}, T_{sp}. The nL increases with I_{ex}, which causes T_{st} much smaller than T_{sp}; and at a certain I_{ex}, T_{st} is similar to the carriers tunneling time T_t, even smaller than T_t. At that time, there are not many carriers accumulated in the WW, and carrier tunneling cannot be blocked strongly. Therefore, most of the carriers excited in the NW tunnel to the WW before recombination in the NW, and they recombine in the WW. Then the lasing arises from the WW and dominates the spectra at high excitation, as shown in Figure 3.55(b). As a result of the fast tunneling process, the carriers excited in the two wells accumulate and recombine in the same well WW, which is the reason, why they think the quantum efficiency can be improved. Further study indicates that the carrier tunneling through the thin barrier is conducive to the lasing from the WW, and the threshold can be lowered by optimizing the structure, which is beneficial to improve the laser performance.

Su et al.[66,67] fabricated ZnO/Zn$_{0.85}$Mg$_{0.15}$O ADQW and multiple QW (MQW) by plasma-assisted molecular epitaxy on *c*-plane sapphire, and their optical properties and optical pumped lasing characteristics studied. The exciton tunneling properties of the ADQWs were studied by means of temperature-dependent PL spectra. The carrier tunneling through the thin barrier is conducive to stimulated emission in the WWs of the ADQWs. The origin of the stimulated emission is exciton-exciton scattering in the WWs of ADQWs.[74] Due to the good crystalline quality, the lasing threshold

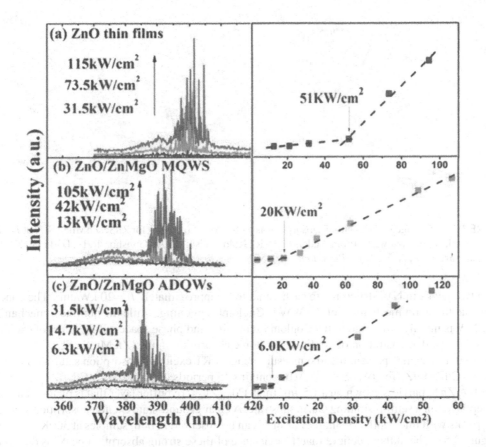

FIGURE 3.56 Room temperature lasing spectra and the integrated intensities of the emission peaks from (a) ZnO thin film, (b) ZnO/MgZnO MQW, and (c) ADQW samples. (From Su, S. C., Zhu, H., Zhang, L. X., He, M., Zhao, L. Z., Yu, S. F., Wang, J. N. and Ling, F. C. C., *Appl. Phys. Lett.*, 103, 131104, 2013. With permission.)

of the MQW is 20 kW/cm². The widths of the NW and the WW of the ADQW were chosen to fascinate rapid LO phonon-assisted carrier tunneling from NW to WW, so as to enhance the exciton density at the WW. Very low lasing threshold of 6 kW/cm² has been achieved (see Figure 3.56).[67]

3.2.4 Nonlinear Optical Properties

3.2.4.1 Optical Nonlinearity of II–VI Group Superlattices

The definition of optical nonlinearity suggests that when the excitation light acts on the material, the variation of the transmitted or reflected light intensity with the incident light intensity is not linear, but nonlinear. The absorption coefficient and refractive index of the material are no longer constant but have an additional change. For semiconductor optical nonlinearity, it can be divided into intrinsic nonlinearity, optoelectronic hybrid nonlinear and carrier transport nonlinearity and so on. Because II–VI group semiconductor SLs have strong exciton effects, the intrinsic nonlinearity is usually related to excitons. Generally speaking, there are five mechanisms that cause the change of absorption coefficient or refractive index: Band filling, bandgap normalization, Coulomb screening of excitons, phase space filling of excitons and collision broadening of exciton band.

In 1986, Andersen et al.[68] investigated the exciton saturation absorption nonlinearity of the ZnSe–ZnMnSe SLs with different well widths at 77K using different pulsed laser intensities, as shown in Figure 3.57. The spectra of two samples clearly show the saturation of the excitonic resonance. The saturation intensity in WW structure corresponds to I_{sat}=1.3 kW/cm², while the saturation intensities

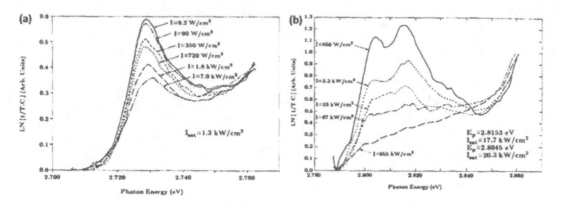

FIGURE 3.57 Change in the excitonic absorption as a function of intensity for ZnSe/ZnMnSe SLS at T=77K. (a) Wide well; (b) narrow well. (From Andersen, D. R., Kolodziejski, L. A., Gunshor, R. L., Datta, S., Kaplan, A. E. and Nurmikko, A. V., *Appl. Phys. Lett.*, 48, 1559, 1986. With permission.)

for the two peaks in NW structure are each found to be approximately I_{sat}=20 kW/cm^2. The conclusion is that the main mechanism of the WW is Coulomb screening, while the nonlinear mechanism of the NW is mainly attributed to the Coulomb screening and phase space-filling effect of excitons. This is the first demonstration of nonlinear excitonic absorption in the (Zn,Mn)Se SLs.

In 1990, Lee et al.[69] present the first measurements of RT excitonic absorption saturation in a II–VI group CdZnTe/ZnTe QWs. Sample A(B) contains 15 periods of 50 (1 0 0) Å Cd$_{0.25}$Zn$_{0.75}$Te wells and 100 Å ZnTe barriers grown on a 1.5-μm-thick Cd$_{0.14}$Zn$_{0.86}$Te buffer layer that is on top of a 1-μm ZnTe epilayer on a semi-insulating GaAs substrate. A clear ground-state exciton absorption showing sharp peaks with line widths as small as 22 meV can be observed in both samples at 300K, as shown in Figure 3.58. The authors believe that the existence of these strong absorption peaks is due to the reduced lattice mismatch for Cd$_{0.25}$Zn$_{0.75}$Te/ZnTe QWs grown on a Cd$_{0.14}$Zn$_{0.86}$Te buffer layer. The peak α for sample A, when normalized to the well thickness alone, is α_w=6×10^4 cm^{-1}. This value is three times larger than the bulk Cd$_{0.14}$Zn$_{0.86}$Te sample in which no exciton peak was observed. It is also noted that since absorption in bulk CdZnTe is higher than that of bulk GaAs, the absorption coefficient per well is larger than that typically found in GaAs/AlGaAs MQWs.

Figure 3.59 shows the absorption coefficient α as a function of effective average intensity I. The effective average intensity is obtained from the incident optical intensity after correcting for reflection at the air-semiconductor interfaces, absorption inside the sample and the Gaussian beam profile. The measured points are shown as open circles. The solid lines show fitted curves to the expression:

$$\alpha(I) = \alpha_0 + \alpha_1 / \left(1 + I/I_{SAT}\right) \tag{3.34}$$

where the fitting parameters α_0 and α_1 represent the nonsaturable and saturable components of the absorption and I_{SAT} is the saturation intensity. From Figure 3.59, strong RT excitonic absorption in CdZnTe/ZnTe QWs is found to saturate at an incident optical intensity that is considerably higher than that for III–V QWs. This is an important advantage for high-power operation of electroabsorptive modulators as well as optical bistable devices such as SEEDs. The authors successfully interpret this high saturation intensity in terms of the smaller II–VI QW exciton Bohr radius. These observations are promising indicators for future device applications of wide-gap II–VI QWs.

3.2.4.2 Optical Bistability of II–VI Group Superlattices

Optical bistability is a kind of all-optical switching effect. It is hoped to be used as the basic element of optical computer-optical switching device. The basic feature is that when an incident light passes through a nonlinear medium, when the incident light intensity changes from strong to weak and

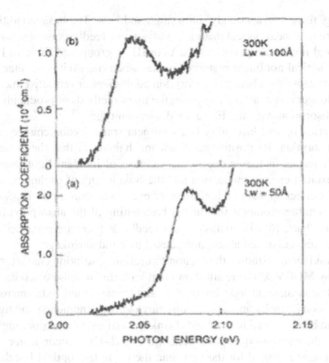

FIGURE 3.58   Room-temperature absorption spectra of $Cd_{0.25}Zn_{0.75}Te/ZnTe$ MQWs showing well-resolved $n=1$ HH exciton features: (a) sample A (50 Å well); (b) sample B (100 Å well). Absorption coefficients refer to the total thickness of CdZnTe wells plus ZnTe barriers, 2250 Å for sample A and 3000 Å for sample B. (From Lee, D., Zucker, J. E., Johnson, A. M., Feldman, R. D. and Austin, R. F., *Appl. Phys. Lett.*, 57, 1132, 1990. With permission.)

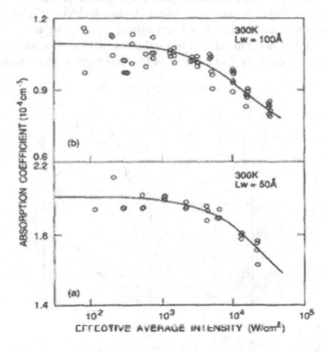

FIGURE 3.59   Room-temperature absorption coefficient as a function of optical intensity for samples A and B. Open circles are experimental data. Solid lines are fits with excitonic saturation intensity ($I_{SAT}=31.6$ and 15.0 kW/cm² for samples A and B, respectively). (From Lee, D., Zucker, J. E., Johnson, A. M., Feldman, R. D. and Austin, R. F., *Appl. Phys. Lett.*, 57, 1132, 1990. With permission.)

from weak to strong, there are two steady states (high and low). The basic conditions for producing optical bistability are nonlinear optical materials and positive feedback mechanism. There are many materials with optical nonlinearity, such as GaAs in III–V group and ZnSe in II–VI group etc. A common feature of optical nonlinear materials is that when the excitation intensity increases to a certain extent, it will cause the change of absorption coefficient or refractive index of the material, so that they are no longer constant. Among them, the main methods of producing positive feedback include enhanced absorption type and FP cavity dispersion type.

Enhanced absorption optical bistability is a nonlinear transmission change caused by the non-linear change of absorption. Its positive feedback mechanism is that the number of excitons in nonlinear optical materials increases due to the increase of incident light intensity, which leads to the increase of exciton-exciton interaction and the broadening of exciton absorption band. The enhanced absorption effect appears on both sides of exciton resonance absorption, and the further increase of exciton number promotes the further broadening of the absorption band. So repeated, forming a positive feedback. Similar to this positive feedback, there are enhanced absorption caused by the thermal effect and enhanced absorption caused by band shrinkage.

In 1988, Gribkovskii et al.[70] studied the exciton saturation absorption characteristics of ZnSe epitaxial layer grown by MOCVD. The result shows that with the increase of excitation light intensity, the relationship between incident light intensity I_0 and transmission light intensity I_t is no longer linear. The transmission intensity appears obvious narrowing phenomenon and light bistability phenomenon, which can be attributed to the band shrinkage and exciton broadening effect. This is the first time to observe the exciton-type optical bistability on II–VI semiconductors.

In 1989, Fan's group[71] reported for the first time the cavityless optical bistability in the ZnSe–ZnS strain SL grown on CaF_2 substrate by MOCVD at 77K. The results are shown in Figure 3.60. Figure 3.60(a) shows the normalized temporal shapes of the incident I_0 and transmitted I_t pulses with a wavelength of 532 nm at RT in the condition of high incident intensities, and Figure 3.60(a′) gives the resulting hysteresis loops $I_t=f(I_i)$. Compared with the incident I_i pulse, the transmitted I_t pulse is obviously compressed. While under low incident light intensities, it was found that the temporal shapes of the incident I_i and transmitted I_t pulses are quite similar, as shown in Figure 3.60(b) and (b′). The abovementioned fact indicates that the dependence of the transmitted intensity I_t on the incident intensity I_i in the case of low and high incident intensities is linear and

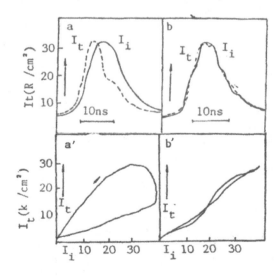

FIGURE 3.60 Time dependence of the temporal shape of the incident (solid curve) I_i and transmitted (dashed curve) I_t pulses (a) and (b) in ZnSe–ZnS SLSs on CaF_2 at 77K. The resulting hysteresis loops (a′) and (b′). (a and a′: under high excitation densities, b and b′: under low excitation densities). (From Shen, D. Z., Fan, X. W., Fan, G. H. and Chen, L. C., *Chin. Acta Optica Sinica*, 10, 643, 1990. With permission.)

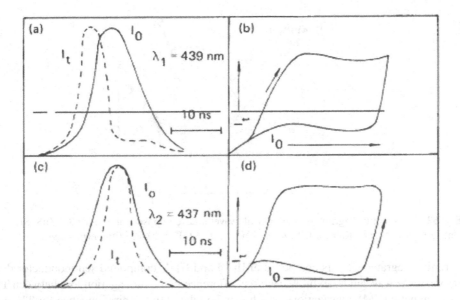

FIGURE 3.61 Time dependence of averaged normalized incident (solid) I_o and transmitted (broken) I_t pulses (a) and (c) in the ZnSe–ZnS/GaAs MQWs at 77K and the resulting hysteresis loops (b) and (d). (From Shen, D. Z., Fan, X. W., Fan, G. H. and Zhang, J. Y., *J. Lumin.*, 48–49, 299, 1991. With permission.)

nonlinear, respectively. According to the change of the loop in the diagram, it is obviously that this is an enhanced absorption optical bistability. Compared with the theoretical calculation based on the band broadening effect, the experimental results are in good agreement with it. Hence, the authors considered that optical bistability obtained on the ZnSe–ZnS/CaF$_2$ is caused by exciton-enhanced absorption. Subsequently, they reported RT picosecond optical bistability in ZnSe–ZnTe MQWs grown by MOCVD on transparent CaF$_2$ (1 1 1) substrates.[72] On the basis of the direction of the hysteresis loop, the absorption and the energy band structure in ZnSe–ZnTe MQWs, the non-linear mechanism for the optical bistability of ZnSe–ZnTe MQWs is due to the effect of increasing absorption, which is ascribed to the bandgap shrinkage of ZnSe–ZnTe MQWs due to the high density of electrons and holes in the MQWs. The ps switching time can be explained by the electrons excited to the CB of ZnTe layer in ZnSe–ZnTe MQWs relaxing quickly to the CB of ZnSe layer in this material system.

In 1992, Shen et al.[73] observed for the first time optical bistability in different loop directions on a ZnSe–ZnS SL grown by MOCVD method on a GaAs substrate with corrosion holes, as shown in Figure 3.61. On the basis of the obtained exciton absorption spectra and the saturation absorption characteristics of excitons, the FP cavity theory and K-K relationship are used to explain the causes of different light bistable return lines. The optical bistability in ZnSe–ZnS/GaAs MQWs is attributed to the dispersive nonlinearity due to the effect of excitonic saturating absorption.

3.2.4.3 Optical Waveguide of ZnSe–ZnS Superlattice

Optical waveguides use the total reflection of light at the interface of different media. The maximum characteristic of waveguide is to concentrate the spatially dispersed beam in a very small region, so it can produce high-power density in the waveguide. The nonlinear guided wave optical bistable device based on semiconductor SL is an effective way to reduce the laser output power and realize the practicality of the device. Yokogawa et al.[74] reported that the ZnSe–ZnS SLS were epitaxially grown on GaAs substrates by LP-MOCVD method, the waveguide results are shown in Figure 3.62. 1.4-μm thick ZnS buffer layer, ZnSe (5–10 nm)–ZnS (5–10 nm) SL waveguide layer and 0.2 μm thick SiO$_2$ upper buffer layer. The transmission loss of the 3D waveguide for the above SL at 0.633 μm in the transverse electric (TE) mode is 0.71 cm^{-1}. This waveguide may be suited for

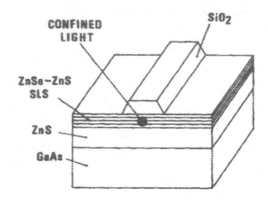

FIGURE 3.62 Schematic diagram of the optical waveguide composed of the ZnSe–ZnS SLS. (From Yokogawa, T., Ogura, M. and Kajiwara, T., *Appl. Phys. Lett.*, 52, 120, 1988. With permission.)

optoelectronic integrated circuits composed of II–VI and III–V compound semiconductor devices. Furthermore, these waveguides exhibited a large difference in the propagation loss between TE and transverse magnetic (TM) polarizations which may be related to the birefringence for TE and TM polarizations due to the slight anisotropy of the refractive index in the ZnSe–ZnS SLS structure. This birefringence effect will be very useful for a polarizing optical device, and it lays a foundation for the further development of nonlinear waveguide optical bistable devices.

REFERENCES

1. Esaki, L. and Tsu, R., *IBM Res. Note*, RC-2418, 1969.
2. Esaki, L. and Tsu, R., *IBM J. Res. Dev.*, 14, 61, 1970.
3. Osbourn, G. C., *J. Appl. Phys.*, 53, 1586, 1982.
4. Fujiyasu, H., Takahashi, H., Shimizu, H., Sasaki, A. and Kuwabara, H., *Proceedings of the 17th International Conference on the Physics of Semiconductors, San Francisco, California*, pp. 539–542, 1985.
5. Esaki, L., Chang, L. L. and Tsu, R., *Proceedings of the 12th International Conference on Low Temperature Physics, Kyoto, Japan*, pp. 262, 1970.
6. Blakeslee, A. E. and Aliotta, C. F., *IBM J. Res. Dev.*, 14, 686, 1970.
7. Chang, L. L., Esaki, L., Howard, W. E. and Rideout, V. L., *Proceedings of the 11th International Conference on the Physics of Semiconductors, Warsaw, Poland*, pp. 52–53, 1972.
8. Gossard, A., *IEEE J. Quantum Electron*, 22, 1649, 1986.
9. Döhler, G. H., *Surf. Sci.*, 73, 97, 1978; J. Vac. Sci. Technol., 16, 851, 1979.
10. Döhler, G. H., Künzel, H., Olego, D., Ploog, K., Ruden, P., Stolz, H. J. and Abstreiter, G., *Phys. Rev. Lett.*, 47, 864, 1981.
11. Nurmikko, A., Gunshor, R. and Kolodziejski, L., *IEEE J. Quantum Electron*, 22, 1785, 1986.
12. Kane, E. O., *J. Phys. Chem. Solids*, 1, 82, 1956 and 1, 249, 1957.
13. Yamada, Y. and Taguchi, T., *J. Cryst. Growth*, 101, 661, 1990.
14. Yang, H., Ishida, A., Fujiyasu, H. and Kuwabara, H., *J. Appl. Phys.*, 65, 2838, 1989.
15. Huang, H. L. and Guo, L. W., *Semicondutor Superlattice*, 277, Shenyang: University of Liaoning, 1992.
16. Miller, R. C., Gossard, A. C., Sanders, G. D., Chang, Y.-C. and Schulman, J. N., *Phys. Rev. B*, 32, 8452, 1985.
17. Anderson, R. L., *Solid State Electron.*, 5, 341, 1962.
18. Harrison, W. A., *J. Vac. Sci. Technol.*, 14, 1016, 1977.
19. Tersoff, J., *Phys. Rev. B*, 30, 4874, 1984.
20. Katnani, A. D. and Margaritondo, G., *J. Appl. Phys.*, 54, 2522, 1983.
21. Ray, B., *II–VI Compounds*, Amsterdam: Pergamon, pp. 54, 1969.
22. Matthews, J. W. and Blakeslee, A. E., *J. Vac. Sci. Technol.*, 14, 989, 1977.
23. Berlincourt, D., Jaffe, H. and Shiozawa, L. R., *Phys. Rev.*, 129, 1009, 1963.
24. Zakharov, O., Rubio, A., Blase, X., Cohen, M. L. and Louie, S. G., *Phys. Rev. B*, 50, 10780, 1994.
25. Asai, H. and Oe, K., *J. Appl. Phys.*, 54, 2052, 1983.

26. Van de Walle, C. G., Shahzad, K. and Olego, D. J., *J. Vac. Sci. Technol.*, 6, 1350, 1988.
27. van der Ziel, J. P., Dingle, R., Miller, R. C., Wiegmann, W. and Nordland, W. A., *Appl. Phys. Lett.*, 26, 463, 1975.
28. Iwamura, H., Saku, T., Ishibashi, T., Otsuka, K. and Horikoshi, Y., *Electron. Lett.*, 19, 181, 1983.
29. Kronig, R. and Penney, W. G., *Proc. Roy. Soc. A*, 130, 499, 1931.
30. Teraguchi, N., Takemura, Y., Kimura, R., Konagai, M. and Takahashi, K., *J. Cryst. Growth*, 93, 720, 1988.
31. Lu, Y. M., Kato, A., Matsumoto-Aoki, T., Sakamoto, Y. and Iida, S., *J. Cryst. Growth*, 214–215, 245, 2000.
32. Yokoyama, M. and Chen, N.-T., *J. Cryst. Growth*, 223, 369, 2001.
33. Taguchi, T., Kawakami, Y. and Yamada, Y., *Physica B*, 191, 23, 1993.
34. Maia, F. F., Freire, J. A. K., Freire, V. N., Farias, G. A. and da Silva, E. F., *Appl. Surf. Sci.*, 237, 261, 2004.
35. Shahzad, K., Olego, D. J. and Van de Walle, C. G., *Phys. Rev. B*, 38, 1417, 1988.
36. Kobayashi, M., Mino, N., Katagiri, H., Kimura, R., Konagai, M. and Takahashi, K., *J. Appl. Phys.*, 60, 773, 1986.
37. Kuwabara, H., Fujiyasu, H., Aoki, M. and Yamada, S., *Jpn. J. Appl. Phys.*, 25, L707, 1986.
38. Rajakarunanayake, Y., Miles, R. H., Wu, G. Y. and McGill, T. C., *J. Vac. Sci. Technol.*, 6, 1354, 1988.
39. Fujiyasu, H., Mochizuki, K., Yamazaki, Y., Aoki, M., Sasaki, A., Kuwabara, H., Nakanishi, Y. and Shimaoka, G., *Surf. Sci.*, 174, 543, 1986.
40. Kobayashi, M., Mino, N., Konagai, M. and Takahashi, K., *Surf. Sci.*, 174, 550, 1986.
41. Fu, Q., Lee, D., Mysyrowicz, A., Nurmikko, A. V., Gunshor, R. L. and Kolodziejski, L. A., *Phys. Rev. B*, 37, 8791, 1988.
42. Guan, Z. P., Fan, G. H., Fan, X. W. and Yu, J. Q., *J. Lumin.*, 53, 355, 1992.
43. Jiang, F. Y., Fan, X. W. and Fan, G. H., *Chin. J. Lumin.*, 11, 174, 1990.
44. O'Donnell, K. P., Parbrook, P. J., Yang, F., Chen, X., Irvine, D. J., Trager-Cowan, C., Henderson, B., Wright, P. J. and Cockayne, B., *J. Cryst. Growth*, 117, 497, 1992.
45. Sack, W., Oberhauser, D., O'Donnell, K. P., Parbrook, P. J., Wright, P. J., Cockayne, B. and Klingshirn, C., *J. Lumin.*, 53, 409, 1992.
46. Schmitt-Rink, S., Chemla, D. S. and Miller, D. A. B., *Adv. Phys.*, 38, 89, 1989.
47. Euceda, A., Bylander, D. M., Kleinman, L. and Mednick, K., *Phys. Rev. B*, 27, 6517, 1983.
48. Haase, M. A., Qiu, J., DePuydt, J. M. and Cheng, H., *Appl. Phys. Lett.*, 59, 1272, 1991.
49. Okuyama, H., Nakano, K., Miyajima, T. and Akimoto, K., *J. Cryst. Growth*, 117, 139, 1992.
50. Kasai, J.-i., Akimoto, R., Hasama, T., Ishikawa, H., Fujisaki, S., Tanaka, S. and Tsuji, S., *Appl. Phys. Express*, 4, 082102, 2011.
51. Che, S.-B., Nomura, I., Kikuchi, A. and Kishino, K., *Appl. Phys. Lett.*, 81, 972, 2002.
52. Nomura, I., Nakai, Y., Hayami, K., Saitoh, T. and Kishino, K., *Phys. Status Solidi B*, 243, 924, 2006.
53. Wu, B. J., Kuo, L. H., DePuydt, J. M., Haugen, G. M., Haase, M. A. and Salamanca-Riba, L., *Appl. Phys. Lett.*, 68, 379, 1996.
54. Chang, L. L., Esaki, L. and Tsu, R., *Appl. Phys. Lett.*, 24, 593, 1974.
55. Esaki, L., *IEEE J. Quantum Electron*, 22, 1611, 1986.
56. Ten, S., Henneberger, F., Rabe, M. and Peyghambarian, N., *Phys. Rev. B*, 53, 12637, 1996.
57. Kim, J. U. and Lee, H. H., *J. Appl. Phys.*, 84, 907, 1998.
58. Tackeuchi, A., Muto, S., Inata, T. and Fujii, T., *Appl. Phys. Lett.*, 58, 1670, 1991.
59. Feng, J. M., Park, J. H., Ozaki, S., Kubo, H., Mori, N. and Hamaguchi, C., *Semicond. Sci. Technol.*, 12, 1116, 1997.
60. Feng, J. M., Ozaki, S., Park, J. H., Kubo, H., Mori, N. and Hamaguchi, C., *Phys. Status Solidi B*, 204, 412, 1997.
61. Krol, M. F., Ten, S., McGinnis, B. P., Hayduk, M. J., Khitrova, G. and Peyghambarian, N., *Phys. Rev. B*, 52, R14344, 1995.
62. Haacke, S., Pelekanos, N. T., Mariette, H., Zigone, M., Heberle, A. P. and Rühle, W. W., *Phys. Rev. B*, 47, 16643, 1993.
63. Hicke, K., Helmbrodt, W., Pier, T., Gumlich, H. E., Rühle, W. W., Nicholls, J. E. and Lunn, B., *J. Cryst. Growth*, 159, 1014, 1996.
64. Yu, G., Fan, X. W., Zhang, J. Y., Zheng, Z. H., Yang, B. J., Zhao, X., Shen, D. and Kong, X., *J. Phys. D Appl. Phys.*, 32, 1506, 1999.
65. Yu, G., Fan, X., Zhang, J., Yang, B., Zhao, X. and Shen, D., *Solid State Commun.*, 110, 127, 1999.
66. Su, S. C., Lu, Y. M., Xing, G. Z. and Wu, T., *Superlattice Microst.*, 48, 485, 2010.

67. Su, S. C., Zhu, H., Zhang, L. X., He, M., Zhao, L. Z., Yu, S. F., Wang, J. N. and Ling, F. C. C., *Appl. Phys. Lett.*, 103, 131104, 2013.
68. Andersen, D. R., Kolodziejski, L. A., Gunshor, R. L., Datta, S., Kaplan, A. E. and Nurmikko, A. V., *Appl. Phys. Lett.*, 48, 1559, 1986.
69. Lee, D., Zucker, J. E., Johnson, A. M., Feldman, R. D. and Austin, R. F., *Appl. Phys. Lett.*, 57, 1132, 1990.
70. Gribkovskii, V. P., Zimin, L. G., Gaponenko, S. V., Malinovskii, I. E., Kuznetsov, P. I. and Yakushcheva, G. G., *Phys. Status Solidi B*, 150, 761, 1988.
71. Shen, D. Z., Fan, X. W., Fan, G. H. and Chen, L. C., *Chin. Acta Optica Sinica*, 10, 643, 1990.
72. Shen, D. Z., Fan, X. W., Piao, Z. S. and Fan, G. H., *J. Cryst. Growth*, 117, 519, 1992.
73. Shen, D. Z., Fan, X. W., Fan, G. H. and Zhang, J. Y., *J. Lumin.*, 48–49, 299, 1991.
74. Yokogawa, T., Ogura, M. and Kajiwara, T., *Appl. Phys. Lett.*, 52, 120, 1988.

3.3 ZnO AND RELATED UV-LUMINESCENCE MATERIALS

Weizhen Liu, Cen Zhang, Youming Lu, Haiyang Xu and Yichun Liu

3.3.1 INTRODUCTION

ZnO and its related materials are II–VI oxide semiconductors with the wurtzite crystal structure and an energy band structure that allows direct interband transitions. Table 3.7 displays a summary of basic physical parameters and properties for ZnO.[1–6] There are some prominent characteristics about ZnO material, such as wide bandgap (3.37 eV), high exciton binding energy (60 meV) and large band-edge absorption coefficient. All these lead to the broad application prospects of ZnO-based materials in many fields, including luminescence/laser, detection, light energy conversion and light-emitting devices, especially in the ultraviolet (UV)/near-UV regions. In principle, a large exciton binding energy should favor efficient excitonic emission process at room temperature (RT). Early research about excitonic stimulated emission from ZnO single crystals at cryogenic temperatures had been observed and found to persist to 280K, which was attributed to the phonon replicas of A-exciton.[7,8] In 1997, an excitonic absorption at RT, as well as near-UV spontaneous and stimulated emission induced by exciton-exciton collision process from ZnO thin film, was observed for the first time (Figure 3.63).[9] Also, the RT free-exciton and bound-exciton emission band had been clearly

TABLE 3.7
Physical Parameters and Properties of ZnO

Property	Value	Reference
Melting point	2248°C	[2]
Stable phase (structure) at 300K	Wurtzite	[2]
Bond length	1.977 nm	[6]
Density (wurtzite)	5.606 g/cm³	[6]
Lattice constant (wurtzite) at 300K		
$a_0 = b_0$	0.32495 nm	[5]
c_0	0.52069 nm	[5]
c_0/a_0	1.602 (ideal hexagonal structure shows 1.633)	[5]
Heat of crystallization (ΔH_{LS})	62 kJ/mol	[6]
Heat capacity (C_P)	9.6 cal/mol K	[6]
Ionicity	62%	[6]
Thermal conductivity	0.6 W/(cm K)	[6]
Linear expansion coefficient	7.2×10^{-6} K^{-1}	[2,6]
Static dielectric constant ($\varepsilon_0/\varepsilon_\infty$)	8.65/4.0	[2,6]
Refractive index (wurtzite)	2.029	[2,6]
Energy gap at 300K (E_g)	3.4 eV, direct	[4]

TABLE 3.7 (*Continued*)
Physical Parameters and Properties of ZnO

Property	Value	Reference
Intrinsic carrier concentration	$<10^6$ cm^{-3} (max n-type doping $>10^{20}$ cm^{-3} electrons; max p-type doping $<10^{17}$ cm^{-3} holes)	[3,4]
Exciton binding energy	60 meV	[4]
Electron effective mass (m^*/m_0)	−0.27	[4]
Electron hall mobility at 300K	200 cm^2/(V s)	[3,4]
Knoop hardness	0.5 N/cm^2	[6]

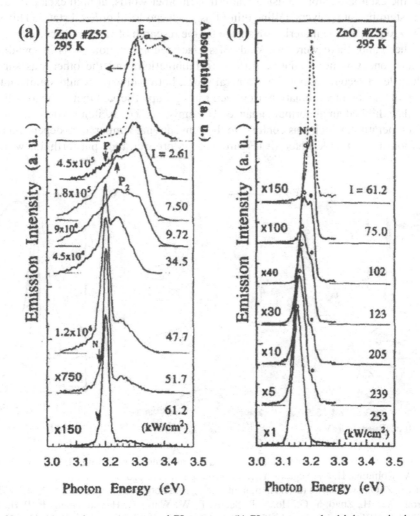

FIGURE 3.63 (a) Absorption spectrum and PL spectra. (b) PL spectra under higher excitation intensities. (From Zu, P., Tang, Z. K., Wong, G. K. L., Kawasaki, M., Ohtomo, A., Koinuma, H. and Segawa, Y., *Solid State Commun.*, 103, 459, 1997. With permission.)

observed by the same research group. Some basic phenomena and processes in luminescence physics were realized in bulk single crystals and thin films of ZnO, such as free-excitonic and bound-excitonic transition processes,[6] an electron-hole plasma emission (P and N bands),[9] DAP transitions, LO-phonon replicas,[10] Auger recombination and other defect-related transitions, including free to bound (electron-acceptor), bound to bound (donor-acceptor), and the so-called yellow/green luminescence.[6] These characteristics greatly promoted the research of ZnO luminescence dynamics.

Subsequently, RT UV laser action due to the radiative recombination process of exciton-exciton collision from the ordered and nano-sized self-assembled ZnO microcrystallite thin films was reported (Figure 3.64(a)).[11] Soon afterward, Cao's team had observed random lasing with coherent feedback in semiconductor powder from ZnO for the first time (Figure 3.64(b)).[12] This phenomenon also provided the direct evidence for the appearance of recurrent scatting of light, which was the key ingredient in Anderson model of photon localization. The success in obtaining random laser from semiconductor powders indicates that localized states with certain order may exist in a disordered system under proper circumstance such as high excitation density. At the same time, Liu's group also observed the optically pumped lasing in the high-quality microcrystallite ZnO films which were prepared by thermal oxidation of ZnS (Figure 3.64(c)).[13] The abovementioned mechanism was attributed to the exciton-exciton collision scattering. In other words, at high excitation density, the disordered system in ZnO polycrystalline thin films produced local ordered states. This was likely to enable ZnO to play a huge material advantage for the research of UV laser devices.

On the other hand, there were two kinds of surface recombination paths in low-dimensional ZnO materials, one was non-radiative surface recombination, and the other was surface trap-mediated deep-level recombination. In a general sense, both pathways would significantly reduce the efficiency of exciton recombination. However, Liu's group observed that some low-dimensional ZnO materials exhibited an abnormal enhanced UV emission, i.e. exciton recombination, with the increase of temperature, which was contrary to the conventional temperature-dependent quenching of UV emission in ZnO bulk materials (Figure 3.65). Moreover, they put forth a new mechanism

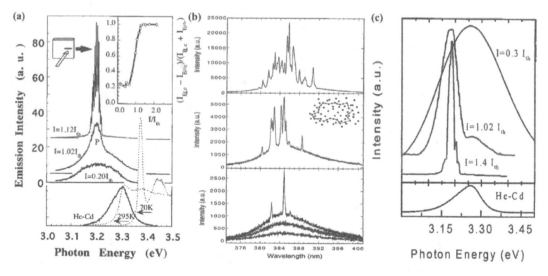

FIGURE 3.64   (a) Absorption spectrum and PL spectra. (From Tang, Z. K., Wong, G. K. L., Yu, P., Kawasaki, M., Ohtomo, A., Koinuma, H. and Segawa, Y., *Appl. Phys. Lett.*, 72, 3270, 1998. With permission.) (b) Spectra of emission from ZnO powder with the excitation intensity (from bottom to top) of 400, 562, 763, 875 and 1387 kW/cm². (From Cao, H., Zhao, Y. G., Ho, S. T., Seelig, E. W., Wang, Q. H. and Chang, R. P. H., *Phys. Rev. Lett.*, 82, 2278, 1999. With permission.) (c) PL spectra of ZnO thin films at below, near and above the threshold of 150 kW/cm². (From Zhang, X. T., Liu, Y. C., Zhang, L. G., Zhang, J. Y., Lu, Y. M., Shen, D. Z., Xu, W., Zhong, G. Z., Fan, X. W. and Kong, X. G., *J. Appl. Phys.*, 92, 3293, 2002. With permission.)

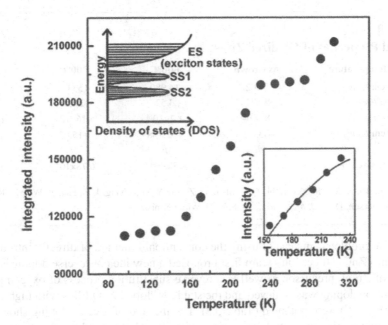

FIGURE 3.65 Temperature dependence of UV PL integrated intensity ranging from 96 to 306K for ZnO nanotubes. Upper left illustration shows schematic diagram of energy against exciton state density and local-ized-state density. Lower right illustration shows temperature dependence of the luminescence intensity from 156 to 231K. (From Tong, Y., Liu, Y., Shao, C., Liu, Y., Xu, C., Zhang, J., Lu, Y., Shen, D. and Fan, X., *J. Phys. Chem. B*, 110, 14714, 2006. With permission.)

for recombination related to surface trapped carriers, i.e., the trapped charge carriers on low-dimensional ZnO surface could be thermally activated to conduction band (CB) and valence band (VB), respectively, and subsequently, get de-excitation via exciton recombination so as to improving UV emission.[14–16] This discovery offered an effective approach for tailoring the nanostructure properties, which was crucial for their practical applications, as well as the potential of using ZnO to fabricate high-temperature UV excitonic laser.

The abovementioned groundbreaking work had important guiding significance for the research of ZnO and related luminescence materials. These basic problems and new discoveries suggested that ZnO was a promising candidate for next generation of semiconductor materials, especially in the field of UV/near-UV light-emitting and lasing devices. Recent studies were described in this section.

3.3.2 p-Type ZnO

The bottleneck of developing ZnO-based UV light-emitting devices is to obtain high-quality p-type ZnO materials. However, ZnO is a "natural" n-type material with strong self-compensation effect, which is extremely difficult to realize the transition to p-type conductance. Therefore, how to achieve stable and controllable p-type doping of ZnO is recognized as a key problem by the scientific community and remains as a major bottleneck for ZnO-based optoelectronics. Nitrogen has a similar ionic radius with oxygen and is considered as an ideal acceptor dopant in ZnO. However, the nitrogen substitution in ZnO lattice has high formation energy, low solubility, and is easy to form the donor defect complexes. As a result, the direct growth of N-doped ZnO suffers from the problems, such as low hole concentration, poor controllability and reproducibility. The self-compensation effect, especially in the low-dimensional ZnO materials, becomes more serious due to the surface defects, which further increases the difficulty in the p-type doping.

TABLE 3.8
Electrical Properties of Oxidized Zn_3N_2

Annealing Temperature	As Grown	600°C	700°C	800°C
Resistivity (Ohm cm)	84.02	14,980	153.05	66
Mobility (cm²/(V s))	39.72	3.52	0.098	56.8
Density (cm⁻³)	1.87×10^{15}	1.2×10^{14}	4.16×10^{17}	1.67×10^{15}
Hall coeff. (cm³/coul)	−3337.3	+52,768	+15.02	−3747
Carriers	E	h	H	e
N concentration (cm³)	l	6.87×10^{21}	6.78×10^{21}	5.48×10^{21}

Source: From Li, B. S., Liu, Y. C., Zhi, Z. Z., Shen, D. Z., Lu, Y. M., Zhang, J. Y., Fan, X. W., Mu, R. X. and Henderson, D. O., *J. Mater. Res.*, 18, 8, 2003. With permission.

To solve these problems, instead of using the conventional method of directly introducing nitrogen impurity into ZnO lattice, Liu's team first proposed a new idea of reverse doping.[17,18] By a thermal oxidation of Zn_3N_2 precursor, as well as a reverse substitution process of oxygen for nitrogen, the effective p-type doping was achieved and the stable N-doped ZnO films with high hole concentration (4.16×10^{17} cm⁻³) were obtained (Table 3.8). The method of reverse doping showed excellent controllability, that is, the nitrogen doping concentration and electrical properties of the resulting N-doped ZnO films could be well controlled by the temperature of thermal treatment. The proposed method of reverse doping had been recognized as an effective way to overcome the crucial problem concerning the low solubility of N acceptors in ZnO (Figure 3.66). This success would have major

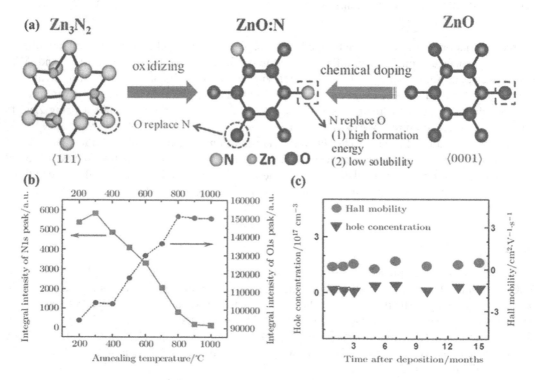

FIGURE 3.66 (a) The preparation process diagram of p-type ZnO. (b) Evolution of N 1s and O 1s core-level bands in X-ray photoelectron spectroscopy with annealing temperature. (c) Evolution of mobility and hole concentration as a function of time after film deposition. (From Li, B. S., Xiao, Z. Y., Ma, J. G. and Liu, Y. C., *Chin. Phys. B*, 26, 117101, 2017. With permission.)

implication for the semiconductor industry, and open a vast new field of applications. When the idea of "reverse doping" was proposed, many research groups around the world cited and followed up this study, and adopted similar reverse doping strategy to prepare ZnO with high hole concentration and good storage stability.[19–26] In 2005, Kawasaki's team used a new repeated temperature modulation technique via nitrogen plasma to reproducibly fabricate p-type ZnO thin films with mobility exceeding that in the bulk and firstly observed the blue-violet luminescence with electric injection from a ZnO homojunction at RT (Figure 3.67).[21] Since this work was reported, more attention had been paid to the preparation and development of p-type ZnO. Many methods were implemented, including Li-doped or Li-N dual-doped p-type ZnO,[27–31] Na-doped or Na-N dual-doped p-type ZnO,[32,33] and Al-N dual-doped p-type ZnO,[34] which had achieved a certain degree of stability improvement. The abovementioned researches further promoted the device-oriented progress of ZnO by optimizing growth conditions and structure designs.

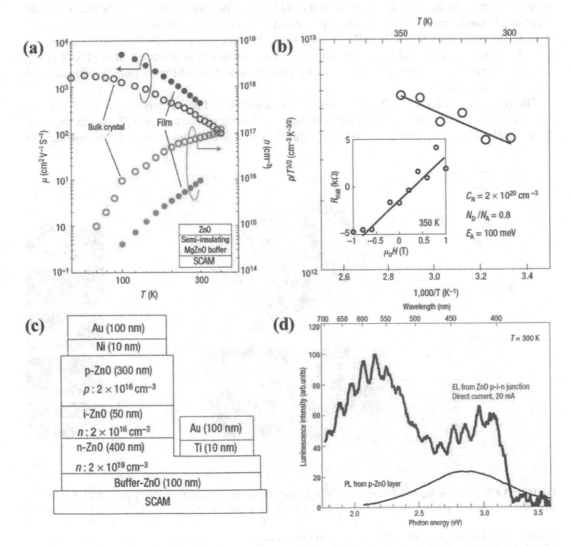

FIGURE 3.67   (a) Temperature evolution of mobility and carrier concentration. (b) Temperature evolution of hole concentration. (c) The structure of a typical p-i-n homojunction LED. (d) EL and PL spectrum from the above homostructure. (From Tsukazaki, A., Ohtomo, A., Onuma, T., Ohtani, M., Makino, T., Sumiya, M., Ohtani, K., Chichibu, S. F., Fuke, S., Segawa, Y., Ohno, H., Koinuma, H. and Kawasaki, M., *Nat. Mater.*, 4, 42, 2005. With permission.)

3.3.3 P-N HOMOJUNCTION AND HETEROJUNCTION LEDs

As mentioned previously, the reproducible and stable p-type ZnO had been obtained in 2005,[21] and the prototype device of ZnO homojunction was first constructed, which had aroused a great interest in the research of ZnO homojunction devices at that time. In the following year, Shen's team became the first research group in China to observe the electric injection emission in ZnO homojunction (Figure 3.68(a)).[35] A ZnO p-n homojunction light-emitting diode with p-ZnO layer via NO plasma activated by a radio frequency (RF) atomic source and natural n-ZnO layer was fabricated on a-plane sapphire substrate. The hole concentration and mobility of p-type ZnO layers were 1.3×10^{17} cm^{-3} and 1.5 cm^2/(V s) at 200K, and with increase of temperature, p-type conduction became unstable gradually. However, the electron concentration and mobility of n-type ZnO layers were stable, fixed on 7×10^{18} cm^{-3} and 40 cm^2/(V s). Then, Park's team tried to further improve the structure properties of n- and p-ZnO (Figure 3.68(b)).[36] An effective thermal annealing process could activate the phosphorus dopants and improve the properties of ZnO layers. Therefore, this device showed excellent rectifying behavior with a lower threshold voltage and UV electroluminescence (EL) emission centered at 380 nm. Later, the electrically pumped Fabry-Perot (FP) waveguide lasing from Sb-doped p-ZnO nanowires and n-ZnO films was first realized in homojunction devices.[37] This FP UV lasing was demonstrated good stability, which could be modeled with finite-different time-domain method (Figure 3.68(c) and (d)).

The representative work of ZnO p-n homojunction mentioned earlier has greatly promoted the practical application prospect of ZnO intrinsic excitonic luminescence and laser. However, the quality and stability of as-prepared p-type ZnO thin films still need to be improved. Meanwhile, the

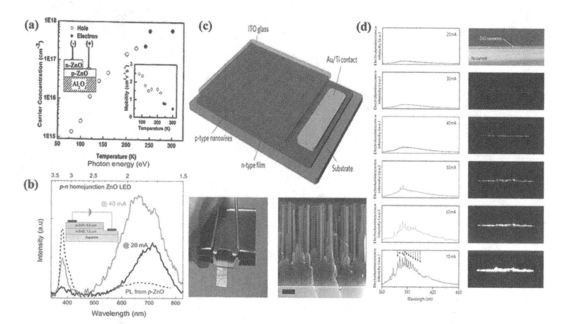

FIGURE 3.68 (a) The dependence of the carrier concentration on temperature. The insets show the device structure and mobility of p-type ZnO film. (From Jiao, S. J., Zhang, Z. Z., Lu, Y. M., Shen, D. Z., Yao, B., Zhang, J. Y., Li, B. H., Zhao, D. X., Fan, X. W. and Tang, Z. K., *Appl. Phys. Lett.*, 88, 031911, 2006. With permission.) (b) EL spectra of ZnO p-n homojunction LED operated at forward currents of 20 and 40 mA. (From Lim, J. H., Kang, C. K., Kim, K. K., Park, I. K., Hwang, D. K. and Park, S. J., *Adv. Mater.*, 18, 2720, 2006. With permission.) (c) Structure and material properties of ZnO nanowire/film laser device. (d) EL spectra and side-view optical microscope images of ZnO p-n homojunction laser device operated between 20 and 70 mA. (From Chu, S., Wang, G., Zhou, W., Lin, Y., Chernyak, L., Zhao, J., Kong, J., Li, L., Ren, J. and Liu, J., *Nat. Nanotechnol.*, 6, 506, 2011. With permission.)

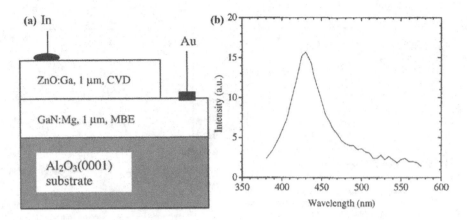

FIGURE 3.69   (a) Heterojunction LED structure diagram. (b) EL spectrum of an n-ZnO/p-GaN heterojunction LED. (From Alivov, Y. I., Nostrand, J. E. V., Look, D. C., Chukichev, M. V. and Ataev, B. M., *Appl. Phys. Lett.*, 83, 2943, 2003. With permission.)

ZnO p-n homojunction devices have low UV-luminescence efficiency. This challenge remains the most important topic and basic question in ZnO research today, and many research efforts are attempting to solve this problem. Logically, an alternative strategy of investigation and development of ZnO-based heterojunction devices using other available p-type materials or design and construction of novel device configurations should be considered. Among the mature p-type semiconductors such as GaN, Si, $CuAlO_2$, NiO, SiC, and even some organics, GaN is taken for the most suitable one due to its identical crystal structure, similar band alignment and small lattice mismatch with ZnO, which favor heteroepitaxial growth and form a high-quality heterointerface. A new round of research upsurge gradually emerged and promoted the continuous development of ZnO-based materials and devices. Actually, a 430-nm blue-violet EL had first already obtained in p-GaN/n-ZnO heterojunction LED in 2003 (Figure 3.69).[38] The layer structure was fabricated by growing an Mg-doped GaN film on c-sapphire substrate firstly, and then depositing a Ga-doped ZnO film on the above p-GaN layer. This heterojunction device had attracted wide attentions in the last 20 years.

However, the origin of the EL in this heterojunction will be mainly determined by the difference in the carrier mobility between n-ZnO and p-GaN. Generally, the electron mobility is higher than that of hole. Hence, electron injection prevails over hole injection under forward bias, that is, electron injection dominates the whole carrier transport process. Thereby, the blue emission usually occupies a large proportion in the EL spectrum, whereas radiative recombination in the n-ZnO side is always suppressed. To address this problem, Liu's team successfully prepared a p-ZnO/n-GaN heterojunction, which showed a typical rectifying characteristic and an EL emission centered at 517 nm (Figure 3.70(a) and (b)).[39] In other words, the carrier's recombination mainly occurred in the deep level probably due to poor quality p-ZnO, whereas ZnO excitonic emission was not fully utilized. To balance the difference between electron and hole mobility, and improve ZnO excitonic emission efficiency, Liu's group first proposed the idea of inserting a thin semi-insulating ZnO (i-ZnO) layer between the p-GaN and n-ZnO layer, which confined parts of carriers to recombine in i-ZnO region (Figure 3.70(c)).[40] Thereby, a ZnO exciton emission at 3.21 eV was observed (Figure 3.70(e)). This heterostructure had the potential applications in UV and laser diodes (LDs) durable at high-temperature operation. Then, a simple method to produce one-chip white LEDs in such a p-i-n heterojunction was proposed (Figure 3.70(f)).[41] Both blue-violet emission and deep-level emission were detected simultaneously, and the intensity ratio was tunable through controlling i-ZnO thickness. Heterojunction with 20 nm-thick i-ZnO layer exhibited good white emission, which were suggested as being more efficient than current phosphor-based double-excitation LEDs.

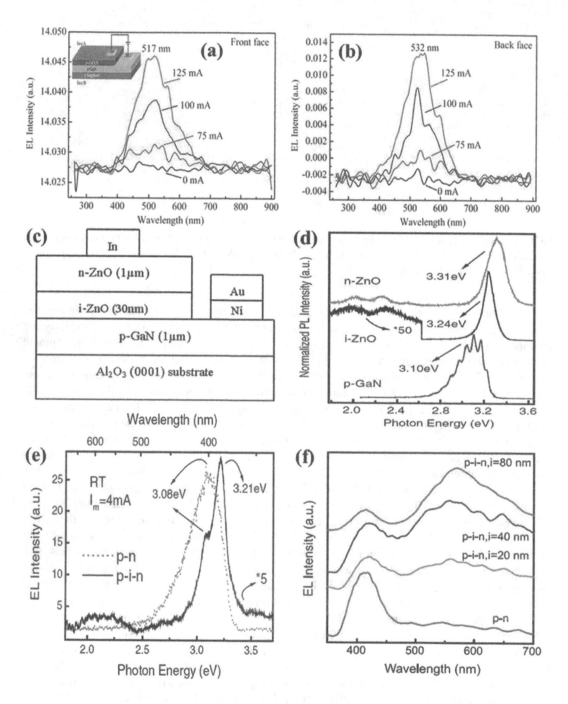

FIGURE 3.70 (a) and (b) EL spectra of p-ZnO/n-GaN LED from the front and back face at forward bias. (From Wang, L., Xu, H., Liu, Y. and Shen, L., *Mater. Res. Express*, 2, 025901, 2015. With permission.) (c) Schematic diagram of p-GaN/i-ZnO/n-ZnO heterojunction LED. (d) PL spectra of p-GaN, i-ZnO and n-ZnO. (e) EL spectra of p-n and p-i-n heterojunction LEDs. (From Xu, H. Y., Liu, Y. C., Liu, Y. X., Xu, C. S., Shao, C. L. and Mu, R., *Appl. Phys. B*, 80, 871, 2005. With permission.) (f) EL spectra of p-n and p-i-n heterojunctions with different i-layer thickness. (From Zhao, L., Xu, C. S., Liu, Y. X., Shao, C. L., Li, X. H. and Liu, Y. C., *Appl. Phys. B*, 92, 185, 2008. With permission.)

3.3.4 CDZNO, BEZNO AND ZNO/BEZNO QUANTUM WELLS

Since high-quality and stable p-type ZnO material can be achieved, a critical strategy for developing high-efficiency ZnO optoelectronic devices is the production of ZnO-based quantum well (QW) and superlattice (SL) structures. It is critical to construct ZnO-based alloy materials with bandgaps lower and larger than that of ZnO to act as active layers for emission in the visible and deep-UV range. In 2001, Kawasaki's group first described the structural and optical properties of $Cd_xZn_{1-x}O$ films grown by pulse laser deposition (Figure 3.71).[42] However, the maximum cadmium concentration ($x=0.07$) was significantly larger than the thermodynamic solubility limit, and the bandgap changed from 3.3 (ZnO) to 3.0 eV ($Cd_{0.07}ZnO_{0.93}O$) at RT with a very small range, probably owing to the large lattice mismatch and the difference in crystal structure between CdO and ZnO. Although Cd had been subsequently used to grow CdZnO alloys and construct ZnO/CdZnO QW structures, the large lattice mismatch made them difficult to extend their bandgaps to the visible region, which hindered their further applications.

Therefore, to expore the growth of different alloys for such a purpose which will have larger bandgaps than that of ZnO motivates the researches' interest to fabricate short-wavelength optoelectronic devices. The ternary II–VI semiconductor compound, BeZnO, is one of the candidates for constructing UV or deep-UV LEDs, because its bandgap varies from 3.3 to 10.6 eV depending on the beryllium mole fraction. In 2006, BeZnO was first proposed and studied by Kim's team (Figure 3.72(a)).[43] The energy bandgap and lattice constants of BeZnO could be continuously modulated by changing Be concentration.

The studies showed that the bandgap of ZnO could be tuned to higher values, and favorable for designing and developing QWs and SLs. In the same year, Kim's team used the same

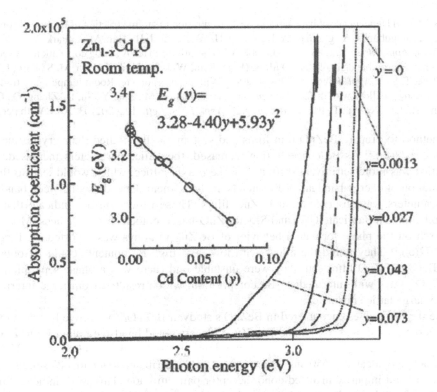

FIGURE 3.71 The dependence of absorption coefficient and energy band at different concentrations (y). (From Makino, T., Segawa, Y., Kawasaki, M., Ohtomo, A., Shiroki, R., Tamura, K., Yasuda, T. and Koinuma, H., *Appl. Phys. Lett.*, 78, 1237, 2001. With permission.)

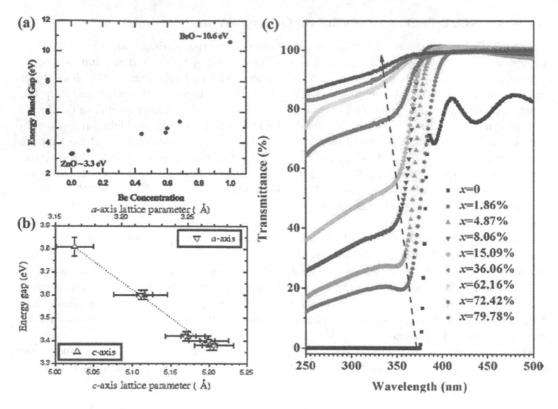

FIGURE 3.72 (a) Dependence of bandgap and Be concentration in atomic fraction of BeZnO. (From Ryu, Y. R., Lee, T. S., Lubguban, J. A., Corman, A. B., White, H. W., Leem, J. H., Han, M. S., Park, Y. S., Youn, C. J. and Kim, W. J., *Appl. Phys. Lett.*, 88, 052103, 2006. With permission.) (b) A strong correlation between lattice parameters and energy bandgap of BeZnO films. (From Kim, W. J., Leem, J. H., Han, M. S., Park, I. W., Ryu, Y. R. and Lee, T. S., *J. Appl. Phys.*, 99, 096104, 2006. With permission.) (c) Room temperature transmission spectra of $Be_xZn_{1-x}O$ alloys with different Be contents. (From Chen, M., Zhu, Y., Su, L., Zhang, Q., Chen, A., Ji, X., Xiang, R., Gui, X., Wu, T., Pan, B. and Tang, Z., *Appl. Phys. Lett.*, 102, 202103, 2013. With permission.)

growth method to prepare BeZnO thin films and systematically studied their crystalline properties (Figure 3.72(b)).[44] As Be concentration increased, the lattice parameters increased, and the c-axis lattice was more sensitively shifted than the a-axis lattice, which would lead to the effective modulation of energy bandgap of BeZnO. Namely, a linear dependency between bandgap and lattice parameters was observed from BeZnO films. These results further indicate that BeZnO was a good choice to fabricate QWs and SLs for ZnO-based optoelectronic devices. More systematic research on the phase formation behavior of $Be_xZn_{1-x}O$ alloys was reported by Tang's team (Figure 3.72(c)).[45] They found the alloys with low- and high-Be contents could be obtained by alloying BeO and ZnO films, but they were unstable and there was a phase separation process of $Be_xZn_{1-x}O$ alloys with intermediate Be composition, which resulted from large internal strain induced by large lattice mismatch.

Despite the difficulties encountered in BeZnO's study, a BeZnO/ZnO active layer comprised of seven QWs between n- and p-ZnO and $Be_{0.3}Zn_{0.7}O$ layers could be also employed into ZnO-based UV LEDs.[46] The multi-QW device showed two dominant EL peaks centered at the UV region, while a broad peak located at 550 nm, which was probably attributed to localized exciton, impurity-bound exciton and impurity-involved donor-acceptor pair emission. This work indicated potential applications of solid-state lighting and antimicrobial lamps with the development of ZnO-based QW structures. Afterward, the properties of QW structures were theoretically explored by using the non-Markovian model with many-body effects.[47] The study found that the emission peak intensity was greatly reduced when Be composition increased due to the increase in the internal field effect. Hence, further investigation on methods to reduce the internal field effect would be necessary for high-efficiency optoelectronic devices.

3.3.5 MgZnO Alloy

Although researchers had made some attempts in CdZnO or BeZnO and their QW structures, the bandgap broadening and narrowing effect of ZnO-alloys was still not obvious due to the large lattice mismatch and phase separation processes. What's more, elements of Cd and Be were rare and highly toxic, which restricted the development of their related materials and devices. Another wide-gap II–VI semiconductor alloy, $Mg_xZn_{1-x}O$, is considered to be one of the promising materials for fabrication of heteroepitaxial UV light-emitting devices based on ZnO, which was first proposed and systematically studied by Ohtomo's team in 1998 (Figure 3.73).[48] The $Mg_xZn_{1-x}O$ films were epitaxially grown on ZnO films and c-sapphire substrates. When the molar content of Mg was not less than 0.36, MgO impurity phase separation would occur, and lattice constants changed slightly. The bandgap and photoluminescence (PL) could be tunable, while maintaining high crystallinity without obvious change of the lattice constants.

Before long, Park's team fabricated high-quality $Mg_xZn_{1-x}O$ films epitaxially grown at 500–600°C on c-sapphire substrates using metal organic vapor-phase epitaxy (Figure 3.74(a)).[49] They found that the c-axis constant decreased by increasing the Mg content in films up to 49 at.% and there was no significant phase separation, as well as the hexagonal wurtzite structure. Because of Mg incorporation, the exciton emission of as-prepared films was blue-shifted from

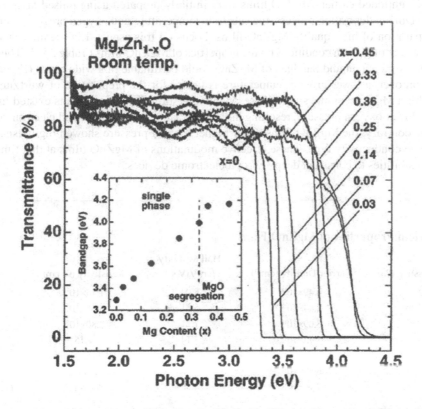

FIGURE 3.73 Transmission spectra of $Mg_xZn_{1-x}O$ films with different Mg contents. (From Ohtomo, A., Kawasaki, M., Koida, T., Masubuchi, K., Koinuma, H., Sakurai, Y., Yoshida, Y., Yasuda, T. and Segawa, Y., *Appl. Phys. Lett.*, 72, 2466, 1998. With permission.)

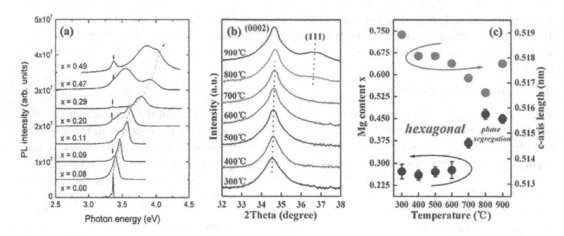

FIGURE 3.74 (a) PL spectra of $Mg_xZn_{1-x}O$ films measured at 15K. (From Park, W. I., Yi, G. C. and Jang, H. M., *Appl. Phys. Lett.*, 79, 2022, 2001. With permission.) (b) XRD patterns of MgZnO films grown at different substrate temperatures. (c) Mg content and *c*-axis length as functions of substrate temperatures. (From Liu, C. Y., Xu, H. Y., Wang, L., Li, X. H. and Liu, Y. C., *J. Appl. Phys.*, 106, 073518, 2009. With permission.)

3.364 to 4.05 eV. This work further optimized the growth method and preparation condition of MgZnO thin films, and obtained high-quality MgZnO alloys, which laid a foundation for further research and is likely to bring the research upsurge of MgZnO material system and related applications. As mentioned earlier, MgZnO films were initially prepared using pulsed laser deposition techniques. During deposition process, the substrate temperature and oxygen pressure were crucial for the preparation of high-quality MgZnO films. Focus on this aspect, Liu's team systematically studied the effect of growth conditions on the properties of MgZnO films (Table 3.9).[50] These results revealed that Mg content and bandgap of MgZnO could be tunable in a wide range (Figure 3.74(b) and (c)). Moreover, a lower growth temperature is useful for the preparation of wurtzite MgZnO alloy with high Mg content and single phase. Many interstitial Zn donor defects existed in MgZnO films under lower oxygen pressure, resulting in the increasing *c*-axis length and electron concentration. On the contrary, the MgZnO films under higher oxygen pressure showed high resistivity and indeterminate conductivity type. These effective modulations of MgZnO films at that time opened up great possibilities to construct deep-UV optoelectronic devices.

TABLE 3.9
The Electrical Properties of MgZnO Films

Oxygen Pressure (Pa)	Carrier Density (cm^{-3})	Hall Mobility (cm^2/(V s))	Resistivity (Ω cm)	Type
10^{-3}	4.49×10^{18}	8.04	1.73×10^{-1}	n
10^{-2}	2.94×10^{18}	5.81	3.65×10^{-1}	n
10^{-1}	3.09×10^{18}	7.21	2.80×10^{-1}	n
1	1.73×10^{18}	3.14	1.15	n
10			High resistance	
20	6.59×10^{13}	7.62	1.30×10^4	Indeterminate
30	7.76×10^{14}	7.58	1.15×10^3	Indeterminate

Source: From Liu, C. Y., Xu, H. Y., Wang, L., Li, X. H. and Liu, Y. C., *J. Appl. Phys.*, 106, 073518, 2009. With permission.

3.3.6 ZnO/MgZnO Multi-quantum Wells

With high-quality MgZnO as the material base, its alloyed QW and SL structures are worth expecting. In 2003, Yi's team first reported the fabrication of ZnO/MgZnO multi-QWs grown on ZnO nanorods without catalyst (Figure 3.75).[51] When the thickness decreased to the monolayer level, these structures showed the obvious quantum confinement and blue-shift with the well layer thickness decreased, which was in good agreement with theoretical predictions. This kind of QW structures opened up an opportunity for developing quantum structures, even probably applied into nanoscale devices, field-effect transistors (FETs) and LEDs. They developed a simple "bottom-up" epitaxial method, which might readily be expanded to produce other heteroepitaxial nanorods.

Based on the emergence of such nanorod multi-QW heterostructures, a new wave of research on the structural properties and applications had spread. The preparation, PL properties and quantum confinement effect of ZnO/MgZnO multi-QWs were systematically studied by Korea research

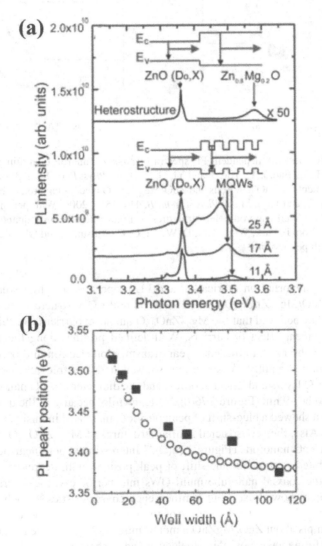

FIGURE 3.75 (a) PL measurements of heterostructure nanorods and multi-quantum well nanorods measured at 10K. (b) Well-width dependence of comparing with PL emission peak positions and theoretically calculated values. (From Park, W. I., Yi, G. C., Kim, M. and Pennycook, S. J., *Adv. Mater.*, 15, 526, 2003. With permission.)

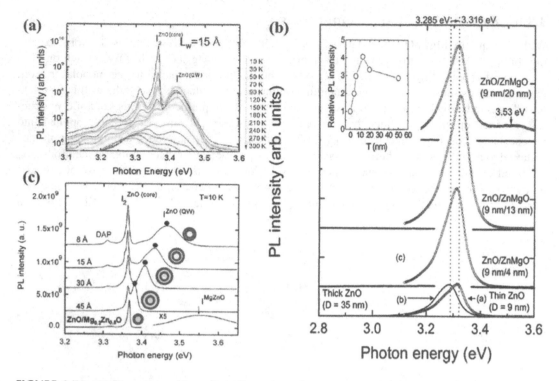

FIGURE 3.76 (a) Temperature-dependent PL spectra of coaxial nanorod quantum wells with well layer thickness of 1.5 nm. (From Bae, J. Y., Yoo, J. and Yi, G. C., *Appl. Phys. Lett.*, 89, 173114, 2006. With permission.) (b) PL measurements of ZnO nanorods and $ZnO/Mg_{0.2}Zn_{0.8}O$ coaxial nanorods. (From Park, W. I., Yoo, J., Kim, D. W., Yi, G. C. and Kim, M., *J. Phys. Chem. B*, 110, 1516, 2006. With permission.) (c) PL spectra of $ZnO/Mg_{0.2}Zn_{0.8}O$ core/shell nanorod heterostructures and multishell nanorod quantum structures. (From Jang, E. S., Bae, J. Y., Yoo, J., Park, W. I., Kim, D. W., Yi, G. C., Yatsui, T. and Ohtsu, M., *Appl. Phys. Lett.*, 88, 023102, 2006. With permission.)

group. In 2006, they reported on fabrication and PL properties, and first studied quantum confinement effect in $ZnO/Mg_{0.2}ZnO_{0.8}O$ coaxial nanorod multi-QW structures with single-QW layer (Figure 3.76(a)).[52] They believed that the $Mg_{0.2}ZnO_{0.8}O$ quantum barrier layer thickness determined the quantum confinement effect of carriers. Well-defined profiles along the multi-QW nanorod radial direction were observed. In the same year, transmission electron microscopy (TEM) images of such coaxial nanorod multi-QW structures showed a defect-free and clean heterointerface between $Mg_{0.2}ZnO_{0.8}O$ layers and ZnO nanorods, and the diameter of ZnO nanorod as a core material was estimated to be ~9 nm (Figure 3.76(b)).[53] Meanwhile, because of the quantum confinement effect, the PL spectra showed a blue-shift of peak position, increased intensity, and notably reduced thermal quenching. Also, they constructed similar structures of $Mg_{0.2}ZnO_{0.8}O/ZnO/Mg_{0.2}ZnO_{0.8}O$ multishell layers on ZnO nanorods (Figure 3.76(c)).[54] Interestingly, both near-field and far-field PL measurements exhibited systematic blue-shift of peak position with decreasing well layer widths. Soon, they turned the coaxial nanotube multi-QWs into periodic vertical arrays arranged in an orderly fashion, which showed excellent luminescence characteristics.[55] Such an effort provided reference for further enhanced the versatility of the components in nano-devicalization process. These research attempts about ZnO/MgZnO nanorod multi-QWs would be useful for exploring and integrating high-performance nanoscale optoelectronic devices.

Since then, some meaningful work had continued to emerge. Cho's team developed a new method with a two-step thermal evaporation procedure to prepare ZnO/MgZnO heterostructure nanorods with high aspect ratio.[56] They demonstrated Mg distribution with shell layers consisting of MgO quantum dots in the upper region and $Mg_{0.35}Zn_{0.65}O$ shell in the middle region. Liu's team reported electrically pumped ZnO QW LD by molecular-beam epitaxy (MBE) (Figure 3.77(a)).[57] Low current density of 10 A/cm² and output power of 11.3 μW were obtained. A series of radial homogeneous ZnO/MgZnO QWs with low-area density nanorod arrays were grown perpendicularly to the substrates by Cao's team (Figure 3.77(c)).[58] The most important finding was that the cathodo-luminescence intensity of axial QW was lower than that of radial QW. In addition, anisotropic strain effects on PL measurements from nonpolar MgZnO/ZnO QW films were studied by Chauveau's team.[59] The anisotropy of lattice parameters resulted in an unusual in-plane strain state, which induced a strong blue-shift of the excitonic transitions. The reduction of structural defects suggested a strong PL improvement, including narrowed full width at half maximum, enhanced emission intensity and improved thermal stability. Park's team introduced a p-type $Mg_{0.15}Zn_{0.85}O$ electron blocking layer between MgZnO/ZnO multiple-QWs and p-type $Mg_{0.1}Zn_{0.9}O$ layers, resulting in increased UV emission and reduced deep-level emission and the total output power was improved

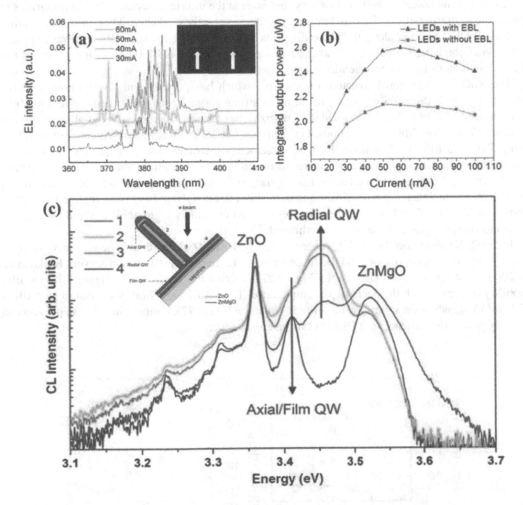

FIGURE 3.77 (a) EL spectra of lasing device with higher injection currents. (From Chu, S., Olmedo, M., Yang, Z., Kong, J. and Liu, J., *Appl. Phys. Lett.*, 93, 181106, 2008. With permission.) (b) Total output power of MgZnO/ZnO multiple-QWs LEDs with and without p-type $Mg_{0.1}Zn_{0.9}O$ electron blocking layer as a function of injection current. (From Choi, Y. S., Kang, J. W., Kim, B. H. and Park, S. J., *Opt. Express*, 21, 31560, 2013. With permission.) (c) Cathodoluminescence spectra (10K) measured at the locations indicated the inset. (From Cao, B. Q., Zúñiga-Pérez, J., Boukos, N., Czekalla, C., Hilmer, H., Lenzner, J., Travlos, A., Lorenz, M. and Grundmann, M., *Nanotechnology*, 20, 305701, 2009. With permission.)

(Figure 3.77(b)).[60] These works greatly promoted the basic research and application research of MgZnO/ZnO QWs from the aspects of material growth, structure optimization and performance improvement.

3.3.7 TWO-DIMENSIONAL ELECTRON GAS CHARACTERISTICS OF ZnO/Mg$_x$Zn$_{1-x}$O HETEROSTRUCTURE

ZnO/Mg$_x$Zn$_{1-x}$O heterostructure and its two-dimensional electron gas (2 DEG) have become a recent research hotspot.[61-63] The ZnO/Mg$_x$Zn$_{1-x}$O heterostructure system, due to its own spontaneous polarization and piezoelectric polarization[64,65] caused by lattice distortion, will induce a layer of 2 DEG with high mobility at the interface.[66-68] It is very advantageous to develop high-performance FET.[69,70] A so-called 2 DEG gas refers to the phenomenon that the electron gas can move freely in the direction of the parallel interface and is confined in the direction of the vertical interface. Since the 2 DEG in the potential well are in a very thin layer at the interface, separated from impurities in space, the scattering effect of ionized impurity centers is greatly weakened. Therefore, the mobility of these 2 DEG moving along the plane direction will be very high, especially at lower temperatures; when the lattice vibration is weakened, its electron mobility is 1000 times higher than that of the normally high-quality bulk semiconductor materials.

The ZnO is a hexagonal structure of wurtzite, which has polarity and low symmetry as the preferred growth direction of c-axis, resulting in strong spontaneous polarization effect of ZnO. Similarly, Mg$_x$Zn$_{1-x}$O ($0<x<0.35$) exhibits a wurtzite hexagonal structure with spontaneous polarization effect at low Mg concentrations. On the other hand, ZnO has a strong piezoelectric effect, using ZnO and Mg$_x$Zn$_{1-x}$O films to form heterostructures, which cause piezoelectric polarization effect due to the strain generated by lattice distortion in the films. High concentration of electron thin layer, i.e., 2 DEG, is induced at the heterojunction interface and sample surface. According to the literature, the integer and fractional quantum Hall effects (QHEs) have been realized in 2 DEG system of MgZnO/ZnO-based heterostructures,[63,71] and extremely high mobilities and remarkable magnetotransport properties have been exhibited.[72,73]

In 2002, Krishnamoorthy et al.[74] prepared a ZnMgO/ZnO/ZnMgO double heterojunction structure by pulsed laser deposition, and the resonant quantum tunneling effect is observed. Koike et al. in 2004[75] first reported two DEG in the MgZnO/ZnO structure. As shown in Figure 3.78, the film mobility increases with the decrease of temperature. The saturation mobility measured at 300K is 170 cm^2/(V s), while the saturation mobility is 400 cm^2/(V s) at 77K temperature. These phenomena directly prove the existence of 2 DEG in this heterostructure.

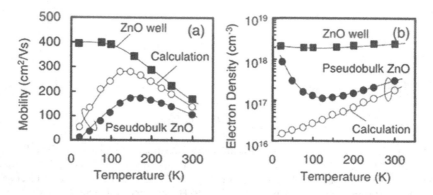

FIGURE 3.78 The relationship between Hall mobility (a) and electron concentration (b) with temperature. (From Koike, K., Hama, K., Nakashima, I., Takada, G.-y., Ozaki, M., Ogata, K.-i., Sasa, S., Inoue, M. and Yano, M., *Jpn. J. Appl. Phys.*, 43, L1372, 2004. With permission.)

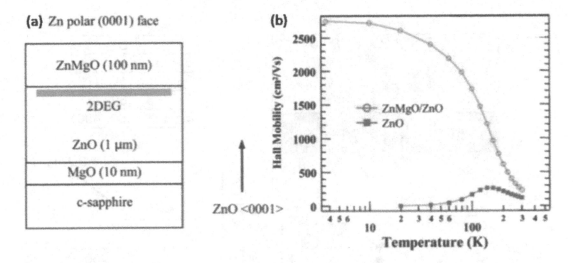

(a) Zn polar (0001) face

ZnMgO (100 nm)

2DEG

ZnO (1 μm)

MgO (10 nm)

c-sapphire

ZnO <0001>

FIGURE 3.79 (a) Cross-sectional schematic of ZnMgO/ZnO heterostructure grown on *c*-sapphire substrate. (b) Temperature dependence of Hall mobility for ZnMgO/ZnO heterostructure and ZnO thin films. (From Tampo, H., Shibata, H., Matsubara, K., Yamada, A., Fons, P., Niki, S., Yamagata, M. and Kanie, H., *Appl. Phys. Lett.*, 89, 132113, 2006. With permission.)

In 2006, Tampo et al.[62] reported the preparation of ZnO/ZnMgO heterostructures Zn polar surfaces using RF MBE and the realization of 2 DEG in this structure. The structure diagram is shown in Figure 3.79(a). Figure 3.79(b) also exhibited the relationship curve of the measured Hall mobility with temperature in ZnO/ZnMgO heterostructures. It can be seen that the electron mobility of heterostructures increases significantly with the increase of Mg components, where the electron mobility reached 250 and 2750 $cm^2/(V\,s)$, at room and 4K temperatures, respectively. The authors point out that ZnMgO/ZnO heterostructures with Zn polar are very suitable for the preparation of top gate thin-film transistor devices. By comparing the dependence of mobility on temperature in ZnO and ZnO/ZnMgO heterojunction, the authors believe that the increasing of mobility with the decrease of temperature is due to the scattering shielding of ionized impurities, and the high electron concentration at the interface makes this shielding particularly significant, leading to high mobility of heterostructures at low temperatures.[62]

In 2007, Tsukazaki et al.[63] reported the observation of Shubnikov-de Haas (SdH) oscillation and the QHE in a high-mobility 2 DEG in polar ZnO/MgZnO heterostructures. Figure 3.80(a) shows the formation mechanism of 2 DEG and the energy band configuration at the interface in ZnO/MgZnO heterostructure. As it can be seen, the electrons move to the interface and surface due to spontaneous polarization and piezoelectric polarization in ZnO and MgZnO layers, forming a 2D electronic layer. For the samples with the growth direction of O-face, both spontaneous and piezoelectric polarization directions point to the surface of the sample, which is based on the fact that the ZnMgO film is completely relaxed and ZnO film is subjected to tensile stress. Figure 3.80(b) shows the theoretical calculated values and experimental results of the 2 DEG concentration with changing Mg compositions. Figure 3.80(c) also gives a schematic diagram of the energy band structure. Because of the band offset of the interface, the electrons under the action of the polarization field are effectively confined to form an electron accumulation layer, and the electron accumulation causes the energy band to bend and form an electron potential well. The authors suggest that the abovementioned results suggest the exciting possibility of realizing the fractional QHE in this polar ZnO/MgZnO heterostructure when the improved mobility can be obtained.

Then, this group achieved a maximum mobility of more than 10^4 $cm^2/(V\,s)$ at 0.5K in ZnO/MgZnO heterostructure with 0.05 Mg components.[66] It was noted that when measurement temperature drops below 50K, the resistivity of the sample changes from semiconductor characteristic to

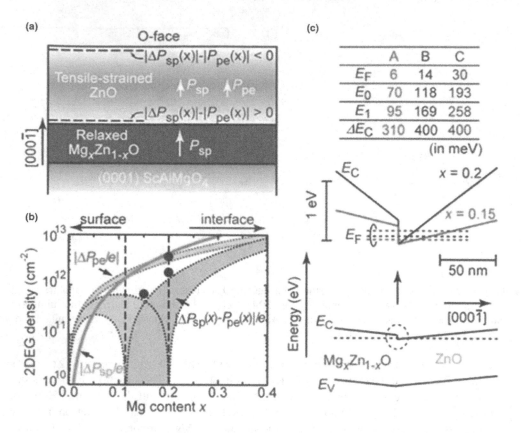

FIGURE 3.80   (a) The schematic diagram of the electron accumulation layer. (b) The calculated value and the measured value of 2-DEG concentration as a function of Mg content. (c) Band calculation parameters and interface band structure diagram. (From Tsukazaki, A., Ohtomo, A., Kita, T., Ohno, Y., Ohno, H. and Kawasaki, M., *Science*, 315, 1388, 2007. With permission.)

metal characteristic, which indicates that 2 DEG has good transportability at low temperature. By 2012, the mobility of the ZnO/MgZnO heterostructure has been continuously improved, reaching 7.7×10^5 cm^2/(V s) at 0.5K temperature.[76] All these studies show that ZnO/MgZnO heterostructure is suitable for preparing electronic devices that can still work normally at very low temperature.

3.3.8   MgZnO/MgO Multiple-quantum Wells

To further shorten the emission wavelength into the deep-UV region, Liu's team considered that the MgZnO and MgO could be chosen as the active layer and barrier material to fabricate MgZnO/MgO QW structures, which would make the carrier confinement stronger than that in ZnO/MgZnO QWs because of the lager energy-band offset.[77] They designed and prepared multiple-QW films and observed a stress-induced transition from hexagonal wurtzite to cubic sphalerite with MgZnO thickness decreasing. Clear alternating light/dark stripes and sharp interface between the MgZnO and MgO layers in TEM images indicated that both layers suffered from in-plane stress. Wavelength-tunable deep-UV emission in the region of 261–314 nm was obtained, and the 261 nm was the shortest emission wavelength reported at the time (Figure 3.81(a)). Furthermore, they analyzed the growth dynamic processes and optical properties of abovementioned multiple-QW structures.[78] Actually, less phonon scattering and higher ballistic electron velocities were expected for rock-salt MgZnO with higher structural symmetry, which was more favorable for deep-UV detectors and emitters (Figure 3.81(b)). Therefore, they grew biaxial rock-salt multiple-QW film structures

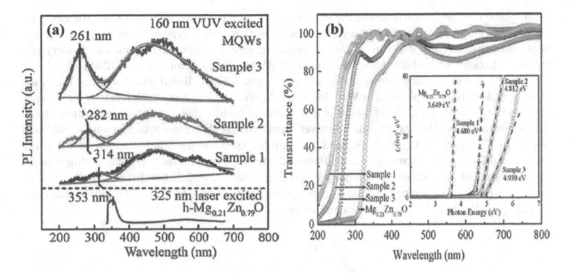

FIGURE 3.81 (a) PL spectra of multiple-quantum-well and film structures. (From Wang, L., Xu, H. Y., Zhang, C., Li, X. H., Liu, Y. C., Zhang, X. T., Tao, Y., Huang, Y. and Chen, D. L., *J. Alloys Compd.*, 513, 399, 2012. With permission.) (b) Transmittance spectra of multiple-quantum-well and film structures. (From Wang, L., Ma, J. G., Xu, H. Y., Zhang, C., Li, X. H. and Liu, Y. C., *Appl. Phys. Lett.*, 102, 031905, 2013. With permission.)

on m-sapphire substrates, and PL measurement showed a relatively broad deep-UV emission and a strong visible emission band. The optical bandgap of the anomalous rock-salt MgZnO was determined by the quantum confinement effect and strain-induced energy shift. A stable rock-salt MgZnO/MgO multiple-QW inherent microcavity without Bragg reflector was also constructed.[79] Obvious FP interference patterns were observed from transmission and PL spectra, which resulted from the fluctuations of thickness and permittivity induced by in-plane strain. These studies provided a new route to develop similar multiple-QW structures for high-performance UV or deep-UV detectors and emitters.

3.3.9 Recent Progress of Novel van der Waals Two-dimensional Semiconductor Materials and Their Heterostructures

Transition metal dichalcogenides (TMDs) as a class of typical 2D materials have drawn significant attentions over the past several years in optoelectronic devices such as LEDs,[80,81] photodetectors,[82,83] and photovoltaic cells,[84,85] relying on their intriguing optical and electronic properties. Different from traditional semiconductors, 2D TMDs materials have dangling bond-free surface and therefore, can be fabricated into a kind of unique heterostructure via stacking different 2D TMD materials on top of each other.[86,87] Herein, strong covalent bonds could supply in-plane stability, whereas the weak van der Waals forces maintain these different 2D TMD layers stack together. The advantage of van der Waals heterostructures is that the constituent materials neither need lattice matching nor satisfy any other stringent conditions. More importantly, the heterostructures constructed by various TMDs crystals with different optical and electronic properties would certainly show various and novel properties and potential applications.

For example, in the van der Waals heterostructure with a suitable band alignment, the photoexcited electrons and holes can reside in the two separate layers, and the transfer of electrons and holes generally exhibited sub-picosecond timescale. This will make the interfacial bandgap can be modulated by selecting different TMDs materials so that the electron-hole recombination in interlayer could generate photons with a frequency determined over a broad range. Nevertheless, a small misalignment or lattice mismatch of the constituent layers can significantly influence the electron-photon coupling and suppress interlayer exciton transitions. To overcome it, recently, Morpurgo et al. had constructed type-II van der Waals heterostructure by assembling atomically thin crystals (Figure 3.82(a)),[88] where both the VB maximum (VBM) and the CB minimum (CBM) locate at the Γ point in k space, avoiding momentum mismatch (Figure 3.82(b)). Based on the strategy, they used multilayer TMDs materials, such as WS_2, MoS_2 and $MoSe_2$, the VBM of which locate at the Γ point, and multilayer InSe the CBM of which is at Γ point, to build a direct interlayer transition in van der Waals interfaces. In this way, the interlayer transitions are direct in k space as long as the band alignment is of type-II (Figure 3.82(c) and (d)). Besides, such van der Waals heterostructure could present radiative optical transitions regardless of the lattice constant, the rotational and/or translational alignment or whether the bandgaps of the constituent materials are direct or indirect. This strategy will extend the range of future optoelectronic applications based on 2D TMDs materials, including LEDs, lasers, photodetectors and so on.

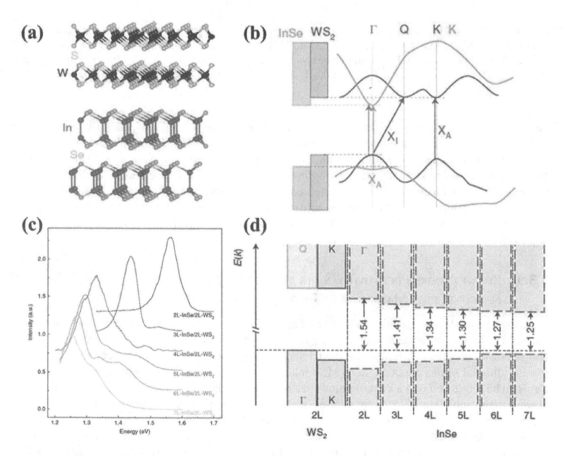

FIGURE 3.82 (a) Schematic diagram of 2D heterostructure. (b) The energy band alignment of 2L-InSe/2L-WS$_2$ heterostructure. (c) PL spectra of (2, 3, 4, 5, 6 and 7) L-InSe/2L-WS$_2$ at 5K. A peak that originates from direct interlayer transition in k point is observed in all spectra. (d) Band alignment of multilayer InSe/2L-WS$_2$ heterostructures, where the CBM is always in the InSe and the VBM situates in the 2L-WS$_2$ in these interfacial bandgaps. Both CBM and VBM are centered around Γ. (From Ubrig, N., Ponomarev, E., Zultak, J., Domaretskiy, D., Zólyomi, V., Terry, D., Howarth, J., Gutiérrez-Lezama, I., Zhukov, A., Kudrynskyi, Z. R., Kudrynskyi, Z. D., Patané, A., Taniguchi, T., Watanabe, K., Gorbachev, R. V., Falko, V. I. and Morpurgo, A. F., *Nat. Mater.*, 19, 299, 2020. With permission.)

For developing 2D spin-photonic devices, the light-induced valley polarization and spin in TMDs materials and their van der Waals heterostructures have also attracted widespread attention recently. Nevertheless, the generation and control of spin polarization reported by most studies were operated at stringent conditions such as low temperature and high-quality samples, to suppress the intervalley scattering and prolong the lifetime of spin-valley polarization. Therefore, acquiring a high degree of spin polarization is very essential from perspective of practical spintronic applications. To solve the problem, Pan et al. had built type-I and type-II van der Waals heterostructures to modulate the carrier lifetime of layered PbI_2 and observed a near-unity degree of polarization.[89]

For previous work, the polarization degree can be expressed as $\rho = \rho_0/(1+\tau_c/\tau_s)$, where ρ_0, τ_c, and τ_s are the polarization without spin relaxation, carrier lifetime and spin relaxation time, respectively. When the carrier lifetime is much longer than the spin relaxation time, namely $\tau_c \gg \tau_s$, the polarization will become very small. In turn, if the carrier lifetime is much shorter than the time of spin relaxation, namely $\tau_c \ll \tau_s$, the polarization will be large. Hence, obtaining a high polarization degree should either prolong the spin relaxation time or reduce the carrier lifetime. However, the spin relaxation time, which depends on the materials and experiment conditions, is not feasible to be controlled and obtained optimum values. In contrast, the carrier lifetime can be easily changed by temperature, strain, defects and van der Waals interfaces. Stacking different 2D materials to form suitable band alignment in van der Waals interface could give rise to interlayer charge transfer, offering an extra decay channel for photogenerated carriers to reduce the carrier lifetime. In the work of Pan et al., the prepared PbI_2/WS_2 heterostructures demonstrate type-I band alignment (Figure 3.83(b)), where the photoexcited electrons and holes can be transferred from the PbI_2 to the monolayer WS_2.[89] Therefore, the carrier lifetime of PbI_2 is reduced, while the spin polarization significantly increases (Figure 3.83(c)). The formed PbI_2/WSe_2 heterostructures show type-II band alignment (Figure 3.83(d)), where the photogenerated electrons transfer from WSe_2 to PbI_2 but the holes transfer from PbI_2 to WSe_2. The type-II alignment between WSe_2 and PbI_2 could also reduce the carrier lifetime of PbI_2 and lead to a high degree of polarization (Figure 3.83(e)). In addition, Pan et al. demonstrate versatile methods of controlling the spin polarization such as modulating the material thickness, temperature, as well as excitation wavelengths.[89] The work supplies a novel strategy via constructing van der Waals interfaces to acquire a high degree of polarization, which is a valuable supplement for developing spintronics in future.

The heterostructures with different twisted angles can also exhibit novel physical phenomena and diverse electronic/optical properties due to changed Coulomb interactions between layers. For instance, the twisted angles between layers could give rise to a larger effect on the emergence of formation time, transition rate and binding energy as well as valley lifetime of interlayer many-body quasiparticles, such as exciton, trion and so forth, which will affect the carrier mobility, linear and nonlinear optical response of TMDs materials.[90] For this reason, Geohegan et al. could be able to take advantage of the optical spectroscopy, such as photoluminescence, absorption spectroscopy, to reveal the interlayer coupling interaction in the WSe_2/WS_2 heterostructures with various twisted angles between layers.[91]

Except for dangling bond-free surface, TMDs materials have unique layer-dependent bandgap structures, where the interlayer coupling between the layers with layer number decreasing could generally result in the transformation of CBM from the Λ valley, namely the middle of K and Γ points, to the K valley and that of VBM from the Γ hill to K hill in the Brillouin zone. When the thickness of TMDs materials thins down to monolayer, the bandgap would crossover from indirect to direct.[92,93] As a result, multilayer TMDs are indirect bandgap materials and optically inactive, restricting their practical applications in light-emitting device fields. However, multilayer TMDs have higher optical density of states and stronger absorbance than the counterpart of monolayer species.[94] In multilayer TMDs system, therefore, if the carriers located at indirect bandgap can redistribute to direct bandgap, the quantum efficiency would be significantly improved. The work provides a new trend for TMD-based optoelectronic applications.

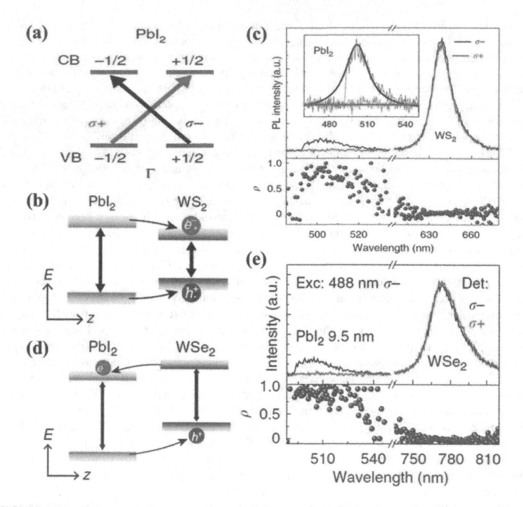

FIGURE 3.83 (a) Schematics of the polarized optical transitions in PbI_2. (b) The energy band alignment and photoexcited carrier behaviors in PbI_2/WS_2 heterostructure. (c) Circularly polarized PL spectra of a typical PbI_2/WS_2 heterostructure at room temperature. (d) The energy band alignment and photoexcited carrier behaviors in PbI_2/WSe_2 heterostructure. (e) Circularly polarized PL spectra of the PbI_2/WSe_2 heterostructure. (From Zhang, D., Liu, Y., He, M., Zhang, A., Chen, S., Tong, Q., Huang, L., Zhou, Z., Zheng, W., Chen, M., Braun, K., Meixner, A. J., Wang, X. and Pan, A., *Nat. Commun.*, 11, 4442, 2020. With permission.)

In recent years, Li et al. have reported that modulating external electric-field achieves the redistribution of injected holes and electrons in multilayer MoS_2.[95] Under the action of external electric field, the injected carriers can acquire kinetic energy to be accelerated, where the electrons transfer from the Λ to the K valley, while the holes transfer from the Γ to K hill (Figure 3.84(a)). Consequently, abundant holes and electrons can recombine at the K point in k space, leading to a considerable EL enhancement in the multilayer MoS_2 the intensity of which is comparable with that of monolayer counterpart, as shown in Figure 3.84(b). In addition, Liu's group confirmed a model of thermally driven intervalley transfer of photogenerated carriers and obtained an improved fluorescence emission from the direct bandgap at high temperature rather than quenching in multilayer MoS_2 (Figure 3.84(c) and (d)).[96] Except for MoS_2, Liu et al. also reported that the above transfers are suitable for other TMDs materials such as multilayer WSe_2 and WS_2 system.[97,98]

Over the last several years, 2D TMDs materials have also attracted momentous research interest from the viewpoint of fundamental physics, such as quantum spin Hall effect, valley polarization, superconductivity and so on, further suggesting potential applications for functional devices in

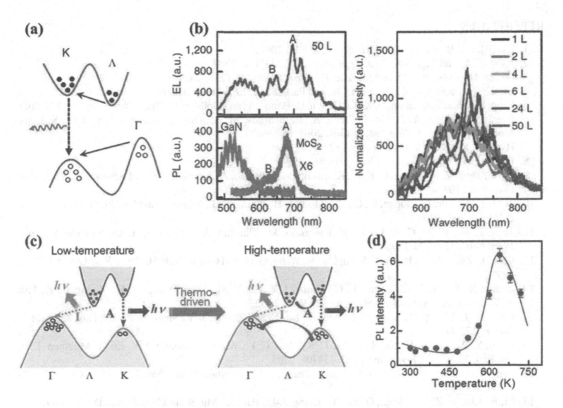

FIGURE 3.84 (a) Schematic diagram of electric-field-induced carrier transfer. (b) Left panel: the EL and PL spectra of 50-layer MoS$_2$ device. Right panel: the EL spectra with different thicknesses of MoS$_2$. (From Li, D., Cheng, R., Zhou, H., Wang, C., Yin, A., Chen, Y., Weiss, N. O., Huang, Y. and Duan, X., *Nat. Commun.*, 6, 7509, 2015. With permission.) (c) Schematic diagram of thermo-driven carrier transfer between different energy valley. (d) PL integrated intensity of A exciton emission as a function of temperature. (From Li, Y., Xu, H., Liu, W., Yang, G., Shi, J., Liu, Z., Liu, X., Wang, Z., Tang, Q. and Liu, Y., *Small*, 13, 1700157, 2017. With permission.)

various areas. The rich varieties of optical and electronic properties of TMDs materials have endowed them with the potential to develop high-performance optoelectronic devices while enabling them downscaling to a feature size with atomically thick. There is a robust reason to believe that TMDs materials could complement the existing silicon complementary metal-oxide-semiconductor technology in future.

3.3.10 Summary

Although the bottleneck problem of ZnO material has not been well solved, a large number of milestone works have emerged in the past nearly two decades of research, including luminescence physics, material modification and device interface control, which has greatly promoted the development prospect of ZnO and related UV-luminescence materials. That is, ZnO-based materials and devices still show considerable promise in the fields of short-wavelength optoelectronic devices, which will continue to attract a lot of attention from researchers. Particularly, some recent efforts to achieve multifunctional expansion of ZnO-based materials and devices are indeed promising, including electrically driven random laser memory, three-dimensional-printed flexible photodetector, ultrasensitive piezotronic/flexible ferroelectric/light-driven FETs, and ultralow-threshold exciton-polariton laser. The cross integration of these emerging fields will certainly bring a new upsurge to further development of ZnO-based materials and related UV-luminescence devices.

REFERENCES

1. Djurišić, A. B. and Leung, Y. H., *Small*, 2, 944, 2006.
2. Ashrafi, A. and Jagadish, C., *J. Appl. Phys.*, 102, 071101, 2007.
3. Özgür, Ü., Hofstetter, D. and Morkoç, H., *Proc. IEEE*, 98, 1255, 2010.
4. Choi, Y. S., Kang, J. W., Hwang, D. K. and Park, S. J., *IEEE Trans. Electron Dev.*, 57, 26, 2010.
5. Fang, X., Zhai, T., Gautam, U. K., Li, L., Wu, L., Bando, Y. and Golberg, D., *Prog. Mater. Sci.*, 56, 175, 2011.
6. Özgür, Ü., Alivov, Y. I., Liu, C., Teke, A., Reshchikov, M. A., Doğan, S., Avrutin, V., Cho, S. J. and Morkoç, H., *J. Appl. Phys.*, 98, 041301, 2005.
7. Hvam, J. M., *Solid State Commun.*, 12, 95, 1973.
8. Hvam, J. M., *Phys. Status Solidi B*, 63, 511, 1974.
9. Zu, P., Tang, Z. K., Wong, G. K. L., Kawasaki, M., Ohtomo, A., Koinuma, H. and Segawa, Y., *Solid State Commun.*, 103, 459, 1997.
10. Teke, A., Özgür, Ü., Doğan, S., Gu, X., Morkoç, H., Nemeth, B., Nause, J. and Everitt, H. O., *Phys. Rev. B*, 70, 195207, 2004.
11. Tang, Z. K., Wong, G. K. L., Yu, P., Kawasaki, M., Ohtomo, A., Koinuma, H. and Segawa, Y., *Appl. Phys. Lett.*, 72, 3270, 1998.
12. Cao, H., Zhao, Y. G., Ho, S. T., Seelig, E. W., Wang, Q. H. and Chang, R. P. H., *Phys. Rev. Lett.*, 82, 2278, 1999.
13. Zhang, X. T., Liu, Y. C., Zhang, L. G., Zhang, J. Y., Lu, Y. M., Shen, D. Z., Xu, W., Zhong, G. Z., Fan, X. W. and Kong, X. G., *J. Appl. Phys.*, 92, 3293, 2002.
14. Chen, S. J., Liu, Y. C., Shao, C. L., Mu, R., Lu, Y. M., Zhang, J. Y., Shen, D. Z. and Fan, X. W., *Adv. Mater.*, 17, 586, 2005.
15. Chen, Y. W., Liu, Y. C., Lu, S. X., Xu, C. S., Shao, C. L., Wang, C., Zhang, J. Y., Lu, Y. M., Shen, D. Z. and Fan, X. W., *J. Chem. Phys.*, 123, 134701, 2005.
16. Tong, Y., Liu, Y., Shao, C., Liu, Y., Xu, C., Zhang, J., Lu, Y., Shen, D. and Fan, X., *J. Phys. Chem. B*, 110, 14714, 2006.
17. Li, B., Liu, Y., Zhi, Z., Shen, D., Lu, Y., Zhang, J.-L., Fan, X., Mu, R. and Henderson, D., *J. Mater. Res.*, 18, 8, 2003.
18. Li, B. S., Xiao, Z. Y., Ma, J. G. and Liu, Y. C., *Chin. Phys. B*, 26, 117101, 2017.
19. Wang, C., Ji, Z., Liu, K., Xiang, Y. and Ye, Z., *J. Cryst. Growth*, 259, 279, 2003.
20. Kaminska, E., Piotrowska, A., Kossut, J., Barcz, A., Butkute, R., Dobrowolski, W., Dynowska, E., Jakiela, R., Przezdziecka, E., Lukasiewicz, R., Aleszkiewicz, M., Wojnar, P. and Kowalczyk, E., *Solid State Commun.*, 135, 11, 2005.
21. Tsukazaki, A., Ohtomo, A., Onuma, T., Ohtani, M., Makino, T., Sumiya, M., Ohtani, K., Chichibu, S. F., Fuke, S., Segawa, Y., Ohno, H., Koinuma, H. and Kawasaki, M., *Nat. Mater.*, 4, 42, 2005.
22. Nakano, Y., Morikawa, T., Ohwaki, T. and Taga, Y., *Appl. Phys. Lett.*, 88, 172103, 2006.
23. Yang, T., Zhang, Z., Li, Y., Lv, M., Song, S., Wu, Z., Yan, J. and Han, S., *Appl. Surf. Sci.*, 255, 3544, 2009.
24. Zou, C. W., Chen, R. Q. and Gao, W., *Solid State Commun.*, 149, 2085, 2009.
25. Zhang, B., Li, M., Wang, J. Z. and Shi, L. Q., *Chin. Phys. Lett.*, 30, 027303, 2013.
26. Jin, Y., Zhang, N. and Zhang, B., *Materials*, 10, 236, 2017.
27. Zhang, Y. Z., Lu, J. G., Ye, Z. Z., He, H. P., Zhu, L. P., Zhao, B. H. and Wang, L., *Appl. Surf. Sci.*, 254, 1993, 2008.
28. Tang, L., Wang, B., Zhang, Y. and Gu, Y., *Mater. Sci. Eng. B*, 176, 548, 2011.
29. Shen, H., Shan C. X., Liu J. S., Li B. H., Zhang, Z. Z. and Shen D. Z., *Phys. Status Solidi B*, 250, 2102, 2013.
30. Shen, H., Shan, C. X., Li, B. H., Xuan, B. and Shen, D. Z., *Appl. Phys. Lett.*, 103, 232112, 2013.
31. Yang, J. J., Fang, Q. Q., Wang, W. N., Wang, D. D. and Wang, C., *J. Appl. Phys.*, 115, 124509, 2014.
32. Ye, Z., Wang, T., Wu, S., Ji, X. and Zhang, Q., *J. Alloys Compd.*, 690, 189, 2017.
33. Swapna R. and Kumar, M. C. S., *Mater. Sci. Eng. B*, 178, 1032, 2013.
34. Kumar, M., Kim, S. K. and Choi, S. Y., *Appl. Surf. Sci.*, 256, 1329, 2009.
35. Jiao, S. J., Zhang, Z. Z., Lu, Y. M., Shen, D. Z., Yao, B., Zhang, J. Y., Li, B. H., Zhao, D. X., Fan, X. W. and Tang, Z. K., *Appl. Phys. Lett.*, 88, 031911, 2006.
36. Lim, J. H., Kang, C. K., Kim, K. K., Park, I. K., Hwang, D. K. and Park, S. J., *Adv. Mater.*, 18, 2720, 2006.
37. Chu, S., Wang, G., Zhou, W., Lin, Y., Chernyak, L., Zhao, J., Kong, J., Li, L., Ren, J. and Liu, J., *Nat. Nanotechnol.*, 6, 506, 2011.

38. Alivov, Y. I., Nostrand, J. E. V., Look, D. C., Chukichev, M. V. and Ataev, B. M., *Appl. Phys. Lett.*, 83, 2943, 2003.
39. Wang, L., Xu, H., Liu, Y. and Shen, L., *Mater. Res. Express*, 2, 025901, 2015.
40. Xu, H. Y., Liu, Y. C., Liu, Y. X., Xu, C. S., Shao, C. L. and Mu, R., *Appl. Phys. B*, 80, 871, 2005.
41. Zhao, L., Xu, C. S., Liu, Y. X., Shao, C. L., Li, X. H. and Liu, Y. C., *Appl. Phys. B*, 92, 185, 2008.
42. Makino, T., Segawa, Y., Kawasaki, M., Ohtomo, A., Shiroki, R., Tamura, K., Yasuda, T. and Koinuma, H., *Appl. Phys. Lett.*, 78, 1237, 2001.
43. Ryu, Y. R., Lee, T. S., Lubguban, J. A., Corman, A. B., White, H. W., Leem, J. H., Han, M. S., Park, Y. S., Youn, C. J. and Kim, W. J., *Appl. Phys. Lett.*, 88, 052103, 2006.
44. Kim, W. J., Leem, J. H., Han, M. S., Park, I. W., Ryu, Y. R. and Lee, T. S., *J. Appl. Phys.*, 99, 096104, 2006.
45. Chen, M., Zhu, Y., Su, L., Zhang, Q., Chen, A., Ji, X., Xiang, R., Gui, X., Wu, T., Pan, B. and Tang, Z., *Appl. Phys. Lett.*, 102, 202103, 2013.
46. Ryu, Y., Lee, T. S., Lubguban, J. A., White, H. W., Kim, B. J., Park, Y. S. and Youn, C. J., *Appl. Phys. Lett.*, 88, 241108, 2006.
47. Park S. H. and Ahn, D., *Physica B*, 441, 12, 2014.
48. Ohtomo, A., Kawasaki, M., Koida, T., Masubuchi, K., Koinuma, H., Sakurai, Y., Yoshida, Y., Yasuda, T. and Segawa, Y., *Appl. Phys. Lett.*, 72, 2466, 1998.
49. Park, W. I., Yi, G. C. and Jang, H. M., *Appl. Phys. Lett.*, 79, 2022, 2001.
50. Liu, C. Y., Xu, H. Y., Wang, L., Li, X. H. and Liu, Y. C., *J. Appl. Phys.*, 106, 073518, 2009.
51. Park, W. I., Yi, G. C., Kim, M. and Pennycook, S. J., *Adv. Mater.*, 15, 526, 2003.
52. Bae, J. Y., Yoo, J. and Yi, G. C., *Appl. Phys. Lett.*, 89, 173114, 2006.
53. Park, W. I., Yoo, J., Kim, D. W., Yi, G. C. and Kim, M., *J. Phys. Chem. B*, 110, 1516, 2006.
54. Jang, E. S., Bae, J. Y., Yoo, J., Park, W. I., Kim, D. W., Yi, G. C., Yatsui, T. and Ohtsu, M., *Appl. Phys. Lett.*, 88, 023102, 2006.
55. Yoo, J., Hong, Y. J., Jung, H. S., Kim, Y. J., Lee, C. H., Cho, J., Doh, Y. J., Dang, L. S., Park, K. H. and Yi, G. C., *Adv. Funct. Mater.*, 19, 1601, 2009.
56. Kong, B. H., Mohanta, S. K., Kim, Y. Y. and Cho, H. K., *Nanotechnology*, 19, 085607, 2008.
57. Chu, S., Olmedo, M., Yang, Z., Kong, J. and Liu, J., *Appl. Phys. Lett.*, 93, 181106, 2008.
58. Cao, B. Q., Zúñiga-Pérez, J., Boukos, N., Czekalla, C., Hilmer, H., Lenzner, J., Travlos, A., Lorenz, M. and Grundmann, M., *Nanotechnology*, 20, 305701, 2009.
59. Chauveau, J. M., Teisseire, M., Chauveau, H. K., Morhain, C., Deparis, C. and Vinter, B., *J. Appl. Phys.*, 109, 102420, 2011.
60. Choi, Y. S., Kang, J. W., Kim, B. H. and Park, S. J., *Opt. Express*, 21, 31560, 2013.
61. Tsukazaki, A., Ohtomo, A. and Kawasaki, M., *Appl. Phys. Lett.*, 88, 152106, 2006.
62. Tampo, H., Shibata, H., Matsubara, K., Yamada, A., Fons, P., Niki, S., Yamagata, M. and Kanie, H., *Appl. Phys. Lett.*, 89, 132113, 2006.
63. Tsukazaki, A., Ohtomo, A., Kita, T., Ohno, Y., Ohno, H. and Kawasaki, M., *Science*, 315, 1388, 2007.
64. Yano, M., Hashimoto, K., Fujimoto, K., Koike, K., Sasa, S., Inoue, M., Uetsuji, Y., Ohnishi, T. and Inaba, K., *J. Cryst. Growth*, 301–302, 353, 2007.
65. Tampo, H., Shibata, H., Maejima, K., Yamada, A., Matsubara, K., Fons, P., Kashiwaya, S., Niki, S., Chiba, Y., Wakamatsu, T. and Kanie, H., *Appl. Phys. Lett.*, 93, 202104, 2008.
66. Brandt, M., von Wenckstern, H., Benndorf, G., Hochmuth, H., Lorenz, M. and Grundmann, M., *Thin Solid Films*, 518, 1048, 2009.
67. Ye, J. D., Pannirselvam, S., Lim, S. T., Bi, J. F., Sun, X. W., Lo, G. Q. and Teo, K. L., *Appl. Phys. Lett.*, 97, 111908, 2010.
68. Tampo, H., Matsubara, K., Yamada, A., Shibata, H., Fons, P., Yamagata, M., Kanie, H. and Niki, S., *J. Cryst. Growth*, 301–302, 358, 2007.
69. Sasa, S., Ozaki, M., Koike, K., Yano, M. and Inoue, M., *Appl. Phys. Lett.*, 89, 053502, 2006.
70. Sasa, S., Maitani, T., Furuya, Y., Amano, T., Koike, K., Yano, M. and Inoue, M., *Phys. Status Solidi A*, 208, 449, 2011.
71. Tsukazaki, A., Akasaka, S., Nakahara, K., Ohno, Y., Ohno, H., Maryenko, D., Ohtomo, A. and Kawasaki, M., *Nat. Mater.*, 9, 889, 2010.
72. Brasse, M., Sauther, S. M., Falson, J., Kozuka, Y., Tsukazaki, A., Heyn, C., Wilde, M. A., Kawasaki, M. and Grundler, D., *Phys. Rev. B*, 89, 075307, 2014.
73. Maryenko, D., Falson, J., Kozuka, Y., Tsukazaki, A. and Kawasaki, M., *Phys. Rev. B*, 90, 245303, 2014.
74. Krishnamoorthy, S., Iliadis, A. A., Inumpudi, A., Choopun, S., Vispute, R. D. and Venkatesan, T., *Solid State Electron.*, 46, 1633, 2002.

75. Koike, K., Hama, K., Nakashima, I., Takada, G.-y., Ozaki, M., Ogata, K.-i., Sasa, S., Inoue, M. and Yano, M., *Jpn. J. Appl. Phys.*, 43, L1372, 2004.
76. Maryenko, D., Falson, J., Kozuka, Y., Tsukazaki, A., Onoda, M., Aoki, H. and Kawasaki, M., *Phys. Rev. Lett.*, 108, 186803, 2012.
77. Wang, L., Xu, H. Y., Zhang, C., Li, X. H., Liu, Y. C., Zhang, X. T., Tao, Y., Huang, Y. and Chen, D. L., *J. Alloys Compd.*, 513, 399, 2012.
78. Wang, L., Ma, J. G., Xu, H. Y., Zhang, C., Li, X. H. and Liu, Y. C., *Appl. Phys. Lett.*, 102, 031905, 2013.
79. Liu, Y., Xue, C., Zhang, J., Gu, D., Shen, L. and Wang, L., *Solid State Sci.*, 102, 106163, 2020.
80. Baugher, B. W. H., Churchill, H. O. H., Yang, Y. and Jarillo-Herrero, P., *Nat. Nanotechnol.*, 9, 262, 2014.
81. Ross, J. S., Klement, P., Jones, A. M., Ghimire, N. J., Yan, J., Mandrus, D. G., Taniguchi, T., Watanabe, K., Kitamura, K., Yao, W., Cobden, D. H. and Xu, X., *Nat. Nanotechnol.*, 9, 268, 2014.
82. Kufer, D. and Konstantatos, G., *Nano Lett.*, 15, 7307, 2015.
83. Wang, X., Wang, P., Wang, J., Hu, W., Zhou, X., Guo, N., Huang, H., Sun, S., Shen, H., Lin, T., Tang, M., Liao, L., Jiang, A., Sun, J., Meng, X., Chen, X., Lu, W. and Chu, J., *Adv. Mater.*, 27, 6575, 2015.
84. Wi, S., Kim, H., Chen, M., Nam, H., Guo, L. J., Meyhofer, E. and Liang, X., *ACS Nano*, 8, 5270, 2014.
85. Pradhan, S. K., Xiao, B. and Pradhan, A. K., *Sol. Energy Mater. Sol. Cells*, 144, 117, 2016.
86. Li, Y., Shi, J., Mi, Y., Sui, X., Xu, H. and Liu, X., *J. Mater. Chem. C*, 7, 4304, 2019.
87. Geim, A. K. and Grigorieva, I. V., *Nature*, 499, 419, 2013.
88. Ubrig, N., Ponomarev, E., Zultak, J., Domaretskiy, D., Zólyomi, V., Terry, D., Howarth, J., Gutiérrez-Lezama, I., Zhukov, A., Kudrynskyi, Z. R., Kudrynskyi, Z. D., Patané, A., Taniguchi, T., Watanabe, K., Gorbachev, R. V., Falko, V. I. and Morpurgo, A. F., *Nat. Mater.*, 19, 299, 2020.
89. Zhang, D., Liu, Y., He, M., Zhang, A., Chen, S., Tong, Q., Huang, L., Zhou, Z., Zheng, W., Chen, M., Braun, K., Meixner, A. J., Wang, X. and Pan, A., *Nat. Commun.*, 11, 4442, 2020.
90. Shi, J., Li, Y., Zhang, Z., Feng, W., Wang, Q., Ren, S., Zhang, J., Du, W., Wu, X., Sui, X., Mi, Y., Wang, R., Sun, Y., Zhang, L., Qiu, X., Lu, J., Shen, C., Zhang, Y., Zhang, Q. and Liu, X., *ACS Photonics*, 6, 3082, 2019.
91. Wang, K., Huang, B., Tian, M., Ceballos, F., Lin, M.-W., Mahjouri-Samani, M., Boulesbaa, A., Puretzky, A. A., Rouleau, C. M., Yoon, M., Zhao, H., Xiao, K., Duscher, G. and Geohegan D. B., *ACS Nano*, 10, 6612, 2016.
92. Splendiani, A., Sun, L., Zhang, Y., Li, T., Kim, J., Chim, C.-Y., Galli, G. and Wang, F., *Nano Lett.*, 10, 1271, 2010.
93. Mak, K. F., Lee, C., Hone, J., Shan, J. and Heinz, T. F., *Phys. Rev. Lett.*, 105, 136805, 2010.
94. Dhall, R., Neupane, M. R., Wickramaratne, D., Mecklenburg, M., Li, Z., Moore, C., Lake, R. K. and Cronin, S., *Adv. Mater.*, 27, 1573, 2015.
95. Li, D., Cheng, R., Zhou, H., Wang, C., Yin, A., Chen, Y., Weiss, N. O., Huang, Y. and Duan, X., *Nat. Commun.*, 6, 7509, 2015.
96. Li, Y., Xu, H., Liu, W., Yang, G., Shi, J., Liu, Z., Liu, X., Wang, Z., Tang, Q. and Liu, Y., *Small*, 13, 1700157, 2017.
97. Chen, H., Li, Y., Liu, W., Xu, H., Yang, G., Shi, J., Feng, Q., Yu, T., Liu, X. and Liu, Y., *Nanoscale Horiz.*, 3, 598, 2018.
98. Li, Y., Liu, W., Xu, H., Chen, H., Ren, H., Shi, J., Du, W., Zhang, W., Feng, Q., Yan, J., Zhang, C., Liu, Y. and Liu, X., *Adv. Opt. Mater.*, 8, 1901226, 2020.

3.4 PHOSPHOR-BASED SINGLE-MODE LASERS FOR OPTICAL COMMUNICATION

Dan Lu, Song Liang, Ming Li and Lingjuan Zhao

The semiconductor lasers and optical fiber invented and developed since the 1960s have revolutionized modern communications. At present, semiconductor lasers are one of the most important light sources in the communication system. To match the low-loss transmission window of the optical fiber, semiconductor lasers for communication mostly work at the wavelength of 0.85, 1.3 and 1.5 μm. The 0.85-μm semiconductor lasers are based on the ternary AlGaAs/GaAs material system, and the 1.3- and 1.5-μm lasers are based on the quaternary InGaAsP/InP or AlGaInAs/InP material systems.

During the last 60 years, the structures and performance of semiconductor lasers keep evolving to accommodate and accelerate the fast development of the optical communication system. The active material of the semiconductor lasers have developed from bulk material to quantum well (QW), quantum dash and quantum dot active structures. They have evolved from pulsed operation cooled by liquid nitrogen to continuous-wave (CW) operation over 100°C, with a lifetime of more than 100,000 hours. The modulation speed has developed from tens of Mbps to more than 400 Gbps[1]; the spectral linewidth has reduced from tens of MHz to the order of kHz[2]; the single-mode output power has evolved from a few milliwatts to hundreds of milliwatts.[3,4]

In this chapter, we will review phosphor-based single-mode lasers for optical communication, with concentration in the distributed feedback (DFB) lasers. The major applications of the single-mode lasers can be categorized into the following aspects:

1. **CW single-wavelength or multiwavelength applications:** The target application is the wavelength division multiplexing (WDM) system, where multiple single-mode lasers with different wavelengths are used in the system. Wavelength control is a key requirement for the lasers, which is mainly achieved by controlling the period of the built-in grating of the laser.

2. **High-speed direct modulation applications:** The 5G technology, data center and access network demand a large number of low-cost optical transceivers, where high-speed directly modulated lasers (DMLs) play a very important role for low-cost, mass deployment scenarios. Uncooled operation (−40–95°C) and high modulation speed (> 25 Gbps) are the technical requirements for this type of application.

3. **High-power applications**: In the area of silicon photonics, phosphor-based high-power lasers are imperative due to the lack of effective lasing and amplification in the silicon platform. The power loss in complex silicon devices needs to be compensated by high-power lasers. Another field is the free-space communication and the eye-safe lidar for autonomous driving, where phosphor-based high-power single-mode lasers working at 1.5 μm are important for long-range transmission and sensing.

4. **Low-noise applications:** Low-noise lasers include narrow linewidth laser and low relative intensity noise (RIN) laser. In coherent optical communication systems, narrow-linewidth single-mode lasers are one of the key devices influencing the overall feasibility of deployment of coherent technology to cost-sensitive short- and medium-reach coherent systems. Low RIN lasers are a crucial component in microwave photonics, which has a stringent requirement on the noise performance of the laser source.

DFB lasers are commercially mature and well developed in most of the above-mentioned fields. In the following sections, the basic structure, design principle and research progress of various DFB lasers are discussed.

3.4.1　Basic Principles of DFB Lasers

Figure 3.85(a) shows the basic structure of a DFB laser. It relies on the built-in grating structure to select the lasing wavelength. Compared with the Fabry-Perot (FP) laser, the DFB laser adds a grating layer above or below the active area. By periodically perturbing the refractive index or gain/loss of the laser waveguide and active area, the grating creates a selectively distributed feedback to the lasing modes, to support a specific lasing wavelength. The concept and theory of DFB lasers were first proposed by H. Kogelnik and C. V. Shank during 1971–1972[5,6] and the experimental realization of a semiconductor DFB laser appeared in 1973.[7] The lasing wavelength λ_B of the DFB laser is determined by the grating period Λ, the grating order m and the effective mode refractive index n_{eff} following the Bragg relationship:

$$\Lambda = m \frac{\lambda_B}{2n_{eff}}$$

(3.35)

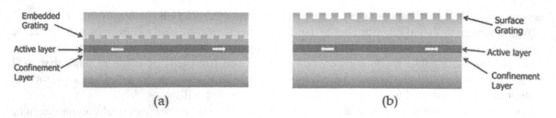

FIGURE 3.85   The basic structure of a distributed feedback (DFB) laser. (a) build-in gratings (b) surface gratings.

DFB lasers working at the 1.3 and 1.5 μm bands normally adopt the first-order grating (m=1), with the corresponding periods of about 200 and 240 nm, respectively. Either holographic or electron-beam lithography can be used to manufacture the gratings. Depending on the nature of perturbation, the DFB lasers can be mainly categorized as gain-coupled (periodic perturbation of the gain) or index-coupled DFB lasers (periodic perturbation of the refractive index). The gain-coupled grating can be realized by directly etching the active region to form a periodic structure. The index-coupled gratings can be realized by etching the separate confinement heterostructure (SCH) layer (such as InGaAsP), followed by the regrowth of material with a different refractive index (such as InP). Besides, there is also a complex-coupled grating structure combining both refractive index and gain coupling mechanism. This structure introduces a periodic current blocking layer or absorption layer in the SCH layer in combination with the index-coupled grating. This structure can take advantage of the gain coupling effect without introducing the potential damage to the active area as compared to the pure gain-coupled grating.

The grating period can be uniform or phase-shifted. For the uniform grating where the gratings have the same period, there are two degenerate modes with the same lowest thresholds gain when there are no additional reflections from the facet. The two modes are equally spaced on each side of the stopband centering around the Bragg wavelength. When the facet reflectivity is not zero, the mode degeneracy may be broken, resulting in a single-mode lasing. There is a single-mode yield problem due to the random relative phase of the laser facets. The phase-shifted grating introduces an additional phase-shifted region in the uniform grating, breaking the mode degeneration of the uniform grating, and single-mode lasing can be realized. Different from the uniform DFB laser, the phase-shifted DFB laser works at a wavelength corresponding to the position inside the stopband of the uniform grating. When the phase shift is 90 degrees—that is, the λ/4 phase shift, the lasing wavelength is located at the center of the stopband and has the best mode selectivity. Its single-mode performance is robust against the variation of the end-face reflectivity and phase, and it can maintain good single-mode performance even under high-speed modulation. Therefore, the λ/4 phase shift grating structure is widely used in high-speed DMLs. However, due to the spatial hole burning (SHB) effect, the single-mode performance of the λ/4 phase-shift DFB laser will be worse when working at high power, and the linewidth will also be broadened. Other phase-adjusted grating structures,[8] multiple phase-shift structures[9] and periodic modulation structures (corrugation pitch modulated grating, CPM)[10] can be used to suppress the hole burning. Gratings with phase-shifted structures have better mode stability against optical feedback, reduced chirp and narrower optical linewidth as compared to uniform gratings. Typical structures include λ/8 phase-shift gratings,[11] periodic modulation gratings[2] and so on.

The gain-coupled DFB laser has an eigenmode with the lowest threshold at the Bragg wavelength, supporting the single-mode lasing. The single-mode operation of the laser is less affected by the end-facet reflectivity,[12] having a low chirp and better mode stability against optical feedback.[13] However, it is difficult to manufacture a pure gain-coupled DFB laser without perturbing the refractive index. Instead, a complex-coupled grating is usually adopted if the gain coupling effect is desired. Overall, gain-coupled DFB lasers are not as widely used as index-coupled DFB lasers.

The DFB laser as shown in Figure 3.85(a) has an embedded grating structure, which requires at least two steps of epitaxy, and a high-precision grating patterning process. Aside from the embedded grating, the surface grating structure, as shown in Figure 3.85(b), can also be used to provide distributed feedback.[14] Surface gratings can be made by etching slots on the side of or the surface of the ridge waveguide. The main advantage of the surface grating structure is its simple manufacturing process, which does not require a second-step epitaxial regrowth to form the embedded grating structure. The surface grating normally adopts a high-order grating structure with a long grating period, which can be made through standard photolithography, without the need for high-precision holographic and electron-beam lithography. Because the surface grating DFB laser only needs one-step epitaxy, this makes it easier to integrate with other photonic devices[15] having potential advantages in photonics integration.

3.4.2 DIRECTLY MODULATED DFB LASERS

Direct modulation is one of the simplest ways to convert electrical signal into optical signal and has been widely used in optical fiber communication systems. Combining the advantages of low cost, low power consumption, small footprint and the capability in mass production, directly modulated semiconductor lasers (DMLs) are most favored by the short- to medium-range optical communication system. Comparing with the FP lasers, the single-mode DFB lasers are immune to intermodal dispersion of the fiber, and compatible with the WDM system. The DMLs have been widely used in situations with a modulation rate above 10 Gbps and the transmission distance exceeds 2 km.

When directly modulated by the electric signal, the DML replicates the electrical waveform in the optical domain, usually accompanied by waveform distortion, frequency chirp, bandwidth limitation etc., which influence the maximum available transmission distance and the modulation rate. By moving the working wavelength from 1.5 μm to the zero-dispersion wavelength of 1.3 μm, the influence of the fiber dispersion and laser frequency chirp can be mitigated. The available single-channel direct-modulation rate can be further increased to more than 25 Gbaud covering a transmission distance of 10–20 km. Further increase of the modulation rate and the coverage reach requires a comprehensive laser design and optimization.

The modulation response of a directly modulated semiconductor laser can be written as[16,17]:

$$|H(f)|^2 = \frac{1}{\left[1+\left(f/f_0\right)^2\right]}\frac{1}{\left[1-\left(f/f_r\right)^2\right]^2+\left(\gamma/2\pi f_r\right)^2\left(f/f_r\right)^2} \qquad (3.36)$$

where $f_0 = 1/(2\pi RC)$ is the parasitic cutoff frequency determined by the device resistance R and capacitance C, γ is the damping factor, and f_r is the laser relaxation oscillation frequency:

$$f_r = \frac{1}{2\pi}\sqrt{\frac{\Gamma v_g a}{qV}\eta_i\left(I-I_{th}\right)} \qquad (3.37)$$

where Γ is the optical confinement factor, v_g is the group velocity, $a=dg/dN$ is the differential gain, g is the gain coefficient, η_i is the injection efficiency, q is the elementary charge of the electron, V is the volume of the active region, I is the injection current and I_{th} is the threshold current.

The design of a high-speed DML usually involves the optimization of the electrical impedance, active material and device structure. The impedance optimization can be achieved by adjusting the doping concentration and adopting electrode structures with lower capacitance. The optimization of the active material and device structure is more complex, typically, including the optimization of the optical confinement factor, the differential gain and the volume of the active area. The major design principles include the following considerations.

3.4.2.1   Quantum Well

The optimization of the QWs can be carried out by introducing strain in the QW structure, selective doping and increasing the number of QWs.

The introduction of the strained QW structure can reduce the density of states and increase the differential gain; the selective doping of Zn in the QW and the top SCH layer can increase the population inversion efficiency in the well and reduce the carrier transport effect, thereby increasing the differential gain. DML with 25 GHz bandwidth using p-doped strained MQW was reported in the 1990s.[18]

Increasing the number of QWs can increase the optical confinement factor and the differential gain.[19,20] In semiconductor lasers, Auger recombination[21] and intervalence band absorption[22] are two major factors affecting the modulation bandwidth of DMLs. With the increase of carrier injection density, the Auger recombination and inter-band absorption will deteriorate the lasing efficiency of the QW. When the multiple-QW structure is adopted, the injected carrier density in each QW can be effectively reduced, which is beneficial to suppress the influence of the Auger recombination and inter-band absorption. A lower threshold carrier density also results in a higher differential gain.[23] The dependence of the differential gain on the number of QWs is shown in Figure 3.86.[20] With 20 pairs of QWs, a bandwidth of 30 GHz was reported for 1.5-μm DFB lasers.[24] However, it should be noted that with the increase of the QWs, the internal loss will also increase, influencing the maximum available laser power.

3.4.2.2   Material System

Both InGaAsP and AlGaInAs can be adopted as the active material for lasers emitting at 1.3 and 1.5 μm. Traditionally, commercial high-speed DMLs mostly use InGaAsP as the QW material. However, due to the small conduction band offset ($\Delta E_c/\Delta E_g$=0.4), InGaAsP has poor electron confinement as compared to AlGaInAs ($\Delta E_c/\Delta E_g$=0.72). When working at high temperature, the modulation performance of InGaAsP-based DML is severely degraded and can hardly be used in the uncooled situation at commercial (0–70°C) or industrial temperature (−40–85°C) while maintaining a high modulation bandwidth. The modulation rate of uncooled InGaAsP-based DML is generally limited to less than 10 Gbps at 85°C.

AlGaInAs has a better temperature characteristic and a higher differential gain. Figure 3.87 shows the comparison of the modulation performance of the two kinds of DMLs based on InGaAsP

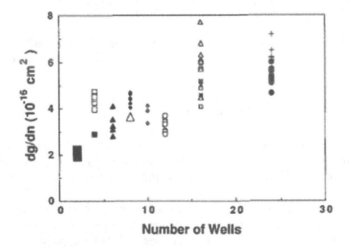

FIGURE 3.86 The influence of the number of quantum wells on the differential gain. (From Tatham, M. C., Lealman, I. F., Seltzer, C. P., Westbrook, L. D. and Cooper, D. M., *IEEE J. Quantum Electron.*, 28, 408, 1992. With permission.)

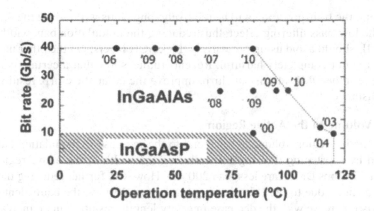

FIGURE 3.87 Comparison of InGaAsP-based and AlGaInAs-based DMLs operated in uncooled condition. (From Fukamachi, T., Adachi, K., Shinoda, K., Kitatani, T., Tanaka, S., Aoki, M. and Tsuji, S., *IEEE J. Sel. Top. Quantum Electron*, 17, 1138, 2011. With permission.)

and AlGaInAs at different temperatures.[25] AlGaInAs-based DMLs show considerably higher modulation bandwidth at high temperatures. After the breakthrough in the high-quality epitaxial growth of Al-containing materials,[26] AlGaInAs-based DMLs with high reliability have been gradually commercialized. Currently, most 1.3-μm DMLs are based on the AlGaInAs material system. The modulation bandwidth of AlGaInAs-based DMLs has reached 34 GHz at 25°C and 23.9 GHz at 80°C in 2019,[27] enabling a 10-km single-mode fiber transmission of 106 Gbps within a temperature of 25–80°C.

3.4.2.3 Gratings and SCH Layer
A well-designed grating can make the DML work at a wavelength with high differential gain while maintaining a stable single mode. By blue-detuning the Bragg wavelength relative to the gain peak, as shown in Figure 3.88, the DML can work in the wavelength region with higher differential gain, enabling a higher modulation bandwidth. To suppress the SHB effect that may influence the dynamic single-mode operation, gain-coupled grating,[28] CPM grating[27] or asymmetric CPM (ACPM) grating can be adopted.

The key consideration in the SCH layer design is the carrier transport effect. The carrier transport refers to the carrier transport time across the SCH layer to the active region. With

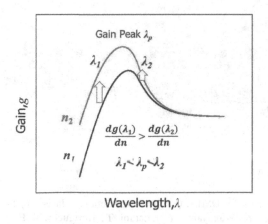

FIGURE 3.88 Influence of wavelength offset on the differential gain.

thick SCH layers, the transport time will be relatively long. This will reduce the differential gain and enhance the low-pass filtering effect, thus reducing the modulation bandwidth. Therefore, a high-speed DML should avoid using excessively thick SCH layers. On the other hand, the carrier transport effect is not completely harmful. Recent studies show that a certain carrier transport effect can suppress the DML's transient chirp, improve the adiabatic chirp, resulting in a longer transmission distance.[29,30]

3.4.2.4 The Volume of the Active Region

Decreasing the active region volume is beneficial to the increase in modulation bandwidth. This can be realized by shortening the length or reducing the width of the active region. The cavity lengths of most 25-Gbps DMLs are less than 200 μm. However, further reducing the active region length will be difficult due to the chip cleaving limitation. Besides, the equivalent mirror loss of the laser will also increase with the decrease in cavity length, resulting in an increase in the laser threshold. The thermal effect will also be more serious in the short device, deteriorating DML's high-temperature performance. This problem with the short cavity can be solved by integrating another passive structure.

Reducing the active region width is another way to reduce the active region volume. Buried heterostructure (BH) can be adopted. This can be realized by using Fe-InP-doped semi-insulating barrier layer to block the lateral diffusion of carriers,[31] to improve the current injection efficiency, reduce the threshold current, lower the parasitic impedance and thereby increase the modulation bandwidth.

3.4.2.5 Passive Structure

When the active region of the DML cannot be further reduced, the integration of a section of passive structure can be used. Distributed reflector (DR)[32,33] and passive waveguide[34] are two common structures for integration, as shown in Figure 3.89. By fabricating passive distributed feedback Bragg reflectors (DBR) or passive waveguides on one or both sides of the active region, the total device length can be long enough for cleaving and effective heat dissipation, while maintaining a short active region length. DR structure can also increase the equivalent facet reflectivity and reduce the threshold current density otherwise faced by short cavity lasers. Using this type of structure, the length of the active region can be reduced to the order of 100 μm. DMLs with such structures can operate beyond 50 Gbps.[34–36]

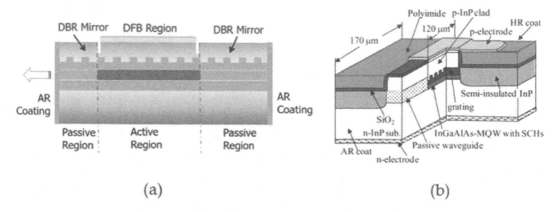

(a)   (b)

FIGURE 3.89 Short-active-cavity DMLs. (a) distributed reflector laser; (b) DML with integrated passive waveguide. (From Nakahara, K., Wakayama, Y., Kitatani, T., Taniguchi, T., Fukamachi, T., Sakuma, Y. and Tanaka, S., *IEEE Photon. Technol. Lett.*, 27, 534, 2015. With permission.)

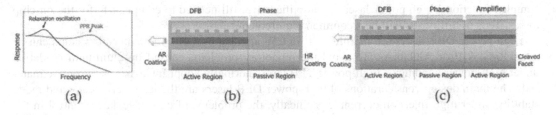

FIGURE 3.90   (a) Photon–photon resonance; (b) passive feedback structure; (c) amplified feedback structure.

3.4.2.6   Photon–Photon Resonance

To further increase the modulation bandwidth, the limitation of the laser's relaxation oscillation has to be overcome. The introduction of the photon–photon resonance (PPR) effect represents a new direction. The PPR effect increases the laser bandwidth by adding another resonance peak beyond the relaxation oscillation peak of the laser, as shown in Figure 3.90(a). This additional response peak can be obtained by introducing a feedback mechanism to the DFB laser. Common structures include passive feedback laser (PFL)[37,38] and amplified feedback laser (AFL).[39,40] The PFL integrates a passive structure with the DFB laser, as shown in Figure 3.90(b). An electrode on the passive waveguide is used to control the feedback phase. The passive region has a high-reflector (HR) coating to increase the amount of feedback. The modulation bandwidth can be extended by three times through the PPR effect, achieving a direct modulation bandwidth of 35 GHz.[41] AFL adds an amplifier section to the PFL, as shown in Figure 3.90(c). This structure is more flexible in controlling the feedback strength and the phase. A bandwidth enhancement factor of two has been reported using the AFL.[42] Besides, the PPR effect can also be achieved by using the DR structure. DR laser with a bandwidth of 55 GHz has been reported in 2017,[43] a sing-mode-fiber transmission of 112-Gbps four-level pulse amplitude modulation (PAM-4) signals was realized. By using DFB laser integrating with a passive waveguide and appropriate reflection coating (DFB+R laser), modulation bandwidth of 67 GHz was also demonstrated by taking advantage of the PPR and detuned loading effect.[1]

3.4.3   High-power DFB Lasers

In optical fiber communications, DFB lasers normally operate at a power level not exceeding 20 mW. When the transmission medium switches from optical fiber to free space, the demand for optical power will be considerably higher, since there is no waveguide in free space, the laser beam tends to diverge under the influence of the atmospheric disturbance and the overall loss will be much higher than that of the optical fiber. Single-lateral-mode high-power FP laser is one solution for single-channel applications. However, with the adoption of WDM technology and coherent communication in free-space communication, single-lateral and single-longitudinal-mode high-power DFB lasers will be required.

With the development of autonomous driving technology, the demand for lidar is also growing. A single-mode high-power light source with high beam quality is desired. Especially in phased array lidar and frequency-modulated CW lidar, a high-power DFB laser is one of the key devices. The wavelength of 1.5-μm lies within the low-loss transmission window of the atmosphere and is also an eye-safe wavelength. This makes the 1.5-μm high-power DFB laser an important light source, with an expected power level of hundreds of milliwatts or even watts.

In the emerging field of silicon photonics, the light source has always been an important issue. Due to the absence of direct lasing and amplification in silicon devices, external light sources have become a realistic choice for the silicon platform. In silicon-based applications, both 1.3- and 1.5-μm light sources are available. Although silicon-based devices are highly integrated and can perform complex functions, high-power lasers or amplifiers are still needed to compensate for the on-chip losses. Tens of milliwatts will be a common requirement.

High-power DFB lasers are mainly used in CW applications, and the material systems can be either InGaAsP or AlGaInAs. To avoid the reliability problems caused by aluminum oxidation under high-power conditions, high-power DFB lasers normally adopt InGaAsP as the active material. The main design considerations of high-power DFB lasers are the lasing efficiency and mode stability under high injection current. Specifically, the problems of carrier leakage control in the active region, internal loss reduction, longitudinal space hole burning suppression and side mode control should be considered in high-power cases.

High-power operation prefers a high slope efficiency under high injection current. At the same time, maintaining single-mode characteristics is also another requirement. The design usually involves the optimization of the following aspects.

3.4.3.1 Quantum Well

For high-power operation, it is desirable to adopt strained QWs to improve the injection efficiency and reduce the number of the wells to reduce the internal losses. Under high-power conditions, the slope efficiency is an important parameter reflecting the performance of the laser. Slope efficiency can be expressed by injection efficiency η_i, internal loss α_i and distributed mirror loss α_m:

$$\eta_{\text{slope}} = \frac{h\nu}{q} \eta_i \frac{1}{1+(\alpha_i / \alpha_m)} \tag{3.38}$$

where h is the Planck constant, ν is the laser frequency. The slope efficiency can be improved by increasing the injection efficiency η_i and reducing α_i/α_m. When designing the QWs, it is necessary to optimize the width and strain of the well barrier, so that the QW has better carrier confinement to increase η_i; on the other hand, it is also necessary to reduce the internal loss α_i of the QW. The internal loss of a semiconductor laser can be written as[44]:

$$\alpha_i = \sum_q \alpha_q = \sum_q \Gamma_q \left(\sigma^n n_q + \sigma^p p_q \right) \tag{3.39}$$

where α_q is the internal loss of the layer q, Γ_q is the optical confinement factor of the layer, σ^n and σ^q are the scattering cross sections of electrons and holes, n_p and n_q, are the free carrier densities, respectively. The internal loss mainly comes from free carrier absorption in the active region and the doped cladding layer.[45,46] The scattering loss is due to epitaxial or material defects. The internal loss can be reduced by reducing the optical confinement factor, which can be realized by reducing the number or the thickness of the QWs. The number of QWs is mostly less than five for DFB lasers with output power beyond 100 mW.

3.4.3.2 Cavity Length

A long cavity is usually adopted in high-power lasers. With the increase of injection carrier density, the Auger effect will play an important role in reducing the lasing efficiency. This effect is particularly serious for long-wavelength lasers such as a 1.5-μm laser. At a high carrier density, the leakage level of the injected carriers from the QWs will also be high, resulting in a reduced injection efficiency. Increasing the length of the laser cavity can effectively suppress the Auger effect and carrier leakage. As the cavity length increases, the carrier density under the same injection current will decrease, thus suppressing the Auger effect; at the same time, the carriers are more effectively confined in the QWs, keeping the injection efficiency at a higher level. Besides, a long cavity also increases the heat dissipation area of the device, thereby reducing the influence of thermal effects and increasing the saturation output power. The cavity lengths of most DFB lasers with output power above 100 mW are generally above 1000 μm. However, it should be noted that the distributed mirror loss α_m is inversely proportional to the laser cavity length. As the cavity length increases, α_m also decreases, and the slope efficiency will be reduced. For a DFB laser, a long cavity means that more energy will be confined by the grating, which will affect the slope efficiency and output power. Therefore, the laser cavity cannot be too long. It is important to reduce the internal loss to maintain a high slope efficiency at a long cavity length.

3.4.3.3 SCH and Cladding Layer

The optimization of the SCH layer and the cladding layer mainly includes the use of a graded-index SCH (GRIN-SCH) structure, doping level control at the interface and large optical cavity (LOC) design. To keep a high injection efficiency, it is necessary to confine the carriers in the active region to reduce the electron-hole recombination outside the QWs. The injected carriers should be easily captured and confined by the SCH layers. The carrier leakage from the SCH layer to the heavily-doped cladding layers should be kept low.

By changing the bandgap/refractive index profile from a stepwise to a graded structure, as shown in Figure 3.91(a), the SCH layer can capture the carriers in a shorter time. This will improve the carrier confinement, increasing the injection efficiency.[47] Another way is to increase the barrier height between the SCH layer and the p-InP cladding layer[48] to block the electron leakage from the top SCH layer to the p-InP cladding layer, so as to maintain a high injection efficiency. The suppression of the leakage current from the SCH/cladding layer can effectively increase the saturation power of the laser under a high injection current.[49] Since the barrier height of the SCH/cladding layer increases as the doping concentration increases, heavy doping of the p-InP cladding layer can suppress the leakage current. Figure 3.91(b) shows the effect of doping level at the interface on the injection efficiency. However, the heavily doped cladding layer will introduce higher absorption loss. To solve the problem, the p-InP cladding layer can be graded doped or delta-doped[49] to balance the injection efficiency and internal loss. Graded doping is realized by doping the SCH/cladding junction with a moderate concentration (such as $5 \times 10^{17}/cm^3$) and gradually increasing the doping concentration (such as $1 \times 10^{18}/cm^3$) away from the junction. Using this method, the leakage current can be suppressed while maintaining a low internal loss level. Delta-doping is to add a heavily doped (such as $2 \times 10^{18}/cm^3$) thin layer at the SCH/cladding interface to form a strong barrier to block the leakage current. Because the heavily doped layer is very thin, its influence on the internal

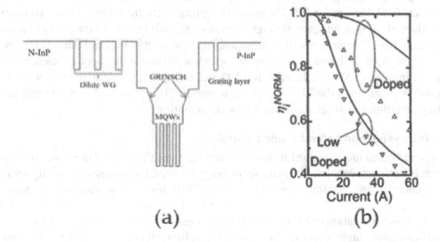

(a) (b)

FIGURE 3.91 (a) Structure of a laser with GRIN-SCH layers and dilute waveguide; (b) influence of doping level on the injection efficiency at the SCH/p-cladding interface. (From Shterengas, L., Menna, R., Trussell, W., Donetsky, D., Belenky, G., Connolly, J. and Garbuzov, D., *J. Appl. Phys.*, 88, 2211, 2000. With permission.)

loss is relatively small. This method can effectively maintain the injection efficiency at a high injection current and increase the saturation output power.[49]

The scattering cross section of holes in long-wavelength lasers is much larger than that of electrons, and the internal loss is more influenced by the p-type cladding layer. It is necessary to reduce the overlap between the mode field and the p-cladding layer and reduce the optical confinement factor of the p-cladding layer. One method is to use a long cavity laser with LOC or broadened waveguides.[45] By increasing the cavity length and the thickness of the SCH layers the degree of optical field penetration in the p-cladding layer can be reduced. The suppression of higher order transverse modes should be considered when using this method. Another way is to introduce an asymmetric dilute waveguide structure,[50–52] as shown in Figure 3.91(a). By using a dilute waveguide layer with a higher refractive index on the n-cladding layer, the mode field is pulled to the n-cladding layer to reduce the mode overlap with the p-cladding layer. The diluted waveguide layer is generally composed of an alternation of thin layers of InP and InGaAsP layers.

In addition to the optimization of injection efficiency and internal loss, the thermal and optical characteristics of the SCH layer and cladding layer also need optimization. For example, to reduce the thermally induced power saturation, the doping concentration of the p-InP cladding layer can be kept at a high level to reduce the series resistance. However, the increase in the doping level will increase internal loss, it is necessary to balance the device series resistance and internal loss. The far-field divergence angle also needs optimization to provide higher coupling efficiency. This can generally be achieved by adopting a LOC or a dilute waveguide structure.

3.4.3.4 Longitudinal Mode Control

Longitudinal mode control mainly involves the design of different grating structures to suppress the SHB effect. SHB is an important factor disturbing the single-mode operating characteristics of high-power lasers. Normally, the optical field of a DFB laser distributes non-uniformly along its longitudinal direction. Under the influence of the non-uniformly optical field and carrier density distribution, the refractive index along the longitudinal axis also becomes non-uniform. The SHB-induced non-uniform refractive index change will counteract the refractive index change introduced by the grating, causing lasing mode switching, jumping or multi-longitudinal mode oscillations in high-power cases. SHB effect becomes severe with the increase of the unevenness of the optical field distribution. The $\lambda/4$ phase-shift DFB laser can be more significantly affected by the SHB since its optical field has a peak distribution at the center of the cavity. To suppress the SHB, multiphase shift gratings, chirped gratings, gain coupled or complex coupled gratings can be used to obtain a more uniform optical field distribution.[53,54] The grating coupling coefficient κ should be well controlled so that the product of κL is in an optimal range. A greater κL will produce a deeper modulation of the optical field, the field distribution will be more uneven, as shown in Figure 3.92(a).[55] The increase in κL will decrease the distributed mirror loss α_m, which will reduce the slope efficiency of the laser according to Equation (3.38). On the other hand, the κL cannot be too small to provide enough mode selection ability to suppress the FP modes.[46] κL is usually chosen around 1 to suppress SHB while maintaining a high-power single-longitudinal-mode operation.

3.4.3.5 Transverse and Lateral Mode Control

Transverse and lateral mode control includes increasing the thickness of the waveguide and using double-trench ridge waveguides. To increase the output power, the waveguides of high-power DFB lasers are generally broader than those of ordinary DFB lasers. The purpose of increasing the transverse dimension of the waveguide is to reduce the internal loss, reduce the beam divergence angle and increase the catastrophic optical damage threshold. The asymmetric LOC structure or the diluted waveguide layer structure can be used to achieve this goal. The purpose of increasing the lateral dimension of the waveguide is to increase the available gain area, reduce the series resistance and increase the saturation power. While increasing the size of the waveguide, it is necessary to maintain the mode stability. As the injection current increases, higher order modes may

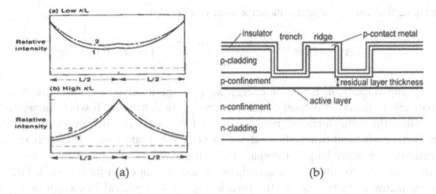

FIGURE 3.92 (a) Optical intensity distribution for low κL (upper curve) and high κL (lower curve) values; (b) a double-trench ridge waveguide laser structure. (From Whiteaway, J. E. A., Thompson, G. H. B., Collar, A. J. and Armistead, C. J., *IEEE J. Quantum Electron.*, 25, 1261, 1989; Wenzel, H., Bugge, F., Dallmer, M., Dittmar, F., Fricke, J., Hasler, K. H. and Erbert, G., *IEEE Photon. Technol. Lett.*, 20, 214, 2008. With permission.)

be excited due to the thermal effect and the change of the lateral refractive index. The single lateral mode operation of the laser can be obtained by increasing the cavity loss to the higher order mode and reducing its modal gain. The double-trench ridge waveguide[4,56,57] structure, as shown in Figure 3.92(b), has a central active region well matching the fundamental mode field, and two passive trenches overlapped with the first-order mode. The fundamental mode can be effectively amplified in the structure, while the first-order mode suffers a high modal loss in the passive trenches.

3.4.4 Low-noise DFB Lasers

Laser noise will influence the optical linewidth and output stability (intensity fluctuation). The broadening of the optical linewidth originates from the contribution of phase noise, while the intensity fluctuation comes from the contribution of intensity noise. Phase noise and intensity noise can be coupled and converted to each other. For semiconductor lasers, noise is related to the phase and amplitude fluctuation of the optical field. Spontaneous emission, the change of the carrier density, external optical feedback, temperature changes, power source ripples etc. will influence the noise of semiconductor lasers. In many applications, low-noise lasers are often required to meet system requirements. Low-noise lasers include narrow linewidth lasers and low RIN lasers.

Narrow linewidth lasers are widely used in coherent optical communications. Higher order modulation demands a laser linewidth on the order of hundreds of kHz or even tens of kHz. In coherent lidar or phased array lidar, the linewidth also needs to be on the order of hundreds of kHz. For laser Doppler radars, the linewidth requirements are on the order of tens of kHz.

Low RIN lasers are mainly used for analog applications. From cable television to the emerging microwave photonics, the demands for the RIN level of the laser are very stringent. In microwave systems, the RIN generally needs to be better than −150 or −160 dB/Hz. For commercial QW DFB laser with a cavity length around 300 μm, the linewidth is normally on the order of several MHz, and the RIN is between −120 and −150 dB/Hz, which is difficult to meet the abovementioned requirements. Special designs have to be adopted to realize the low-noise laser.

The spectral linewidth originates from the phase noise of the light field and is influenced by the driver noise and environment fluctuation. The laser linewidth can be decomposed into the intrinsic linewidth and technical noise. Intrinsic linewidth is related to the quantum noise of the laser, with a lineshape of Lorenzen, manifesting as white noise in the noise spectrum. The typical quantum noise is the phase perturbation coupled from the spontaneous emission to the stimulated emission. Technical noise is related to external perturbation, the lineshape is usually a Gaussian distribution, which appears as the low-frequency $1/f$ noise and small irregular undulations on the noise spectrum.

The intrinsic linewidth of a semiconductor laser can be written as[58-60]:

$$\Delta v = \left(1 + \alpha^2\right)\frac{v_g^2 h v n_{sp}\left(\alpha_m + \alpha_i\right)\alpha_m}{8\pi P} = \left(1 + \alpha^2\right)\delta f_{ST} \tag{3.40}$$

where v_g is group velocity, h_v is photon energy, n_{sp} is spontaneous emission factor, α is the linewidth enhancement factor, P is laser output power, δf_{ST} is Schawlow-Townes linewidth.[61,62] This expression takes the same form for both DFB and FP lasers. The difference is that α_m of the DFB laser is the equivalent mirror loss related to κL, while that of the FP laser is related to the facet reflectivity and cavity length. Compared with the classic Schawlow-Townes laser linewidth formula, the linewidth formula of semiconductor lasers has an extra term $(1+\alpha^2)$. The linewidth enhancement factor α contributes to the broadening of the spectral linewidth of semiconductor lasers comparing with other types of lasers. The linewidth enhancement factor α can be written as:

$$\alpha = \frac{dn'/dN}{dn''/dN} = \frac{4\pi}{\lambda}\frac{dn'/dN}{dg/dN} \tag{3.41}$$

where n' and n'' are the real and imaginary parts of the refractive index, and N is the carrier density. In a semiconductor laser, the phase of the optical field is not only perturbed by the spontaneous emission, but also by an additional contribution from the relaxation oscillation. Relaxation oscillation is a result of the energy exchange process between the carriers and photons. The fluctuation of the carrier relative to its steady state causes the instantaneous change of the laser gain (imaginary part of the refractive index). This gain change will affect the real part of the refractive index, thereby introducing an additional phase perturbation to the laser field, which will bring an additional broadening of the laser linewidth. The linewidth enhancement factor reflects the degree of coupling between the real and imaginary parts of the refractive index. In addition to the relaxation oscillation, power supply noise, SHB, mode distribution noise and nonlinear gain will also cause additional broadening of the semiconductor laser linewidth.

The RIN is also related to the spontaneous emission and relaxation oscillation process of the semiconductor laser. It describes the intensity fluctuation of the laser output, which is defined as:

$$\text{RIN} = \frac{\langle|\Delta P(\omega)|\rangle^2}{\langle P\rangle^2} \tag{3.42}$$

where $<|\Delta P(\omega)|>$ is the noise spectral density, and $<P>$ is the average optical power. The spectral distribution of laser RIN can be written as[63,64]:

$$\text{RIN}(f) = \frac{4}{\pi}\delta f_{ST}\frac{f^2 + \left(\gamma^*/2\pi\right)^2}{\left(f_r^2 - f^2\right)^2 + f^2\left(\gamma/2\pi\right)^2} \tag{3.43}$$

where γ^* is the damping factor by taking the nonlinear gain term into account. It can be seen from Equation (3.43) that RIN is related to the Schawlow-Townes linewidth and the laser relaxation oscillation frequency and damping factor. From the RIN spectrum, δf_{ST}, f_r, γ can be extracted using parameters fitting.

The origins of the laser linewidth and RIN are closely related. The design principles are similar, and the optimization usually focuses on the intrinsic linewidth. In practice, it is often possible to obtain a low RIN output while maintaining a narrow linewidth.

It can be seen from Equation (3.40) that the low-noise DFB laser can be optimized by reducing the linewidth enhancement factor, reducing the equivalent mirror loss and internal loss and increasing the output power. The device design mainly includes the following aspects.

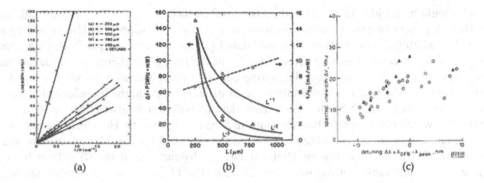

FIGURE 3.93 The dependence of a DFB laser's linewidth on (a) output power; (b) cavity length; (c) wavelength detuning. (From Liou, K., Dutta, N. and Burrus, C., *Appl. Phys. Lett.*, 50, 489, 1987; Ogita, S., Kotaki, Y., Kihara, K., Matsuda, M., Ishikawa, H. and Imai, H., *Electron. Lett.*, 24, 613, 1988. With permission.)

3.4.4.1 Quantum Well

The optimization of the active region includes the use of the strained QW and an appropriate increase in the number of QWs. The use of strained QW can significantly reduce the linewidth enhancement factor α[65] and suppress SHB,[66] which is a result of higher differential gain dg/dN. The minimum linewidth can decrease from MHz to the order of hundreds of kHz[67,68] even kHz[2] after the introduction of the strained QWs.

By increasing the number of QWs, the differential gain can be improved. However, with the increase of the number of QWs, the internal loss also increases, resulting in limited maximum achievable output power. Equation (3.40) reveals the linear relationship between the laser linewidth and the inverse of the laser power $1/P$, as shown in Figure 3.93(a). The decrease in output power corresponds to the increase in the linewidth of the laser. Therefore, the number of QWs should be balanced between high differential gain and high output power.

Increasing the output power is also an important method to obtain a low-noise laser. The structural design is similar to that of high-power lasers. However, it should be noted that the linewidth of the laser cannot monotonically decrease as the output power increases. Instead, the laser linewidth will keep unchanged or become re-broadened above a certain output power level. It is generally believed that this is related to the increase of the linewidth enhancement factor α due to the SHB under high injection current.[69] The suppression of SHB is another important design consideration.

3.4.4.2 Cavity Length

The intrinsic linewidth is related to the cavity Q, which can be increased by increasing the cavity length and equivalent reflectivity. For low-noise DFB lasers, the optimization can be translated to the control of κL. The longer the cavity length, the narrower the linewidth. It can be proved that the DFB laser's linewidth is proportional to $1/\kappa^2 L^3$ when the end-facet reflection is ignored for a large κL.[71] When considering the end-facet reflectivity, the laser linewidth will be proportional to $1/L^2$ ($\alpha_m \gg \alpha_i$) or $1/L$ ($\alpha_m \ll \alpha_i$).[59] Figure 3.93(b) shows the dependence of the laser linewidth on the cavity length. Comparing to FP lasers, cavity length has a more significant influence on the linewidth of DFB lasers.[58] Increasing the cavity length can be more effective in reducing the linewidth for DFB lasers ($\propto 1/L^3$). The cavity lengths of narrow-linewidth DFB lasers are mostly between 1000 and 2000 μm. However, the potential problems accompanying the long cavity are the decrease in slope efficiency and grating uniformity,[59] resulting in the deterioration of the operation of the DFB laser.

3.4.4.3 Gratings

Grating optimization mainly includes controlling of κ, choice of grating type and detuning of the Bragg wavelength. With a long cavity, κ needs to be controlled at a sufficiently small value. A large κL should be avoided in reducing the linewidth[70], which will broaden the linewidth due to the SHB.[55] A variety of grating structures have been proposed to suppress SHB, including multiphase-shift gratings,[72] corrugation-pitch modulated gratings,[2,73] chirped gratings, gain-coupled grating or complex-coupled grating. Blue-detuning of the Bragg wavelength is similar to that of the high-speed DML. By controlling the grating period to make the laser working at the shorter wavelength side of the gain peak to obtain a higher differential gain, the linewidth enhancement factor α will be reduced.[2,74] Studies also show that a smaller α is also effective in suppressing SHB,[66,73] which can suppress the linewidth re-broadening effects. However, the wavelength detune should not be too large, otherwise, the gain of the DFB laser mode will be too small to suppress the FP cavity mode. Figure 3.93(c) shows the dependence of the DFB laser linewidth on the wavelength detuning of the Bragg wavelength. The blue-detuning generally needs to be controlled in a range of 20 nm.[59]

The increase in facet reflectivity may be helpful in reducing the laser linewidth.[75] However, the increase in facet reflectivity will increase the competition between the FP cavity modes and the DFB mode. Even if the FP cavity modes are below the lasing threshold, the beating of the amplified spontaneous cavity modes will still influence the carrier density distribution, resulting in an increase in the linewidth. Therefore, narrow-linewidth DFB lasers tend to adopt anti-reflection/HR coating, which can ensure high output power while suppressing the FP modes.

By using comprehensive designs, the linewidth and RIN of DFB lasers continue to decrease. By using compressive strain QWs combined with CPM grating and wavelength detuning technology, a minimum linewidth of 3.6 kHz and a power of 55 mW was demonstrated for a DFB laser with a cavity length of 1200 μm.[2] The lowest RIN of some DFB lasers have reached the level of −170 dB/Hz.[76]

3.4.5 Summary and Outlook

After 60 years of development, the theory of single-mode lasers has been well established. Single-mode DFB lasers with various functionalities have been developed to meet the demands of a wide variety of applications. In the pursuit of extreme performance, the researches on QW-based DFB lasers have gradually shifted from academia to industry. It can be predicted that in the next 5–10 years, for commercial products, the modulation rate of the single-wavelength high-speed DML will increase to 100–200 Gbps, and the power of high-power DFB lasers will increase to the watts level and the commercialize narrow linewidth and low RIN DFB lasers will reach kHz and −170 dB/Hz, respectively. For academic research, on the one hand, the focus will shift to large-scale functional photonic integration based on QW DFB lasers; on the other hand, quantum-wire-based and quantum-dot-based single-mode lasers with better performance and more diversified functionalities will be widely adopted for new applications.

REFERENCES

1. Matsui, Y., Schatz, R., Che, D., Khan, F., Kwakernaak, M. and Sudo, T., 2020 Optical Fiber Communications Conference and Exhibition (OFC), 1, pp. 3–5, 2020.
2. Okai, M., Suzuki, M. and Taniwatari, T., *Electron. Lett.*, 29, 1696, 1993.
3. Doussiere, P., Shieh, C.-L., Demars, S. and Dzurko, K., *Proc. SPIE*, 6485, 64850G, 2007.
4. Menna, R., Komissarov, A., Maiorov, M., Khalfin, V., DiMarco, L., Connolly, J. and Garbuzov, D. Summaries of Papers Presented at the Conference on Lasers and Electro-Optics. Postconference Technical Digest, pp. CPD12-CP1, 2001.
5. Kogelnik, H. and Shank, C. V., *Appl. Phys. Lett.*, 18, 152, 1971.
6. Kogelnik, H. and Shank, C. V., *J. Appl. Phys.*, 43, 2327, 1972.
7. Nakamura, M., Yariv, A., Yen, H. W., Somekh, S. and Garvin, H. L., *Appl. Phys. Lett.*, 22, 515, 1973.

8. Soda, H., Kotaki, Y., Sudo, H., Ishikawa, H., Yamakoshi, S. and Imai, H., *IEEE J. Quantum Electron.*, 23, 804, 1987.
9. Agrawal, G. P., Geusic, J. E. and Anthony, P. J., *Appl. Phys. Lett.*, 53, 178, 1988.
10. Okai, M., Chinone, N., Taira, H. and Harada, T., *IEEE Photon. Technol. Lett.*, 1, 200, 1989.
11. Huang, Y., Sato, K., Okuda, T., Suzuki, N., Ae, S., Muroya, Y., Mori, K., Sasaki, T. and Kobayashi, K., *IEEE J. Quantum Electron.*, 38, 1479, 2002.
12. Luo, Y., Nakano, Y., Tada, K., Inoue, T., Hosomatsu, H. and Iwaoka, H., *Appl. Phys. Lett.*, 56, 1620, 1990.
13. Nakamura, K., Miyamura, S., Sekikawa, R., Shimura, D., Nakaya, S., Ori, T., Yaegashi, H. and Ogawa, Y. 2007 Conference on Optical Fiber Communication and the National Fiber Optic Engineers Conference, pp. 1–3, 2007.
14. Miller, L. M., Verdeyen, J. T., Coleman, J. J., Bryan, R. P., Alwan, J. J., Beernink, K. J., Hughes, J. S. and Cockerill, T. M., *IEEE Photon. Technol. Lett.*, 3, 6, 1991.
15. Guo, W.-H., Lu, Q., Nawrocka, M., Abdullaev, A., O'Callaghan, J., Lynch, M., Weldon, V. and Donegan, J. F., *IEEE Photon. Technol. Lett.*, 24, 634, 2012.
16. Coldren, L. A. and Corzine, S. W., *Diode Lasers and Photonic Integrated Circuits*, Hoboken, NJ: Wiley, 1995.
17. Thibeault, B. J., Bertilsson, K., Hegblom, E. R., Strzelecka, E., Floyd, P. D., Naone, R. and Coldren, L. A., *IEEE Photon. Technol. Lett.*, 9, 11, 1997.
18. Morton, P. A., Logan, R. A., Tanbun-Ek, T., Sciortino, P. F., Sergent, A. M., Montgomery, R. K. and Lee, B. T. *Electron. Lett.*, 28, 2156, 1992.
19. Shimizu, J., Yamada, H., Murata, S., Tomita, A., Kitamura, M. and Suzuki, A., *IEEE Photon. Technol. Lett.*, 3, 773, 1991.
20. Tatham, M. C., Lealman, I. F., Seltzer, C. P., Westbrook, L. D. and Cooper, D. M., *IEEE J. Quantum Electron.*, 28, 408, 1992.
21. Chiu, L. and Yariv, A., *IEEE J. Quantum Electron.*, 18, 1406, 1982.
22. Childs, G., Brand, S. and Abram, R. A., *Semicond. Sci. Technol.*, 1, 116, 1986.
23. Yamamoto, T., Optical Fiber Communication Conference, Los Angeles, CA, pp. OTh3F.5, 2012.
24. Matsui, Y., Murai, H., Arahira, S., Kutsuzawa, S. and Ogawa, Y., *IEEE Photon. Technol. Lett.*, 9, 25, 1997.
25. Fukamachi, T., Adachi, K., Shinoda, K., Kitatani, T., Tanaka, S., Aoki, M. and Tsuji, S., *IEEE J. Sel. Top. Quantum Electron*, 17, 1138, 2011.
26. Tsuchiya, T., Takemoto, D., Taike, A., Aoki, M. and Uomi, K., Conference Proceedings. Eleventh International Conference on Indium Phosphide and Related Materials, pp. 47–50, 1999.
27. Sasada, N., Nakajima, T., Sekino, Y., Nakanishi, A., Mukaikubo, M., Ebisu, M., Mitaki, M., Hayakawa, S. and Naoe, K., *J. Lightwave Technol.*, 37, 1686, 2019.
28. Lu, H., Blaauw, C., Benyon, B. and Makino, T., Proceedings of IEEE 14th International Semiconductor Laser Conference, pp. 51–52, 1994.
29. Chang, F., *Datacenter Connectivity Technologies: Principles and Practice*, River Publishers, Gistrup, 2018.
30. Matsui, Y., Li, W., Roberts, H., Bulthuis, H., Deng, H., Lin, L. and Roxlo, C., 2016 Optical Fiber Communications Conference and Exhibition (OFC), pp. 1–3, 2016.
31. Bang, D., Shim, J., Kang, J., Um, M., Park, S., Lee, S., Jang, D. and Eo, Y., *IEEE Photon. Technol. Lett.*, 14, 1240, 2002.
32. Shim, J., Komori, K., Arai, S., Arima, I., Suematsu, Y. and Somchai, R., *IEEE J. Quantum Electron.*, 27, 1736, 1991.
33. Yamamoto, T., Uetake, A., Otsubo, K., Matsuda, M., Okumura, S., Tomabechi, S. and Ekawa, M., 22nd IEEE International Semiconductor Laser Conference, pp. 193–194, 2010.
34. Nakahara, K., Wakayama, Y., Kitatani, T., Taniguchi, T., Fukamachi, T., Sakuma, Y. and Tanaka, S., *IEEE Photon. Technol. Lett.*, 27, 534, 2015.
35. Simoyama, T., Matsuda, M., Okumura, S., Uetake, A., Ekawa, M. and Yamamoto, T., European Conference on Optical Communication (ECOC), pp. 1–3, 2012.
36. Kobayashi, W., Ito, T., Yamanaka, T., Fujisawa, T., Shibata, Y., Kurosaki, T., Kohtoku, M., Tadokoro, T. and Sanjoh, H., *IEEE J. Sel. Top. Quantum Electron*, 19, 1500908, 2013.
37. Tager, A. A. and Petermann, K., *IEEE J. Quantum Electron.*, 30, 1553, 1994.
38. Radziunas, M., Glitzky, A., Bandelow, U., Wolfrum, M., Troppenz, U., Kreissl, J. and Rehbein, W., *IEEE J. Sel. Top. Quantum Electron*, 13, 136, 2007.
39. Bauer, S., Brox, O., Kreissl, J., Sahin, G. and Sartorius, B., *Electron. Lett.*, 38, 334, 2002.

40. Brox, O., Bauer, S., Radziunas, M., Wolfrum, M., Sieber, J., Kreissl, J., Sartorius, B. and Wunsche, H., *IEEE J. Quantum Electron.*, 39, 1381, 2003.
41. Troppenz, U., Kreissl, J., Moehrle, M., Bornholdt, C., Rehbein, W., Sartorius, B., Woods, I. and Schell, M., *Proc. SPIE*, 7953, 79530F, 2011.
42. Yu, L., Guo, L., Lu, D., Ji, C., Wang, H. and Zhao, L., *Chin. Opt. Lett.*, 13, 051401, 2015.
43. Matsui, Y., Schatz, R., Pham, T., Ling, W. A., Carey, G., Daghighian, H. M., Adams, D., Sudo, T. and Roxlo, C., *J. Lightwave Technol.*, 35, 397, 2017.
44. Slipchenko, S., Vinokurov, D., Pikhtin, N. A., Sokolova, Z. N., Stankevich, A., Tarasov, I. and Alferov, Z., *Semiconductors*, 38, 1430, 2004.
45. Garbuzov, D., Xu, L., Forrest, S. R., Menna, R., Martinelli, R. and Connolly, J. C., *Electron. Lett.*, 32, 1717, 1996.
46. Tirong, R. C. and Wei, H., *Proc. SPIE*, 3547, 24, 1998.
47. Yong, Y. S., Wong, H. Y., Yow, H. K. and Sorel, M., *Laser Phys.*, 20, 811, 2010.
48. Shterengas, L., Menna, R., Trussell, W., Donetsky, D., Belenky, G., Connolly, J. and Garbuzov, D., *J. Appl. Phys.*, 88, 2211, 2000.
49. Han, I. K., Cho, S. H., Heim, P. J. S., Woo, D. H., Kim, S. H., Song, J. H., Johnson, F. G. and Dagenais, M., *IEEE Photon. Technol. Lett.*, 12, 251, 2000.
50. Nagashima, Y., Onuki, S., Shimose, Y., Yamada, A. and Kikugawa, T., 2004 IEEE 19th International Semiconductor Laser Conference, pp 47–48, 2004.
51. Shih, M. H., Choa, F.-S., Kapre, R. M., Tsang, W. T., Logan, R. A. and Chu, G., *Electron. Lett.*, 31, 1058, 1995.
52. Faugeron, M., Tran, M., Parillaud, O., Chtioui, M., Robert, Y., Vinet, E., Enard, A., Jacquet, J. and Dijk, F. V., *IEEE Photon. Technol. Lett.*, 25, 7–10, 2013.
53. Borchert, B., *Proc. SPIE*, 1523, 194, 1992.
54. Stegmüller, B. and Borchert, B., Conference on Lasers and Electro-Optics, Baltimore, MD, 15, pp. CThK1, 1995.
55. Whiteaway, J. E. A., Thompson, G. H. B., Collar, A. J. and Armistead, C. J., *IEEE J. Quantum Electron.*, 25, 1261, 1989.
56. Wenzel, H., Bugge, F., Dallmer, M., Dittmar, F., Fricke, J., Hasler, K. H. and Erbert, G., *IEEE Photon. Technol. Lett.*, 20, 214, 2008.
57. Dmitri, Z. G., Mikhail, A. M., Raymond, J. M., Anatoly, V. K., Khalfin, V., Igor, V. K., Alexander, V. L., Louis, A. D. and John, C. C., *Proc. SPIE*, 4651, 92, 2002.
58. Kojima, K. and Kyuma, K., *Electron. Lett.*, 20, 869, 1984.
59. Liou, K., Dutta, N. and Burrus, C., *Appl. Phys. Lett.*, 50, 489, 1987.
60. Kunii, T. and Matsui, Y., *Opt. Quant. Electron.*, 24, 719, 1992.
61. Schawlow, A. and Townes, C., *Phys. Rev.*, 112, 1940, 1958.
62. Fleming, M. W. and Mooradian, A., *Appl. Phys. Lett.*, 38, 511, 1981.
63. Yokouchi, N., Yamanaka, N., Iwai, N., Nakahira, Y. and Kasukawa, A., *IEEE J. Quantum Electron.*, 32, 2148, 1996.
64. Kikuchi, K. and Okoshi, T., *IEEE J. Quantum Electron.*, 21, 1814, 1985.
65. Dutta, N., Wynn, J., Sivco, D. and Cho, A., *Appl. Phys. Lett.*, 56, 2293, 1990.
66. Aoki, M., Uomi, K., Tsuchiya, T., Sasaki, S., Okai, M. and Chinone, N., *IEEE J. Quantum Electron.*, 27, 1782, 1991.
67. Yamazaki, H., Sasaki, T., Kida, N., Kitamura, M. and Mito, I., Optical Fiber Communication, pp. PD33, 1990.
68. Kotaki, Y., Fujii, T., Ogita, S., Matsuda, M. and Ishikawa, H., Optical Fiber Communication, pp. THE3, 1990.
69. Pan, X., Olesen, H. and Tromborg, B., 12th IEEE International Conference on Semiconductor Laser, pp. 118, 1990.
70. Ogita, S., Kotaki, Y., Kihara, K., Matsuda, M., Ishikawa, H. and Imai, H., *Electron. Lett.*, 24, 613, 1988.
71. Kojima, K., Kyuma, K. and Nakayama, T., *J. Lightwave Technol.*, 3, 1048, 1985.
72. Ogita, S., Kotaki, Y., Matsuda, M., Kuwahara, Y. and Ishikawa, H. *Electron. Lett.*, 25, 629, 1989.
73. Okai, M., Tsuchiya, T., Uomi, K., Chinone, N. and Harada, T., *IEEE Photon. Technol. Lett.*, 2, 529, 1990.
74. Kikuchi, K., *Electron. Lett.*, 24, 80, 1988.
75. Ogita, S., Hirano, M., Soda, H., Yano, M., Ishikawa, H. and Imai, H., *Electron. Lett.*, 23, 347, 1987.
76. Zhao, Y., Luo, X., Tran, D., Hang, Q., Weber, P., Hang, T., Knust-Graichen, R., Nuttall, N., Cendejas, R., Nikolov, A. and Dutt, R., 2012 IEEE Avionics, Fiber-Optics and Photonics Digest CD, pp. 66–67, 2012.

3.5 NEAR-INFRARED MATERIALS AND LASERS ON GaAs SUBSTRATES

Yugang Zeng, Yongqiang Ning and Lijun Wang

3.5.1 STRUCTURES AND BANDGAP ENERGIES

III–V compound semiconductor materials based on GaAs substrates, which comprise Group III (Al/Ga/In) and Group V (As/P), including AlGaAs, GaAsP, GaInP, InGaAs and AlGaInP, are used for lasers in the visible wavelength region and near-infrared region. These materials are all zinc blende structure (cubic β-ZnS structure).

In a bulk semiconductor, both direct and indirect energy gaps in semiconductor materials are temperature-dependent quantities, with the functional form often fitted to the empirical Varshni Equation (3.44)[1]:

$$E_g(T) = E_g(0K) - \frac{\alpha T^2}{\beta + T} \qquad (3.44)$$

The lattice constants at room temperature (RT) (300K), thermal expansion coefficients and band parameters for III–V binary semiconductors are listed in Table 3.10.[2]

The lattice constants $a_{A_x B_{1-x}}$ of ternary alloys are determined by their composition and generally vary with the Vegard's law as the following equation[3]:

$$a_{A_x B_{1-x}} = x a_A + (1-x) a_B \qquad (3.45)$$

But the dependence of the energy gap on alloy composition is not linear, generally assumed to fit a simple quadratic form as Equation (3.46):

$$E_g(A_x B_{1-x}) = x E_A + (1-x) E_B - x(1-x)C \qquad (3.46)$$

Where the so-called bowing parameter C accounts for the deviation from a linear interpolation between the two binaries A and B, the bowing parameter for III–V alloys is typically positive (i.e., the alloy bandgap is smaller than the linear interpolation result) and can in principle be a function of temperature. The physical origin of the bandgap bowing can be traced to disorder effects created by the presence of different cations and anions.

TABLE 3.10
Lattice Constants and Band Parameters for III–V Binary Materials

Material	Lattice Constant (nm)	Coefficient of Thermal Expansion (10^{-6} nm/K)	Bandgap (eV) @RT	α (meV/K)	β (K)	Remarks
GaAs	0.56533	3.88	$E_g^{\Gamma} = 1.519$	0.5405	204	
AlAs	0.56611	2.90	$E_g^{X} = 2.44$	0.885	530	Indirect bandgap
			$E_g^{\Gamma} = 3.099$			
InAs	0.60583	2.74	$E_g^{\Gamma} = 0.417$	0.276	93	
GaP	0.54505	2.92	$E_g^{X} = 2.35$	0.5771	372	Indirect bandgap
			$E_g^{\Gamma} = 2.886$			
AlP	0.54672	2.92	$E_g^{X} = 2.52$	0.5771	372	Indirect bandgap
			$E_g^{\Gamma} = 3.63$			
InP	0.58679	2.79	$E_g^{\Gamma} = 1.4236$	0.363	162	

TABLE 3.11

The Bandgap Energy Bowing Parameters for Ternary Alloys (in eV)

Material	E_g^Γ	E_g^X	E_g^L	Δ_{so}	VBO (Valence Band Offset)	a_c
AlGaAs	$-0.127+1.310x$	0.055	0	0	–	–
InGaAs	0.477	1.4	0.33		−0.38	2.61
InAlAs	0.7	0	0	0.15	−0.64	−1.4
GaInP	0.65	0.2	1.03	–	–	–
AlInP	−0.48	0.38	−0.19	–	–	–
AlGaP	–	0.13	–	–	–	–
GaAsP	0.19	0.24	0.16	–	–	–
InAsP	0.1	0.27	0.27	0.16	–	–
AlAsP	0.22	0.22	0.22	–	–	–

The bandgap energy bowing parameters for ternary alloys are listed in Table 3.11.[2]

As marked in Table 3.10, the bandgap of AlAs, AlP and GaP are indirect. Therefore, the band structures for the alloys such as $Ga_{1-x}Al_xAs$, $In_{0.5}(Ga_{1-x}Al_x)_{0.5}P$ and $GaAs_{1-x}P_x$ will vary between direct bandgap and indirect bandgap depending on the fraction x.[4] Because direct bandgap crystals have high radiative efficiency and narrow emission spectrum, the quantum wells (QWs) of direct bandgap materials are mandatary in laser diodes (LDs).

For the quaternary alloys, $(Al_zGa_{1-z})_{0.51}In_{0.49}P$ (or, more accurately, $(Ga_{0.51}In_{0.49}P)_{1-z}(Al_{0.52}In_{0.48}P)_z$) and $(GaAs)_{1-z}(Ga_{0.51}In_{0.49}P)_z$ are lattice-matched to GaAs. $(Al_zGa_{1-z})_{0.51}In_{0.49}P$ is often employed as the barrier and cladding material in GaInP/AlGaInP and at $z=0.55$, the bandgap changes from direct to indirect. $(GaAs)_{1-z}(Ga_{0.51}In_{0.49}P)_z$ is direct bandgap for any z. The bandgaps as a function of composition for z are plotted in Figure 3.94.[2]

3.5.2 CRYSTAL GROWTHS

High-quality material growth is crucially important for lasers. Usually GaAs-based thin films for lasers using III–V compound semiconductors are grown by various methods, including liquid phase epitaxy (LPE), molecular-beam epitaxy (MBE) and vapor phase epitaxy (VPE).

LPE was invented in 1963 by Nelson utilizing the recrystallization of the solute from a super-saturated solution. Because of the fast growth rate, LPE is difficult to obtain epitaxial materials with nanometer-scale thickness, and the surface morphology of the epitaxial layer is usually not good and the roughness is high. It is also difficult to control the composition of epitaxial layers. For example, the segregation coefficient (defined as the ratio of atoms incorporated from the liquid solution to those in the solid crystal) of Al is relatively large in the case of LPE growth for GaAlAs. This causes a gradual decrease in the Al amount in the solution, resulting in a graded composition structure for thick GaAlAs growth. The problem of segregation is even more serious for InGaAlP growth. It was very difficult to obtain high-quality InGaAlP crystals by the LPE methods because of the extremely large segregation coefficient of Al.[5,6] However, previous to the introduction of MBE and metalorganic chemical vapor deposition (MOCVD), semiconductor heterostructures were typically grown by LPE.

During the 1970s and early 1980s, the epitaxial systems improved greatly, mainly, including MBE and MOCVD. Both MBE and MOCVD made it realistic that precise compositional layers were grown at the nanometer scale to tailor the bandgap and lattice constant.

MBE is an advanced technology for the growth of high-quality epitaxial materials on crystal substrates. In ultrahigh vacuum chamber, the various molecular or atomic beams produced by

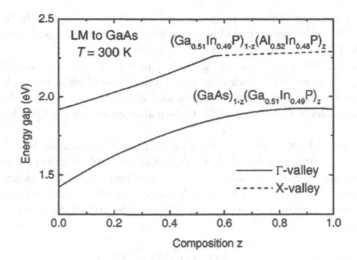

FIGURE 3.94 Bandgaps as a function of composition z for $(Al_{0.52}In_{0.48}P)_z/(Ga_{0.51}In_{0.49}P)_{1-z}$ and $(Ga_{0.51}In_{0.49}P)_z/(GaAs)_{1-z}$ quaternary alloys. (From Vurgaftman, I., Meyer, J. R. and Ram-Mohan, L. R., *J. Appl. Phys.*, 89, 5815, 2001. With permission.)

heating the furnaces, spraying directly onto the single-crystal substrate and scanning the substrate at an appropriate temperature, the films can be grown on the substrate precisely. MBE can control and grow ultrathin crystals for the III–V systems that can be applied to form quantum dots (QDs), multi-QWs (MQWs), doping layers and structures for light-emitting devices.[7] MBE is mainly used for researches due to its low production and high cost.

VPE is an important epitaxy growth method. Especially, MOCVD, also called organic-metal VPE (OMVPE), is the most widely used VPE method. MOCVD uses metal organic (MO) compounds such as trimethyl gallium (TMGa), trimethyl aluminum (TMAl) and trimethyl indium (TMIn) as source gases for Group III materials. Hydrides such as arsine (AsH_3) and phosphorane (PH_3) are used as source gases for Group V. MO sources and hydrides are introduced into the reactor, decomposed at high temperature. Chemical reactions take place on the surface of the substrate and the exceptionaly solid films are deposited finally. Because of its characteristics of precise control of material growth and uniform growth in a large area, MOCVD has greatly promoted the research and development of semiconductor materials and devices and has now become an important and dominant growth means for III–V layers and lasers.[8] In particular, for manufacturing considerations such as high throughput, the low-pressure MOCVD (LP-MOCVD) process has become the preferred method for the growth of the LDs.

To realize an LD device, both n-type and p-type semiconductor layers are necessary. In general, Group VI elements, such as S and Se, due to one more valence electron than Group V, will act as donors for the III–V system and thus are used as n-type dopants. Group II elements such as Zn, Mg and Be behave as acceptors and are used as p-type dopants due to one less valence electron than Group III. Group IV dopants such as C and Si have four valence electrons, thus they can act both as donors and acceptors when depositing III–V layers. The substitution site depends on the growth condition. Generally speaking, Si is used as an n-type dopant and C as a p-dopant in depositing III–V films by LP-MOCVD.[9]

3.5.3 Strained InGaAs QWs and Growths by LP-MOCVD

Strained QWs are the most important active region for lasers. Strained QWs have improved the performance of near-infrared diode lasers revolutionarily. The strained QW materials consist of III–V alloys mentioned previously. These alloys have miscibility bandgaps so we can expand the span of emission wavelengths through precise compositional layers. Especially, the compressively strained InGaAs QW has become the workhorse for highly reliable, high-performance lasers in the important 0.90- to 1.2-μm wavelength regime.[10]

Diode lasers employing such active regions have demonstrated the highest total power conversion efficiency (PCE) up to 76% for wavelengths <1 μm.[11] Nevertheless, many challenges remain to expanding the accessible emission wavelengths and addressing the associated strain-relaxation limitations of such material systems. And the high-quality growth of the QWs becomes the main challenge.

To reach near infrared, for example, 980-nm LD on the GaAs substrate, InGaAs QWs with the fraction of In ~20% and strains ($\Delta a/a$, where a is the lattice constant) of up to 1.5% would be needed. Because of the stress, the epilayer quality could become bad when the thickness is higher than the critical thickness. Matthews and Blakeslee developed the M-B model to calculate the critical thickness using elastic accommodation of strain in layers. The critical thickness H is shown in the following equation[12]:

$$H = \frac{b\left(1 - v\cos^2\theta\right)}{8\varepsilon\pi\left(1+v\right)\cos\lambda}\ln\frac{\alpha H}{b} \qquad (3.47)$$

Because the M-B model is based on ideal elastic accommodation of strain at thermodynamic equilibrium state, the real critical thickness is much higher than M-B's by LP-MOCVD, especially at low temperature, where it is usually non-thermodynamic equilibrium. So, people grow ultrahigh indium (>35% at wavelength of 1150 nm or longer) InGaAs QW at the temperature lower than 550°C.[13]

To obtain good-quality QW, it is a must that there is no relaxation in the layers that means the thickness is lower than the real critical thickness. For MQWs, people utilize GaAsP for the strain compensation so as to increase the total critical thickness.[14,15] InGaAs/GaAsP QWs are usually grown by LP-MOCVD at a temperature lower than 650°C because indium will diffuse when the temperature is high, which could cause bad interface quality. To obtain high quality, people utilize 0.5–2 nm GaAs as an insert layer between InGaAs and GaAsP.

3.5.4 808-NM HIGH-POWER LDS AND ARRAYS

808-nm high-power semiconductor lasers are extensively used in a variety of applications, such as solid-state laser and fiber laser pumping, material processing, as well as health-care and military applications. Depending on the application, single emitters and different arrays are needed. The continuous-wave (CW) output power of the single device has reached from 8.1 W in 1990s to nearly 30 W now (Figure 3.95). And the power of the bar has reached 1010 W (CW), while pulse output power up to 2800 W (Figure 3.96).[16]

Many methods have been applied to achieve high-power 808-nm LDs. Generally speaking, researchers make efforts to achieve high-power devices through the following means.

3.5.4.1 In(Al)GaAs Stained QWs

The biaxial compression of the strained In(Al)GaAs QWs defeats the cubic symmetry of the semiconductor. It splits the degeneracy of the heavy and light hole valence band edges and reduces the in-plane effective mass of the heavy-hole band. It will effectively reduce the threshold current of the laser and improve the quantum efficiency and differential gain compared with traditional double heterostructures (DHs).[17]

3.5.4.2 Al-Free QWs

Aluminum is easy to be oxidized leading to catastrophic optical mirror damage (COMD). Al-free QWs have high COMD than traditional InAlGaAs QWs. It will increase the power of the LDs.

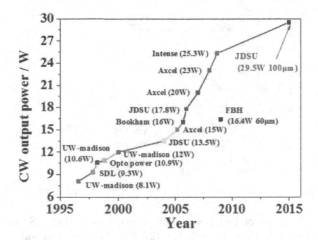

FIGURE 3.95 Evolution of the CW single emitter at wavelength of 808 nm. (From Ning, Y. Q., Chen, Y. Y., Zhang, J., Song, Y., Lei, Y. X., Qiu, C., Liang, L., Jia, P., Qin, L. and Wang, L. J., *Acta Optica Sinica*, 41, 0114001, 2021. With permission.)

People usually use GaAsP/GaInP and InGaAsP/GaInP as Al-free active region.[18] Therefore, high quality of (Al)GaInP growth is necessary. It is known that a disordered alloy has larger bandgap energy than that of an ordered one. (Al)GaInP is apt to be ordered deposited by LP-MOCVD. Researchers utilize tilted substrates to decrease the orderliness. Figure 3.97 shows the dependence of the bandgap energy, obtained by photoluminescence (PL) measurement of InGaAlP alloys on the substrate orientation. Usually 15 degree toward [0 1 1] substrates is preferred in LP-MOCVD.[19]

3.5.4.3 Structural Improvements

The photon is easier to be absorbed in p-layer. Researchers utilize asymmetric waveguide where the p-layer is thinner than n-layer. This leads to much more light distribution in n-layer to decrease the power loss.[20]

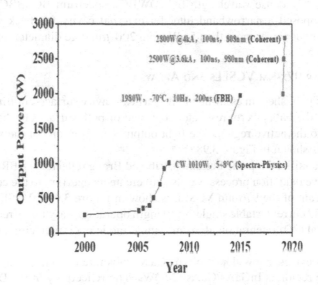

FIGURE 3.96 The output power of the 808-nm LD arrays. (From Ning, Y. Q., Chen, Y. Y., Zhang, J., Song, Y., Lei, Y. X., Qiu, C., Liang, L., Jia, P., Qin, L. and Wang, L. J., *Acta Optica Sinica*, 41, 0114001, 2021. With permission.)

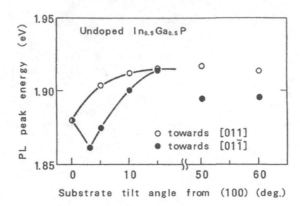

FIGURE 3.97 PL peak energy of InGaP vs. substrate orientation. (From Suzuki, M., Nishikawa, Y., Ishikawa, M. and Kokubun, Y., *J. Cryst. Growth*, 113, 127, 1991. With permission.)

To increase the COMD threshold, it is necessary to reduce the energy density of the light field in the active region. This requires increasing the size of the waveguide, increasing the size of the spot and broadening the light field distribution. This is called the large optical cavity (LOC) technology that is used widely nowadays.[21]

3.5.4.4 Facet Passivation

Facet passivation is very important for high-power LDs. Facet surface passivation can effectively remove impurities such as contamination and oxide layer on the cavity surface, reducing the surface state density, effectively improving the thermal stability of the device and enhancing COMD.[22]

3.5.4.5 Beam Combination

The semiconductor LD has poor beam quality. The beam combination (BC) technology can achieve high-quality beam by LD arrays. This makes it possible to approach high-power density and total power sufficient for direct application of LDs for materials processing. The BC mainly includes conventional BC (CBC), dense wavelength BC (DWBC), spectrum BC (SBC) and coherent BC (CBC).[16] Trumpf proposed a narrowband filter for external cavity feedback wavelength locking structure with the output power more than 5 kW from 200-μm core diameter optical fiber.[23,24]

3.5.5 High-power 97x-nm VCSELs and Arrays

In 1979, Kenichi Iga published an essay about the vertical-cavity surface-emitting lasers (VCSELs) firstly. In this essay, the Fabry-Perot resonator consisted of both surfaces of the wafer and its axis was perpendicular to the active region. The light output was from one of the wafer surfaces. The schematic diagram is shown in Figure 3.98.[25]

Later, Kenichi Iga utilized top and bottom distributed Bragg reflectors (DBR) to cover the active region.[26] The selective oxidation process was used to create an aperture to the current confinement. The schematic diagram of the 976-nm VCSEL is show in Figure 3.99.[27] VCSELs have the advantages of low-threshold current, stable single-wavelength operation, easy high-frequency modulation, easy two-dimensional (2D) integration, dynamic single-mode operation, circular symmetrical spot and high-fiber coupling efficiency.

For the 97X-nm devices, a low-doped n-type GaAs substrate is used to minimize absorption of the light. The active region is InGaAs/GaAsP MQWs. The reflectivity of the DBRs is higher than 98% for maximum PCE. For low-resistivity DBRs with low-absorption losses, gradual AlGaAs layers and modulation doping are necessary. A 98% Al fraction AlGaAs layer is grown near the first pair of the P-DBR for the oxide aperture. The materials are grown usually by LP-MOCVD.

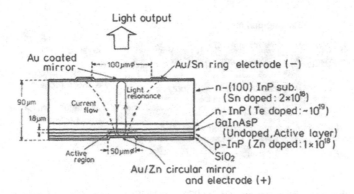

FIGURE 3.98 Schematic structure of GaInAsP/InP surface injection laser. (From Soda, H., Motegi, Y. and Iga, K., *IEEE J. Quantum Electron.*, 19, 1035, 1983. With permission.)

Ti/Pt/Au is evaporated to form the p-type contacts. Mesas are formed by reactive ion etching (RIE) and are deep enough to expose the 98% Al fraction AlGaAs layer. Then the samples are placed in a high humidity furnace to form the oxidation aperture. The substrate is thinned to less than 150 μm. The antireflection Si_3N_4 coating is grown by plasma enhances chemical vapor deposition. Ge/Au/Ni/Au is evaporated to form the n-type contacts. Finally, the devices are cleaved and packaged on heat sink for testing.[28]

The VCSEL was firstly considered low-power operation, i.e., mW level, because of its short resonant cavity. However, with the improvements in the epitaxial growth, processing, device design and packaging, researchers obtained high-power VCSELs emitting from single large aperture device. In 2001, Michael Miller and Karl Joachim Ebeling obtained 890 mW output VCSEL.[29] L. J. Wang and Y. Q. Ning reported 1.95 W output CW VCSEL in 2007 and achieved 92 W peak output in 2011.[30,31]

The VCSEL is very fit for high-power 2D arrays. High-power VCSEL arrays will play an important role in LiDAR and three-dimensional (3D) sensing. Figure 3.100 is the schematic diagram of the VCSEL arrays.[32] The technical challenges for the development of a high-power VCSEL arrays can be summarized as follows: (1) Develop a high-efficiency single emitter with low thermal impedance; (2) improve the growth and process uniformity for cm-scale chips; (3) develop the packaging for reliability and kW-level heat-removal.[28]

For properly designed arrays, there is little drop in PCE compared to the single device results. For higher power and power density, arrays with more closely spaced elements were fabricated for CW and QCW operation. The results are shown in Figure 3.101. At a 320-A drive current, as high

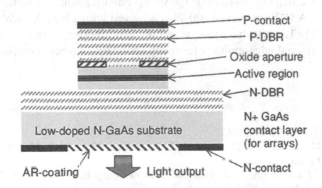

FIGURE 3.99 Schematic of the VCSEL single device for bottom-emitting structure. (From Jean-Francois, S., Chuni, L. G., Viktor, K., Aleksandr, M., Guoyang, X., James, D. W., Prachi, P. and D'Asaro, L. A., *Proc. SPIE*, 6908, 2008. With permission.)

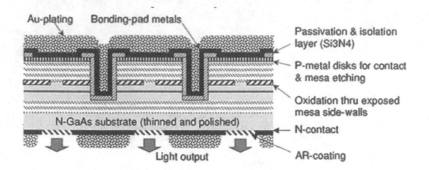

FIGURE 3.100 Schematic cross section of a 2D 97X-nm VCSEL array. (From Jean-Francois, S., Chuni, L. G., Viktor, K., Aleksandr, M., Guoyang, X., James, D. W., Prachi, P. and D'Asaro, L. A., *Proc. SPIE*, 6908, 2008. With permission.)

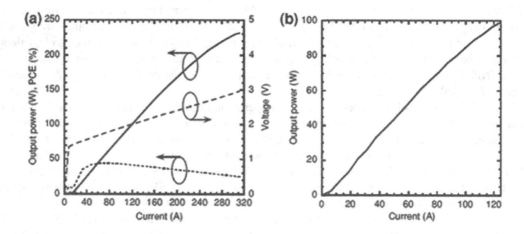

FIGURE 3.101 (a) High-power 976-nm VCSEL array achieving 231 W of CW output power at 320 A. (b) High-power-density 976-nm QCW VCSEL array. At 125 A, the output power is 100 W, corresponding to a 3.5 kW/cm²-power density. (From Jean-Francois, S., Chuni, L. G., Viktor, K., Aleksandr, M., Guoyang, X., James, D. W., Prachi, P. and D'Asaro, L. A., *Proc. SPIE*, 6908, 2008. With permission.)

as 231 W output power was reached corresponding to a power density of 1 kW/cm². This array has a peak conversion efficiency >44%. The QCW hip-on-submount was tested directly without a micro-channel cooler. A QCW power of 100 W is achieved with a high 3.5 kW/cm² power density at 125 A. Because VCSEL arrays have unique and obvious advantages in performance and cost, it could replace the traditional edge-emitters for many applications where compact high-power laser sources are required.[28,32]

REFERENCES

1. Varshni, Y. P., *Physica*, 34, 149, 1967.
2. Vurgaftman, I., Meyer, J. R. and Ram-Mohan, L. R., *J. Appl. Phys.*, 89, 5815, 2001.
3. Vegard, L., *Z. Phys.*, 5, 17, 1921.
4. Prins, A. D., Sly, J. L., Meney, A. T., Dunstan, D. J., O'Reilly, E. P., Adams, A. R. and Valster, A., *J. Phys. Chem. Solids*, 56, 349, 1995.
5. William, M. Y., Shigeo, S. and Hajime, Y., *Phosphor Handbook*, Second Edition, 2007, Boca Raton, FL: CRC Press Taylor&Francis Group.
6. Kazumura, M., Ohta, I. and Teramoto, I., *Jpn. J. Appl. Phys.*, 22, 654, 1983.

7. Arthur, J. R., *Surf. Sci.*, 500, 189, 2002.
8. Dupuis, R. D., *IEEE J. Sel. Top. Quantum Electron.*, 6, 1040, 2000.
9. Uchida, K., Bhunia, S., Sugiyama, N., Furiya, M., Katoh, M., Katoh, S., Nozaki, S. and Morisaki, H., *J. Cryst. Growth*, 248, 124, 2003.
10. Mawst, L. J., Kim, H., Smith, G., Sun, W. and Tansu, N., *Prog. Quant. Electron.*, 79, 6727, 2020.
11. Lauer, C., König, H., Grönninger, G., Hein, S., Gomez-Iglesias, A., Furitsch, M., Maric, J., Kissel, H., Wolf, P., Biesenbach, J. and Strauss, U., *Proc. SPIE*, 8241, 824111, 2012.
12. Matthews, J. W. and Blakeslee, A. E., *J. Cryst. Growth*, 27, 118, 1974.
13. Sato, S. and Satoh, S., *Jpn. J. Appl. Phys.*, 38, L990, 1999.
14. Won-Jin, C., Dapkus, P. D. and Jewell, J. J., *IEEE Photon. Technol. Lett.*, 11, 1572,1999.
15. Tansu, N. and Mawst, L. J., *IEEE Photon. Technol. Lett.*, 13, 179, 2001.
16. Ning Y. Q., Chen Y. Y., Zhang J., Song Y., Lei Y. X., Qiu C., Liang. L., Jia P., Qin L. and Wang L. J., *Acta Optica Sinica*, 41, 0114001, 2021.
17. Laidig, W. D., Lin, Y. F. and Caldwell, P. J., *J. Appl. Phys.*, 57, 33, 1985.
18. Mawsi, L. J., Rusli, S., Al-Muhanna, A. and Wade, J. K., *IEEE J. Sel. Top. Quantum Electron*, 5, 785, 1999.
19. Suzuki, M., Nishikawa, Y., Ishikawa, M. and Kokubun, Y., *J. Cryst. Growth*, 113, 127, 1991.
20. Crump, P., Pietrzak, A., Bugge, F., Wenzel, H., Erbert, G. and Tränkle, G., *Appl. Phys. Lett.*, 96, 131110, 2010.
21. Knauer, A., Erbert, G., Staske, R., Sumpf, B., Wenzel, H. and Weyers, M., *Semicond. Sci. Technol.*, 20, 621, 2005.
22. Lu, Z., Yunhua, W., Baoshan, J., Duanyuan, B., Jing, X., Xin, G. and Baoxue, B., 2011 Academic International Symposium on Optoelectronics and Microelectronics Technology, pp. 35, 2011
23. Zimer, H., Haas, M., Ried, S., Tillkorn, C., Killi, A., Heinemann, S., Negoita, V., An, H. and Schmidt, B., *2014 IEEE Photonics Conference, San Diego, CA*, pp. 230, 2014.
24. Strohmaier, S., Erbert, G., Meissner-Schenk, A. H., Lommel, M., Schmidt, B., Kaul, T., Karow, M. and Crump, P., *Proc. SPIE*, 10514, 1051409, 2018.
25. Soda, H., Motegi, Y. and Iga, K., *IEEE J. Quantum Electron.*, 19, 1035, 1983.
26. Sakaguchi, T., Koyama, F. and Iga, K., *Electron. Lett.*, 24, 928, 1988.
27. Jean-Francois, S., Chuni, L. G., Viktor, K., Aleksandr, M., Guoyang, X., James, D. W., Prachi, P. and D'Asaro, L. A. *Proc. SPIE*, 6908, 690808, 2008.
28. Michalzik R., *Springer Series in Optical Sciences*, 166, Berlin, Heidelberg: Springer, 2013.
29. Miller, M., Grabherr, M., King, R., Jager, R., Michalzik, R. and Ebeling, K. J., *IEEE J. Sel. Top. Quantum Electron.*, 7, 210, 2001.
30. Sun, Y., Ning, Y., Li, T., Qin, L., Yan, C. and Wang, L., *J. Lumin.*, 122–123, 886, 2007.
31. Zhang, L., Ning, Y., Zeng, Y., Qin, L., Liu, Y., Zhang, X., Liu, D., Xu, H., Zhang, J. and Wang, L., *Appl. Phys. Express*, 4, 052102, 2011.
32. Jean-Francois, S., Guoyang, X., Viktor, K., Alexander, M., James, D. W., Prachi, P., Chuni, L. G. and D'Asaro, L. A. *Proc. SPIE*, 7229, 722903, 2009.

3.6 GaN-BASED LEDs ON SI SUBSTRATE

Zhijue Quan, Jianli Zhang and Fengyi Jiang

3.6.1 INTRODUCTION

The LEDs, which greatly influence people's lives, have been widely used in general lighting, signage lighting, display, automotive application and etc. The prosperity of LEDs should be attributed to the success of GaN-based LEDs. The GaN-based materials, including GaN and its alloys with AlN and InN, are III–V nitride compound semiconductors with wurtzite crystal structure. GaN and AlN have a bandgap energy of 3.4 and 6.2 eV at room temperature, respectively. There was some controversy with the bandgap energy of InN. Prior to the year of 2002, the generally accepted value for the bandgap energy of InN was 1.9 eV. However, Wu[1,2] showed by luminescence measurements that the bandgap energy of InN is lower. To this day, the bandgap energy of 0.7 eV for InN[3] is generally accepted. Therefore, the wavelengths of GaN-based LEDs, in theory, can span a very wide range covering deep ultraviolet (UV), near UV, visible light and even near infrared, with great potential to provide us a colorful artificial vision world. However, it is generally difficult to synthesize In-rich InGaN alloys with high internal quantum efficiencies (IQEs). As a result, the AlGaInN material system is exclusively used for UV, blue and green LEDs in a long term and rarely for longer wavelengths. In the orange and red wavelength range, AlGaInP LEDs play the dominant role owing to the high luminous efficiency and the lower manufacturing cost compared with AlGaInN LEDs. Until recently, efficient InGaN-based yellow and red LEDs on Si substrate have been fabricated.[4,5] A significant breakthrough is made on the yellow and red wavelength by improving the quality of InGaN, reducing the strain in quantum wells (QWs) and utilizing the 3D p-n junction of V-pits.

The substrate used for GaN-based materials epitaxial growth has great impact on the crystal quality, the strain, the luminescence behavior and the light extraction mode of LEDs. To obtain high-quality and crack-free GaN-based materials, homoepitaxial GaN on GaN substrate is preferable. However, in the initial stages of GaN LED, it is lack of the bulk GaN substrate with low cost and high quality. Even now, the bulk GaN substrate is still too expensive to be used to fabricate commercial LEDs. Instead, sapphire, SiC and Si substrates are commonly used.

Among them, Si is not considered as an excellent substrate choice for GaN epitaxy. It is mainly attributed to the mismatches in lattice constant and thermal expansion coefficient (TEC). As shown in Figure 3.102, the lattice mismatch and thermal mismatch between GaN and Si (1 1 1) substrate are 16.9 and 57%, respectively, which are much larger than those of GaN/Sapphire and GaN/SiC.[6] The large mismatches bring greate challenge to the epitaxial growth of GaN. The GaNs grown on Si substrate (GaN/Si) are more prone to high dislocation density, film cracking and wafer bending.

However, despite abovementioned drawbacks, Si substrate also possesses many attractive advantages. First, the cost of LEDs goes down due to the cheap price and large-size (up to 300-mm diameter) of Si substrates. Second, as the processing technology of Si is very mature in microelectronic industry, GaN/Si LED can be fabricated very flexibly and have the possibility in optoelectronic integration. Third, GaN/Si LEDs can get a good reliability since Si has good thermal and electrical conductivity. Finally, the tensile-type thermal strain and lattice strain raised by the Si substrate will contribute to indium incorporation in InGaN QWs. It is favorable to the growth of long wavelength LEDs as the growth temperature of InGaN QWs can be raised. The enhanced growth temperature can improve the quality of InGaN QWs.

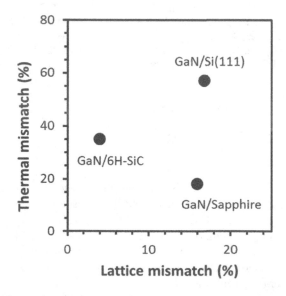

FIGURE 3.102 The lattice and thermal mismatch of GaN epitaxy on different substrates. (From Jiang, F., Zhang, J., Sun, Q. and Quan, Z., *Light-Emitting Diodes*, Springer International Publishing, Cham, pp. 133–170, 2019. With permission.)

Since the 1970s, when the first GaN/Si technology was reported,[7] great efforts have been made by many researchers to industrialize the GaN/Si LEDs. The GaN/Si blue LEDs were realized in 2004,[8] and it took until 2009 for the first bright LED on Si substrate to be made.[9] By 2014, the performance of blue GaN/Si LEDs is comparable to the top-level LEDs on sapphire and SiC substrates.[10] Nowadays, the LEDs on Si substrate have already been commercially available and widely spread in our lives. In this chapter, we will summarize the development of GaN/Si LEDs in our laboratory, mainly focusing on works performed since 2004.

3.6.2 A Glance at Performance of GaN/Si LED

A series of GaN/Si LED with various wavelengths ranging from blue to red was successfully developed in our laboratory. Table 3.12 shows the data of wall-plug efficiency (WPE), voltage (V_f) and full-width-at-half maxima (FWHM) for GaN/Si LED with various dominant wavelengths (WLD), respectively, measured at 20 A/cm^2 and J_{max} (the current density corresponding to the peak WPE).

Figure 3.103 shows the dependence of WPE, voltage and FWHM on wavelength for GaN/Si LEDs. At 20 A/cm^2, the WPE is 68% when wavelength is 457 nm and 3.2% when wavelength is 620 nm, which deceases approximately linearly with increasing wavelength. Fitting the data shows that WPE decays around 0.41% for each 1nm increment in wavelength. The voltage of the devices shows a negative relationship with wavelength, which follows the nature that bandgap energy decreases with increasing wavelength. The FWHM is found to be narrow for short wavelength and broad for long wavelength, this could be attributed to indium fluctuation and piezoelectric effect in the QWs. More indium content will lead to sever indium fluctuation and stronger piezoelectric field.

3.6.3 Epitaxial Structures of GaN/Si LED

GaN/Si LEDs are normally grown by metalorganic chemical vapor deposition (MOCVD) with mutilayer structure. Figure 3.104(a) shows the typical cross-sectional view of GaN/Si LED epilayer. The epitaxial structure from the bottom to top is described as follow: A buffer layer is deposited on a Si (1 1 1) substrate firstly, then a Si-doped n-type GaN is grown on the buffer layer, and followed by a preparation layer. InGaN/GaN multiple QWs (MQWs) are grown on the preparation layer, followed by an Mg-doped AlGaN electron blocking layer (EBL) and an Mg-doped p-GaN layer. The structure uses GaN-based materials with different Al or In compositions, doped with species like

TABLE 3.12

The Data of WPE, Voltage and FWHM for GaN/Si LED in Our Laboratory

@20 A/cm^2				@J_{max}			
WLD (nm)	FWHM (nm)	V_f (V)	WPE (%)	WLD (nm)	FWHM (nm)	V_f (V)	WPE Peak (%)
457	16.8	2.83	68.0	459	15.9	2.72	75.8
475	19.7	2.72	62.5	480	18.6	2.59	71.3
495	22.1	2.61	54.8	501	21.3	2.48	66.5
520	28.9	2.51	46.1	521	25.8	2.37	62.5
540	33.4	2.42	38.3	540	31.3	2.30	62.8
555	36.7	2.37	32.3	563	35.7	2.21	51.7
565	38.5	2.35	27.9	578	38.3	2.15	46.9
578	43.3	2.32	19.2	590	41.3	2.09	39.9
598	49.5	2.30	9.5	603	45.6	2.03	30.2
620	55.1	2.28	3.2	615	48.0	1.97	22.5

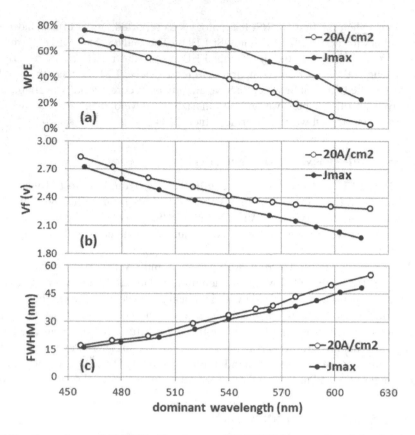

FIGURE 3.103   Dependence of (a) WPE, (b) voltage and (c) FWHM on wavelength for GaN/Si LED.

Mg and Si, and suffering from contamination like H, C and O. The layer structure, interface quality, doping concentration and impurity contamination can be characterized by the second ion mass spectrum (SIMS), providing depth profiles with excellent detection limits for Al, In matrix species, and Mg, Si dopants, together with H, C, O impurities. The SIMS profile of a typical GaN/Si LED sample is shown in Figure 3.104(b).

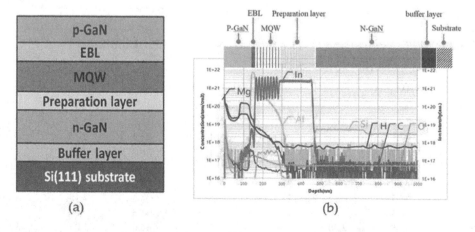

FIGURE 3.104   (a) Schematic cross-sectional structure of InGaN MQW LED on Si (1 1 1), (b) SIMS depth profile of elements in epitaxial film for GaN LED grown on Si.

In this section, the key technologies of GaN/Si LED structure will be described, including the technology of buffer layer, preparation layer and three-dimensional (3D) p-n junction embedded in MQWs.

3.6.3.1 The Buffer Layer

The buffer layer is used to reduce dislocations and cracks for the GaN/Si epitaxial growth. Various buffer technologies have been studied. Up to now, two types of buffer technology are more successful than others and have been widely used in the epitaxial growth of GaN/Si LEDs.

One is the graded AlGaN buffer technology. Enough compressive strain in the GaN film at epitaxial growth temperature can be build up by utilizing the positive lattice mismatch of +2.4% between GaN (a=0.3189 nm) and AlN (a=0.3112 nm) in the Al-composition graded AlGaN/AlN buffer layers. The compressive strain can compensate for the tensile stress induced by the TEC difference during the cool-down process[11,12] and avoid film cracking. Figure 3.105 shows a typical structure of Al-composition graded AlGaN/AlN buffer layers from reference.[13] The buffer layer consisted of a 280-nm-thick AlN, a 180-nm-thick $Al_{0.35}Ga_{0.65}N$ layer and a 310-nm-thick $Al_{0.17}Ga_{0.83}N$ layer. The threading dislocation (TD) density in the as-grown 3-μm GaN film grown on the buffer layer was estimated to be around 5.8×10^8 cm^{-2}.

The other one is thin AlN buffer technology. In the last two decades, the technical route has been studied and developed in our laboratory. In 2006, LEDs based on the technical route has been commercialized in company established by the technical team. In this section, the thin AlN buffer technology will be introduced in detail.

For the thin AlN technology, the grid-patterned Si substrate (GPSS) is needed. The size of the unit patterns depends on the size of the chips to be made. Figure 3.106 shows a schematic view of a GPSS. Figure 3.106(a) is a vertical view, in which the Si substrate is divided into unit patterns isolated from each other by the boundaries.[6] There are two different process methods for the formation of boundary. One is the dielectric film boundary, as shown in Figure 3.106(b). The dielectric film boundary is obtained by growing layer of dielectric film (such as SiN_x, SiO_2) and then lithographic etching off the maskless region. Another is the trench boundary, as shown in the sectional view of Figure 3.106(c). The trenches are produced by photolithography and dry etching on the Si substrate. During the process of epitaxial growth, GaN almost cannot be grown on the boundaries. Therefore, the GaN thin film is also divided into small sizes isolated from each other by the boundaries, which significantly reduces the thermal stress between GaN thin film and Si substrate.

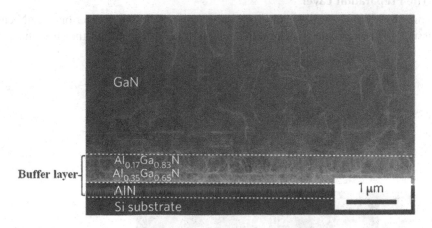

FIGURE 3.105   TEM image of GaN film on Si substrate with an AlN/$Al_{0.35}Ga_{0.65}$N/$Al_{0.17}Ga_{0.83}$N buffer layer. (From Sun, Y., Zhou, K., Sun, Q., Liu, J. P., Feng, M. X., Li, Z. C., Zhou, Y., Zhang, L. Q., Li, D. Y., Zhang, S. M., Ikeda, M., Liu, S. and Yang, H., *Nat. Photonics*, 10, 595, 2016. With permission.)

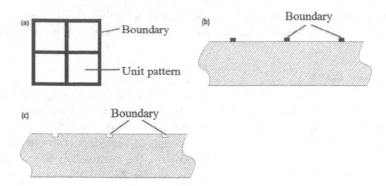

FIGURE 3.106 Scheme for patterned Si substrate in (a) vertical view; (b) sectional view of Si substrate with dielectric film boundary and (c) sectional view of Si substrate with trench boundary. (From Jiang, F., Zhang, J., Sun, Q. and Quan, Z., *Light-Emitting Diodes*, Springer International Publishing, Cham, pp. 133–170, 2019. With permission.)

When the GPSS is placed into the MOCVD reactor, a thermal treatment at around 1100°C is conducted to remove the native oxide layer on the surface of Si substrates and obtain a flat surface for epitaxial growth. And then, a few monolayer of metallic aluminum (Al) was deposited on the Si substrate prior to the growth of the AlN intermediate layer. If AlN is directly grown on Si substrates, Si surface easily reacts with NH_3 to form SiN_x. Finally, the AlN buffer layer with a thickness of 100 nm is grown.

AlN is an ideal intermediate material, which not only serves as seed layer or nucleation layer to prevent Ga melt-back etching[14] but also facilitates the subsequent growth of GaN.[15]

Figure 3.107 shows the transmission electron microscopy image of crack-free GaN film with low dislocation density grown on Si (1 1 1) substrate, using a 100-nm thick AlN buffer layer. As shown in Figure 3.107, in the AlN buffer layer, there exists a large number of dislocations (dislocation density $>10^{10}$ cm^{-2}), and GaN dislocation density can be reduced to a range of $3×10^8$–$1.5×10^9$ cm^{-2} by controlling the growth conditions.

Compare to the graded AlGaN buffer, the thin AlN buffer with simple structure was more reliable and less time-consuming, and hence facilitates the rapid efficient batch production of GaN/Si LEDs.

3.6.3.2 The Preparation Layer

Similar to the growth of GaN on Si substrate, the growth of the active layer on n-GaN encounters the same issue of lattice mismatch. The active region of LEDs consists of the structure of multiple

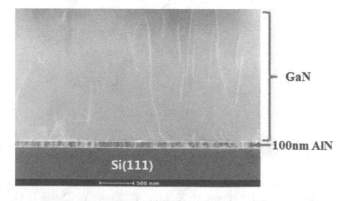

FIGURE 3.107 TEM image of GaN film on Si substrate with a 100-nm-AlN buffer layer.

QW (MQW), including the InGaN well and GaN barrier. The lattice parameter of InN a-axis is 0.3548 nm, which has ~10% lattice mismatch to GaN. For InGaN QW and GaN QB, the value of mismatch is 2–3%. The large lattice mismatch will introduce huge compressive stress to the QWs and lead to producing new dislocations and indium segregation, thereby reduce the efficiency of LEDs.

To improve LED's efficiency, the preparation layers (PLs) are frequently inserted into between n-GaN and the active layer. The PLs are generally InGaN/GaN superlattices (SLs), or MQWs with less In composition than the active MQWs. Some reports claimed that the SLs embedded under MQWs in the LED structure could serve as strain relaxation layers, which are beneficial to the strain relaxation in the MQW layer and the enhancement of efficiency due to the reduction of the quantum confinement stark effects (QCSE).[16–18]

The other physical mechanism about the PLs has also attracted special interest for its significant improvement on the emission efficiency of InGaN/GaN QWs. Some researchers reported that the SLs can act as current spreading layers,[19,20] or electron cooler.[21,22] The current spreading layers are helpful for LED to obtain uniform current spreading and hence to achieve high-efficiency performances. The electron cooler increases electron capture rate and improves hole transport across the MQW and then leads to enhance optical output power and reduce efficiency droop. In addition, it has also been suggested that the SLs embedded under MQWs provide templates for V-pits formation and can vary the V-pits in terms of the SLs' growth temperature, well and barrier composition and period number.[23,24] The V-pits can promote the injection of holes in addition to screening the dislocations, and then improves the quantum efficiency of the LEDs.[25]

The detail about V-pits will be described in the next section. Figure 3.108 shows the STEM images of two samples grown on Si in our laboratory. Sample A has the same structure as sample B except no PLs before the MQWs. The PLs consist of 32 periods of alternating 5-nm thick $In_xGa_{1-x}N$ wells with x~0.1 and 1-nm thick GaN barriers. The STEM clearly reveals that the MQWs of sample B are divided into two parts (the tilted MQWs and the flat MQWs) by the large V-pits around dislocations, while the MQWs of sample A are only typically flat. The tilted MQWs are the sidewalls of V-pits.

3.6.3.3   The Three-Dimensional p-n Junction of V-Pits

In general, heterojunction, polarization field and built-in electric field were considered to be perpendicular to c-planes. These are attributed to the traditional design ideas of planar devices. However, in InGaN-based LEDs with c-plane MQWs containing large V-pits, the situations will be different.

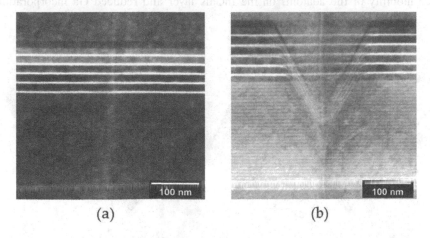

(a)   (b)

FIGURE 3.108   Cross-sectional STEM images of the active regions of (a) sample A without preparation layer and (b) sample B with the SLs preparation layer.

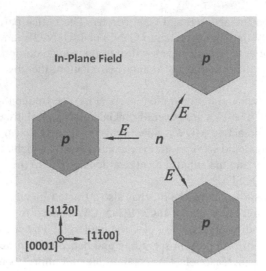

FIGURE 3.109 Schematics of horizontal p-n junctions in the LEDs with V-pits, where hexagon represents outline of V-pits in the top view along [0 0 0 1] growth direction. (From Gao, J. D., Zhang, J. L., Quan, Z. J., Pan, S., Liu, J. L. and Jiang, F. Y., *J. Phys. D Appl. Phys.*, 53, 335103, 2020. With permission.)

Normally, the V-pits are formed in the preparation layer with low growth temperature and high growth rate. The subsequent grown MQWs can have two different shapes: The tilted MQWs on the V-pit side and the flat MQWs on the terrace. And then, during the growth progress of p-GaN with high temperatures and low growth rates, the V-pits will be filled up by p-GaN. Thus, in addition to the regular vertical p-n junction along [0 0 0 1], a lot of horizontal p-n junctions along ⟨1–100⟩ created by n-type MQWs and the p-GaN filled in the pits are existed as shown in Figure 3.109. Together the horizontal p-n junction and the regular vertical p-n junction form the 3D p-n junction in InGaN/GaN MQW LEDs.[26] It can be seen from above that the changes in geometric shapes of p-n junctions are caused by V-pits.

The V-pits with six {10–11}-oriented sidewalls, induced from dislocations, are inverted hexagonal pits embedded in MQW structures.[27] Figure 3.110 shows the structure of V-pits. As a typical characteristic in GaN-based LEDs, the V-pits have attracted attention since 1998.[28–32] These works mainly focused on the formation mechanism of V-pits. Several suggestions, such as strain release, low surface mobility of the adatoms on the InGaN layer and reduced Ga incorporation on the

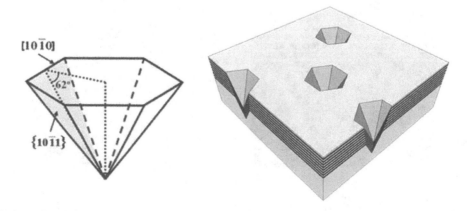

FIGURE 3.110 Schematic structure of the V-shaped pits.

{10–11} planes in comparison to the 0 0 0 1 surface, have been put forward to explain the generation of such V-pits.

In the early research, the V-pits are considered to be a kind of harmful defect to device performance. The purpose of research is to reduce and eliminate the V-pits. Until 2000, Takahashi[33] investigated the effect of V-pits on the emission efficiency of GaInN-based single QW (SQW). It was demonstrated that the V-pits increase the luminescence intensity and reduces the dependence of photoluminescence (PL) intensity on the TD density of underlying GaN. The explanation for this is that the partial isolation of the well layer from TDs due to the V-pits formation improves the emission efficiency. That is an embryonic form of the "V-pits screening TDs" model. However, this work was ignored at that time by other researchers. In 2005, Hangleiter found[34,35] that the thickness and In concentrations in sidewall structure of V-pits are much lower compared to those of the flat region. It provides an energy barrier around each dislocation and prevents carriers from the dislocation, which is considered as the non-radiative recombination center. Consequently, the V-pits induced from dislocations can effectively screen the dislocations themselves and improve the emission efficiency of LEDs. Thereafter, the "V-pits screening TDs" model has been confirmed and accepted widely.[36–41] Further studies have indicated that the better shielding effect can be obtained with the larger V-pits size.[42,43]

As is known to all, there is a high density of TDs in InGaN/GaN LED grown on foreign substrates, due to high lattice and thermal mismatch between the epilayer and substrate. But the luminous efficiency of GaN-based LEDs seems to be insensitive to the TDs, unlike to that of the traditional III–V semiconductor LEDs. The "V-pits screening TDs" model can partly explain the unexpectedly high emission efficiency of InGaN-based LEDs with high dislocation density. However, it is worth noting that the role of V-pits is only screening TDs in the model. As long as there are no dislocations, the V-pits will not be needed. Moreover, the V-pits occupy part of the MQWs and reduce the luminous area. In this point, it will lead to decline emission efficiency of LED. Therefore, in the logical framework of "V-pits screening TDs" model, it is still considered that the density of TDs should be as less as possible to improve the emission efficiency of LED. The homoepitaxial LED grown on GaN substrate is considered to have the potential to achieve high luminous efficiency for the density of TDs that is about less than 10^6 cm^{-2}. However, it is not the truth. So far, no research on the homoepitaxial GaN LED with high efficiency has been reported. The density of TDs is about 10^8–10^9 cm^{-2} for LED grown on sapphire or SiC substrate, 10^9–10^{10} cm^{-2} for LED grown on Si substrate. But the GaN/sapphire or GaN/SiC LED with high efficiency have been achieved and commercially produced. Moreover, the GaN/Si LEDs with the higher density of TDs have also been developed and achieve the efficiency comparable to the GaN/sapphire LED. The contradiction between the model and practice implies that the "V-pits screening TDs" model has a flaw. Besides screening TDs, does the V-pits play other roles?

In 2014, a physical model, called the "V-pits enhancing hole injection" model, was established by our group.[25] In this model, the V-pits can promote the injection of holes in addition to screening the dislocations, and then improves the quantum efficiency of LEDs. It is the reason that the injection of carrier into the flat MQW is easier via the sidewalls of V-pits than via the flat region, which attributes to the lower polarization charge densities at both InGaN/GaN and AlGaN/GaN interfaces of sidewalls structure with the lower In concentrations and {10–11}-oriented semipolar face. Figure 3.111 illustrates clearly the process of hole injection into c-plane QW. There are two ways for hole injection into c-plane QW. One way is hole injection via flat region, in which holes are directly injected from p-type layer into c-plane QW. Another one is hole injection via the side wall of V-pits, in which most of the holes in the V-pits are injected into sidewalls QW laterally rather than vertically from p-type layer, run along the sidewall QW and then flow into c-plane QW.

This physical model can well explain the experimental phenomena that the V-pits have a significant influence on the hole injection depth[44] and the experimental results that the photoelectric performance of GaN-based LEDs varies with the size of the V-pits.[45] In addition, our group also observed the phenomenon of holes injection from the sidewall of V-pits into the c-plane MQW

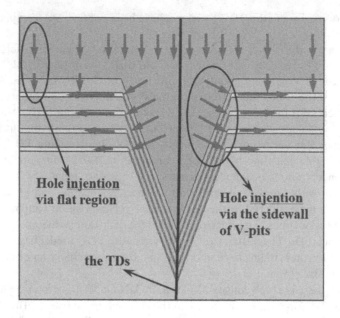

FIGURE 3.111 The diagram about the process of hole injection into *c*-plane QW. The red arrows denote the direction of hole transport.

at low temperature experimentally.[46] At present, the "V-pits enhancing hole injection" model has been accepted by many researchers. Li et al. used the model to calculate the influence of different V-pits sizes on carrier transport and radiative recombination rate.[47] Zhou et al. also used the model to explain the experimental results of LED devices with different V-pits sizes and distribution densities.[48]

According to this model, the density and size of V-pits are closely related to the hole injection, thereby affecting the quantum efficiency of LED. Blue LEDs with different density and size of V-pits are numerically investigated.[49] The results are shown in Figure 3.112. The IQE increases

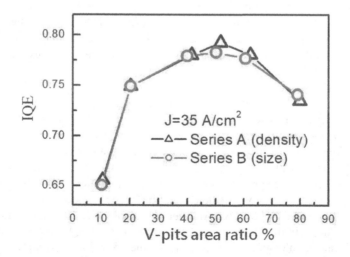

FIGURE 3.112 The dependence of IQE on the area ratio of pits from the calculated results of series A (density) and series B (size) at the current density of 35 A/cm² for blue LED. The area ratio of V-pits here means the area ratio of the V-pits accounting for the last QW. (From Quan, Z. J., Liu, J. L., Fang, F., Wang, G. X. and Jiang, F. Y., *Opt. Quant. Electron.*, 48, 1, 2016. With permission.)

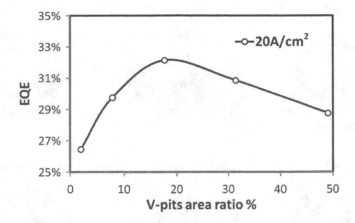

FIGURE 3.113 The experimental dependence of EQE on the area ratio of pits from yellow GaN/Si LED in our laboratory.

firstly and then decreases with an increase in the density and size of pits. Moreover, the dependences of the calculated IQE on area ratio are almost the same for both two series of calculations. It indicates that the area ratio is the key factor. That is, the IQE is firstly improved with an increase in the area ratio of pits due to the improvement of the hole injection. Nevertheless, the excessive increase of the area ratio could lead to the smaller area of emissive region, consequently resulting in poor optical output of LEDs. A balance point must be found between V-pits enhancing hole injection and enough emissive area to maximize the IQE. There lies an optimal value for pits density and the V-pits are induced from dislocations, there lies an optimal value for the density of dislocations.

It should be noted that the abovementioned results are obtained from the models based two hypotheses in the simulation. The one is that the V-shaped pits are uniformly distributed. Another is that all of the pits have the same size. It is very difficult to realize the two hypotheses for a real LED. Therefore, the enhancing effect of V-pits on hole injection will be weakened in the real devices, which resulting in the real optimal data is much smaller than theoretical value (less than 50%), as is shown in Figure 3.113. However, the results point out the future development direction for boosting the LED performance: How to get the same size and uniformly distributed V-pits in the device.

In conclusion, the "V-pits enhancing hole injection" model is established on the basis of the "V-pits screening TD" model. The essential difference of the former from the latter is there lays an optimal value for the density of dislocations, which is a new concept contrary to conventional viewpoints. It is the reason for the unexpectedly high emission efficiency of InGaN-based LEDs with high dislocation density, especially for LED grown on Si substrate. This new concept also eliminates the prejudice that the efficiency upper limit of GaN LED on Si substrate is lower than that on other substrates.

3.6.4 GaN/Si LEDs

After the epitaxial growth, a vertical structure LED chip with the n-GaN upward is fabricated, as the substrate can be easily removed by wet etching for GaN on Si substrate. The chip fabrication process of vertical LEDs grown on Si substrate is simply described in Figure 3.114.

After the growth of (1) GaN LED epi-structure on Si substrate finished, (2) Ag metal layers are deposited on the p-surface of the wafer as p-contact and light reflector, (3) then the p-surface is bonded to new electric conductive carrier via metal bonding technique, (4) and the original Si substrate is etched off by wet etching, leaving n-surface exposed, (5) the n-surface is then roughed by wet etching to enhance light extraction, and finally,(6) n-contact is prepared, completing the chip

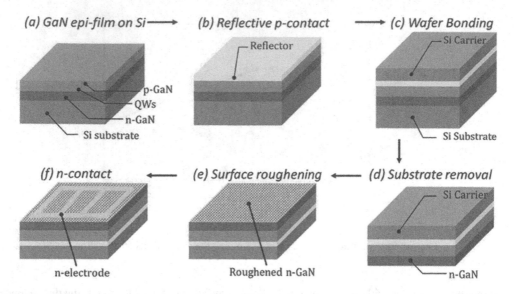

FIGURE 3.114   Device fabrication processes of vertical LEDs grown on Si substrate. (From Jiang, F., Zhang, J., Sun, Q. and Quan, Z., *Light-Emitting Diodes*, Springer International Publishing, Cham, pp. 133–170, 2019. With permission.)

process. The details of the chip fabrication process have been reported[6] and not described here. A luminous chip of green GaN/Si LED is shown in Figure 3.115.

In this section, the electroluminescence (EL) performances of GaN/Si LEDs with different luminous wavelength, which have been fabricated into the vertical LEDs with size of 1×1 mm² in our laboratory, will be shown. The fabricated LED chips are packaged as LUXEON structure with silicone for LED encapsulation. The EL measurements for packaged LEDs are performed by photoelectric test system, which contains Keithley Instruments 2635A Sourcemeter, Instrument Systems CAS140CT spectrometer and IPS 250 integrating sphere.

3.6.4.1   The Blue LEDs

Since 2003, our group began a study on the blue GaN/Si LEDs, and then the blue GaN/Si LEDs were realized in 2004,[8] and it took until 2009 for the first bright LED on Si substrate to be present.[9] By 2013, the performance of our blue GaN/Si LEDs is comparable to the best LEDs on sapphire

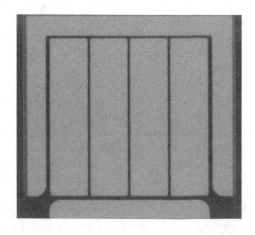

FIGURE 3.115   A Luminous chip of green GaN/Si LEDs fabricated by our group.

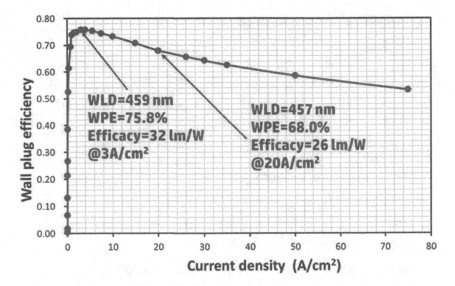

FIGURE 3.116 Plot of WPE as function of current density for the blue GaN/Si LEDs fabricated by our group.

or SiC substrates. The WPE is plotted as functions of current density in Figure 3.116. At 3 A/cm², WPE reaches its peak, where the value is 75.8% and the corresponding WLD is 459 nm. Increasing current density to 20 A/cm², WPE drops down to 68%, where the WLD is 457 nm.

3.6.4.2   The Green LEDs

After the technology of the blue GaN/Si LED was successfully transformed to commercialized production, our group turned to focus on the GaN/Si LED with longer wavelengths.[50] The LED lighting based on phosphor conversion from blue LEDs (PC-LEDs) has a limited efficiency and insufficient illumination quality. By mixing high-efficient LEDs with different colors to form white light, there will be great potential on efficiency improvement and better illumination quality. The key point of realizing color mixing white light is to enhance the efficiency of LEDs in long wavelength range.

As shown in Figure 3.117,[51] the efficiencies of AlGaInN-based LEDs drops rapidly as wavelength increases from 450 to 550 nm, and the efficiencies of AlGaInP-based LEDs drops even faster as wavelength decreases from 650 to 600 nm, the WPE of LEDs is relatively low within the wavelength range from 500 to 600 nm, which is also known as "green gap". The peak of CIE eye response curve just lies in the center of the gap, implies low efficiencies for the sensitive colors. Note that the luminous efficacy (lm/W) is the product of the WPE and the eye response, hence violet LEDs have a poor luminous efficacy despite their excellent WPE. The purpose of developing GaN-based LEDs within long wavelength range is not only to increase the luminous efficiency of LEDs, but also to fill a blank emission region of visible light for solid state lighting. In Figure 3.117, the high WPE of the long wavelengths LEDs fabricated by our group is also displayed with square symbols.

Nowadays, the performance of our green GaN/Si LEDs is comparable to the top-level green LEDs on sapphire or SiC substrates. The WPE is plotted as functions of current density in Figure 3.118. At 1 A/cm², WPE reaches its peak, where the value is 62.4% and the corresponding WLD is 529 nm. Increasing current density to 20 A/cm², WPE drops down to 46.1%, where the WLD is 520 nm.

3.6.4.3   The Yellow LEDs

For many decades, the WPE of direct yellow LEDs is always lower than 10%.[52] In 1971, Craford[53] fabricated the first yellow LED using with WPE of 0.01% nitrogen-doped GaAsP. In the 1990s,

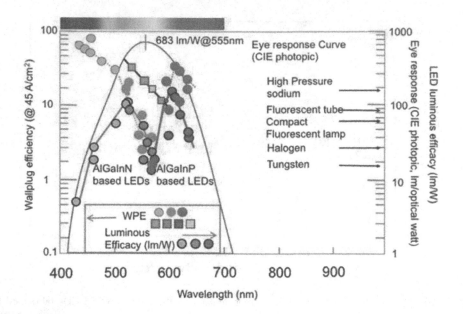

FIGURE 3.117 WPE, luminous efficacy and eye response curve (full line curve) of visible LEDs. This figure come from the reference, and the high WPE of the long wavelengths LEDs fabricated by our group is displayed with square symbols. (From Weisbuch, C., *ECS J. Solid State Sci. Technol.*, 9, 016022, 2020. With permission.)

HP Company developed the AlGaInP-based yellow LEDs with double heterostructure,[54,55] and the WPE was increased to 1.3%. In 2008, to reduce piezoelectric field of InGaN QWs, Nakamura fabricated InGaN-based yellow LEDs on semipolar GaN substrate.[56] However, the voltage is as high as 5.4 V, even if external quantum efficiency (EQE) is increased to 13.4%, WPE is only 5.4%. In 2013, the Toshiba Company reported their InGaN-based yellow LEDs grown on *c*-faced sapphire substrate,[57] with the key technology of inserting AlGaN interlayer between InGaN well and

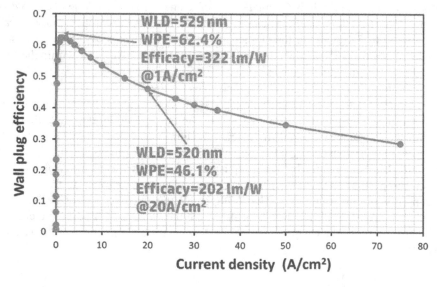

FIGURE 3.118 Plot of WPE as function of current density for the green GaN/Si LEDs fabricated by our group.

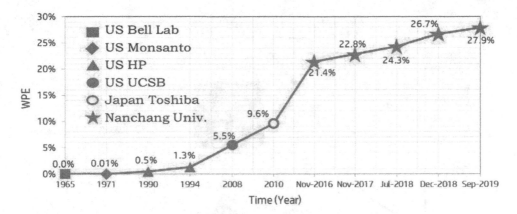

FIGURE 3.119 The efficiency development progress of yellow LEDs in the last 50 years. (From Jiang, F. Y., Liu, J. L., Zhang, J. L., Xu, L. Q., Ding, J., Wang, G. X., Quan, Z. J., Wu, X. M., Zhao, P., Liu, B. Y., Li, D., Wang, X. L., Zheng, C. D., Pan, S., Fang, F. and Mo, C. L., *Acta Physica Sinica*, 68, 168503, 2019. With permission.)

GaN barrier. The EQE was increased to 18.9%, but again the voltage was as high as 4.34 V, thus finally, the WPE was only 9.4%. In 2014, high-power yellow LEDs on silicon substrate were first fabricated by our group, with 74-mW light output power and the EQE of 9.4% at 350 mA-driven current.[58] After that, we made a significant breakthrough on the "yellow gap" by improving the quality of InGaN, reducing the strain in QWs and utilizing the 3D p-n junction of V-pits. Efficient InGaN-based yellow LEDs on Si substrate have been fabricated by using our home-made MOCVD.[4] Figure 3.119 shows the efficiency development progress of yellow LEDs in the last 50 years.[59]

The latest results for the yellow LED of our group are shown in Figure 3.120. A yellow LED has a WPE of 27.9% with dominant wavelength of 565 nm at 20 A/cm^2, and the WPE peak reaches 43.9% with dominant wavelength of 579 nm at 1 A/cm^2. This is currently highest efficiency data for the yellow LED, mainly due to the improved material quality, reduced compressive strain of InGaN QWs by prestrained layer and substrate and enhanced hole injection by 3D p-n junction with V-pits.

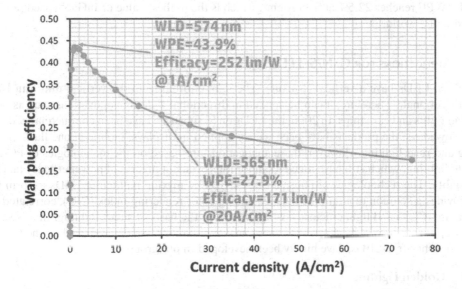

FIGURE 3.120 Plot of WPE as function of current density for the yellow GaN/Si LEDs fabricated by our group.

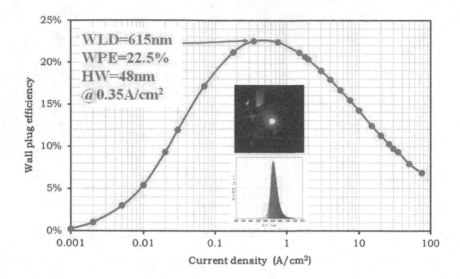

FIGURE 3.121   Plot of WPE as function of current density for the red GaN/Si LEDs fabricated by our group.

3.6.4.4 The Red LEDs

Currently, most commercial red LED chips are made of AlGaInP material with luminous efficiency above 50% in normal size. But when the size of chip is reduced to microlevel, the efficiency drops down dramatically as it has high surface recombination due to the high minority carrier diffusion length.[60–62] Meanwhile, it is reported that the size effect of InGaN-based LED is smaller than that of AlGaInP-based LED.[63–65] Thus, InGaN could be a promising material for micro-red LED, which can be used in the micro-LED full-colors displays. The micro-LED displays are deemed as a promising technology that can replace LCD and organic LED (OLED) in some applications.

Based on the previous work on InGaN-based yellow LEDs, we have extended the efficient emission of InGaN QW from yellow to red and made a significant progress.[5]

The latest results for the red LED of our group is shown in Figure 3.121. For a 615-nm red LED, the WPE reaches 22.5% at 0.35 A/cm^2, which is the highest value of InGaN-based red LEDs reported in the world.

3.6.5 Applications for GaN/Si LEDs

The GaN/Si LEDs have a vertical structure. The light emitted from the vertical thin-film LED is highly directional.[6] Therefore, it can be applied in directional lighting applications such as projectors, head lights and car lamps. In 2006, the LEDs on Si substrate have been commercialized by our group. Nowadays, the products have been widely applied in displays, mobile phone flash, portable lighting and road lighting. However, these applications are based on the technologies of PC-LEDs for white lighting, which still have much room for improvement in term of power efficacy and spectrum quality. The technologies of phosphor-free LEDs by mixing multicolor LEDs to form white light provides a solution to attain the balance among color rendering index (CRI), correlated color temperature (CCT) and light efficiency, which would be the basis of smart lighting. In this section, two new applications will be described. These new applications, which are based on the technologies of phosphor-free LEDs, have highly been developed in our group, recently.

3.6.5.1 Golden Lighting

The "Golden Lighting" is the mixed light source of the red AlGaInP LEDs and yellow GaN/Si LEDs (R&Y-mixed LEDs). The packaging of high-power R&Y-mixed LEDs uses the ceramic

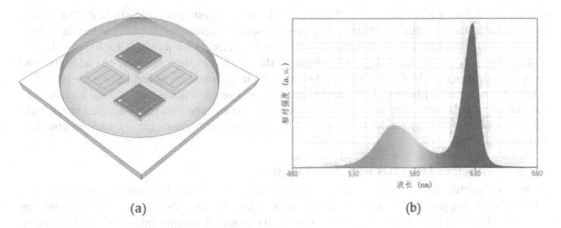

(a)   (b)

FIGURE 3.122 (a) The structure schematic diagrams and (b) EL spectra of golden light source with yellow GaN/Si LEDs and red AlGaInP LEDs.

packaging structure. Four vertical thin-film LED chips with the size of 1 mm×1 mm×0.2 mm are bonded at the bottom of the ceramic substrate. The four LED chips consist of two yellow LED chips with a dominant wavelength of 560 nm and two red LED chips with a dominant wavelength of 620 nm. Each color chips are electronically connected by gold wire bonding process. The hemispherical silicone lens was fabricated by molding process to reduce the reflection loss at the chip surface and enhance light extraction. Figure 3.122 shows the structure schematic diagrams and the EL spectra of golden light source with yellow and red LEDs. The "Golden Lighting" is measured at 20 A/cm² under continuous-wave (CW) mode. Typically, the sample gets a CCT of 2000K, a CRI of Ra=77 and an efficacy of 160 lm/W.

The "Golden Lighting" is mainly applied to the outdoor street lights, as shown in Figure 3.123(a). On the one hand, it has the advantages of low color temperature and strong penetrability as the traditional high-pressure sodium lamp. On the other hand, it has still the merits of high light efficiency and long life as the current mainstream LED street lamps.

The indoor atmosphere lamp is the other application of "Golden Lighting" as shown in Figure 3.123(b). Currently, light pollution has become a non-negligible issue in our daily life. Artificial light sources with high color temperature are considered to be the major pollution source, which could induce several adverse effects on human's health. In our previous

FIGURE 3.123 Applications of golden light LEDs as (a) outdoor street lights for ZhouShan Sea-Crossing bridge of Zhejiang province in china, and (b) indoor atmosphere lamps.

research,[66] the biological effects of "Golden Lighting" are systematically studied. The results indicated that the "Golden Lighting" could provide several significant benefits in human's daily life. The "Golden Lighting" presented several positive effects on human's biological rhythm and health. It is an "artificial light of harmony" that promotes the secretion of melatonin and improves sleeping quality. What's more, because the large amounts of red light were included, it is also an "artificial light of life", which can protect eyes and accelerate wound healing and hair growth. This promising indoor lighting source is thus expected to be applied in the field of advanced biomedical products such as sleep management, cosmetology and precise hair regeneration.

3.6.5.2 Phosphor-Free LEDs for General Lighting

Despite its merits, the "Golden Lighting" is not suitable for general lighting because of the low CRI. To settle the problem, our group developed a technology of phosphor-free LEDs by mixing five-color LEDs to form white light.[4] The white LEDs made by mixing InGaN-based blue, cyan, green and yellow LEDs on silicon substrate, together with AlGaInP red LEDs, are successfully developed and commercialized. For the packaging of five-color LEDs, chip-on-board packaging structure was used and the multicolor LED chips were directly bonded on the substrate. The vertical structure LED chips with the size of 1 mm×1 mm×0.2 mm and 0.6 mm×0.6 mm×0.2 mm were used. For the LED chips, blue GaN/Si LED chips with a dominant wavelength of 455 nm, cyan GaN/Si LED chips with a dominant wavelength of 490 nm, green GaN/Si LED chips with a dominant wavelength of 530 nm, yellow LED chips with a dominant wavelength of 570 nm and red AlGaInP LED chips with a dominant wavelength of 629 nm were used. Each color chip is electronically connected by gold wire bonding process. The hemispherical silicone lens was fabricated by molding process to reduce the reflection loss at the chip surface and enhance light extraction.

The spectrum of a five-color LED module measured at 20 A/cm^2 under CW mode is presented in Figure 3.124. With color coordinate (x=0.4401, y=0.4033) lays right on Planckian locus, the sample (CCT=2941K) gets a high CRI (Ra: 97.5 and R9: 96.2) and a high efficacy of 121.3 lm/W, which is comparable with those PC-LEDs with similar CCT and CRI.

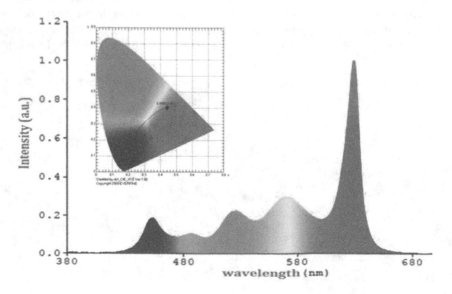

FIGURE 3.124 Spectrum of a color mixing white LED without phosphor made up by blue, cyan, green and yellow GaN/Si LEDs and AlGaInP red LEDs.

3.6.6 SUMMARY

In summary, a significant breakthrough was made on the "green and yellow gap" by improving the quality of InGaN, and reducing the strain in QWs and utilizing the 3D p-n junction of V-pits. InGaN-based blue, cyan, green and yellow LEDs on silicon substrate have been developed and commercialized. The technology of phosphor-free LEDs by mixing multicolors LEDs to form white light provides a promising way to attain the balance among CRI, CCT and light efficiency. Two new applications based on the technologies of phosphor-free LEDs have been developed rapidly and were being successfully or gradually commercialized by our group.

REFERENCES

1. Wu, J., Walukiewicz, W., Yu, K. M., Ager, J. W., Haller, E. E., Lu, H., Schaff, W. J., Saito, Y. and Nanishi, Y., *Appl. Phys. Lett.*, 80, 3967, 2002.
2. Wu, J., Walukiewicz, W., Yu, K. M., Ager, J. W., Haller, E. E., Lu, H. and Schaff, W. J., *Appl. Phys. Lett.*, 80, 4741, 2002.
3. Wu, J. and Walukiewicz, W., *Superlattice Microst.*, 34, 63, 2003.
4. Jiang, F. Y., Zhang, J. L., Xu, L. Q., Ding, J., Wang, G. X., Wu, X. M., Wang, X. L., Mo, C. L., Quan, Z. J., Guo, X., Zheng, C. D., Pan, S. and Liu, J. L., *Photonics Res.*, 7, 144, 2019.
5. Zhang, S., Zhang, J., Gao, J., Wang, X., Zheng, C., Zhang, M., Wu, X., Xu, L., Ding, J., Quan, Z. and Jiang, F., *Photonics Res.*, 8, 1671, 2020.
6. Jiang, F., Zhang, J., Sun, Q. and Quan, Z., *Light-Emitting Diodes*, Cham: Springer International Publishing, pp. 133–170, 2019.
7. Chu, T. L., *J. Electrochem. Soc.*, 118, 1200, 1971.
8. Mo, C. L., Fang, W. Q., Pu, Y., Liu, H. C. and Jiang, F. Y., *J. Cryst. Growth*, 285, 312, 2005.
9. Jiang F., Wang, L., Wang, X., et al., *Abstract Book: The 8th International Conference on Nitride Semiconductors*, 1, pp. 82–83, 2009.
10. Onomura, M., *Proc. SPIE*, 8986, 898620, 2014.
11. Cheng, K., Leys, M., Degroote, S., Van Daele, B., Boeykens, S., Derluyn, J., Germain, M., Van Tendeloo, G., Engelen, J. and Borghs, G., *J. Electron. Mater.*, 35, 592, 2006.
12. Leung, B., Han, J. and Sun, Q., *Phys. Status Solidi C*, 11, 437, 2014.
13. Sun, Y., Zhou, K., Sun, Q., Liu, J. P., Feng, M. X., Li, Z. C., Zhou, Y., Zhang, L. Q., Li, D. Y., Zhang, S. M., Ikeda, M., Liu, S. and Yang, H., *Nat. Photonics*, 10, 595, 2016.
14. Ishikawa, H., Yamamoto, K., Egawa, T., Soga, T., Jimbo, T. and Umeno, M., *J. Cryst. Growth*, 189–190, 178, 1998.
15. Calleja, E., Sanchez-Garcia, M. A., Sanchez, F. J., Calle, F., Naranjo, F. B., Munoz, E., Molina, S. I., Sanchez, A. M., Pacheco, F. J. and Garcia, R., *J. Cryst. Growth*, 201, 296, 1999.
16. Huang, C.-F., Tang, T.-Y., Huang, J.-J., Shiao, W.-Y., Yang, C. C., Hsu, C.-W. and Chen, L. C., *Appl. Phys. Lett.*, 89, 051913, 2006.
17. Huang, C.-F., Liu, T.-C., Lu, Y.-C., Shiao, W.-Y., Chen, Y.-S., Wang, J.-K., Lu, C.-F. and Yang, C. C., *J. Appl. Phys.*, 104, 123106, 2008.
18. Liu, L., Wang, L., Li, D., Liu, N. Y., Li, L., Cao, W. Y., Yang, W., Wan, C. H., Chen, W. H., Du, W. M., Hu, X. D. and Feng, Z. C., *J. Appl. Phys.*, 109, 073106, 2011.
19. Ryu, H.-Y. and Choi, W. J., *J. Appl. Phys.*, 114, 173101, 2013.
20. Jang, C.-H., Sheu, J.-K., Tsai, C. M., Chang, S.-J., Lai, W.-C., Lee, M.-L., Ko, T. K., Shen, C. F. and Shei, S. C., *IEEE J. Quantum Electron.*, 46, 513, 2010.
21. Chang, S.-J. and Lin, Y.-Y., *J. Disp. Technol.*, 10, 162, 2014.
22. Zhang, Z.-H., Liu, W., Tan, S. T., Ju, Z., Ji, Y., Kyaw, Z., Zhang, X., Hasanov, N., Zhu, B., Lu, S., Zhang, Y., Sun, X. W. and Demir, H. V., *Opt. Express*, 22, A779, 2014.
23. Kim, J., Kim, J., Tak, Y., Chae, S., Kim, J. Y. and Park, Y., *EEE Electron Device Lett.*, 34, 1409, 2013.
24. Kim, J., Cho, Y. H., Ko, D. S., Li, X. S., Won, J. Y., Lee, E., Park, S. H., Kim, J. Y. and Kim, S., *Opt. Express*, 22, A857, 2014.
25. Quan, Z. J., Wang, L., Zheng, C. D., Liu, J. L. and Jiang, F. Y., *J. Appl. Phys.*, 116, 183107, 2014.
26. Gao, J. D., Zhang, J. L., Quan, Z. J., Pan, S., Liu, J. L. and Jiang, F. Y., *J. Phys. D Appl. Phys.*, 53, 335103, 2020.
27. Le, L. C., Zhao, D. G., Jiang, D. S., Li, L., Wu, L. L., Chen, P., Liu, Z. S., Li, Z. C., Fan, Y. M., Zhu, J. J., Wang, H., Zhang, S. M. and Yang, H., *Appl. Phys. Lett.*, 101, 252110, 2012.

28. Kim, I. H., Park, H. S., Park, Y. J. and Kim, T., *Appl. Phys. Lett.*, 73, 1634, 1998.
29. Chen, Y., Takeuchi, T., Amano, H., Akasaki, I., Yamano, N., Kaneko, Y. and Wang, S. Y., *Appl. Phys. Lett.*, 72, 710, 1998.
30. Mahanty, S., Hao, M., Sugahara, T., Fareed, R. S. Q., Morishima, Y., Naoi, Y., Wang, T. and Sakai, S., *Mater. Lett.*, 41, 67, 1999.
31. Cho, H. K., Lee, J. Y., Yang, G. M. and Kim, C. S., *Appl. Phys. Lett.*, 79, 215, 2001.
32. Son, K. S., Kim, D. G., Cho, H. K., Lee, K. H., Kim, S. W. and Park, K. S., *J. Cryst. Growth*, 261, 50, 2004.
33. Takahashi, H., Ito, A., Tanaka, T., Watanabe, A., Ota, H. and Chikuma, K., *Jpn. J. Appl. Phys.*, 39, L569, 2000.
34. Hangleiter, A., Hitzel, F., Netzel, C., Fuhrmann, D., Rossow, U., Ade, G. and Hinze, P., *Phys. Rev. Lett.*, 95, 127402, 2005.
35. Netzel, C., Bremers, H., Hoffmann, L., Fuhrmann, D., Rossow, U. and Hangleiter, A., *Phys. Rev. B*, 76, 155322, 2007.
36. Abell, J. and Moustakas, T. D., *Appl. Phys. Lett.*, 92, 091901, 2008.
37. Fang, Z., *J. Appl. Phys.*, 106, 023517, 2009.
38. Tomiya, S., Kanitani, Y., Tanaka, S., Ohkubo, T. and Hono, K., *Appl. Phys. Lett.*, 98, 181904, 2011.
39. Cho, Y.-H., Kim, J.-Y., Kim, J., Shim, M.-B., Hwang, S., Park, S.-H., Park, Y.-S. and Kim, S., *Appl. Phys. Lett.*, 103, 261101, 2013.
40. Weidlich, P. H., Schnedler, M., Eisele, H., Dunin-Borkowski, R. E. and Ebert, P., *Appl. Phys. Lett.*, 103, 142105, 2013.
41. Wu, X. M., Liu, J. L., Quan, Z. J., Xiong, C. B., Zheng, C. D., Zhang, J. L., Mao, Q. H. and Jiang, F. Y., *Appl. Phys. Lett.*, 104, 221101, 2014.
42. Chang, C. Y., Li, H., Shih, Y. T. and Lu, T. C., *Appl. Phys. Lett.*, 106, 091104, 2015.
43. Okada, N., Kashihara, H., Sugimoto, K., Yamada, Y. and Tadatomo, K., *J. Appl. Phys.*, 117, 025708, 2015.
44. Li, Y., Yun, F., Su, X., Liu, S., Ding, W. and Hou, X., *J. Appl. Phys.*, 116, 123101, 2014.
45. Quan, Z. J., Liu, J. L., Fang, F., Wang, G. X. and Jiang, F. Y., *J. Appl. Phys.*, 118, 193102, 2015.
46. Wu, X., Liu, J. and Jiang, F., *J. Appl. Phys.*, 2015, 118, 164504.
47. Li, C. K., Wu, C. K., Hsu, C. C., Lu, L. S., Li, H., Lu, T. C. and Wu, Y. R., *AIP Adv.*, 6, 055208, 2016.
48. Zhou, S. J. and Liu, X. T., *Phys. Status Solidi A*, 214, 1600782, 2017.
49. Quan, Z. J., Liu, J. L., Fang, F., Wang, G. X. and Jiang, F. Y., *Opt. Quant. Electron.*, 48, 1, 2016.
50. Liu, J. L., Zhang, J. L., Wang, G. X., Mo, C. L., Xu, L. Q., Ding, J., Quan, Z. J., Wang, X. L., Pan, S., Zheng, C. D., Wu, X. M., Fang, W. Q. and Jiang, F. Y., *Chin. Phys. B*, 24, 067804, 2015.
51. Weisbuch, C., *ECS J. Solid State Sci. Technol.*, 9, 016022, 2020.
52. Damilano, B. and Gil, B., *J. Phys. D Appl. Phys.*, 48, 403001, 2015.
53. Groves, W. O., Herzog, A. H. and Craford, M. G., *Appl. Phys. Lett.*, 19, 184, 1971.
54. Kuo, C. P., Fletcher, R. M., Osentowski, T. D., Lardizabal, M. C., Craford, M. G. and Robbins, V. M., *Appl. Phys. Lett.*, 57, 2937, 1990.
55. Kish, F. A., Steranka, F. M., Defevere, D. C., Vanderwater, D. A., Park, K. G., Kuo, C. P., Osentowski, T. D., Peanasky, M. J., Yu, J. G., Fletcher, R. M., Steigerwald, D. A., Craford, M. G. and Robbins, V. M., *Appl. Phys. Lett.*, 64, 2839, 1994.
56. Sato, H., Chung, R. B., Hirasawa, H., Fellows, N., Masui, H., Wu, F., Saito, M., Fujito, K., Speck, J. S., DenBaars, S. P. and Nakamura, S., *Appl. Phys. Lett.*, 92, 221110, 2008.
57. Saito, S., Hashimoto, R., Hwang, J. and Nunoue, S., *Appl. Phys. Express*, 6, 207, 2013.
58. Zhang, J. L., Xiong, C. B., Liu, J. L., Quan, Z. J., Wang, L. and Jiang, F. Y., *Appl. Phys. A* 114, 1049, 2014.
59. Jiang, F. Y., Liu, J. L., Zhang, J. L., Xu, L. Q., Ding, J., Wang, G. X., Quan, Z. J., Wu, X. M., Zhao, P., Liu, B. Y., Li, D., Wang, X. L., Zheng, C. D., Pan, S., Fang, F. and Mo, C. L., *Acta Physica Sinica*, 68, 168503, 2019.
60. Bulashevich, K. A. and Karpov, S. Y., *Phys. Status Solidi RRL*, 10, 480, 2016.
61. Royo, P., Stanley, R. P., Ilegems, M., Streubel, K. and Gulden, K. H., *J. Appl. Phys.*, 91, 2563, 2002.
62. Boroditsky, M., Gontijo, I., Jackson, M., Vrijen, R., Yablonovitch, E., Krauss, T., Cheng, C. C., Scherer, A., Bhat, R. and Krames, M., *J. Appl. Phys.*, 87, 3497, 2000.
63. Wong, M. S., Lee, C., Myers, D. J., Hwang, D., Kearns, J. A., Li, T., Speck, J. S., Nakamura, S. and DenBaars, S. P., *Appl. Phys. Express*, 12, 097004, 2019.
64. Oh, J. T., Lee, S. Y., Moon, Y. T., Moon, J. H., Park, S., Hong, K. Y., Song, K. Y., Oh, C. H., Shim, J. I., Jeong, H. H., Song, J. O., Amano, H. and Seong, T. Y., *Opt. Express*, 26, 11194, 2018.

65. Hwang, D., Mughal, A., Pynn, C. D., Nakamura, S. and DenBaars, S. P., *Appl. Phys. Express*, 10, 032101, 2017.
66. Lin, J. Q., Ding, X. W., Hong, C., Pang, Y. L., Chen, L. M., Liu, Q. W., Zhan, X., Xin, H. B. and Wang, X. L., *Sci. Rep.*, 9, 7560, 2019.

3.7 STRONG EXCITON-PHOTON COUPLING AND POLARITON LASING IN InGaN QUANTUM WELL MICROCAVITIES

Hao Long and Baoping Zhang

3.7.1 INTRODUCTION

Strong coupling between exciton and photon induced a novel quasiparticle: Exciton-polariton, which was firstly proposed by Prof. Hopfield in 1950s.[1] Exciton-polariton is a half matter-half light bosonic particle. Since the existence of photon in this quasi particle, the effective mass could be tuned and substantially small (10^{-4} to the mass of electron or exciton). In the early stage, the research on exciton-polariton was based on the bulk material without photon confinement by cavity. Since the photon density was low, exciton-polariton didn't attract so much attention at that time. In recent decades, the quality of semiconductor microcavity was substantially improved by advanced material epitaxy and fabrication technique, including metal-organic vapor-phase epitaxy (MOVPE), molecular-beam epitaxy (MBE) and e-beam deposition etc. The quality of Fabry-Perot (FP) microcavity exceeded 10^3, which enabled strong coupling between exciton in active regions and photon in microcavity. In 1992, C. Weisbuch firstly introduced the semiconductor microcavity into the exciton-polariton community and observed anti-crossing in dispersion curves by reflection spectrum, as shown in Figure 3.125.[2]

Following this work, Houdré et al. group investigated the in-plane dispersion relation by angle-resolved photoluminescence (PL) measurement.[3] Two PL bands named upper polariton branch and lower polariton branch were observed, as shown in Figure 3.126.

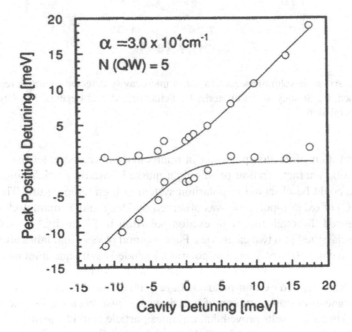

FIGURE 3.125 Dispersion relation of exciton-polariton of GaAs micro cavity. (From Weisbuch, C., Nishioka, M., Ishikawa, A. and Arakawa, Y., *Phys. Rev. Lett.*, 69, 3314, 1992.)

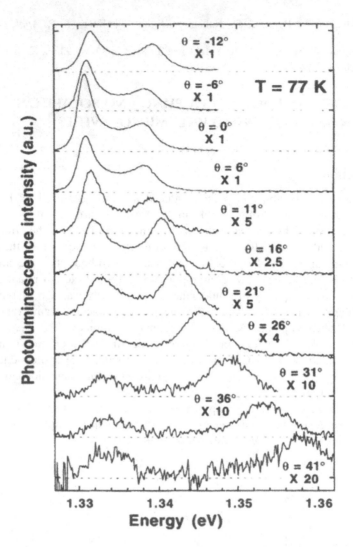

FIGURE 3.126 Polariton photoluminescence in GaAs microcavity detected by angle resolved PL. (From Houdré, R., Weisbuch, C., Stanley, R. P., Oesterle, U., Pellandini, P. and Ilegems, M., *Phys. Rev. Lett.*, 73, 2043, 1994. With permission.)

The clear observation of exciton-polariton in microcavity attracted extensive attention in this area. Since the exotic characterization of this light-massed bosonic particle, many fundamental physic phenomena could be observed in polariton system at high temperature. The Bose-Einstein condensation (BEC) of exciton-polariton was observed by Deng and Yamamoto.[4] Alexey Kavokin et al. also investigated thorough theory of exciton-polariton in FP cavity.[5] Research on exciton-polariton could be classified into two categories: Fundamental physics and fabrication of polaritonic devices. The fundamental physics in exciton-polariton included cavity quantum mechanism, polariton trapping and transportation, and BEC etc. In GaAs[6] and perovskite (CsPbBr$_3$)[7] material systems, the "s" and "p" states of tapped exciton-polariton have been realized by artificially localized potential design. The momentum transportation of polaritons has also been realized within micrometer length scale.[8] Experimental results proved that this quasiparticle could be artificially controlled and be promising for the future electrical and optoelectrical devices. Since the effective mass of exciton-polariton is typically only 10^{-8} to the He atom, the BEC threshold temperature approaches or even exceeds room temperature (RT), because that the condensation temperature is inverse ratio to the effective mass of bosons. Therefore, scientists could realize BEC without the expensive and cumbersome cryogenic equipment. In 1996, Imamoglu et al. proposed that exciton-polariton would be beneficial to the BEC condensation and spontaneous coherent emission.[9] In 2006, Le Si Dang et al. experimentally proved the BEC phase transition in CdTe microcavity.[10] The threshold temperature in CdTe system was about 20K, which was greatly elevated, compared with the $10^{-6}-10^{-7}$K of cold atoms. In 2008, Lagoudakis et al. observed vortex in CdTe polariton system.[11] Yamamoto group proved the existence of Bogoliubov effect in GaAs polariton.[12] The super fluid in polariton system has also been observed in 2009.[13] Therefore, in 2008, "Nature" journal anticipated that polariton would play pivot role in coherent bosonic system and low-threshold lasing devices.[14] Since then, some substantial milestones have been achieved: Coherent scattering between exciton-polaritons,[15] coherent interaction between polariton and phonon,[16] polariton lasing at RT,[17] polariton superlattice[18] and etc.

3.7.1.1 Strong Exciton-Photon Coupling and Experimental Observations

Among the semiconductor material community, GaAs, CdTe, ZnO and GaN etc. have already been successfully utilized into polariton. Compared with conventional GaAs and CdTe, nitride material yields large exciton binding energy (>26 meV of GaN).[19,21] Since the 5- to −6-meV exciton binding energy of GaAs and CdTe was smaller than the RT thermal broadening, exciton-photon strong coupling only exists at cryogen temperatures in these materials. However, GaN, especially InGaN quantum wells (QWs) structure could bind electron-hole pair with 40 meV binding energy,[19] qualifying RT polariton's research. Moreover, unlike CdTe and ZnO, with available p- and n-dopants and sophisticated wafer epitaxy, device fabrication and chip package techniques, nitride semiconductor seems to be more promising for the RT electrically injected polariton devices, e.g. RT non-threshold polariton lasing. However, due to the serious InN/GaN phase separation in InGaN alloy and relatively low Q-quality of nitride FP cavity, the polariton research in InGaN was very rare.

Research on exciton-polariton in GaN systems started from bulk GaN material. Until 2006, I.R. Sellers et al. introduced GaN bulk material into a $\lambda/2$ resonant cavity consisted of hybrid bottom AlInN/AlGaN distributed Bragg reflector (DBR) and upper oxide dielectric DBR and observed polariton photoluminescence by angle-resolved reflection and emission spectrums.[22] In 2007, S. Christopoulos et al. utilized GaN bulk material as active layer to realize RT polariton lasing in a $3/2\lambda$ microcavity.[23] They used bottom nitride DBR, precise cavity length controlled by MBE to eliminate the polarization field in active QWs and elevated DBR reflectivity. Their polariton lasing threshold was one magnitude lower than the one of photonic vertical-cavity surface-emitting laser (VCSEL) lasing. Pallab Bhattacharya et al. from Michigan Univ. achieved electrically injected polariton lasing at RT based on GaN bulk active layer and lateral dielectric DBR structures, as shown in Figure 3.127.[24]

Following the electrically polariton lasing in GaN bulk material, the polariton in GaN nanowires (NWs) also attracted the researchers' interest since the extremely large exciton binding energy. In 2011, Ayan Das et al. achieved super-low lasing threshold (three order of magnitude lower than the photonic lasing) by placing single GaN NW on the antinode of a 1λ FP cavity with both side dielectric SiO_2/TiO_2 DBR.[25] Their polariton lasing threshold was not only lower than the photonic lasing threshold, but also two orders of magnitude lower than the polariton lasing threshold in GaN bulk material-based microcavity. In 2013, vertical upright GaN NWs array was placed between the dielectric DBR structure by Junseok Heo et al. exciton-polariton lasing was also observed (Figure 3.128).[26]

In nitride system, multi-QW structure played essential role in optoelectrical devices, including light-emitting diodes (LEDs), laser diodes (LDs), and VCSEL etc. In 2004, T. Tawara firstly introduced the InGaN/GaN QW structure into dielectric DBR with 4λ cavity length. They grew the InGaN/GaN QWs structure on SiC substrate followed by dry etching of SiC and bonding processes.[27]

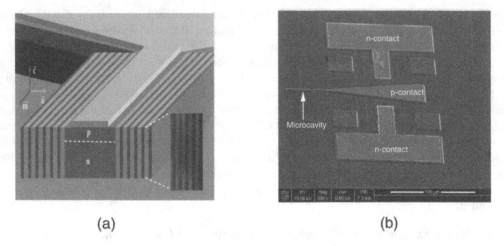

(a)   (b)

FIGURE 3.127   (a) Schematic of electrically injection of exciton-polariton from Michigan Univ.; (b) SEM image of devices. (From Bhattacharya, P., Frost, T., Deshpande, S., Baten, M. Z., Hazari, A. and Das, A., *Phys. Rev. Lett.*, 112, 236802, 2014. With permission.)

Anti-crossing of upper branch (UPB) and lower branch (LPB) of polariton was observed by reflection spectrum, confirming the strong coupling between exciton and photon. The Rabi splitting value in this system was also investigated: 6 meV with 3 pairs QWs and 17 meV with 10 pairs QWs, as shown in Figure 3.129.

In 2007, G. Christmann et al. realized the 345-nm polariton lasing in GaN/AlGaN MQWs confined by a 3λ length cavity.[28] In 2011, T. C. Lu et al. from Taiwan National Chiao Tung Univ. utilized hybrid DBR structure to achieve the InGaN/GaN MQW-based exciton-polariton LED at RT.[29] The dispersion relation of exciton-polariton was studied by angle-resolved electroluminescence and temperature dependent experiments, confirming the exciton photon strong coupling in electrically injected InGaN/GaN MQWs.

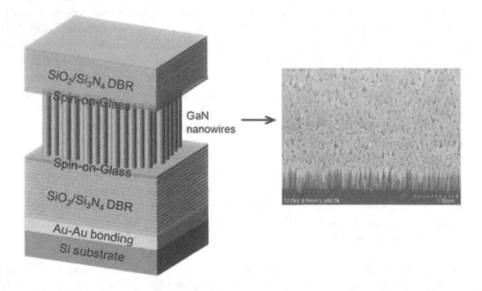

FIGURE 3.128   Microcavity utilizing nanowire array and SEM image. (From Heo, J., Jahangir, S., Xiao, B. and Bhattacharya, P., *Nano Lett.*, 13, 2376, 2013. With permission.)

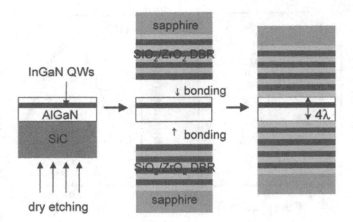

FIGURE 3.129 Structure and fabrication process of InGaN microcavity with dielectric DBRs. (From Tawara, T., Gotoh, H., Akasaka, T., Kobayashi, N. and Saitoh, T., *Phys. Rev. Lett.*, 92, 256402, 2004. With permission.)

3.7.1.2 Difficulty in InGaN Materials

To study the exciton-polariton in nonlinear region at RT, GaN with high exciton binding energy exhibited extraordinary superiority. The RT optically pumped polariton lasing has been observed in GaN bulk, NW and MQWs structures.[30–32] Among this active layers' geometry, multi-QWs (MQWs) structure yields not only larger exciton binding energy, but also great potential in electrically injected devices. Nitride MQWs included GaN/AlGaN MQWs and InGaN/GaN MQWs, with ultraviolet and visible emission, respectively. Both MQWs have large exciton binding energy, but InGaN/GaN suffered indium atom phase separation, which was detrimental to strong coupling. In exciton photon strong coupling, the inhomogeneous broadening and oscillator strength are both essential factors. The inhomogeneous broadening by indium's concentration non-uniformity reduced the excitons' lifetime and destructs the oscillator strength. Compared with binary GaN compound, InGaN layer exhibits very large exciton inhomogeneous broadening since the indium's phase separation. Meanwhile, quantum confined stark effect (QCSE) also spatially separate electrons and holes and reduced the excitons' oscillator strength. Therefore, from the viewpoint of optically pumping polariton, GaN/AlGaN MQWs are preferred, while for the electrically pumping and visible spectrum, InGaN/GaN MQWs seem to be the only choice. The InGaN/GaN MQWs light-emitting diodes were also very mature now, which may benefit the InGaN electrical polariton lasing.

3.7.2 DESIGN AND FABRICATION PROCESSES

In this section, we will introduce our work on exciton-polariton based on InGaN/GaN MQWs, including the coupled QWs design, FP cavity fabrication process, exciton-polariton characterization and polariton lasing.

3.7.2.1 Coupled QWs

When exciton photon interacts with strong coupling, there was a Rabi splitting value Ω to character the coupling strength[33]:

$$\Omega = \sqrt{\Omega_{max}^2 - \left(\gamma_c^2 + \gamma_e^2\right)/2}$$ (3.48)

where γ_c and γ_e represented the width of cavity mode and exciton oscillator; Ω_{max} was the maximum Rabi value without broadening of cavity mode and exciton.

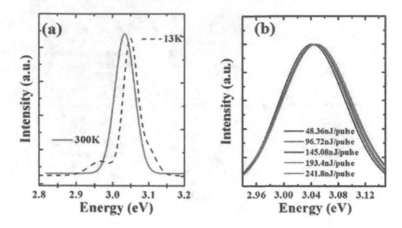

FIGURE 3.130   (a) PL of InGaN/GaN MQWs at 13 and 300K (b) PL spectrums under different excitation.

At low temperature, the width of exciton was dominated by inhomogeneous broadening without exciton thermal broadening. To realize the strong coupling, the Rabi splitting value should be as high as possible. The cavity mode width γ_c was determined by the lifetime of photon in cavity (i.e. quality factor of FP cavity). In high-quality FP cavity, the γ_c could be negligible. The exciton inhomogeneous broadening γ_e was a key factor to determine the coupling strength. If γ_e was too high, the exciton-photon interaction will weaken, no strong coupling could exist. Since the inhomogeneous broadening of InGaN/GaN MQWs mainly came from the non-uniformity of indium atoms, in this work, we utilized the 4 nm/4 nm $In_{0.1}Ga_{0.9}N$/ GaN coupled QWs structures with average 10% indium concentration. Figure 3.130 shows the PL spectrum of this structure at 13 and 300K. At low temperature, the luminescent peak was 3.048 eV with 43 meV inhomogeneous broadening. At 300K, the width was 54 meV, which was the convolution result of inhomogeneous broadening and homogeneous broadening. From the PL width, we could deduce that the homogeneous broadening at RT was 25 meV. From the theoretical anticipation, strong coupling could survive when inhomogeneous broadening was narrower than 46 meV.[34]

The QCSE in InGaN/GaN MQWs induced by large polarization fields inclined the energy bands. Separation of electrons and holes reduced the oscillator strength.[35,36] The coupled QWs structure in this work enabled the carriers tunneling through the barrier layers, which significantly reduced the QCSE effect. Figure 3.130(b) showed that the peak energy blue shifted by increasing the excitation power. In InGaN/GaN MQWs, the blueshift of peak energy with excitation may be caused by two effects: The screening of QCSE and band filling effect. The screening of QCSE causes the peak energy blueshift without bandwidth broadening, while the band filling effect could also cause the width increasing by broadening of high energy side. From Figure 3.130(b), with increasing the excitation power, the band filling effect dominated in our sample. Therefore, we utilized the coupled QWs structure to reduce the QCSE effect under the premise of low inhomogeneous broadening. Our sample structure not only satisfied the prerequisite of exciton-photon strong coupling but also enhanced the oscillator strength. In our work, we calculated the effect of polarization field of InGaN/GaN MQWs structure on the oscillator strength.[37] The Rabi splitting value was determined by the oscillator strength as:

$$\hbar\Omega \approx 2\hbar\left(\frac{2\Gamma_0 c N_{QW}}{n_c L_{eff}}\right)^{1/2} \tag{3.49}$$

where c is light velocity, N_{QW} is QWs number, n_c is the refractive index of cavity material, $L_{eff}=L_c+L_{DBR}$ means the effective cavity length (L_{DBR} is the penetration depth of light into DBR). Γ_0 relies on the oscillator strength per area f_{ex} as:

$$\Gamma_0 = \frac{\pi}{n_c}\frac{e^2}{4\pi\varepsilon_0}\frac{\hbar}{m_e c}f_{ex} \qquad (3.50)$$

with

$$f_{ex} = \frac{2M_{QW}M^2}{m_0 E_{ex}}\left|\int_{\infty}^{+\infty}dz f_e(z)f_h(z)\right|^2|\varphi(0)|^2 \qquad (3.51)$$

where M represents the optical transition element; M_{QW} is the polarization index with 3/2 in transverse electric case and 1/2 in transverse magnetic case. E_{ex} is the transition energy, m_0 is the mass of free electron; f_e and f_h are the wave functions of electron and hole along z direction; $|\varphi(0)|^2$ represents the probability of electron and hole at same xy planar position.

From the equations above, the strong coupling between exciton and photons strongly depends on the overlap between electrons and holes. In our work, coupled QWs structure and high-quality microcavity ($Q>3000$) were utilized to implemented extremely high Rabi splitting value of 130 meV.[38] Based on this achievement, polariton lasing was further investigated.

3.7.2.2　Fabrication Procedure

In this work, five pairs of 4-nm/4-nm $In_{0.1}Ga_{0.9}N$/GaN coupled MQWs were firstly grown on sapphire substrate with 2-μm-thick undoped GaN buffer layer and followed by 50-nm-thick GaN capping layer. Then, 16.5 pairs of Ti_3O_5/SiO_2 were deposited on the wafer surface to form bottom DBR. After that, the fabricated structure was glue bonded onto silica glass, followed by laser lift off (LLO) to peel off the sapphire substrate. The chemical mechanism polishing (CMP) was utilized to grind the GaN buffer layer to form 4.5λ thick resonant cavity length. The $In_{0.1}Ga_{0.9}N$/GaN coupled MQWs were also designed to place at the antinodes of resonant electrical field to maximum the coupling strength. It was worth noting that the CMP process was very important in this fabrication, because it rendered the thinner cavity length and placed the MQWs' location at the electric field's antinode. At last, the 13.5 pairs of Ti_3O_5/SiO_2 were deposited onto the grinded side of samples to form up DBR. Short cavity length of 4.5λ and antinode location of MQWs assured the strong coupling between exciton and photon. Figure 3.131(b) presented the location of coupled MQWs at the cavity.

Considering the PL width of 15 and 300K, the inhomogeneous and homogeneous broadening could be derived. In Figure 3.131(c), 3.03 and 2.94 eV corresponded to the exciton energy and LO-assisted peak.

3.7.3　Characteristics of Exciton-Polaritons

In our work,[39] we used two measurement geometry to character exciton-polariton. In Figure 3.132, the YAG:Nd laser emitted 355-nm pulse with 50 Hz frequency and 8-ns pulse width. The prism "P" was used to disperse 532 nm from YAG laser. Two RND attenuation lens were utilized to weaken the light intensity to the sample and spectrometer. The micro-objective "MO1" focused the excitation laser spot to 20 μm, irradiating onto the sample surface. After excitation, "MO2" was used to collect the polariton PL from sample. Prior to the MO2, a pin aperture restricted the incident angle of light from sample, improving the angle resolution. The MO2 objective and pin aperture rotated around the excitation spot to derive the angle-resolved spectrum. An optical fiber was used to collect the spectrum from MO2 objective. Spectrum was recorded by monochromator and charge coupled device. A BS mirror was used to split half excitation energy into the energy probe and power meter.

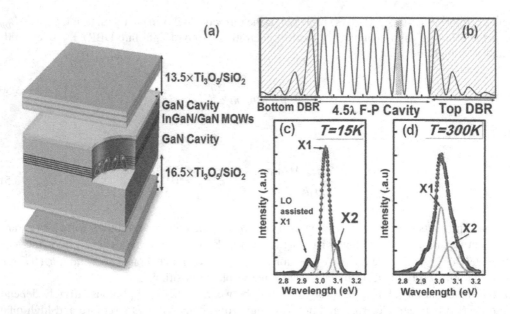

FIGURE 3.131   (a) Sample structure of InGaN/GaN MQWs microcavity for strong coupling. (b) The optical field in the microcavity with MQWs placed on the antinode. (c) and (d) Photoluminescence spectrums of bare wafer at 15 and 300K.

In our second measurement geometry, Fourier angle-resolved optical system was set up, as shown in Figure 3.133. In this system, 20 kHz, 3-ns pulse width, 355-nm citation was chosen. 2-μm excitation spot was focused onto the sample by lens. The optical path included Fourier space—real space—Fourier space transition by a high NA (0.75) objective and two confocal optical lenses. A pin aperture was located at the real plane between f_1 and f_2 to restrict the collected area and enhance the accuracy. The angle detective capability was determined by the numerical aperture value of objective lens. Objective lens with NA=0.75 give about ±48.6° detective angle range.

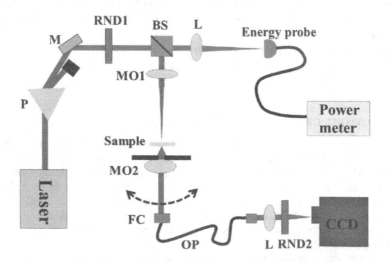

FIGURE 3.132   Angle resolve spectrum measurement setup.

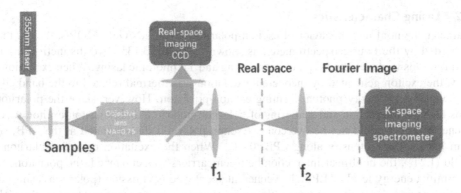

FIGURE 3.133   Schematic of angle-resolved PL Fourier image setup.

3.7.3.1   PL vs. Emitting Angle

In exciton-polariton, the dispersion relation, detuning energy and anti-crossing phenomena could be studied by tuning the energies of photon and exciton. When excitation light incidents to the cavity with inclined angle, the cavity mode also changed. Therefore, the dispersion relation and anti-crossing phenomena could be investigated by modulating the excitation angle. In Figure 3.134(a), the polariton spectrum changed by detective angle from 0° (vertical incidence) to 16°. Red vertical dash line presented the exciton energy. Two luminescent bands (LPB and UPB) were detected. From Figure 3.134(b), the peak energies of LPB and UPB changed by detection angle. Dashed red and black lines represent the exciton and photon energies. The experimental LPB and UPB could be well fitted by the transfer matrix model with the cavity mode energy E_{ph} related with detect angel by:

$$E_{ph}\left(\theta_{\text{out}}\right)=\frac{E_0}{\sqrt{1-\left(\sin^2\theta_{\text{out}}/n_{cav}^2\right)}} \tag{3.52}$$

where θ_{out} was the emitting angle; E_0 was the cavity mode at vertical direction; n_{cav} was the effective refraction index. Along vertical direction, the detuning value between exciton and photon was negative (–3 meV); where it changed to positive when angle increased. At the zero-detuning point, the energy divergence between LPB and UPB was called Rabi splitting value, which represents the coupling strength between exciton and photon. A Rabi splitting of 45 meV was deduced from our simulation.

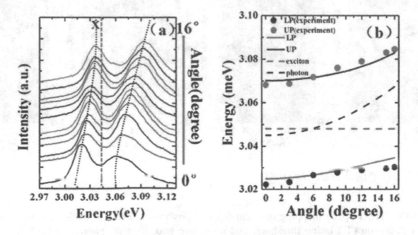

FIGURE 3.134   (a) PL under different detecting angle; (b) peak energies of LPB and UPB.

3.7.3.2 Lasing Characteristics

To character the nonlinear behavior of exciton-polariton in InGaN/GaN MQWs, the PL mapping was recorded by the Fourier spectrometer, as shown in Figure 3.135. Two distinctive and typical lasing thresholds were detected, T_1: Polariton lasing and T_2: photonic lasing. When excitation power was low, the exciton generated by incoherent excitation was thermal relaxed to the band edge and strong coupled with cavity photon, forming exciton-polariton. However, since the polariton density was low, the interaction and scattering of polariton-polariton and polariton-exciton was weak. "Bottleneck" effect could not be overcome. Exciton-polariton disassociated at the LPB, sending uniform luminescent intensity along LPB ($0.8T_1$). When the excitation exceeds polariton lasing threshold ($1.2T_1$), the nonlinear interaction between carriers was enhanced, the polaritons relaxed to the minima energy level of LPB. The coherent spontaneous emission (polariton lasing) occurs. Further increasing the excitation power, the exciton binding energy was reduced by the Colombia screening effect between electrons and holes. Exciton tended to dissociate. When excitation power increased approaching T_2 (Mott transition), carrier population inversion occurred, and the conventional photonic lasing emitted. From T_1 to T_2, the peak energy was gradually blue shifted from the LPB minima to cavity mode. The strong coupling was also degraded into weak coupling.

Figure 3.136 was the PL intensity, peak energies and width evolution with the excitation power from 40 to 800 μW. Two clear nonlinear regions corresponding to the T_1 and T_2 in Figure 3.135 were observed. Both nonlinear regions exhibited nonlinear intensity increasing, narrowing of luminescent width and blueshift of peaks. The blueshift of peak energy around the T_1 excitation power proved the setup of coherent states. The polariton distribution of k-space and real space started to concentrate. Meanwhile, the phenomena of width broadening before T_1 and width shrinkage after T_1 also confirmed the built-up of coherent polariton states. The polariton density at LPB minima was increased by this condensation, enhancing the polariton-polariton scattering and peak blueshift, which is the benchmark of polariton condensation. After the swift blueshift and further increasing of excitation, the peak energy stays stable, which was coincided with the GaN polariton data.[40] The scattering between polariton and polariton could also induce the width broadening. Since in our system, the detuning value was large and the LPB minima was very close to the cavity mode, the magnitude of blueshift from T_1 to T_2 was relatively small. Although this large negative detuning

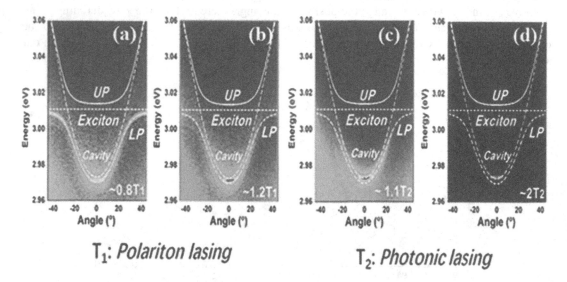

FIGURE 3.135 The k-space mapping of exciton-polaritons under different excitation powers. (a) before and (b) after polariton (T_1) lasing threshold; and (c) before and (d) after photonic (T_2) lasing thresholds, respectively.

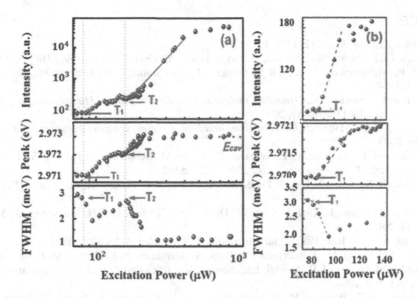

FIGURE 3.136   (a) The dependence of luminous intensity, peak position and half-width of LPB on the excitation power with double nonlinear regimes. (b) Enlarged polariton lasing region with linear coordinates. (T_1 and T_2 denote the polariton lasing threshold and photonic lasing threshold.)

was detrimental to the strong coupling, our coupled MQWs and high-quality FP cavity in this work could increase the excitons' oscillator strength, coupling strength and Rabi splitting value.

It is worth noting that although the polariton lasing in InGaN QWs was observed here in our study, the polariton lasing threshold was about half of the photonic lasing threshold. In previous reports on different materials with small exciton broadening, the polariton lasing thresholds were one or two order of magnitudes smaller than the photonic lasing threshold. Therefore, the polariton lasing threshold observed in our study was still higher than expection. This large polariton lasing threshold may originate from the large negative detuning value and severe exciton broadening in InGaN/GaN MQWs. As reported in GaN/AlGaN MQWs structures, an optimum negative detuning value for the lowest polariton lasing threshold resulted from the trade-off between thermodynamics and relaxation kinetics.[41,42] LPB with large negative detuning yields large portion of photons and very light effective mass, which inhibits the LPB condensation and increases the polariton lasing threshold due to the reduced polariton scattering. Meanwhile, the half-width broadening of InGaN/GaN MQWs is also essential in polariton lasing action. Only those excitons the energies of which match the cavity photon energy could be strong coupled and contributed to the strong coupling and polariton lasing. Therefore, the large inhomogeneous broadening of 43 meV in our sample should also significantly increase the polariton lasing threshold.

3.7.4 SUMMARY

In this chapter, we introduced the strong coupling between exciton and photon in semiconductor, especially the exciton-polariton based on nitride semiconductor system. Although exciton-polariton has been extensively studied in many material systems, the strong coupling in InGaN/GaN MQWs was still big challenge. In our work, strong coupling and polariton lasing were observed in InGaN/GaN MQWs embedded in 4.5λ microcavity at RT. Dual thresholds corresponding, respectively, to the polariton lasing and photonic lasing were clearly observed. The polariton lasing threshold was half of the photonic lasing threshold. Large negative detuning value and severe inhomogeneous broadening of exciton in InGaN/GaN MQWs are important factors in determining the lasing threshold of polariton. Nevertheless, we believe that our work will pave a way for ultralow-threshold InGaN QWs lasers and BEC.

REFERENCES

1. Hopfield, J. J., *Phys. Rev.*, 112, 1555, 1958.
2. Weisbuch, C., Nishioka, M., Ishikawa, A. and Arakawa, Y., *Phys. Rev. Lett.*, 69, 3314, 1992.
3. Houdré, R., Weisbuch, C., Stanley, R. P., Oesterle, U., Pellandini, P. and Ilegems, M., *Phys. Rev. Lett.*, 73, 2043, 1994.
4. Deng, H. and Yamamoto, Y., *Dynamic Condensation of Semiconductor Microcavity Polaritons*, Ph.D dissertation, Stanford University, 2006.
5. Kavokin, A. V. and Malpuech, G., *Thin Films and Nanostructures: Cavity Polaritons*, San Diego, CA: Elsevier, 2003.
6. Gao, T., Egorov, O. A., Estrecho, E., Winkler, K., Kamp, M., Schneider, C., Höfling, S., Truscott, A. G. and Ostrovskaya, E. A., *Phys. Rev. Lett.*, 2018, 121, 225302.
7. Su, R., Ghosh, S., Wang, J., Liu, S., Diederichs, C., Liew, T. C. H. and Xiong, Q., *Nat. Phys.*, 16, 301, 2020.
8. Su, R., Wang, J., Zhao, J., Xing, J., Zhao, W., Diederichs, C., Liew, T. C. H. and Xiong, Q., *Sci. Adv.*, 4, eaau0244, 2018.
9. Imamoglu, A., Ram, R. J., Pau, S. and Yamamoto, Y., *Phys. Rev. A*, 53, 4250, 1996.
10. Kasprzak, J., Richard, M., Kundermann, S., Baas, A., Jeambrun, P., Keeling, J. M. J., Marchetti, F. M., Szymańska, M. H., André, R., Staehli, J. L., Savona, V., Littlewood, P. B., Deveaud, B. and Dang, L. S., *Nature*, 443, 409, 2006.
11. Lagoudakis, K. G., Wouters, M., Richard, M., Baas, A., Carusotto, I., André, R., Dang, L. S. and Deveaud-Plédran, B., *Nat. Phys.*, 4, 706, 2008.
12. Utsunomiya, S., Tian, L., Roumpos, G., Lai, C. W., Kumada, N., Fujisawa, T., Kuwata-Gonokami, M., Löffler, A., Höfling, S., Forchel, A. and Yamamoto, Y., *Nat. Phys.*, 4, 700, 2008.
13. Amo, A., Sanvitto, D., Laussy, F. P., Ballarini, D., Valle, E. D., Martin, M. D., Lemaître, A., Bloch, J., Krizhanovskii, D. N., Skolnick, M. S., Tejedor, C. and Viña, L., *Nature*, 457, 291, 2009.
14. Deveaud-Plédran, B., *Nature*, 453, 297, 2008.
15. Liu, W., Xie, W., Guo, W., Xu, D., Hu, T., Ma, T., Yuan, H., Wu, Y., Zhao, H., Shen, X. and Chen, Z., *Phys. Rev. B*, 89, 201201, 2014.
16. Xu, D., Xie, W., Liu, W., Wang, J., Zhang, L., Wang, Y., Zhang, S., Sun, L., Shen, X. and Chen, Z., *Appl. Phys. Lett.*, 104, 082101, 2014.
17. Zhang, L., Xie, W., Wang, J., Poddubny, A., Lu, J., Wang, Y., Gu, J., Liu, W., Xu, D., Shen, X., Rubo, Y. G., Altshuler, B. L., Kavokin, A. V. and Chen, Z., *Proc. Natl. Acad. Sci. U.S.A.*, 112, E1516, 2015.
18. Wang, J., Xie, W., Zhang, L., Xu, D., Liu, W., Lu, J., Wang, Y., Gu, J., Chen, Y., Shen, X. and Chen, Z., *Phys. Rev. B*, 2015, 91, 165423.
19. Malpuech, G., Di Carlo, A., Kavokin, A., Baumberg, J. J., Zamfirescu, M. and Lugli, P., *Appl. Phys. Lett.*, 81, 412, 2002.
20. Bajoni, D., Senellart, P., Wertz, E., Sagnes, I., Miard, A., Lemaître, A. and Bloch, J., *Phys. Rev. Lett.*, 100, 047401, 2008.
21. Tsintzos, S. I., Pelekanos, N. T., Konstantinidis, G., Hatzopoulos, Z. and Savvidis, P. G., *Nature*, 453, 372, 2008.
22. Sellers, I. R., Semond, F., Leroux, M., Massies, J., Zamfirescu, M., Stokker-Cheregi, F., Gurioli, M., Vinattieri, A., Colocci, M., Tahraoui, A. and Khalifa, A. A., *Phys. Rev. B*, 74, 193308, 2006.
23. Christopoulos, S., von Högersthal, G. B. H., Grundy, A. J. D., Lagoudakis, P. G., Kavokin, A. V., Baumberg, J. J., Christmann, G., Butté, R., Feltin, E., Carlin, J. F. and Grandjean, N., *Phys. Rev. Lett.*, 98, 126405, 2007.
24. Bhattacharya, P., Frost, T., Deshpande, S., Baten, M. Z., Hazari, A. and Das, A., *Phys. Rev. Lett.*, 112, 236802, 2014.
25. Das, A., Heo, J., Jankowski, M., Guo, W., Zhang, L., Deng, H. and Bhattacharya, P., *Phys. Rev. Lett.*, 107, 066405, 2011.
26. Heo, J., Jahangir, S., Xiao, B. and Bhattacharya, P., *Nano Lett.*, 13, 2376, 2013.
27. Tawara, T., Gotoh, H., Akasaka, T., Kobayashi, N. and Saitoh, T., *Phys. Rev. Lett.*, 92, 256402, 2004.
28. Christmann, G., Butté, R., Feltin, E., Carlin, J.-F. and Grandjean, N., *Appl. Phys. Lett.*, 93, 051102, 2008.
29. Lu, T.-C., Chen, J.-R., Lin, S.-C., Huang, S.-W., Wang, S.-C. and Yamamoto, Y., *Nano Lett.*, 11, 2791, 2011.
30. Hsieh, D. H., Tzou, A. J., Kao, T. S., Lai, F. I., Lin, D. W., Lin, B. C., Lu, T. C., Lai, W. C., Chen, C. H. and Kuo, H. C., *Opt. Express*, 23, 27145, 2015.

31. Chang, T.-C., Kuo, S.-Y., Lian, J.-T., Hong, K.-B., Wang, S.-C. and Lu, T.-C., *Appl. Phys. Express*, 10, 112101, 2017.
32. Jewell, J. L., Huang, K. F., Tai, K., Lee, Y. H., Fischer, R. J., McCall, S. L. and Cho, A. Y., *Appl. Phys. Lett.*, 55, 424, 1989.
33. Hayashi, Y., Mukaihara, T., Hatori, N., Ohnoki, N., Matsutani, A., Koyama, F. and Iga, K., *Electron. Lett.*, 31, 560, 1995.
34. Yang, G. M., MacDougal, M. H. and Dapkus, P. D., *Electron. Lett.*, 31, 886, 1995.
35. Liu, W., Chen, S., Hu, X., Liu, Z., Zhang, J., Ying, L., Lv, X., Akiyama, H., Cai, Z. and Zhang, B., *IEEE Photon. Technol. Lett.*, 25, 2014, 2013.
36. Zhang, J.-Y., Cai, L.-E., Zhang, B.-P., Li, S.-Q., Lin, F., Shang, J.-Z., Wang, D.-X., Lin, K.-C., Yu, J.-Z. and Wang, Q.-M., *Appl. Phys. Lett.*, 93, 191118, 2008.
37. Shi, X., Long, H., Wu, J., Chen, L., Ying, L., Zheng, Z. and Zhang, B., *Superlattice Microst.*, 128, 151, 2019.
38. Wu, J., Shi, X., Long, H., Chen, L., Ying, L., Zheng, Z. and Zhang, B., *Mater. Res. Express*, 6, 076204, 2019.
39. Wu, J. Z., Long, H., Shi, X. L., Luo, S., Chen, Z. H., Feng, Z. C., Ying, L. Y., Zheng, Z. W., and Zhang, B. P., *OEA*, 2, 190014, 2019.
40. Vurgaftman, I. and Meyer, J. R., *J. Appl. Phys.*, 94, 3675, 2003.
41. Butté, R., Levrat, J., Christmann, G., Feltin, E., Carlin, J.-F. and Grandjean, N., *Phys. Rev. B*, 80, 233301, 2009.
42. Levrat, J., Butté, R., Feltin, E., Carlin, J.-F., Grandjean, N., Solnyshkov, D. and Malpuech, G., *Phys. Rev. B*, 81, 125305, 2010.

3.8 UV EMISSION FROM AlGaN MATERIALS AND RELATED QUANTUM STRUCTURES

Xinqiang Wang

3.8.1 INTRODUCTION

Over the last decades, GaN and related materials such as AlGaN and InGaN have received significant research attention and have been developed dramatically since they are widely used in both optoelectronic and electronic devices such as light-emitting diodes (LEDs), laser diodes (LDs), ultraviolet (UV) emitters and detectors, solar cells, high-temperature and high-frequency field-effect transistors (FETs), heterojunction bipolar transistors (HBT) and so on.[1-4] This kind of optoelectronic device is required for solid-state lighting, full-color display, laser printers, high-density information storage and communication while the electronic devices are needed for automobile engines, advanced power distribution systems and all electronic vehicles.

As members of the III-nitride family, wurtzite GaN and AlN have direct room temperature (RT) bandgap energy (E_g) of 3.4 and 6.2 eV. Since the E_g of InN was believed to be 1.9 eV previously, the bandgap of AlGaInN varies between 1.9 and 6.2 eV, which makes the III-nitrides attractive for the optoelectronic device application in the green, blue and UV ranges. Around 2002, InN has been proved to be a narrow bandgap semiconductor with E_g of about 0.7 eV at low temperature and 0.65 eV at RT.[5-9] This observation has a big impact because the cover range of the III-nitrides and their ternary and quaternary alloys is not 1.9–6.2 eV any more but 0.65–6.2 eV, i.e. from infrared to deep UV (as shown in Figure 3.137).[10]

3.8.2 FUNDAMENTAL PROPERTIES OF AlGaN

3.8.2.1 Bandgap and Bowing Parameter

By alloying GaN with AlN the emission of quantum structure can be tuned to cover almost the entire UV spectral range, making it suited to applications like such as water/air purification, non-line-of-sight communications, surface modification, biological/chemical analysis and so on.

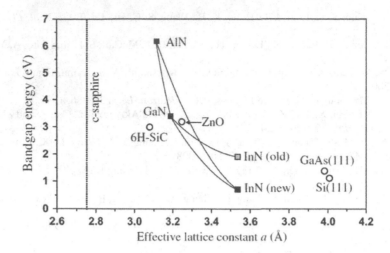

FIGURE 3.137 Bandgaps and lattice constants of III-nitrides (solid rectangles). Effective lattice constants considered as substrates for hetero-epitaxy of III-nitrides are also shown (hollow circles). (From Wang, X. and Yoshikawa, A., *Prog. Cryst. Growth Ch.*, 48–49, 42, 2004. With permission.)

The bandgap energy of ternary alloys $Al_xGa_{1-x}N$ has the following parabolic dependence on the molar fraction x:

$$E_g^{AlGaN}(x) = xE_g^{AlN} + (1-x)E_g^{GaN} - bx(1-x) \tag{3.53}$$

where E_g^{AlN} and E_g^{GaN} are bandgap energies of AlN and GaN, respectively, and b is the bowing parameter, which is determined by experiments. Here, the bowing parameter, which is also called nonlinear parameter, accounts for the downward deviation of the bandgap energy of ternary compounds compared to the linear relation between the bandgap energy of binary compounds.

Initial studies of the compositional dependence of the bandgap energy of AlGaN reported downward, upward, and negligible bowing.[11] Subsequent early photoluminescence (PL)[12] and absorption[13] measurements found a bowing parameter of 1.0 eV, which continues to be widely used in band structure calculations even though a number of more recent investigations question the conclusions of the early work. At the beginning of 21st century, with the improvement of growth technology and crystal quality, bandgap energy of AlGaN over entire composition range has been carefully investigated by multiple characterization. F. Yun et al. have grown whole composition range AlGaN on c-plane sapphire by plasma-induced molecular-beam epitaxy (MBE).[14] The molar Al fraction was deduced from X-ray diffraction and the bandgap of the alloy was extracted from low-temperature optical reflectance measurements, as shown in Figure 3.138. They found a downward bowing with a bowing parameter $b=1.0$ eV. Afterward, AlGaN films were grown by metalorganic chemical vapor deposition (MOCVD) using a high-temperature AlN interlayer on a thick GaN template.[15] The Al composition (x) of the $Al_xGa_{1-x}N$ was varied in the range from 0.13 to 0.8. RT cathodoluminescence spectra show pronounced near-bandedge emission from these AlGaN alloys. The optical bandgaps (E_g) are found to deviate from linear interpolation between E_g (GaN) and E_g (AlN) with a bowing parameter $b=0.89$, which is close to the values from Refs. [11,16].

Combining experimental measurements with an analysis of sample strain, Paduano et al. have determined the "unstrained" bandgap energy of AlGaN.[16] Two different AlGaN layer types were grown by low-pressure MOCVD. For AlGaN random alloys on GaN films deposited on sapphire, the strain-corrected bandgap dependence on alloy composition is fit well by a bowing parameter of $b=0.70\pm0.05$ eV. In contrast, AlGaN films deposited directly on sapphire had much higher X-ray linewidths and their E_g's are not fit well by one value of the bowing parameter. This suggests that

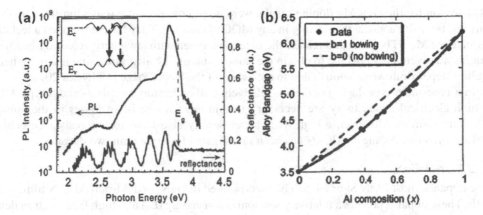

FIGURE 3.138 (a) Low-temperature (15 K) reflectance and PL spectra of AlGaN (x~0.13) sample. The inset is a schematic energy band diagram showing the effects of any local AlGaN fluctuation. (b) Experimental data of energy bandgap of AlGaN ($0 \leq x \leq 1$) plotted as a function of Al composition (solid circle), and the least-squares fit (solid line) giving a bowing parameter of b=1.0 eV. The dashed line shows the case of zero bowing. (From Yun, F., Reshchikov, M. A., He, L., King, T., Morkoç, H., Novak, S. W. and Wei, L., *J. Appl. Phys.*, 92, 4837, 2002. With permission.)

material quality may be a significant reason for the large range of bowing parameters reported in the literature. Figure 3.139 reveals that the strain correction noticeably reduces the scatter about the b=0.7 eV fitted line.

3.8.2.2 Doping of GaN and AlGaN

For all III-nitrides, obtaining well p-doping is a key issue for its development. For many years, researchers struggled to get any p-conductivity at all, until H. Amano and I. Akasaki[17] discovered

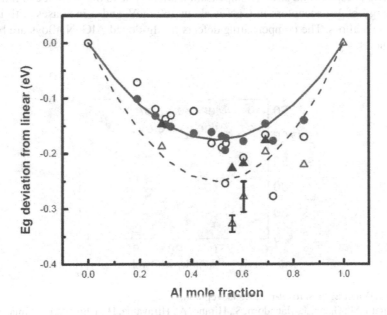

FIGURE 3.139 Deviation of E_g from linear interpolation between E_g^{GaN} and E_g^{AlN}. Open symbols denote E_g without stain correction; solid symbols are for strain-corrected E_g. Solid line is calculated with b=0.7 eV. Dashed line represents b=1.0 eV. (From Paduano, Q. S., Weyburne, D. W., Bouthillette, L. O., Wang, S.-Q. and Alexander, M. N., *Jpn. J. Appl. Phys.*, 41, 1936, 2002. With permission.)

that the reason for the poor Mg doping results were due to the dopant passivation from hydrogen impurities, typically abundantly existing in any MOCVD reactor. They have developed a technique to dissolve the Mg–H bonding and release the excess hydrogen with low-energy electron-beam irradiation. It was eventually possible to obtain the same result until Nakamura using a simple thermal annealing step.[18] Both works contributed to the award of the Nobel Prize in Physics 2014.

Both n-type and p-type doping of GaN is relatively well understood, while n- and p-type AlGaN with high electrical conductivity are necessary to minimize the resistive losses in the epilayers of UV light sources. Moreover, a high free carrier density also leads to low-resistance contacts. Therefore, effective doping of AlGaN has been investigated for more than two decades.

3.8.2.2.1 n-AlGaN

N-type dopants, most likely Si or Ge can be incorporated in GaN- and Al-rich AlGaN films during growth. These dopants provide a relatively low ionization energy (E_A) and high free electron density (>10^{19} cm^{-3}) in AlGaN films. A low E_A of Si donors in (Al)GaN yields a free electron density in the range of ~10^{19} cm^{-3} and conductivity of <10 mΩ·cm in n-Al$_x$Ga$_{1-x}$N films (x<0.8).[19,20] However, the donor E_A undergoes a steady rise in Al-rich AlGaN films when the Al mole fraction goes above 80%, as shown in Figure 3.140.[21] This sharp increase in E_A is attributed to the formation of Si-DX centers acting as acceptor-type compensating point defects. While the low Si doping density is responsible for the formation of carbon and vacancy-oxygen complexes, a high Si doping leads to self-compensation and vacancy-related complexes (e.g., VIII+Si).[22,23] All these acceptor-type point defects compensate the free electron density in Al-rich AlGaN films.

3.8.2.2.2 p-AlGaN

Although lots of efforts have been made in development of AlGaN alloy and their application in UV devices, achieving p-type conductivity in Al-rich AlGaN is still highly challenging. Mg is the only known viable acceptor doping source that can occupy a substitutional site in (Al)GaN epitaxial films. The main reasons of difficulty to achieve p-type conductivity are the large activation energy of the Mg acceptors and strong compensation effects due to the presence of intrinsic defects. Activation energy of Mg acceptors in GaN is about 160 meV, and it increases with increasing Al content in AlGaN alloys. The compensating defects in Mg-doped AlGaN alloys are believed to be nitrogen vacancies.

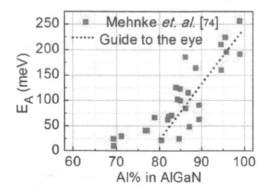

FIGURE 3.140 Donor E_A of Si in Al-rich AlGaN epitaxial films. (From Amano, H., Collazo, R., Santi, C. D., Einfeldt, S., Funato, M., Glaab, J., Hagedorn, S., Hirano, A., Hirayama, H., Ishii, R., Kashima, Y., Kawakami, Y., Kirste, R., Kneissl, M., Martin, R., Mehnke, F., Meneghini, M., Ougazzaden, A., Parbrook, P. J., Rajan, S., Reddy, P., Römer, F., Ruschel, J., Sarkar, B., Scholz, F., Schowalter, L. J., Shields, P., Sitar, Z., Sulmoni, L., Wang, T., Wernicke, T., Weyers, M., Witzigmann, B., Wu, Y.-R., Wunderer, T. and Zhang, Y., *J. Phys. D Appl. Phys.*, 53, 503001, 2020. With permission.)

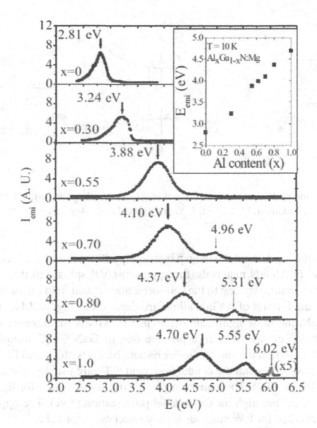

FIGURE 3.141 Low temperature (10 K) PL spectra of Mg-doped $Al_xGa_{1-x}N$ alloys with varying Al contents (x) from 0 to 1. Inset shows the PL emission peak position (E_{emi}) of the group of deep impurity transitions in $Al_xGa_{1-x}N$ alloys (highlighted by the bold arrows) as a function of the Al content. (From Nakarmi, M. L., Nepal, N., Lin, J. Y. and Jiang, H. X., *Appl. Phys. Lett.*, 94, 091903, 2009. With permission.)

Figure 3.141 shows the low-temperature (10 K) PL spectra of Mg-doped $Al_xGa_{1-x}N$ alloys of varying x.[24] In all samples, a group of impurity transitions, which are highlighted by the bold arrows, is dominant over that of the band edge. The spectral peak positions of this group of impurity transitions are blue-shifted from 2.81 eV in GaN to 4.7 eV in AlN. For Mg-doped AlN, a weak I_1 (exciton bound to neutral Mg) emission peak at 6.02 eV and an additional impurity peak at 5.55 eV are evident. The emission peak at 5.55 eV is attributed to the transition of electrons in the conduction band (or bound to shallow donors) to neutral Mg acceptors.[25] This assignment provides an energy level of Mg acceptors in AlN of about 0.5 eV, which is consistent with previously reported results.[25,26] The emission peaks at about 4.96 and 5.31 eV in $Al_xGa_{1-x}N$ with $x=0.7$ and 0.8 are of the same origin as the 5.55 eV line in AlN.

3.8.2.3 Polarization

In addition, the light extraction efficiency (LEE) is also a challenge for UV emitters due to its strong dependence on the optical polarization of the emitted light. The emission pattern for transverse magnetic (TM, **E//c**) polarized light is in-plane and thus it's harder to extract TM polarized light through the surface than transverse electric (TE, **E⊥c**) polarized light.[27] The band structure of AlN near Γ point was found to be very different from GaN, the dominant PL emission of GaN is with polarization of **E⊥c** while that of AlN is with polarization of **E//c**.[28] It was shown that as the Al component of AlGaN quantum wells (QWs) increased, TM-polarized emission became predominant, which was attributed to a reordering of the valence band (VB) in AlGaN at higher Al mole fractions.

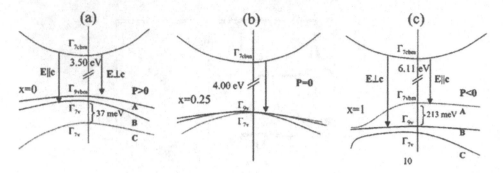

FIGURE 3.142   The band structures near the Γ point of $Al_xGa_{1-x}N$ alloys for (a) $x=0$, (b) $x=0.25$, and (c) $x=1$. (From Nam, K. B., Li, J., Nakarmi, M. L., Lin, J. Y. and Jiang, H. X., *Appl. Phys. Lett.*, 84, 5264, 2004. With permission.)

The behavior around the fundamental bandgap is strongly influenced by the VB order at the Γ point of the Brillouin zone. In AlGaN materials, the uppermost VB splits into three energetically close subbands with different symmetry due to the spin-orbit and crystal field interaction. The illustrative band structures near the Γ point of AlGaN alloys are depicted in Figure 3.142. Compared with the band structure of GaN, the most significant difference in AlN is the negative crystal-field energy Δ_{cf} of about -220 meV[29] compared with a positive one in GaN.[29,30] Consequently, one finds an uppermost VB with Γ_{7+}^y symmetry and two lower remote bands with Γ_9^y and Γ_{7-}^y symmetry. Optical transitions between these states and the conduction band (CB) are usually labeled as A ($\Gamma_{7+}^y \leftrightarrow \Gamma_7^c$), B ($\Gamma_9^y \leftrightarrow \Gamma_7^c$) and C ($\Gamma_{7-}^y \leftrightarrow \Gamma_7^c$). The transition probability for A is low for light with the perpendicular polarization ($E \perp c$) but high for the parallel polarization ($E // c$). The opposite applies for C, while B is strictly forbidden for $E // c$ and thus is only observable for $E \perp c$.

The optical properties of AlGaN alloys grown on c-plane sapphire substrates have been studied by PL spectroscopy.[28] Figure 3.143 shows the low-temperature (10 K) PL spectra for $Al_xGa_{1-x}N$

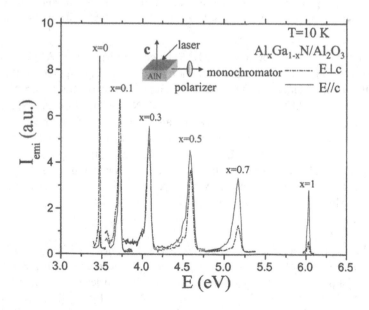

FIGURE 3.143   Low-temperature (10 K) PL spectra of $Al_xGa_{1-x}N$ alloys of varying x, for from $x=0$ to 1. The experimental geometry was depicted in the inset, where the electrical field of PL emission can be selected either parallel or perpendicular to the c-axis. (From Nam, K. B., Li, J., Nakarmi, M. L., Lin, J. Y. and Jiang, H. X., *Appl. Phys. Lett.*, 84, 5264, 2004. With permission.)

alloys ($0 \leq x \leq 1$). The dotted (solid) lines indicate the emission spectra, collected with the polarization of **E**⊥**c** (**E**//**c**). The degree of polarization (P) is defined by $P=(I_\perp-I_\parallel)/(I_\perp+I_\parallel)$, where $I_\perp$ and $I_\parallel$ are the integrated PL intensities for the polarization components of **E**⊥**c** and **E**//**c**, respectively. K. B. Nam et al. found that the emission intensity of TE-polarized light as well as P decreases almost linearly with increasing x, and $P=0$ at $x=0.25$.

This transition depends on various factors that include the strain in the QW as well as composition and thickness of the QW. The polarization of the in-plane electroluminescence (EL) of (0 0 0 1) orientated (In)(Al)GaN MQWs LEDs in the UVA and UVB spectral range has been investigated by T. Kolbe et al.[31] They found that the degree of polarization depends nearly linearly on wavelength and appears largely independent of the composition of the (In)(Al)GaN active layers for a given certain emission wavelength. Zero polarization was found for LEDs emitting near 300 nm. J. E. Northrup et al. reported model calculations that quantify the dependence of polarization of light emission on Al barrier composition and strain for AlGaN-based LEDs.[32] The results show that as the strain in the QW becomes more compressive, it is more likely to achieve TE-polarized light at shorter wavelength. T. M. Al Tahtamouni et al. have investigated the effects of QW thickness (t_{QW}) on the polarization of AlN/AlGaN SQW with Al content fixed at 0.65 and the t_{QW} varying from 1 to 3 nm.[33] The predominant polarization switched from TM to TE with the QW thickness decreasing and $P=0$ at $t_{QW}=2$ nm. These findings above will serve as a guideline for designing and optimizing the active region of deep UV emitters.

C. Reich et al. have studied the influence of design parameters for the active region of ultraviolet C (UVC) LEDs on the polarization of the emitted light, using model calculations based on **k·p** perturbation theory.[34] Compressive strain in the growth-plane and high quantum barriers are beneficial for enhanced TE-polarized emission, even at short wavelengths in the UVC spectral range. The composition range used for the fabricated UVC LEDs is indicated by dotted yellow lines in Figure 3.144(a). Strongly TE-polarized light emission was obtained from UVC LEDs with MQWs pseudomorphically grown on compressively strained epitaxially laterally overgrown (ELO) AlN/sapphire templates. Figure 3.144(b) shows the polarization-dependent in-plane emission spectra of the five LEDs grown on ELO AlN/sapphire templates. All spectra are normalized to the maximum of the TE-polarized spectrum. As expected, the fraction of the TM-polarized emission increases when decreasing the wavelength. However, the TE polarization is dominant in all LEDs, even down to emission wavelengths of 239 nm.

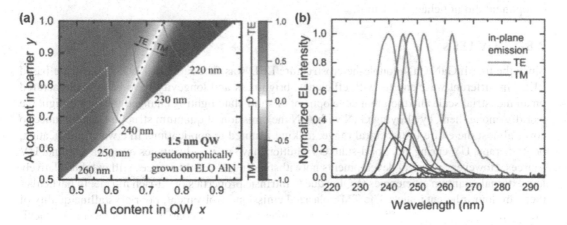

FIGURE 3.144 (a) Degree of polarization q of quantum well structures pseudomorphically grown on ELO AlN/sapphire, obtained by **k·p**-perturbation theory. (b) Polarization-dependent in-plane emission spectra of selected UVC LEDs. (From Reich, C., Guttmann, M., Feneberg, M., Wernicke, T., Mehnke, F., Kuhn, C., Rass, J., Lapeyrade, M., Einfeldt, S., Knauer, A., Kueller, V., Weyers, M., Goldhahn, R. and Kneissl, M., *Appl. Phys. Lett.*, 107, 142101, 2015. With permission.)

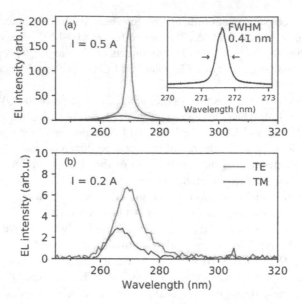

FIGURE 3.145   Edge emission spectra with TE and TM polarization at (a) 0.5 A and (b) 0.2 A forward current. The inset figure in part (a) shows the spectrum of the TE mode at 0.5 A with the highest wavelength resolution. (From Zhang, Z., Kushimoto, M., Sakai, T., Sugiyama, N., Schowalter, L. J., Sasaoka, C. and Amano, H., *Appl. Phys. Express*, 12, 124003, 2019. With permission.)

Very recent, edge-emitting LDs emitting in the UVC spectral range at 271.8 nm have been realized.[35] A polarization-induced doping cladding layer was employed to achieve hole conductivity and injection without intentional impurity doping. Even with this undoped layer, they were still able to achieve a low operation voltage of 13.8 V at a lasing threshold current of 0.4 A. Figure 3.145 shows the measured edge emission spectra of the TE and TM polarization components at currents (a) above and (b) below threshold. The spontaneous emission of the TE and TM components, observed at a current of 0.2 A, showed a full-width-at-half maximum (FWHM) value of 6.6 and 11 nm, respectively. The TE component was sharp and dominant with an FWHM value of 0.41 nm at 0.5 A forward current, while the FWHM value of the TM component did not change (11 nm).

3.8.3   UV LEDs

Since the first InGaN/GaN double-heterostructure LED was fabricated in 1993,[36] III-nitride-based LEDs in different wavelengths with efficiency, brightness and longevity have been manufactured on an industrial scale and the great development of solid-state lighting changed the way to light the world. Among them, by alloying GaN with AlN the emission of quantum structure can be tuned to cover almost the entire UV spectral range, making it suited to applications like water purification, phototherapy, UV curing etc. Solid-state UV emitters have many advantages over conventional UV sources. However, technical developments notwithstanding, UV light sources still exhibit relatively low external quantum efficiencies (EQEs) due to intrinsic properties of Al-rich nitride. In particular, there are four inherent issues: The TM-polarized emission involvement, poor crystalline quality of the AlGaN active region, difficulty in p-type doping for high-Al-content AlGaN layers and inevitable light absorption in the commonly used p-type GaN injection layers.[27,37–39]

Depending on the specific wavelength, the effects of UV light on materials and organisms will vary significantly. In general, UV radiation can be divided into three spectral bands: UVA (400–320 nm), UVB (320–280 nm) and UVC (280–100 nm). In the UVA wavelength range, the key applications include UV curing of polymers, inks, coatings, resins and adhesives.[40] UVA emitters are

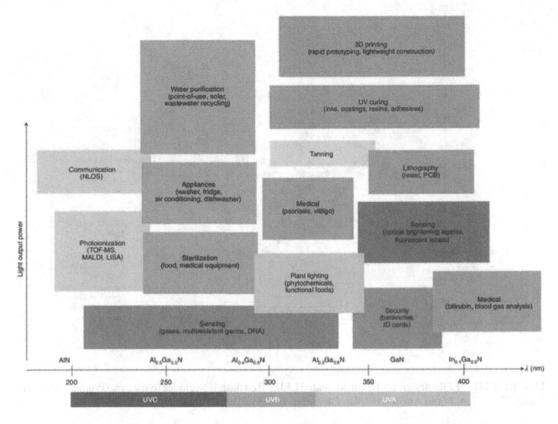

FIGURE 3.146 Applications of UVA (400–320 nm), UVB (320–280 nm) and UVC (280–200 nm) LEDs. (From Kneissl, M., Seong, T.-Y., Han, J. and Amano, H., *Nat. Photonics*, 13, 233, 2019. With permission.)

also employed for sensing applications, like blood gas analysis and light therapy. Phototherapy, in particular, the treatment of skin diseases like vitiligo and psoriasis, is one of the key applications for LEDs in the UVB band. UVB LEDs are also of great interest for the curing of surfaces since the penetration depth of UVB light in polymers is much smaller than that of UVA emitters. In addition, plant growth lighting constitutes a large-scale application for UVB LEDs. Disinfection and sterilization are certainly the highest volume applications for LEDs emitting in the UVC spectral band. Figure 3.146 presents some of the key applications for UV LEDs. As can be seen, the wavelength and power requirements vary greatly for these applications. While sensing applications require low power but pure spectra, for UV curing and disinfection LED modules delivering many Watts of UV light are crucial.

As shown in Figure 3.147, the EQE of UV LEDs varies greatly with the emission wavelength. One trend is that when the wavelength goes shorter, the EQE of UV LEDs goes lower. Whereas UVA LEDs with emission longer than 315 nm exhibit EQEs exceeding 50%, LEDs in the UVB and UVC band lag significantly behind. Although record EQEs of more than 20% for LEDs emitting near 275 nm have been reported,[41] the performance levels of commercial devices in these wavelength bands are still less than 10%.

3.8.3.1 UVA LEDs

Since the bandgap of GaN locates in the center of UVA spectral band, the UVA region can be divided into two parts based on the emission wavelength longer/shorter than 365 nm. UVA LEDs emit longer than 365 nm typically have GaN or InGaN QWs in the active region, and those with shorter wavelengths in general have AlGaN QWs. In 1998, J. Han et al. presented the first UV LED based on $Al_{0.2}Ga_{0.8}N/GaN$ multiple-QW (MQW) light-emitting diodes.[42] As shown in

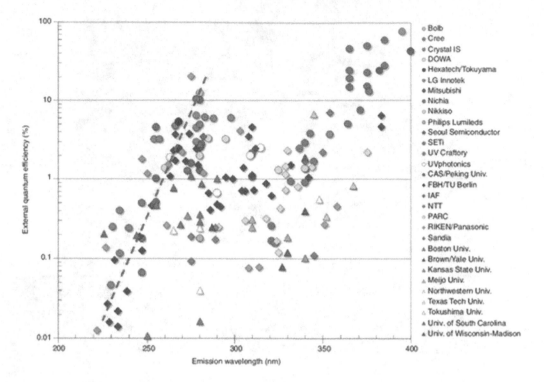

FIGURE 3.147 EQEs for group-III-nitride-based LEDs. Red dash line guides to the eye. (From Kneissl, M., Seong, T.-Y., Han, J. and Amano, H., *Nat. Photonics*, 13, 233, 2019. With permission.)

Figure 3.148(a), The RT EL emission is peaked at 353.6 nm with a narrow linewidth of 5.8 nm. In the simple planar devices, without any efforts to improve LEE, an output power of 13 μW at 20 mA was measured. Pulsed EL data demonstrate that the output power does not saturate up to current densities approaching 9 kA/cm^2 (seen in Figure 3.148(b)).

By introducing thick bulk GaN as a substrate, T. Nishida et al. improved the performance of AlGaN-based UV LED.[43] The output power of the 352-nm UVA LED exceeds 3 mW at the injection current of 100 mA with internal quantum efficiency (IQE) estimated as more than 80%. A UVA LED emits at 340 nm is reported by M. A. Khan's group.[44] The active layers were comprised of quaternary AlInGaN/AlInGaN MQWs, which were deposited over sapphire substrates

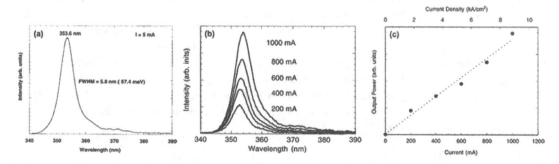

FIGURE 3.148 (a) Room-temperature electroluminescence spectrum of a 120-μm diameter LED at *I*=5 mA. (b) Electroluminescence spectra from a 120-μm diameter LED under various levels of pulsed current injection. The current levels are indicated. (c) Peak UV emission from the LED as a function of injected current. (From Han, J., Crawford, M. H., Shul, R. J., Figiel, J. J., Banas, M., Zhang, L., Song, Y. K., Zhou, H. and Nurmikko, A. V., *Appl. Phys. Lett.*, 73, 1688, 1998. With permission.)

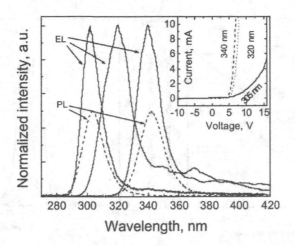

FIGURE 3.149 Room temperature dc EL spectra of LEDs with three different MQW configurations. Dotted line shows RT PL spectra at 305 and 340 nm. The I–V curves for LEDs are depicted in the inset. (From Khan, M. A., Adivarahan, V., Zhang, J. P., Chen, C. Q., Kuokstis, E., Chitnis, A., Shatalov, M., Yang, J. W. and Simin, G., *Jpn. J Appl. Phys.*, 40 (12A), L1308–L1310, 2001.)

using a pulsed atomic-layer epitaxy process. By varying the alloy compositions of the quaternary AlInGaN active layers, the peak emission wavelength can be tuned from 340 to 305 nm, as shown in Figure 3.149.[45] It was the first time that the UVB LED has been reported. At 340 nm, for a 20 μm×1000 μm stripe geometry device an output power as high as 1 mW was measured from the sapphire substrate side.

D. Morita et al. have made a high output power 365-nm UVA LED of GaN-free structure in 2002.[46] The GaN templates and the sapphire substrates were removed by using laser-induced liftoff and polishing techniques to reduce the absorption of UV light by the GaN layer. When this UV LED was operated at a forward-bias direct current of 500 mA at RT, it emitted at 365 nm with the output power of 118 mW and the operating voltage of 4.9 V. The EQE of this UV LED can be determined as 6.9%. Recently, benefit from the developments in crystal growth, chip processing and packaging technologies, there have been significant improvements in the EQE of the UVA LEDs specifically in the spectral range between 400 and 365 nm, such as 30% at 365 nm and 50% at 385 nm. A 365-nm UVA LED with an output power of up to 12 W (14 A) has been reported.[47]

3.8.3.2 UVB LEDs

For the UVB LEDs, the EQE values reached so far are lower than those record efficiencies of LEDs emitting around 275 nm and UVA LEDs, shown as "UVB gap". Nevertheless, commercial UVB LEDs emitting at 305 nm have been launched by LG Innotek, with an emission power of 100 mW at a current of 350 mA (EQE 6%) and an operation voltage of 7 V. Recently, to mitigate the effect of threading dislocations (TDs), the UVB LEDs have been fabricated on n-AlGaN templates with a TD density larger than 5×10^8 cm^{-2}, which were grown on (0 0 0 1) sapphire with a 1.0° miscut relative to the m-plane with a step bunched surface.[48] It achieved a record high EQE of 6% at about 300 nm. Figure 3.150(a) shows the EL output as a function of forward current for the deep-ultraviolet (DUV) LEDs with the highest EQEs. EQEs of 3.5, 3.9, 6.1 and 6.0% were obtained at wavelengths of 266, 271, 283 and 298 nm at 40 mA, respectively.

Very recent, an AlGaN UVB LD at 298 nm was realized at RT using pulse operation.[49] The LD has a lattice-relaxed $Al_{0.6}Ga_{0.4}N$ layer from the underlying AlN/sapphire template and a composition-graded p-AlGaN cladding layer. The multimodal laser spectrum with proper polarization properties at 298 nm was obtained over the threshold current at 0.90 A corresponding to 67 kA/cm^2. By broadening the width of the p-electrode to 11.5 μm, the threshold current density decreased to 41 kA/cm^2. As shown in Figure 3.151, the emission spectrum has been performed

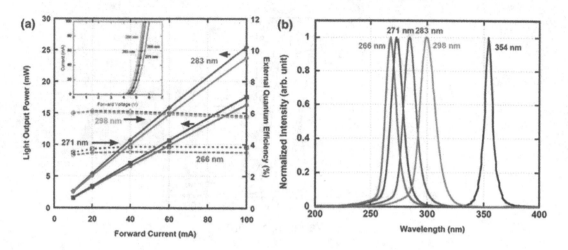

FIGURE 3.150 (a) CW output power (solid lines) and EQE (dashed lines) characteristics of AlGaN-based DUV LEDs. The inset shows the I–V characteristics of these LEDs. (b) EL spectra of these LEDs shown in comparison with that of a 354-nm LED with a GaN well. (From Kaneda, M., Pernot, C., Nagasawa, Y., Hirano, A., Ippommatsu, M., Honda, Y., Amano, H. and Akasaki, I., *Jpn. J. Appl. Phys.*, 56, 061002, 2017. With permission.)

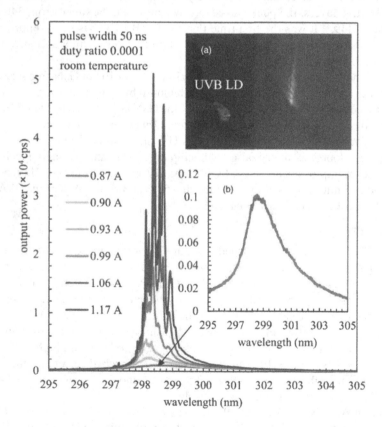

FIGURE 3.151 Emission spectra of the UVB LD at 0.87, 0.90, 0.93, 0.99, 1.06 and 1.17 A. Inset (a) is FFP of UVB laser irradiated to orange fluorescent coated on a paper. Inset (b) is magnified emission spectrum at 0.87 A. (From Sato, K., Yasue, S., Yamada, K., Tanaka, S., Omori, T., Ishizuka, S., Teramura, S., Ogino, Y., Iwayama, S., Miyake, H., Iwaya, M., Takeuchi, T., Kamiyama, S. and Akasaki, I., *Appl. Phys. Express*, 13, 031004, 2020. With permission.)

under different current conditions from 0.87 to 1.17 A for the UVB LD with a p-electrode width of 4.5 μm. Multimodal laser emission from the mirror facet became clear once the current exceeded 0.93 A.

3.8.3.3 UVC LEDs

In 2002, A. Yasan et al. demonstrated the first UVC LED emitting at 280 nm with AlInGaN/AlInGaN MQWs.[50] In the following years, M. A. Khan's group have reported a series of UVC LEDs toward short wavelength, e.g., 269-nm LEDs grown on sapphire with EQE as high as 0.32% under 10 mA direct current,[51] 254 nm LEDs with a novel micro-pixel design reaching pulsed output power as high as 1 mW (1 A)[52] and 250 nm LEDs.[53] Later on, Y. Taniyasu et al. demonstrated an AlN p-i-n homojunction LED with the successful control of both n-type and p-type doping in AlN.[54] As shown in Figure 3.152, The AlN LED had an emission wavelength of 210 nm, which is the shortest reported to date for any kind of LED. The emission is attributed to an exciton transition and represents an important step toward achieving exciton-related LEDs as well as replacing gas light sources with solid-state light sources.

At a wavelength of 275 nm, EQEs above 10% were achieved by using a p-reflective contact and unknown encapsulation.[55] In the case of using a p-GaN contact layer, EQEs of 4.9% at 270 nm with encapsulation[56] and 5.5–6.0% at 265 nm with distinctive roughening[57] were obtained using an AlN bulk substrate. Recently, record EQEs of more than 20% (Figure 3.153) have been reported for UVC LEDs by enhancing LEE.[41] Figure 3.153(a) shows the schematics of novel UVC-LED structures. They introduced several features to improve the light extraction: A transparent AlGaN:Mg contact layer, a Rh mirror electrode, an AlN template on a patterned sapphire substrate and encapsulation resin. The combination of the AlGaN:Mg contact layer and the Rh mirror electrode significantly improved the output power and the EQE of UVC LEDs.

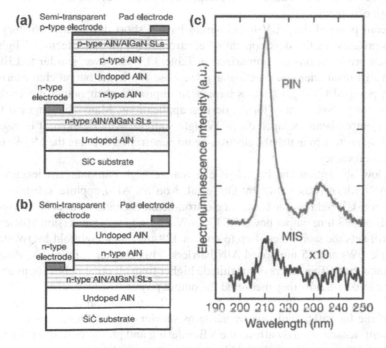

FIGURE 3.152   AlN LEDs. (a) Schematic illustration of the PIN LED. (b) Schematic illustration of the MIS LED. (c) Electroluminescence spectra of the LEDs under a DC bias condition. (From Taniyasu, Y., Kasu, M. and Makimoto, T., *Nature*, 441, 325, 2006. With permission.)

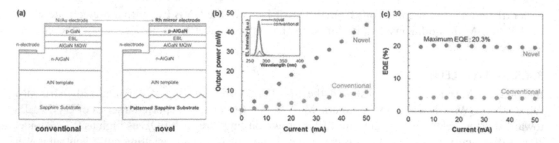

FIGURE 3.153 (a) Schematics of conventional and novel UV-LED structures. (b) Output power-current and (c) EQE-current characteristics. Blue and red dots show the characteristics of the conventional and novel UV-LED structures, respectively. Inset in part (b) shows the EL spectra of the UV LEDs at a DC of 20 mA. (From Takano, T., Mino, T., Sakai, J., Noguchi, N., Tsubaki, K. and Hirayama, H., *Appl. Phys. Express*, 10, 031002, 2017. With permission.)

3.8.4 E-BEAM PUMPED EMITTERS

The challenges of realizing effective p-type doping in AlGaN materials limit the performance of UV emitters. An alternative approach to bypass the efficiency-limiting key issues of UV LEDs and realize efficient deep-UV light emission is using electron-beam pumping of an (Al)GaN/(Ga)AlN active region. The device can be made very compact by using a miniature electron gun and electrostatic optics. When using e-beam excitation, electron-hole pairs are created through a sequence of scattering events within the semiconductor material that obviate the need for p-type doping. Therefore, without p-type doping layer, the resistive electrical losses from contact and sheet resistances become negligible. Also, there is no need for an electron blocking layer or other means to balance the electrical properties between electrons and holes that influence the effectiveness of carrier injection. Moreover, high LEE can be realized without p-doping layer. All of these aspects make it suitable for fabrication of large-scale light sources with both, high power and high efficiency.

Electron-beam pumped deep-UV light sources have a short development history of less than 10 years. The progress in the development of electron-beam pumped deep-UV light sources by various research organizations is summarized in Table 3.13. However, similar to LEDs, their performances are still insufficient for practical applications. While the output characteristics of these electron-beam pumped UV light sources have been improved, their output powers and emission efficiencies are not yet satisfactory for the desired applications. Many challenges still exist in the fabrication of electron-beam pumped deep-UV light sources. Among them, it is essential to optimize the MQWs structure to match the electron-beam penetration depth in the MQWs to obtain high carrier injection efficiency.[58]

QWs with low absorption and high IQE can realize high emission efficiencies. In 2010, an $Al_{0.69}Ga_{0.31}N$/AlN MQW was grown by Oto et al.[59] on an AlN/sapphire substrate, and a high-performance deep-UV light source based on electron-beam excitation was achieved with an operating wavelength of 238 nm, output power of 100 mW, and a power conversion efficiency (PCE) of 40%, which still sets the world record up to now. A 128-nm-thick eightfold MQW stack was used with 1-nm-thick QWs and 15-nm doped AlN barriers. This chapter is in intense discussion in the community, since the PCE is orders of magnitude higher than all other publications and the authors do not describe in detail, how they measured the output power.

In thick $Al_xGa_{1-x}N$ coherently grown on AlN, the VB ordering prevents surface emission from the (0 0 0 1) plane for $x>0.6$ (emission wavelengths shorter than 250 nm). The authors found that the quantum confinement strongly affects the VB ordering and promotes (0 0 0 1) surface emission, which motivated the use of a 1-nm-thick QW.

Tabataba-Vakili et al. from Park (Palo Alto/United States)[65] are working on design and characterization of electron-beam pumped AlGaN UVC emitters to realize a high-power vertical-cavity

TABLE 3.13
Recent Progress in Development of Electron-Beam Pumped UV Light Sources

Material and Structure	Institution	Excitation Condition	Output Power (mW)	PCE (%)	Wavelength (nm)	Year
AlGaN MQW	Kyoto University (Japan)	8 keV, 45 μA	100	40	238	2010[59]
Si-doped AlGaN film	Mie University (Japan)	10 keV, 100 μA	2.2	0.24	247	2011[60]
AlGaN MQW	Stanley Electric Corporation (Japan)	7.5 keV, 80 μA	20	4	240	2012[61]
Si-doped AlGaN MQW	Mie University (Japan)	10 keV, 200 μA	15	0.75	256	2013[62]
AlGaN MQW	Ioffe Institute (Russia)	20 keV, 100 μA	3.2	0.16	270	2015[63]
Quasi-2D GaN inserted in AlGaN MQW	Peking University (China)	15 keV, 700 μA	27	0.3	285	2016[64]
AlGaN MQW	Palo Alto Research Center, Inc. (United States)	12 keV, 4.4 mA (pulsed)	230	0.43	246	2016[65]
AlN/GaN MQW	Ioffe Institute (Russia)	20 keV, 1 mA (pulsed)	150	0.75	235	2018[66]
AlN/GaN MQW	Peking University (China)	20 keV, 1.25 mA	179	0.56	267	2019[67]
AlN/GaN MQW	Peking University (China)	20 keV, 1 mA	122.5	0.56	246	2019[67]
AlN/GaN MQW	Peking University (China)	20 keV, 1 mA	24	0.12	232	2019[67]
AlN/GaN MQW	Peking University (China)	18 keV, 37 mA	2200	0.3	254	2019[67]

laser emitting between 220 and 240 nm with good-quality beam profile and narrow linewidth. They succeeded in demonstrating pulsed cathodoluminescence (CL) around 240 nm of thinned samples that employed a thick stack of AlGaN/AlN MQWs (up to 20 QWs) grown by MOCVD. They achieved >200 mW peak output power with a wall-plug efficiency (WPE) of 0.43% with $Al_{0.56}Ga_{0.44}N/Al_{0.9}Ga_{0.1}N$ MQWs grown on ELO AlN/sapphire template. Chen et al.[68] are working on an electron-beam pumped external-cavity structure AlGaN-based UVC vertical-cavity surface-emitting laser emitting at 280 nm. In particular, LD devices require a relatively high injection to reach an exciton above the threshold for lasing, which leads to high energy consumption in the end.

Dimkou et al. report about embedded stacks of AlGaN/GaN quantum disks in a dense array of non-coalesced nanowires grown by MBE on Si substrate.[69] They achieved different emission wavelengths, ranging from 340 (IQE>60%), 286 nm (IQE ~44%) down to 258 nm (IQE ~22%), respectively. Y. Wang et al. have achieved high-output-power electron-pumped deep UV light sources, operating at 230–270 nm, by adjusting the well thickness of binary ultrathin AlN/GaN MQWs (Figure 3.154).[67] These structures are fabricated on high-quality thermally annealed AlN templates by MOCVD. Owing to the reduced dislocation density, large electron-hole overlap, and efficient carrier injection by electron beam, the UV light sources demonstrated high output powers of 24.8, 122.5 and 178.8 mW, where the estimated IQE values are 11.4, 61.6 and 51.7% at central wavelengths of 232, 243 and 265 nm, respectively. Further growth optimization and employing an electron-gun

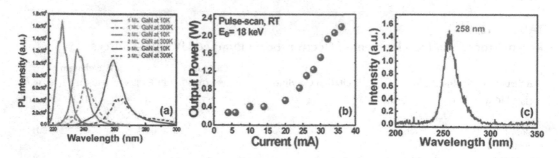

FIGURE 3.154 (a) PL spectra measured at 10 (solid line) and 300K (dashed line) for sample with nominally one monolayer (ML) GaN (red lines), two MLs GaN (green lines), and three MLs GaN (blue lines). (b) Dependence of output power on electron-beam current under pulse mode for sample with 150 periods of GaN/AlN MQWs with a well width of two MLs. The optimized electron energy for the maximum output power is 18 keV. (c) Corresponding CL spectrum with a peak wavelength of 258 nm for the optimized sample. (From Wang, Y., Rong, X., Ivanov, S., Jmerik, V., Chen, Z., Wang, H., Wang, T., Wang, P., Jin, P., Chen, Y., Kozlovsky, V., Sviridov, D., Zverev, M., Zhdanova, E., Gamov, N., Studenov, V., Miyake, H., Li, H., Guo, S., Yang, X., Xu, F., Yu, T., Qin, Z., Ge, W., Shen, B. and Wang, X., *Adv. Opt. Mater.*, 7, 1801763, 2019. With permission.)

with increased beam current lead to a record output power of ≈2.2 W at emission wavelength of ≈260 nm, the key wavelength for water sterilization. Whereas the state-of-the-art output power can reach up to 2 W, the WPE of electron-beam pumped UV devices (emission wavelength 200–280 nm) is still under 1% (neglecting still disputed publications).

3.8.5 CONCLUSION

In this section, we introduced the basic physical mechanisms behind and fundamental properties of the AlGaN-based materials and related UV light sources. Since the first UV LEDs were demonstrated, almost two decades ago, lots of efforts have been put into the research field of UV emitters. However, there are still many challenges existed for the development of high-efficiency, high-power UV LEDs. Nevertheless, the improvement of device performance will be gained by cumulative breakthrough for each of the key points about the EQE and LEE. We could expect that the UVB- and UVC LEDs will soon be commercially available in large scale and will trigger the rapid development in public health.

REFERENCES

1. Nakamura, S. and Fasol, G., *The Blue Laser Diode: GaN Based Light Emitters and Lasers*, New York: Springer-Verlag Berlin Heidelberg GmbH, 1997.
2. Morkoç, H., *Nitride Semiconductors and Devices*, New York: Springer-Verlag Berlin Heidelberg GmbH, 1999.
3. Pearton, S. J., *GaN and Related Materials II*, Boca Raton: CRC Press, 2000.
4. Gil, B., *Group III Nitride Semiconductor Compounds: Physics and Applications*, Midsomer Noton: Clarendon Press, Oxford University Press, 1998.
5. Wu, J., Walukiewicz, W., Yu, K. M., Ager, J. W., Haller, E. E., Lu, H., Schaff, W. J., Saito, Y. and Nanishi, Y., *Appl. Phys. Lett.*, 80, 3967, 2002.
6. Xu, K. and Yoshikawa, A., *Appl. Phys. Lett.*, 83, 251, 2003.
7. Wu, J. and Walukiewicz, W., *Superlattice Microst*, 34, 63, 2003.
8. Bhuiyan, A. G., Hashimoto, A. and Yamamoto, A., *J. Appl. Phys.*, 94, 2779, 2003.
9. Nanishi, Y., Saito, Y. and Yamaguchi, T., *Jpn. J. Appl. Phys.*, 42, 2549, 2003.
10. Wang, X. and Yoshikawa, A., *Prog. Cryst. Growth Ch.*, 48–49, 42, 2004.
11. Vurgaftman, I., Meyer, J. R. and Ram-Mohan, L. R., *J. Appl. Phys.*, 89, 5815, 2001.

12. Khan, M. R. H., Koide, Y., Itoh, H., Sawaki, N. and Akasaki, I., *Solid State Commun.*, 60, 509, 1986.
13. Koide, Y., Itoh, H., Khan, M. R. H., Hiramatu, K., Sawaki, N. and Akasaki, I., *J. Appl. Phys.*, 61, 4540, 1987.
14. Yun, F., Reshchikov, M. A., He, L., King, T., Morkoç, H., Novak, S. W. and Wei, L., *J. Appl. Phys.*, 92, 4837, 2002.
15. Liu, B., Zhang, R., Xie, Z. L., Liu, Q. J., Zhang, Z., Li, Y., Xiu, X. Q., Yao, J., Mei, Q., Zhao, H., Han, P., Lu, H., Chen, P., Gu, S. L., Shi, Y., Zheng, Y. D., Cheung, W. Y., Ke, N. and Xu, J. B., *J. Cryst. Growth*, 310, 4499, 2008.
16. Paduano, Q. S., Weyburne, D. W., Bouthillette, L. O., Wang, S.-Q. and Alexander, M. N., *Jpn. J. Appl. Phys.*, 41, 1936, 2002.
17. Amano, H., Kito, M., Hiramatsu, K. and Akasaki, I., *Jpn. J. Appl. Phys.*, 28, L2112, 1989.
18. Nakamura, S., Mukai, T., Senoh, M. and Iwasa, N., *Jpn. J. Appl. Phys.*, 31, L139, 1992.
19. Mehnke, F., Trinh, X. T., Pingel, H., Wernicke, T., Janzén, E., Son, N. T. and Kneissl, M., *J. Appl. Phys.*, 120, 145702, 2016.
20. Pampili, P. and Parbrook, P. J., *Mat. Sci. Semicon. Proc.*, 62, 180, 2017.
21. Amano, H., Collazo, R., Santi, C. D., Einfeldt, S., Funato, M., Glaab, J., Hagedorn, S., Hirano, A., Hirayama, H., Ishii, R., Kashima, Y., Kawakami, Y., Kirste, R., Kneissl, M., Martin, R., Mehnke, F., Meneghini, M., Ougazzaden, A., Parbrook, P. J., Rajan, S., Reddy, P., Römer, F., Ruschel, J., Sarkar, B., Scholz, F., Schowalter, L. J., Shields, P., Sitar, Z., Sulmoni, L., Wang, T., Wernicke, T., Weyers, M., Witzigmann, B., Wu, Y.-R., Wunderer, T. and Zhang, Y., *J. Phys. D Appl. Phys.*, 53, 503001, 2020.
22. Harris, J. S., Baker, J. N., Gaddy, B. E., Bryan, I., Bryan, Z., Mirrielees, K. J., Reddy, P., Collazo, R., Sitar, Z. and Irving, D. L., *Appl. Phys. Lett.*, 112, 152101, 2018.
23. Washiyama, S., Reddy, P., Sarkar, B., Breckenridge, M. H., Guo, Q., Bagheri, P., Klump, A., Kirste, R., Tweedie, J., Mita, S., Sitar, Z. and Collazo, R., *J. Appl. Phys.*, 127, 105702, 2020.
24. Nakarmi, M. L., Nepal, N., Lin, J. Y. and Jiang, H. X., *Appl. Phys. Lett.*, 94, 091903, 2009.
25. Nakarmi, M. L., Nepal, N., Ugolini, C., Altahtamouni, T. M., Lin, J. Y. and Jiang, H. X., *Appl. Phys. Lett.*, 89, 152120, 2006.
26. Nam, K. B., Nakarmi, M. L., Li, J., Lin, J. Y. and Jiang, H. X., *Appl. Phys. Lett.*, 83, 878, 2003.
27. Ryu, H.-Y., Choi, I.-G., Choi, H.-S. and Shim, J.-I., *Appl. Phys. Express*, 6, 062101, 2013.
28. Nam, K. B., Li, J., Nakarmi, M. L., Lin, J. Y. and Jiang, H. X., *Appl. Phys. Lett.*, 84, 5264, 2004.
29. Li, J., Nam, K. B., Nakarmi, M. L., Lin, J. Y., Jiang, H. X., Carrier, P. and Wei, S.-H., *Appl. Phys. Lett.*, 83, 5163, 2003.
30. Ishii, R., Kaneta, A., Funato, M., Kawakami, Y. and Yamaguchi, A. A., *Phys. Rev. B*, 81, 155202, 2010.
31. Kolbe, T., Knauer, A., Chua, C., Yang, Z., Einfeldt, S., Vogt, P., Johnson, N. M., Weyers, M. and Kneissl, M., *Appl. Phys. Lett.*, 97, 171105, 2010.
32. Northrup, J. E., Chua, C. L., Yang, Z., Wunderer, T., Kneissl, M., Johnson, N. M. and Kolbe, T., *Appl. Phys. Lett.*, 100, 021101, 2012.
33. Al Tahtamouni, T. M., Lin, J. Y. and Jiang, H. X., *Appl. Phys. Lett.*, 101, 042103, 2012.
34. Reich, C., Guttmann, M., Feneberg, M., Wernicke, T., Mehnke, F., Kuhn, C., Rass, J., Lapeyrade, M., Einfeldt, S., Knauer, A., Kueller, V., Weyers, M., Goldhahn, R. and Kneissl, M., *Appl. Phys. Lett.*, 107, 142101, 2015.
35. Zhang, Z., Kushimoto, M., Sakai, T., Sugiyama, N., Schowalter, L. J., Sasaoka, C. and Amano, H., *Appl. Phys. Express*, 12, 124003, 2019.
36. Nakamura, S., Senoh, M. and Mukai, T., *Jpn. J. Appl. Phys.*, 32, L8, 1993.
37. Hirayama, H., Maeda, N., Fujikawa, S., Toyoda, S. and Kamata, N., *Jpn. J. Appl. Phys.*, 53, 100209, 2014.
38. Kneissl, M., Seong, T.-Y., Han, J. and Amano, H., *Nat. Photonics*, 13, 233, 2019.
39. Bao, X., Sun, P., Liu, S., Ye, C., Li, S. and Kang, J., *IEEE Photonics J.*, 7, 1, 2015.
40. Kneissl, M. and Rass, J., *III-Nitride Ultraviolet Emitters*, Springer International Publishing, eBook, 2016.
41. Takano, T., Mino, T., Sakai, J., Noguchi, N., Tsubaki, K. and Hirayama, H., *Appl. Phys. Express*, 10, 031002, 2017.
42. Han, J., Crawford, M. H., Shul, R. J., Figiel, J. J., Banas, M., Zhang, L., Song, Y. K., Zhou, H. and Nurmikko, A. V., *Appl. Phys. Lett.*, 73, 1688, 1998.
43. Nishida, T., Saito, H. and Kobayashi, N., *Appl. Phys. Lett.*, 79, 711, 2001.
44. Adivarahan, V., Chitnis, A., Zhang, J. P., Shatalov, M., Yang, J. W., Simin, G., Khan, M. A., Gaska, R. and Shur, M. S., *Appl. Phys. Lett.*, 79, 4240, 2001.

45. Khan, M. A., Adivarahan, V., Zhang, J. P., Chen, C. Q., Kuokstis, E., Chitnis, A., Shatalov, M., Yang, J. W. and Simin, G., *Jpn. J Appl. Phys.*, 40 (12A), L1308-L1310, 2001.
46. Morita, D., Sano, M., Yamamoto, M., Murayama, T., Nagahama, S.-I. and Mukai, T., *Jpn. J. Appl. Phys.*, 41, L1434, 2002.
47. Muramoto, Y., Kimura, M. and Nouda, S., *Semicond. Sci. Technol.*, 29, 084004, 2014.
48. Kaneda, M., Pernot, C., Nagasawa, Y., Hirano, A., Ippommatsu, M., Honda, Y., Amano, H. and Akasaki, I., *Jpn. J. Appl. Phys.*, 56, 061002, 2017.
49. Sato, K., Yasue, S., Yamada, K., Tanaka, S., Omori, T., Ishizuka, S., Teramura, S., Ogino, Y., Iwayama, S., Miyake, H., Iwaya, M., Takeuchi, T., Kamiyama, S. and Akasaki, I., *Appl. Phys. Express*, 13, 031004, 2020.
50. Yasan, A., McClintock, R., Mayes, K., Darvish, S. R., Kung, P. and Razeghi, M., *Appl. Phys. Lett.*, 81, 801, 2002.
51. Adivarahan, V., Wu, S., Zhang, J. P., Chitnis, A., Shatalov, M., Mandavilli, V., Gaska, R. and Khan, M. A., *Appl. Phys. Lett.*, 84, 4762, 2004.
52. Wu, S., Adivarahan, V., Shatalov, M., Chitnis, A., Sun, W.-H. and Khan, M. A., *Jpn. J. Appl. Phys.*, 43, L1035, 2004.
53. Adivarahan, V., Sun, W. H., Chitnis, A., Shatalov, M., Wu, S., Maruska, H. P. and Khan, M. A., *Appl. Phys. Lett.*, 85, 2175, 2004.
54. Taniyasu, Y., Kasu, M. and Makimoto, T., *Nature*, 441, 325, 2006.
55. Shatalov, M., Sun, W., Jain, R., Lunev, A., Hu, X., Dobrinsky, A., Bilenko, Y., Yang, J., Garrett, G. A., Rodak, L. E., Wraback, M., Shur, M. and Gaska, R., *Semicond. Sci. Technol.*, 29, 084007, 2014.
56. Grandusky, J. R., Chen, J., Gibb, S. R., Mendrick, M. C., Moe, C. G., Rodak, L., Garrett, G. A., Wraback, M. and Schowalter, L. J., *Appl. Phys. Express*, 6, 032101, 2013.
57. Inoue, S.-I., Naoki, T., Kinoshita, T., Obata, T. and Yanagi, H., *Appl. Phys. Lett.*, 106, 131104, 2015.
58. Li, D., Jiang, K., Sun, X. and Guo, C., *Adv. Opt. Photonics*, 10, 43, 2018.
59. Oto, T., Banal, R. G., Kataoka, K., Funato, M. and Kawakami, Y., *Nat. Photonics*, 4, 767, 2010.
60. Shimahara, Y., Miyake, H., Hiramatsu, K., Fukuyo, F., Okada, T., Takaoka, H. and Yoshida, H., *Appl. Phys. Express*, 4, 042103, 2011.
61. Matsumoto, T., Iwayama, S., Saito, T., Kawakami, Y., Kubo, F. and Amano, H., *Opt. Express*, 20, 24320, 2012.
62. Fukuyo, F., Ochiai, S., Miyake, H., Hiramatsu, K., Yoshida, H. and Kobayashi, Y., *Jpn. J. Appl. Phys.*, 52, 01AF03, 2013.
63. Ivanov, S. V., Jmerik, V. N., Nechaev, D. V., Kozlovsky, V. I. and Tiberi, M. D., *Phys. Status Solidi A*, 212, 1011, 2015.
64. Rong, X., Wang, X., Ivanov, S. V., Jiang, X., Chen, G., Wang, P., Wang, W., He, C., Wang, T., Schulz, T., Albrecht, M., Jmerik, V. N., Toropov, A. A., Ratnikov, V. V., Kozlovsky, V. I., Martovitsky, V. P., Jin, P., Xu, F., Yang, X., Qin, Z., Ge, W., Shi, J. and Shen, B., *Adv. Mater.*, 28, 7978, 2016.
65. Tabataba-Vakili, F., Wunderer, T., Kneissl, M., Yang, Z., Teepe, M., Batres, M., Feneberg, M., Vancil, B. and Johnson, N. M., *Appl. Phys. Lett.*, 109, 181105, 2016.
66. Jmerik, V. N., Nechaev, D. V., Toropov, A. A., Evropeitsev, E. A., Kozlovsky, V. I., Martovitsky, V. P., Rouvimov, S. and Ivanov, S. V., *Appl. Phys. Express*, 11, 091003, 2018.
67. Wang, Y., Rong, X., Ivanov, S., Jmerik, V., Chen, Z., Wang, H., Wang, T., Wang, P., Jin, P., Chen, Y., Kozlovsky, V., Sviridov, D., Zverev, M., Zhdanova, E., Gamov, N., Studenov, V., Miyake, H., Li, H., Guo, S., Yang, X., Xu, F., Yu, T., Qin, Z., Ge, W., Shen, B. and Wang, X., *Adv. Opt. Mater.*, 7, 1801763, 2019.
68. Chen, Y. R., Zhang, Z. W., Miao, G. Q., Jiang, H., Li, Z. M. and Song, H., *J. Alloy Compd.*, 820, 153415, 2020.
69. Dimkou, I., Harikumar, A., Ajay, A., Donatini, F., Bellet-Amalric, E., Grenier, A., den Hertog, M. I., Purcell, S. T. and Monroy, E., *Phys. Status Solidi A*, 217, 1900714, 2020.

3.9 SILICON CARBIDE (SiC) AS A LUMINESCENCE MATERIAL

Yan Peng and Xiangang Xu

As a member of the third-generation semiconductor material, silicon carbide (SiC) has plenty of advantages imposed by the fundamental material properties, such as large bandgap, high thermal conductivity and high breakdown electric field strength. The last 20 years have witnessed a remarkable development in SiC crystal growth and now 2–8 inch SiC single crystal can be grown.[1-4] Thus SiC has been widely applied in various fields, such as electric vehicle, aerospace, microwave communications and so on.[5-7]

3.9.1 POLYTYPES

SiC shows polytypism arising from different stacking possibilities. In hexagonal close packing of the Si–C pair, the positions of the pair in the first and second layers are uniquely determined (A and B) as shown in Figure 3.155(a). However, in the third layer, there are two possibilities, either A or C as shown. In the former case, the stacking order becomes ABAB..., giving a wurtzite (hexagonal) structure, and the latter becomes ABCABC..., giving a zinc-blende (cubic) structure.

In SiC crystals, there can exist various combinations of these two structures, which give different stacking orders called polytypes. Among the many polytypes, 3C–, 6H– and 4H–SiC appear frequently: These structures are shown in Figures 3.155(b)–(d) together with 2H–SiC (Figure 3.155(e)). Here, the number indicates the period of stacking order and the letter gives its crystal structure: C=cubic, H=hexagonal, R=rhombohedral. Since the position of each atom has a different configuration of nearest-neighbor atoms, the sites are crystallographically different; that is, they have cubic or hexagonal site symmetry. Hence, when an impurity atom substitutes into the position of Si or C, it gives rise to different energy levels depending upon the number of inequivalent sites present in the material. In 3C–SiC and 2H–SiC, only one cubic or one hexagonal site exists, respectively, whereas in 6H–SiC, there exist one hexagonal and two cubic sites and in 4H–SiC one hexagonal and one cubic site.

3.9.2 BAND STRUCTURE AND OPTICAL ABSORPTION

Figure 3.156 shows the absorption spectra of different polytypes of SiC at 4.2K.[8] The spectra contain shoulder features related to phonon-assisted transitions, which are characteristics of indirect

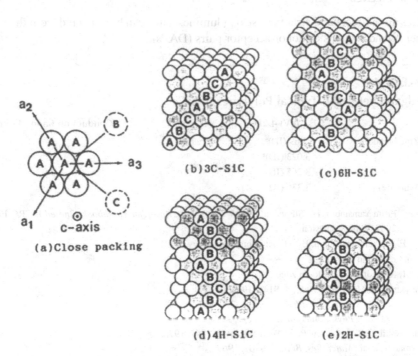

FIGURE 3.155 Position of Si–C pair in typical SiC polytypes. (a) Close packing of equal spheres (Si–C pair), (b) 3C–SiC, (c) 6H–SiC, (d) 4H–SiC, and (e) 2H–SiC. (From Yamamoto, H., Shionoya, S. and Yen, W. M., *Phosphor Handbook (2nd ed.)*, CRC Press, 2007. With permission.)

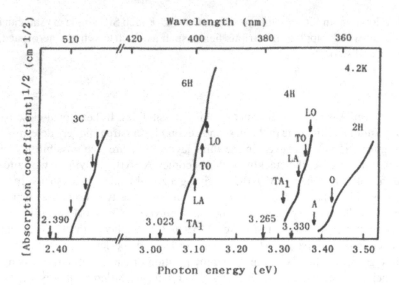

FIGURE 3.156 Absorption spectra for typical SiC polytypes. Exciton bandgap is shown for each polytype. (From Choyke, W., *Mater. Res. Bull.*, 4, S141, 1969. With permission.)

band structures. In the figure, the positions of the exciton bandgaps are shown. In Table 3.14, the values of exciton bandgaps and exciton binding energies are tabulated.[8] The characteristics near the fundamental absorption edge have quite similar structure for all the polytypes except 2H–SiC. This is due to the similarity of the phonons involved in optical absorption in the different polytypes.

3.9.3 Luminescence[8,9]

Since SiC has an indirect band structure, strong luminescence can be expected from the recombination of either bound excitons or donor-acceptor pairs (DAPs).

TABLE 3.14
Bandgap Energies in Typical Polytypes of SiC

	E_{GX} (eV) 4.2K	E_{exc} (meV)	Conduction Band Minimum
3C (Zinc blende)	2.69 (ID)[a]	13.5[b]	X[e]
6H	3.023 (ID)[a]	78[c]	U[f]
4H	3.265 (ID)[a]	20[d]	M[g]
2H (Wurtzite)	3.330 (ID)[a]		K[e]

Source: From Yamamoto, H., Shionoya, S. and Yen, W. M., *Phosphor Handbook (2nd ed.)*, CRC Press, 2007. With permission.

Note: E_{GX}: Exciton bandgap, E_{exc}: Exciton binding energy.

ID: indirect band structure.

X, U, M, K: position in Brillouin zone.

[a] Choyke, W.J., *Mater. Res. Bull.*, 4, S141, 1969.
[b] Nedzvetskii, D.S. et al., *Sov. Phys. Semicon.*, 2, 914, 1969.
[c] Sankin, V.I., *Sov. Phys. Solid State.*, 17, 1191, 1975.
[d] Dubrovskii, G.B. et al., *Sov. Phys. Solid State.*, 17, 1847, 1976.
[e] Herman, F. et al., *Mater. Res. Bull.*, 4, S167, 1969.
[f] Choyke, W.J., unpublished result, 1995.
[g] Patrick, L. et al., *Phys. Rev.*, 137, A1515, 1965.

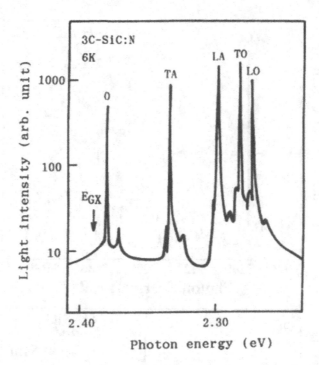

FIGURE 3.157 Photoluminescence spectrum of excitons bound at N donors in 3C–SiC. EGX indicates the exciton bandgap. (0: zero phonon; TA: transverse acoustic; LA: longitudinal acoustic; TO: transverse optic; and LO: longitudinal optic). (From Choyke, W., *Mater. Res. Bull.*, 4, S141, 1969. With permission.)

3.9.3.1   Luminescence from Excitons

Figure 3.157 depicts the photoluminescence spectrum from excitons bound at N donors in 3C–SiC.[8] From the energy difference between the exciton bandgap and the peak energy corresponding to the zero-phonon line, the exciton binding energy for N donors is estimated to be 10 meV. Since the resolution of peak energies is much better than that in the absorption spectra, the exact value of phonon energies can be obtained from the photoluminescence spectra.

In the photoluminescence spectrum of 6H–SiC, there exists a zero-phonon peak due to the recombination of excitons bound at N donors substituted into hexagonal C sites and two zero-phonon peaks due to those located in cubic C sites.[10] Since the energy levels of N donors in inequivalent (hexagonal, cubic) sites are different, the photoluminescence peaks have different energies. Figure 3.158 shows PL for 4H–SiC and 6H–SiC at low temperature. P_0, Q_0 in Figure 3.158(a) (P_0, R_0, S_0 in Figure 3.158(b)) denotes the zero-phonon lines, associated to the direct recombination of excitons to neutral nitrogen donors.[11]

3.9.3.2   Luminescence from Donor-Acceptor Pairs

In SiC, N atoms belonging to the fifth column of the periodic table work as donors, and B, Al and Ga in the third column work as acceptors. When donors and acceptors are simultaneously incorporated in a crystal, electrons bound at donors and holes at acceptors can create a pair due to the Columbic force between electrons and holes. This interaction leads to strong photoluminescence through recombination and is known as DAP luminescence.

The photon energy of DAPs is given by[12]:

$$E = E_g - E_d - E_A + \frac{e^2}{4\pi\varepsilon R}$$

(3.54)

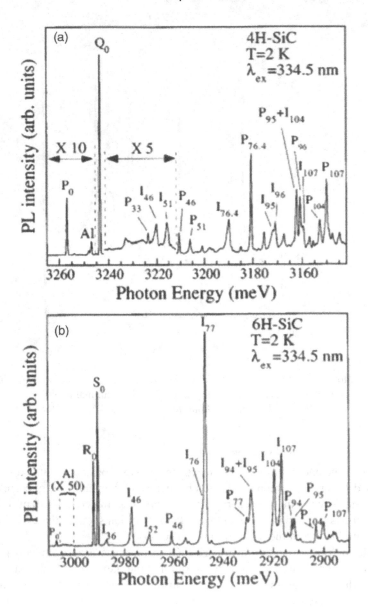

FIGURE 3.158 Low-temperature PL spectra of high-quality CVD-grown epilayers of 4H–SiC (a) and 6H–SiC (b).

where E_g is the bandgap, E_d is the energy level of donor, E_A is related to ionization energy of acceptor and $e^2/4\pi\varepsilon R$ is Columbic interaction energy between the donor and the acceptor. Here, R is the function of the donor-acceptor distance. e and ε are the elementary charge and dielectric constant. Columbic interaction energy is always ignored due to unnoticeable value.

DAPs luminescence has been detailed characterized to identify the physical properties of doping in early studies. And the ionization energies of nitrogen donor, aluminum acceptors have been obtained from PL spectrum.[13] As shown in Figure 3.159, DAP spectra of Al-doped, Ga-doped and B-doped crystal of (a) 4H, (b) 4H and (c) 15R SiC at 4.2 and 77K. Since the value of E_A is calculated from free-to-acceptor recombination as in Figure 3.158, E_d can also be determined.

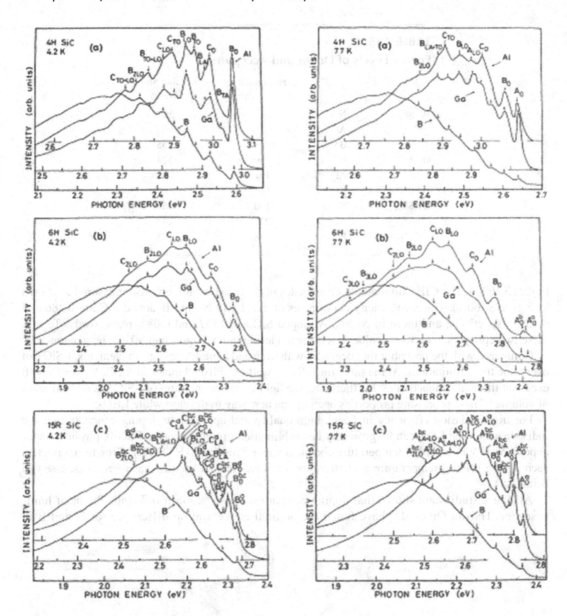

FIGURE 3.159 DAP spectra of Al-doped, Ga-doped and B-doped crystal of (a) 4H, (b) 4H and (c) 15R SiC at 4.2 (left) and 77K (right). (From Ikeda, M., Matsunami, H. and Tanaka, T., *Phys. Rev. B*, 22, 2842, 1980. With permission.)

From a similar analysis, the Al, B- and Ga-acceptor levels for 3C, 4H and 6H–SiC can also be determined, as listed in Table 3.15.[13]

The luminescence application of SiC is limited due to its indirect bandgap and lower light emission efficiency. A power efficiency of early blue LED produced by DAP recombination of 6H–SiC is extremely low ~0.05–0.07%.[14] And in the last 20 years, researchers turn attention back to N–B and N–Al co-doped SiC, called fluorescent SiC (f-SiC), which are excellent candidate of phosphor materials for white LEDs.[12,15–17]

The N- and B-doped SiC emits a yellow-orange light, while the N- and Al-doped SiC emits blue-green light under the excitation of ultraviolet (UV) light. To combine these two spectra, a full-range of visible spectrum similar to the sun-light spectrum can be produced, as shown in

TABLE 3.15

Energy Levels of Donor and Acceptors

		Hexagonal Site (meV)	Cubic Site (meV)
3C–SiC	N		56.5
	Al		254
	B		735
4H–SiC	N	66	124
	Al	191	
	B	647	
6H–SiC	N	100	155
	Al	239	249
	B	698	723

Figure 3.160(a). The CIE Chromaticity coordinates of these two f-SiCs are also measured as shown in Figure 3.160(b). The chromaticity coordinates of x and y in N- and B-doped SiC are 0.486 and 0.465, respectively, and those in N- and Al-doped SiC are 0.137 and 0.085, respectively. To mix these two epilayers, pure white color with color rendering index larger than 90 can be generated.[18]

Compared with the phosphor that degrades with time and contains rear earth elements, f-SiC can overcome these limitations. At the same time, SiC is well-established material for GaN growth with excellent thermal conductivity.[19] Furthermore, the light-emission medium of f-SiC has no limitation of volume.[7] These promising properties open up the new way to develop white LEDs.

For high emission efficiency in f-SiC, high-quality and appropriate doping concentration are indispensable. The f-SiC can be grown by fast sublimation growth process (FSGP), physical vapor deposition (PVT) and high-temperature chemical vapor deposition (HTCVD). Growth parameters such as the source, temperature and structure and electrical properties have been discussed in detail.[19–23]

All these studies are shown that doping concentration is the main influence factor of luminescence. Haiyan Ou et al.[19] have found the optimal concentration difference for N and B in

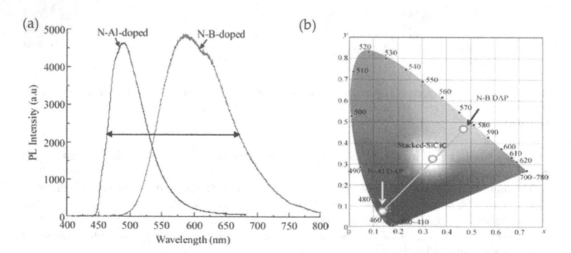

FIGURE 3.160 Photoluminescence spectra of f-SiC epilayers (a); CIE chromaticity coordinate plot of f-SiCs (b). (From Kamiyama, S., Iwaya, M., Takeuchi, T., Akasaki, I., Syväjärvi, M. and Yakimova, R., *J. Semicond.*, 32, 013004, 2011. With permission.)

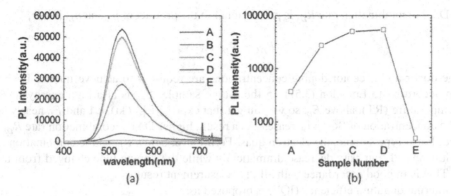

FIGURE 3.161 PL spectra (a) and PL intensities (b) of five DA co-doped 4H–SiC substrates.

n-type 4H–SiC is of 4.6×10^{18} cm^{-3}. Rusheng Wei reported that with the increase of Boron concentration, the PL intensity increased and strongest DAP emission occurred with the least concentration difference of 6.8×10^{18} cm^{-3}.[23] While the conduction type becomes to p-type, that means the concentration of Boron is more than nitrogen, the DAP emission decrease distinctly, as shown in Figure 3.161. Despite the intensity differences, all the DAP emission spectra show the same peak wavelength at 530 nm (2.34 eV) and the same full-width-at-half maximum (FWHM) of 82 nm. Both the doping type and the concentration difference affect the DAP emission intensity but do not introduce changes in the peak wavelength and FWHM at this doping level.[24–26]

All abovementioned results are explained by using band diagram with Fermi-Dirac statistics.

Figure 3.162 shows a schematic band diagram of n-type doped SiC under excitation condition. It is seen that all the acceptor states (E_A) are occupied by photoexcited holes due to the large ionization energy of acceptor states (E_A-E_V), so the hole density at the acceptor states $pA \cong N_A$. On the other hand, the electron density at donor states (E_D) under excitation is given by:

$$n_D \cong N_D \cdot F_D \tag{3.55}$$

due to the small ionization energy (E_C-E_D). Here F_D is the occupancy probability of an electron on donor states. It can be represented by:

$$F_D = \frac{1}{1+(1/2)\exp\left(\dfrac{E_D-E_{FC}}{kT}\right)} \tag{3.56}$$

where E_{FC} is the electron quasi-Fermi level under excitation, k is the Boltzmann constant and T is the absolute temperature.

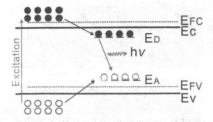

FIGURE 3.162 Band diagram of n-type SiC under excitation (black dots represent electrons and unfilled circles represent holes).

The DAP recombination rate R_{DA} is proportional to the product of p_A and n_D, hence,

$$R_{DA} \propto N_D \cdot F_D \cdot N_A \tag{3.57}$$

Large donor and acceptor doping concentrations are required to achieve intense DAP recombination according to Equation (3.57). In the n-type sample, E_{FC} is at a high energy level under room temperature (RT), above E_D, so we can get that $\exp(E_D-E_{FC}/kt) \ll 1$ and F_D nearly constant while the concentration of Boron increases. As a result, higher DAP recombination rate R_{DA} will be achieved. For p-type sample, the electron quasi-Fermi level below E_D, so the combination rate R_{DA} is very low and PL intensity decrease dramatically while the doping type changed from n-type to p-type. That is in good accordance with all PL measurement results.

The internal quantum efficiency (IQE) is proposed as:

$$IQE = \frac{1}{1+\left(\tau_r / \tau_{nr}\right)} \tag{3.58}$$

where τ_r is radiation lifetime, and τ_{nr} is non-radiative lifetime. The estimate IQEs is affected by doping concentration and growth method. The high-quality f-SiC shows almost 100% IQE at B concentration $+10^{19}$ cm^{-3} (Figure 3.163).[18] These results are consistent with experimental results reported by Satoshi Kamiyama in 2006 which found that nitrogen (N) and boron (B) co-doped SiC had high DAPs emission efficiency with IQE up to 95%.[27]

However, the luminescence is not only influenced by optimal doping concentration and crystalline quality, but also effected by excitation wavelength, excitation power and emission angle which has been clarified by the research of GaN, ZnO and ZnTe, and other semiconductor luminescent materials.[28–30]

The photoluminescence excitation (PLE), angle-resolved PL and the excitation power dependence of DAPs emission have been carried out.

The PLE spectrum was obtained by monitoring the DAPs luminescence peak while scanning the whole laser excitation energy to find the optimal excitation wavelengths. There is a slight difference in threshold excitation energy of different samples, but the values are basically maintained at around 3.2 eV. It also proves that the photons absorbed in the SiC can only generate the excited states of the DAP by the processes of band-to-band excitation.[30] The process is shown in Figure 3.164. When the photon energy of the UV light is greater than the bandgap energy, the electrons are excited to the conduction band, leaving holes in the valence band. These electrons and

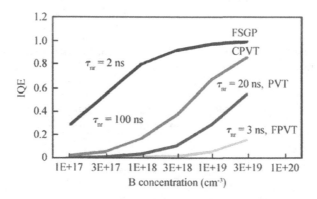

FIGURE 3.163 IQEs as a function of B concentration for f-SiCs with a variation of non-radiative lifetime. (From Kamiyama, S., Iwaya, M., Takeuchi, T., Akasaki, I., Syväjärvi, M. and Yakimova, R., *J. Semicond.*, 32, 013004, 2011. With permission.)

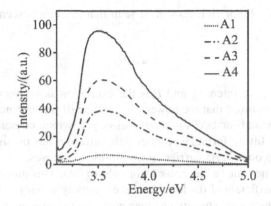

FIGURE 3.164 Photoluminescence excitation spectra of the DAPs emission. (From Ci, S., Peng, Y., Xie, X., Hu, X., Duan, P., Yu, J., Xu, X., Chen, X., Yang, X. and Wang, R., *Semicond. Sci. Technol.*, 35, 055009, 2020. With permission.)

holes are trapped by the donor and the acceptor separately. Then these electrons and holes recombine to produce a DAP and emit photons.

Figure 3.164 shows that the optimal excitation wavelengths ranged from 340 to 360 nm. Carrier concentration has no affection on the optimal excitation wavelength. When the excitation photon energy is between 3.5 and 3.7 cV, the luminescence intensity is the strongest. However, when the excitation photon energy is greater than 3.7 eV, the luminescence intensity decreases as the excitation energy increases. It's different from the convention observed in other semiconductor DAPs luminescence such as ZnO,[29] where the luminescence intensity is substantially constant when the excitation energy is greater than the bandgap energy. As reported, the PLE spectroscopy can also prove optical absorption properties. I. G. Ivanov et al.[31] studied the absorption characteristics of SiC near the absorption edge using PLE spectroscopy. Therefore, the decrease in luminescence intensity of PLE spectrum is affected by the absorption properties. For light with photon energy greater than 3.7 eV, the greater the photon energy, the weaker the samples absorb photons, so the luminescence intensity decreases.

The photoluminescence spectra under different power excitations were collected by a 325-nm wavelength He–Cd laser at RT. The relationship between luminous intensity and excitation power is shown in Figure 3.165, where the excitation power varies from 2 to 44 mW. It was fitted to nonlinear

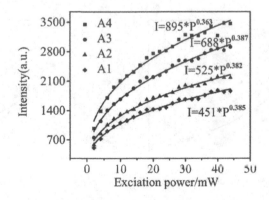

FIGURE 3.165 The power dependence of DAPs intensity of A1–A4. The data was fitted nonlinear curve by the function of Allometric1 in origin. (From Ci, S., Peng, Y., Xie, X., Hu, X., Duan, P., Yu, J., Xu, X., Chen, X., Yang, X. and Wang, R., *Semicond. Sci. Technol.*, 35, 055009, 2020. With permission.)

curve by the function of Allometric1. As can be seen that the luminescence intensity follows the power dependence:

$$I = aP^\gamma \qquad (3.59)$$

where I is the luminescence intensity and P is the excitation laser intensity, γ is a coefficient. Schmidt et al.[32,33] have reported that the power dependence of the luminescence intensity can be described by a $I \propto P^\gamma$ law and for DAPs recombination $\gamma<1$, which coincide with the experiment result. According to the fitting formulas, carrier concentration has no significant influence on γ but has a great influence on "a". As carrier concentration decreases, "γ" is basically maintained at around 0.38, at the same time "a" increased from 451 to 895. This means the carrier concentration does not affect γ coefficient of the luminescence intensity to vary with excitation power, but it affects the slope at which the intensity changes with excitation power. With the increase of B doping concentration, the carrier concentration gets lower and the intensity gets more sensitive to the excitation power change.

This is due to the fact that DAPs luminescence is achieved by tunneling, and transition probability is exponentially attenuated as the spacing between atoms increases. When the excitation power increases, the average spacing between donor and acceptor decreases, so the transition probability increases, resulting in an increase in PL intensity.[34] Therefore, changing the excitation power, the average spacing between the donor and acceptor of high B doping concentration samples decreases rapidly. As a consequence, the luminescence intensity is more sensitive to the excitation power change.

Figure 3.166(c) and (d) shows that the luminescence peak position of sample 3 and 4 varies with the excitation power. As the excitation power increases, the luminescence peak shifts toward the

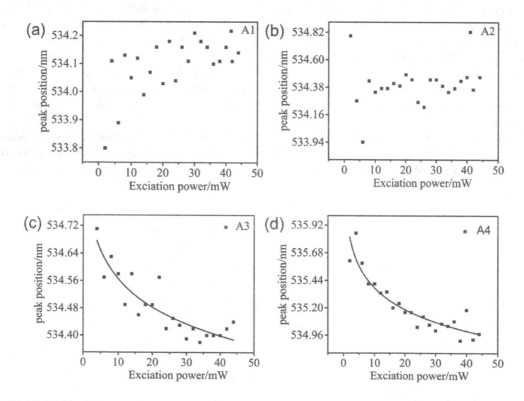

FIGURE 3.166 The power dependence of DAPs peak position of each sample (a) A1, (b) A2, (c) A3, (d) A4. (From Ci, S., Peng, Y., Xie, X., Hu, X., Duan, P., Yu, J., Xu, X., Chen, X., Yang, X. and Wang, R., *Semicond. Sci. Technol.*, 35, 055009, 2020. With permission.)

high energy. When the excitation power increases from 2 to 44 mW, the luminescence peak position of samples A3 and A4 shift by 1.3 and 3.5 meV, respectively. As discussed earlier, the photon energy of DAPs is affected by the energy level of donor and acceptor and Coulombic interaction energy. When the excitation power increases, the average spacing between the donor and the acceptor decreases. It results in the luminescence peak move to the high energy peak position.[34] This phenomenon is evident in samples A4 and A3 and does not appear in samples A1 and A2, as shown in Figure 3.160(a) and (b). This proves that the influence of Coulombic interaction energy on the luminescence peak position cannot be negligible under the conditions of high doping concentration and high excitation power.

Furthermore, angle-resolved PL was measured on sample by using a goniometer. The excitation light was fixed to a certain direction, and the observation (emission) angle varied from 0 (normal to the sample surface) to 80 degrees. From the acquired spectra shown in Figure 3.167(a), one can see that the PL intensity decreases and its peak shifts toward shorter wavelength as the emission angle θ increases.

The PL intensity becomes lower because the internal reflection increases as the emission angle increases, hence, the emission light extracted out of the SiC sample decreases. From Figure 3.167(b), it is also seen that the luminescence intensity at an angle of 80 degrees is still around 57% of the one normal to the surface, which is quite promising among most of commercial LEDs (less than 30%). In addition, this angular-dependent PL peak wavelength is attributed to the Fabry-Perot (FP) microcavity interference effect, and similar phenomena have also been observed in Refs. [35,36].

The FP equation

$$m\lambda = 2dn\cos\psi \qquad (3.60)$$

gives the condition for constructive interference of light undergoing multiple internal reflections in a solid film or an etalon. In Equation (3.60), m is the interference order, λ is the propagating wavelength, d is the etalon thickness, n is the refractive index of the etalon and ψ is the wavefront propagation angle with respect to the surface normal. The relationship between ψ and θ can be represented by Snell's law:

$$n\sin\psi = \sin\theta \qquad (3.61)$$

Thus, the FP equation can be rewritten as

$$m\lambda = 2dx\sqrt{n^2 - \sin^2\theta} \qquad (3.62)$$

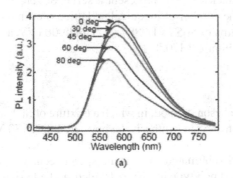

(a)

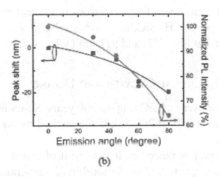

(b)

FIGURE 3.167   (a) Angular-dependent PL spectra, (b) emission angle vs. peak shift and normalized PL intensity (%). (From Ou, Y., Jokubavicius, V., Kamiyama, S., Liu, C., Berg, R. W., Linnarsson, M., Yakimova, R., Syväjärvi, M. and Ou, H., *Opt. Mater. Express*, 1, 1439, 2011. With permission.)

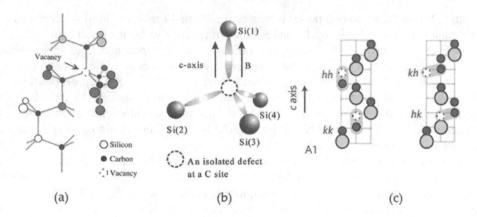

(a)   (b)   (c)

FIGURE 3.168 (a) 12 NNN Si atoms of a silicon vacancy. (From Mizuochi, N., Yamasaki, S., Takizawa, H., Morishita, N., Ohshima, T., Itoh, H., Umeda, T. and Isoya, J., *Phys. Rev. B*, 72, 235208, 2005. With permission.); (b) An isolated defect at a C site with the magnetic field parallel to the *c*-axis. (From Son, N. T., Hai, P. N. and Janzén, E., *Phys. Rev. B*, 63, 201201, 2001. With permission.); (c) The proposed atomic structure of the defects $C_{Si}V_C$. (From Castelletto, S., Johnson, B. C., Ivády, V., Stavrias, N., Umeda, T., Gali, A. and Ohshima, T., *Nat. Mater.*, 13, 151, 2014. With permission.)

The N-B doped SiC film grown on the 6H–SiC substrate has large refractive index difference at the SiC–air interface. Also due to its smooth surface, it acts as an FP etalon and thus optical interference occurs. The DAP emission in SiC is broad as shown in the PL spectra. From Equation (3.62), it is clear that the interference wavelength decreases as θ increases. The PL peak shifts toward shorter wavelengths as the emission angle increases and this is consistent with the observed results.

3.9.3.3 Other Luminescence Centers

SiC substrates, which are irradiated with high-energy (>MeV) electrons and annealed at various temperatures, are robust and cheap light sources emitting single photons. There are different kinds of isolate single defects, such as Si vacancy, C vacancy, $C_{Si}V_C$, N_CV_{Si} and so on.[37–40] Figure 3.168 shows the atomic structure of defects. The major challenge is to identify and form isolated defects due to complex defects existed during growth process.[39]

The PL of Si vacancy and $C_{Si}V_C$ has been detailed researched. Photoluminescence spectra when excited using either 660 or 532 nm are shown in Figure 3.169. Typical maps of photoluminescence raster scans using RT confocal microscopy are obtained by exciting either with 532- or 660-nm light.

The photoluminescence (PL) spectra of Si vacancies in SiC represent a series of zero-phonon lines the number of which agrees in general with the number of inequivalent lattice sites in the SiC polytype. In 6H–SiC, these lines arise at wavelengths of 865 (V1), 887 (V2) and 906 (V3) and for 4H–SiC at 862 (V1) and 917 nm (V2), as shown in Figure 3.170.[40]

3.9.4 CRYSTAL GROWTH AND DOPING

The discovery of SiC is up to 200 years. So-called Acheson method, in which a mixture of SiO_2 and C is heated to about 2000°C, is the method proposed in 1855 to synthesis the powdered SiC.[41] 70 Years later, in 1955, the Lely method is designed to grow the SiC crystal. In a specially designed crucible, SiC power is placed near side wall of crucible and sublimated to nucleate spontaneously at inner porous graphite.[42] The relationship of temperature and polytype has been detailed studied using Lely method. Modified Lely process has been carried out by placing the small size slice of SiC obtained by Lely method on lid of crucible as the seed.[43–45] Now the modified Lely process also called physical vapor transport (PVT) method is most widely used method for commercial SiC substrate.

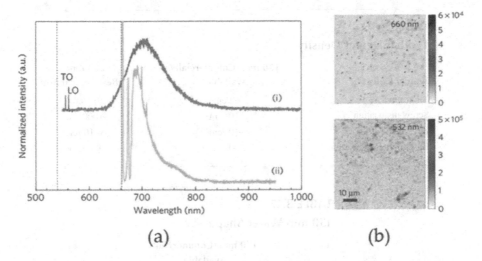

FIGURE 3.169 Normalized room-temperature photoluminescence of a single-photon source in the irradiated sample excited with 532-nm (i) and 660-nm (ii) lasers (a); confocal maps of the low electron fluence sample excited with 660- and 532-nm lasers (b). The diffraction-limited spots all correspond to single-photon emitters. (From Castelletto, S., Johnson, B. C., Ivády, V., Stavrias, N., Umeda, T., Gali, A. and Ohshima, T., *Nat. Mater.*, 13, 151, 2014. With permission.)

High-quality 150-mm crystals and substrates that reduce device manufacturing costs have obtained since 2010.[3,46–48] The first 200-mm diameter SiC substrate is exhibited.[3] However, the transition from 4 to 6 inch wafers, coupled with increasing wafer demand, led to a wafer short supply. The commercial 200-mm SiC substrate is not available nowadays. Expect for supply chain, there are still challenges considered as an industrial development problem.

The first challenge is defect density. In comparison to silicon, there still exist large defects in SiC substrate. Especially with the enlargement of diameter, the radial and axial temperature gradients increase and result in crystal stresses and formation of defects. Typical defect density is listed in Table 3.16.[3,46–48] Defects cause reduction of carrier lifetime and increase of on-resistance and

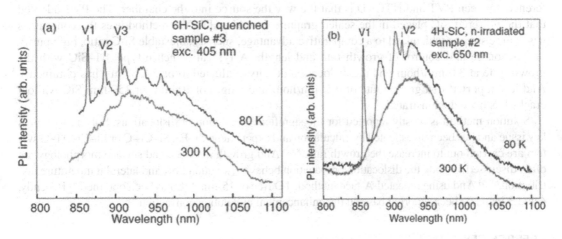

FIGURE 3.170 (a) PL spectra recorded in neutronirradiated 6H–SiC (sample no. 1) at 77K and room temperature with excitation at 405 nm. (b) PL spectra recorded in neutron-irradiated 4H–SiC (sample no. 2) at 77K and room temperature with excitation at 650 nm. (From Baranov, P. G., Bundakova, A. P., Soltamova, A. A., Orlinskii, S. B., Borovykh, I. V., Zondervan, R., Verberk, R. and Schmidt, J., *Phys. Rev. B*, 83, 125203, 2011. With permission.)

TABLE 3.16
Typical Defect Density

Typical Defect	150 mm (Commercially Available)	200 mm (Best Reported)
Micropipe	<2/cm^2	<0.1/cm^2
Screw dislocation	~10^2/cm^2	~1000/cm^2
Edge dislocation	~10^3/cm^2	~5×10^3/cm^2
Basal plane dislocation	10^2–10^3/cm^2	~900/cm^2

TABLE 3.17
150 mm Wafer Shape

	150 mm (Commercially Available)
TTV	<10 μm
Bow	<10 μm
Warp	<30 μm

leakage current, thereby affect the reliability and performance of SiC power devices. KOH etching, white beam X-ray topography and transmission electron microscopy are mostly common methods for dislocation observation.[49–51] Many new method, such as PL mapping with deep convolutional neural networks, photo-induced currents measurements, mirror projection electron microscopy and so on, have been proposed.[52–54]

Wafer shape control is another problem. Processing accommodated the hardness of SiC, mainly include cutting, grinding and polishing. Damage during processing must be properly managed to main a suitable wafer shape. Excellent shape has been demonstrated at 150 mm due to precise control of wafer processing, which is shown in Table 3.17.[3,47,48]

Beyond PVT method, HTCVD and solution growth are two promising candidates.[55–64] The difference between PVT and HTCVD is that the way the source into the chamber. The PVT adopted that the powered SiC placed in the sealed graphite crucible. HTCVD method uses the continuous gas source supply, which lead to a comparative advantage, such as adjustable Si/C ratio, high purity of gas source and control of growth rate and length. A typical 4-inch n-type 4H–SiC with the growth rate at 3.0 mm/hour and the dislocation density evaluated to be 3020 cm^{-1} has obtained.[56] And it is reported that growth rate of 0.5 mm/hour and length of 10 mm for 150 mm SiC without cracks has been demonstrated.[57]

Solution method is mostly reported for the significant reduction of dislocations. Solvent system is key issue and has been investigated by calculation and experiment. Si–Fe, Si–Cr–C or Si–Cr–Al–C system are carried out to increase the growth rate.[58–60] And growth interface and surface morphology are often discussed to study the dislocation propagation behavior, solution flow and lateral temperature distribution.[61–63] And using repeated-A-face method, TD free of 15-mm SiC has been obtained.[64] Recently, 4 and 3 inch has been reported base on the enhancement of solution convection.[65–67]

REFERENCES

1. Gupta, A., Semenas, E., Emorhokpor, E., Chen, J., Zwieback, I., Souzis, A. E. and Anderson, T., *Mater. Sci. Forum*, 527, 43, 2006.
2. Nakabayashi, M., Fujimoto, T., Katsuno, M., Ohtani, N., Tsuge, H., Yashiro, H., Aigo, T., Hoshino, T., Hirano, H. and Tatsumi, K., *Mater. Sci. Forum*, 600, 3, 2009.

3. Powell, A. R., Sumakeris, J. J., Khlebnikov, Y., Paisley, M. J., Leonard, R. T., Deyneka, E., Gangwal, S., Ambati, J., Tsevtkov, V., Seaman, J., McClure, A., Horton, C., Kramarenko, O., Sakhalkar, V., O'Loughlin, M., Burk, A. A., Guo, J. Q., Dudley, M. and Balkas, E., *Mater. Sci. Forum*, 858, 5, 2016.
4. Manning, I., Matsuda, Y., Chung, G., Sanchez, E., Dudley, M., Ailihumaer, T. and Raghothamachar, B., *Mater. Sci. Forum*, 1004, 37, 2020.
5. Itoh, A. and Matsunami, H., *Crit. Rev. Solid State*, 22, 111, 1997.
6. Bhatnagar, M. and Baliga, B. J., *IEEE Trans. Electron Devices*, 40, 645, 1993.
7. Casady, J. B. and Johnson, R. W., *Solid-State Electron.*, 39, 1409, 1996.
8. Choyke, W., *Mater. Res. Bull.*, 4, S141, 1969.
9. Marshall, R., Faust, J. and Ryan, C., *Silicon Carbide*, Columbia: University of South Carolina Press, pp. 261–283, 1973.
10. Ivanov, I. G., Hallin, C., Henry, A., Kordina, O. and Janzén, E., *J. Appl. Phys.*, 80, 3504, 1996.
11. Choyke, W. J. and Patrick, L., *Phys. Rev.*, 127, 1868, 1962.
12. Sun, J. W., Kamiyama, S., Jokubavicius, V., Peyre, H., Yakimova, R., Juillaguet, S. and Syväjärvi, M., *J. Phys. D Appl. Phys.*, 45, 235107, 2012.
13. Ikeda, M., Matsunami, H. and Tanaka, T., *Phys. Rev. B*, 22, 2842, 1980.
14. Edmond, J., Kong, H., Suvorov, A., Waltz, D. and Carter, J. C., *Phys. Status Solidi A*, 162, 481, 1997.
15. Syväjärvi, M., Müller, J., Sun, J. W., Grivickas, V., Ou, Y., Jokubavicius, V., Hens, P., Kaisr, M., Ariyawong, K., Gulbinas, K., Hens, P., Liljedahl, R., Linnarsson, M. K., Kamiyama, S., Wellmann, P., Spiecker, E. and Ou, H., *Phys. Scripta*, T148, 014002, 2012.
16. Ivanov, I. G., Magnusson, B. and Janzén, E., *Phys. Rev. B*, 67, 165211, 2003.
17. Yang, A., Murata, K., Miyazawa, T., Tawara, T. and Tsuchida, H., *J. Phys. D Appl. Phys.*, 52, 10LT01, 2019.
18. Kamiyama, S., Iwaya, M., Takeuchi, T., Akasaki, I., Syväjärvi, M. and Yakimova, R., *J. Semicond.*, 32, 013004, 2011.
19. Ou, H., Ou, Y., Argyraki, A., Schimmel, S., Kaiser, M., Wellmann, P., Linnarsson, M. K., Jokubavicius, V., Sun, J., Liljedahl, R. and Syväjärvi, M., *Eur. Phys. J. B*, 87, 58, 2014.
20. Jokubavicius, V., Kaiser, M., Hens, P., Wellmann, P. J., Liljedahl, R., Yakimova, R. and Syväjärvi, M., *Mater. Sci. Forum*, 740–742, 19, 2013.
21. Kaiser, M., Hupfer, T., Jokubavicius, V., Schimmel, S., Syväjärvi, M., Ou, Y. Y., Ou, H. Y., Linnarsson, M. K. and Wellmann, P. J., *Mater. Sci. Forum*, 740–742, 39, 2013.
22. Ci, S., Peng, Y., Xie, X., Hu, X., Duan, P., Yu, J., Xu, X., Chen, X., Yang, X. and Wang, R., *Semicond. Sci. Technol.*, 35, 055009, 2020.
23. Wei, R., Chen, X., Wang, L., Song, M., Yang, K., Hu, X., Peng, Y. and Xu, X., *Int. J. Electrochem. Sci.*, 8, 7099, 2013.
24. Ou, Y., Jokubavicius, V., Kamiyama, S., Liu, C., Berg, R. W., Linnarsson, M., Yakimova, R., Syväjärvi, M. and Ou, H., *Opt. Mater. Express*, 1, 1439, 2011.
25. Ohishi, M., *Jpn. J. Appl. Phys.*, 25, 1546, 1986.
26. Bäume, P., Kubacki, F. and Gutowski, J., *J. Cryst. Growth*, 138, 266, 1994.
27. Kamiyama, S., Maeda, T., Nakamura, Y., Iwaya, M., Amano, H., Akasaki, I., Kinoshita, H., Furusho, T., Yoshimoto, M., Kimoto, T., Suda, J., Henry, A., Ivanov, I. G., Bergman, J. P., Monemar, B., Onuma, T. and Chichibu, S. F., *J. Appl. Phys.*, 99, 093108, 2006.
28. Rhee, S. J., Kim, S., Reuter, E. E., Bishop, S. G. and Molnar, R. J., *Appl. Phys. Lett.*, 73, 2636, 1998.
29. Wang, L. and Giles, N. C., *J. Appl. Phys.*, 94, 973, 2003.
30. Nakashima, S. and Yamaguchi, Y., *J. Appl. Phys.*, 50, 4958, 1979.
31. Ivanov, I. G., Egilsson, T., Henry, A. and Janzén, E., *Mater. Sci. Eng. B*, 61–62, 265, 1999.
32. Schmidt, T., Lischka, K. and Zulehner, W., *Phys. Rev. B*, 45, 8989, 1992.
33. Zielinski, M., Balloud, C., Juillaguet, S., Boyer, B., Soulière, V. and Camassel, J., *Mater. Sci. Forum*, 483–485, 449, 2005.
34. Chang, Y., Chu, J.-H., Tang, W.-G. and Tang, D.-Y., *Acta Phys. Sin.*, 46, 959, 1997.
35. Sotta, D., Hadji, E., Magnea, N., Delamadeleine, E., Besson, P., Renard, P. and Moriceau, H., *J. Appl. Phys.*, 92, 2207, 2002.
36. Pan, S. S., Ye, C., Teng, X. M. and Li, G. H., *J. Phys. D Appl. Phys.*, 40, 4771, 2007.
37. Mizuochi, N., Yamasaki, S., Takizawa, H., Morishita, N., Ohshima, T., Itoh, H., Umeda, T. and Isoya, J., *Phys. Rev. B*, 72, 235208, 2005.
38. Son, N. T., Hai, P. N. and Janzén, E., *Phys. Rev. B*, 63, 201201, 2001.
39. Castelletto, S., Johnson, B. C., Ivády, V., Stavrias, N., Umeda, T., Gali, A. and Ohshima, T., *Nat. Mater.*, 13, 151, 2014.

40. Baranov, P. G., Bundakova, A. P., Soltamova, A. A., Orlinskii, S. B., Borovykh, I. V., Zondervan, R., Verberk, R. and Schmidt, J., *Phys. Rev. B*, 83, 125203, 2011.
41. Zhu, W. Z. and Yan, M., *Scr. Mater.*, 39, 1675, 1998.
42. Lely, J. A., *Ber. Deut. Keram. Ges.*, 32, 229, 1955.
43. Tairov, Y. M. and Tsvetkov, V. F., *J. Cryst. Growth*, 43, 209, 1978.
44. Rengarajan, V., Brouhard, B. K., Nolan, M. C. and Zwieback, I., Axial gradient transport growth process and apparatus utilizing resistive heating, United States, II-VI Incorporated (Saxonburg, PA, US), 9228274, 2016. https://www.freepatentsonline.com/9228274.html
45. Zwieback, I., Anderson, T. E., Souzis, A. E., Ruland, G. E., Gupta, A. K., Rengarajan, V., Wu, P. and Xu, X., Large diameter, high quality SiC single crystals, method and apparatus, United States, II-VI Incorporated (Saxonburg, PA, US), 8741413, 2014. https://www.freepatentsonline.com/8741413.html
46. Kondo, H., Takaba, H., Yamada, M., Urakami, Y., Okamoto, T., Kobayashi, M., Masuda, T., Gunjishima, I., Shigeto, K., Ooya, N., Sugiyama, N., Matsuse, A., Kozawa, T., Sato, T., Hirose, F., Yamauchi, S. and Onda, S., *Mater. Sci. Forum*, 778–780, 17, 2014.
47. Quast, J., Hansen, D., Loboda, M., Manning, I., Moeggenborg, K., Mueller, S., Parfeniuk, C., Sanchez, E. and Whiteley, C., *Mater. Sci. Forum*, 821–823, 56, 2015.
48. Manning, I., Zhang, J., Thomas, B., Sanchez, E., Hansen, D., Adams, D., Chung, G. Y., Moeggenborg, K., Parfeniuk, C., Quast, J., Torres, V. and Whiteley, C., *Mater. Sci. Forum*, 858, 11, 2016.
49. Ohno, T., Yamaguchi, H., Kuroda, S., Kojima, K., Suzuki, T. and Arai, K., *J. Cryst. Growth*, 260, 209, 2004.
50. Ellison, A., Sörman, E., Sundqvist, B., Magnusson, B., Yang, Y., Guo, J. Q., Goue, O., Raghothamachar, B. and Dudley, M., *Mater. Sci. Forum*, 858, 376, 2016.
51. Sakwe, S. A., Müller, R. and Wellmann, P. J., *J. Cryst. Growth*, 289, 520, 2006.
52. Leonard, R., Conrad, M., Van Brunt, E., Giles, J., Hutchins, E. and Balkas, E., *Mater. Sci. Forum*, 1004, 321, 2020.
53. Privitera, S., Litrico, G., Camarda, M., Piluso, N. and La Via, F., *Mater. Sci. Forum*, 858, 380, 2016.
54. Isshiki, T. and Hasegawa, M., *Mater. Sci. Forum*, 858, 371, 2016.
55. Tokuda, Y., Kojima, J., Hara, K., Tsuchida, H. and Onda, S., *Mater. Sci. Forum*, 778–780, 51, 2014.
56. Tokuda, Y., Hoshino, N., Kuno, H., Uehigashi, H., Okamoto, T., Kanda, T., Ohya, N., Kamata, I. and Tsuchida, H., *Mater. Sci. Forum*, 1004, 5, 2020.
57. Okamoto, T., Kanda, T., Tokuda, Y., Ohya, N., Betsuyaku, K., Hoshino, N., Kamata, I. and Tsuchida, H., *Mater. Sci. Forum*, 1004, 14, 2020.
58. Harada, S., Yamamoto, Y., Xiao, S., Tagawa, M. and Ujihara, T., *Mater. Sci. Forum*, 778–780, 67, 2014.
59. Kado, M., Daikoku, H., Sakamoto, H., Suzuki, H., Bessho, T., Yashiro, N., Kusunoki, K., Okada, N., Moriguchi, K. and Kamei, K., *Mater. Sci. Forum*, 740–742, 73, 2013.
60. Yoshikawa, T., Kawanishi, S., Morita, K. and Tanaka, T., *Mater. Sci. Forum*, 740–742, 31, 2013.
61. Kawanishi, S., Yoshikawa, T. and Morita, K., *Mater. Sci. Forum*, 740–742, 35, 2013.
62. Alexander, Seki, K., Kozawa, S., Yamamoto, Y., Ujihara, T. and Takeda, Y., *Mater. Sci. Forum*, 679–680, 24, 2011.
63. Komatsu, N., Mitani, T., Takahashi, T., Okamura, M., Kato, T. and Okumura, H., *Mater. Sci. Forum*, 740–742, 23, 2013.
64. Danno, K., Yamaguchi, S., Kimoto, H., Sato, K. and Bessho, T., *Mater. Sci. Forum*, 858, 19, 2016.
65. Kusunoki, K., Yashiro, N., Okada, N., Moriguchi, K., Kamei, K., Kado, M., Daikoku, H., Sakamoto, H., Suzuki, H. and Bessho, T., *Mater. Sci. Forum*, 740–742, 65, 2013.
66. Kusunoki, K., Kamei, K., Okada, N., Moriguchi, K., Kaido, H., Daikoku, H., Kado, M., Danno, K., Sakamoto, H., Bessho, T. and Ujihara, T., *Mater. Sci. Forum*, 778–780, 79, 2014.
67. Kusunoki, K., Kishida, Y. and Seki, K., *Mater. Sci. Forum*, 963, 85, 2019.

3.10 PEROVSKITE MATERIALS AND EMISSION PROPERTIES

Wenna Du and Xinfeng Liu

3.10.1 INTRODUCTION

Perovskite is named after the Russian mineralogist L. A. Perovski, originally referring to a calcium titanium oxide mineral consisting of calcium titanate ($CaTiO_3$). Thereafter, the terminology of perovskite was extended to describe the class of compounds that have a crystal structure similar to that of $CaTiO_3$. Nowadays, perovskite usually refers to metal halide perovskite, which has the generic chemical formula ABX_3 or A_2MX_4, where A means the monovalent inorganic or organic cation, B represents a divalent metallic cation and X is a halide anion. By using either a mixture/substitution of halide anion or the monovalent cation, the emission region of perovskites can be extended from ultraviolet (UV) to near-infrared (IR) region.

Since Miyasaka et al.[1] first successfully used two organolead halide perovskite nanocrystals (NCs) as a visible photosensitizer of photoelectrochemical cells in 2009, the research on the preparation and application of perovskite materials has flourished. In the process of the rapid development, metal halide perovskites have exhibited many excellent properties, such as large optical absorption coefficients,[2] high carrier mobilities,[3,4] long carrier lifetimes and diffusion lengths,[3,4] and low defect state density as well as low-cost facile processing method.[5,6] As direct bandgap semiconductors, perovskites also show wideband tunable emission color with high photoluminescence quantum yield (PLQY). Owing to these outstanding properties, metal halide perovskite materials have shown great potential for applications in solar cells,[7,8] light-emitting diode (LED) devices,[9–12] lasers,[13–16] photodetectors[17–19] etc.

3.10.2 DIFFERENT GROWTH TECHNIQUES

The main used perovskite growth methods so far can be basically divided into the conventional solution growth method,[20–23] the "new solution method"—spin coating,[12,24–26] and the chemical vapor deposition (CVD).[6,27–35]

The conventional solution growth method: With the advantages of operation simplicity, versatility and low cost, the conventional solution method plays a critical role in the preparation of metal halide perovskites. In addition, it is more facile to obtain the high-quality single crystals compared to other methods. According to the difference of basic principles, the conventional solution growth approach can be mainly classified into the solution temperature lowering (STL), inverse temperature crystallization (ITC) and the anti-solvent vapor-assisted crystallization (AVC) methods.

STL growth method obtains the driving force, that is, the solution super-saturation, through gradually lowering the solution temperature. It is first to prepare the saturated solution at a relatively high temperature and subsequently, cooling the solution gradually down to a target temperature within a certain time for the precipitation of small seed crystals, which, in turn, grow large-sized bulk single crystals. Based on the different positions of seed during the cooling process, the STL method includes the bottom-seeded solution growth and the top-seeded solution growth methods (Figure 3.171(a) and (b)).[20,23] STL methods provide a feasible and practical technique for the synthesis of bulk perovskite single crystals. Nevertheless, such methods are time-consuming.

ITC growth method is a method utilizing the abnormal phenomenon that the solubility of $APbX_3$ perovskites is inversely proportional to temperature in certain solvents. It is an approach of rapid crystal growth, and its apparatus schematic is demonstrated in Figure 3.171(c).[21] For the ITC method, it is crucial to choose the appropriate solvent for the preparation of high-quality single crystals. For the growth of perovskite crystals with different halogens, the used solvents and the growth temperatures are different. In the ITC method, the crystallization is actuated by the solubility that decreases with increasing the solution temperature in a specific selected organic solvent. The whole crystal growth process of ITC method will take a few hours, which is much faster than the STL method; therefore, the ITC method is mainly utilized to rapidly grow larger sized perovskite single crystals.

AVC growth method is based on the different solubility of perovskite compounds in different solvents. For the AVC method (Figure 3.171(d)),[22] dimethylsulfoxide (DMSO), dimethyl formamide (DMF) and gamma-butyrolactone (GBL) are commonly regarded as effective solvents, whereas toluene, benzene, chlorobenzene, xylene, chloroform, 2-propanol and acetonitrile usually acted as anti-solvents for the growth of metal halides. Therefore, it is pivotal to choose the right solvent/anti-solvent group for the perovskite crystal growth, according to the specific material component. Based on the driven force of solubility contrast, it shows a relatively weak dependence on temperature during the crystal growth of the AVC method. However, large-sized single crystals

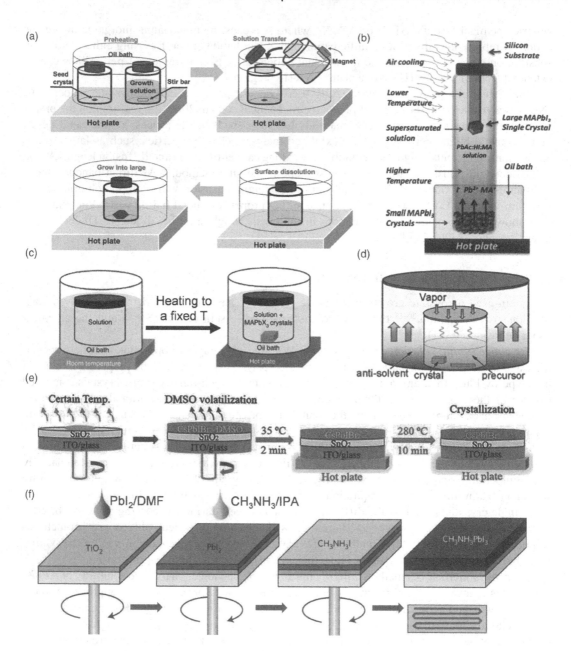

FIGURE 3.171 (a and b) Schematic of the two solution temperature lowering methods to grow large MAPbI$_3$(Cl) bulk single crystals. (From Lian, Z., Yan, Q., Gao, T., Ding, J., Lv, Q., Ning, C., Li, Q. and Sun, J.-l., *J. Am. Chem. Soc.*, 138, 9409, 2016; Dong, Q., Fang, Y., Shao, Y., Mulligan, P., Qiu, J., Cao, L. and Huang, J., *Science*, 347, 967, 2015. With permission.) (c) Schematic of the inverse temperature crystallization method, whers the crystallization vial is immersed into a heating bath and crystal begins to grow. (From Saidaminov, M. I., Abdelhady, A. L., Murali, B., Alarousu, E., Burlakov, V. M., Peng, W., Dursun, I., Wang, L., He, Y. and Maculan, G., *Nat. Commun.*, 6, 7586, 2015. With permission.) (d) Schematic of the anti-solvent vapor-assisted crystallization method. (From Hang, H., Liu, X., Dong, J., Yu, H., Zhou, C., Zhang, B., Xu, Y. and Jie, W., *Cryst. Growth Des.*, 17, 6426, 2017. With permission.) (e) One-step and (f) two-step spin coating method. (From Guo, Y., Yin, X., Liu, J. and Que, W., *J. Mater. Chem. A*, 7, 19008, 2019; Im, J.-H., Jang, I.-H., Pellet, N., Grätzel, M. and Park, N.-G., *Nat. Nanotechnol.*, 9, 927, 2014. With permission.)

are generally difficult to obtain by the AVC method compared to the STL and ITC methods. The as-grown perovskite single crystals achieved by the AVC method are usually small and numerous.

Spin coating method: This method is a low-cost film production method and is mainly implemented to prepare high-performance perovskite polycrystalline or microcrystalline films applied to solar cells. Once spinning, the evaporation and convective self-assembly processes will rapidly induce the formation of well-crystallized perovskite materials because of the strong ionic interaction between halogen anions and metal ions. Summarily, the spin coating of perovskite crystals includes three processes: the supersaturated solution forms through the rapid evaporation of the solvent once the precursor solution is dropped on the substrate, nucleation occurs when Gibbs free energy overcomes the energy barrier, and more crystal nuclei comes into being leading to the growth toward a large crystal. The centrifugal force caused by spreading diffusion flow was applied to the slowly evaporating solvent, but it is hard to produce a uniform large-area perovskite layer by the simple spin coating. Hence, numerous optimized methods have been developed to improve the quality of perovskite films, including one-step, two-step, multistep spin coating and other methods of spin coating (Figure 3.171(e) and (f)).[24,25]

CVD method: CVD is an approach widely utilized in both industry and research and is one of the numerous methods implemented for the growth of perovskites. The perovskite materials obtained by the CVD process have fewer defects and higher crystal qualities compared to perovskite films grown by solution method. With the progress of perovskite materials accompanied by the development of CVD technology, different morphologies, such as nanoplatelets (NPs), microplatelets (MPLs), films, microwires and nanowires (NWs), microspheres and pyramids, could be obtained by modulating the growth conditions (e.g., growth temperature, pressure, source ratio, time, substrate etc.). For pure inorganic perovskite growth, it is commonly used a CVD equipment with a single source and heating zone, as shown in Figure 3.172(a).[27] Generally, the CsX and PbX_2 (X=Cl, Br, I) mixture powder in the heated zone is utilized to grow the $CsPbX_3$ (X=Cl, Br, I) single crystal, including ultrathin NPs, microspheres, oriented NWs,[29] and even single-crystal thin films.[28] According to the different melting temperatures of precursors as well as the target morphologies of the products, the growth of all-inorganic perovskite may also require dual sources and single heated zone in the CVD systems (Figure 3.172(b)).[6,30] A double-temperature zone tube furnace, depicted in Figure 3.172(c), was used as a reactor to synthesize all-inorganic $CsPbBr_3$ MPLs with high quality.

The CVD method has also been adopted in several studies for the preparation of uniform perovskite films with large areas, such as low vacuum one step process,[32] sequential deposition process,[33] hybrid CVD (HCVD),[31,34] hybrid physical CVD method[35] and growth under atmospheric conditions. Figure 3.172(d) illustrates the HCVD furnace and methylammonium iodide (MAI) deposition as an example. Additionally, the toxicity of lead in metal halide perovskites has blocked their developments and practical applications. It is more practical to explore the preparation and the associated methods of lead-free perovskites in the future.

3.10.3 CHARACTERISTICS OF TYPICAL PEROVSKITES

3.10.3.1 Crystal Structure

The typical crystal structure of perovskite is depicted in Figure 3.173(a) with the general chemical formula of AMX_3, where A is a small monovalent inorganic or organic cation (e.g., $CH_3NH_3^+$, abbreviated with MA, $HC(NH_2)_2^+$, abbreviated with FA, and Cs^+); B is a divalent metallic cation (usually Pb^{2+} or Sn^{2+}); and X is a halide anion such as $Cl^-/Br^-/I^-$. Here, in a typical perovskite lat tice consisting of a three-dimensional (3D) framework, A, M and X are, respectively, situated at the eight corners, the body center, and the six face centers of the unit cell. Among, M is surrounded by six X atoms to form an $[MX_6]^{4-}$ octahedral, which thereupon constructs a 3D octahedral network of corner-sharing MX_6 octahedral. A locates at the center of eight octahedrals in the 3D network. To maintain the stability of perovskite's crystalline structure, the relationship of the radii of A, M

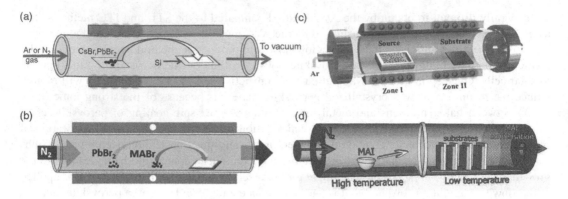

FIGURE 3.172 Different chemical vapor deposition (CVD) methods. (a) CVD growth with single source and one heated zone.[27] (b) Schematic of growing MAPbBr$_3$ by a dual source in a single heated zone CVD system.[30] (c) Schematic of growing CsPbBr$_3$ microplatelets with single source and dual heated zone. (From Li, Y., Shi, Z., Lei, L., Zhang, F., Ma, Z., Wu, D., Xu, T., Tian, Y., Zhang, Y., Du, G., Shan, C. and Li, X., *Chem. Mater.*, 30, 6744, 2018. With permission.) (d) Diagram of the hybrid CVD system with methylammonium iodide (MAI) deposited onto metal halide seeded substrates. (From Leyden, M. R., Ono, L. K., Raga, S. R., Kato, Y., Wang, S. and Qi, Y., *J. Mater. Chem. A*, 2, 18742, 2014. With permission.)

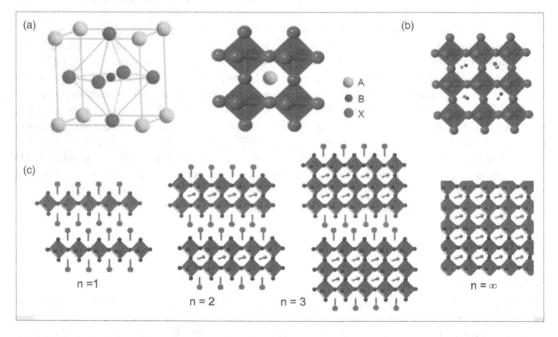

FIGURE 3.173 Schematic of crystal structure. (a) The unit cell (left panel) and crystalline structure (right panel) of representative perovskite. (b) Three-dimensional (3D) perovskite crystalline structure with a molecular formula of AMX$_3$. (From Zhang, Q., Su, R., Du, W., Liu, X., Zhao, L., Ha, S. T. and Xiong, Q., *Small Methods*, 1, 1700163, 2017. With permission.) (c) Two-dimensional (2D) layered perovskite crystalline structure with chemical formula of (LA)$_2$(A)$_{n-1}$M$_n$X$_{3n+1}$, where n corresponds to number of octahedral layer in the inorganic layers and $n=\infty$ corresponds to 3D perovskite AMX$_3$. (From Cohen, B.-E., Wierzbowska, M. and Etgar, L., *Sustain. Energy Fuels*, 1, 1935, 2017. With permission.)

and X atoms should meet tolerance formula. The empirical Goldschmidt tolerance factor can be expressed as[36,37]

$$t = \frac{r_A + r_X}{\sqrt{2(r_A + r_B)}} \tag{3.63}$$

where r_A, r_B, and r_X are the effective radii of A, M and X atoms, respectively. Empirically, when $0.813 < t < 1.107$, it is favorable for the 3D crystalline structure (Figure 3.173(b)). In particular, when $t=1$, the perovskite tends to form an ideal cubic lattice, whereas when $t<1$, the symmetry of the lattice decreases. Meanwhile, when $t>1$ or $t<0.8$, nonperovskite structures occur; these structures adapt different types of connectivity among the MX_6 octahedra. To maintain the structure of the perovskite, the radius of A should be large enough. A could also be an organic cation with several functional groups, the length of which mainly depends on the length of the carbon chain. When the organic cation has a short carbon chain, such as in MA^+, it is large enough for the spacing between the $[MX_6]^{4-}$ octahedrals to place the organic cation. However, when the radius of A is too large, meaning that the organic cation has a long carbon chain, e.g., $C_5H_6CH_3NH_{2+}$ (BA), 3D perovskite's crystalline structure could not maintain stably leading that the $[MX_6]^{4-}$ octahedral chains separate to provide enough spacing placing the organic cation. Under the circumstance, the organic cation is sandwiched between two corner-sharing $[MX_6]^{4-}$ octahedral layers. 3D perovskite structure changes to a two-dimensional (2D) layered structure.

The 2D layer perovskite can be written with a chemical formula of $(LA)_2(A)_{n-1}M_nX_{3n+1}$ (Figure 3.173(c)),[38] where LA^+ represents a long chain cation and n is an integer. The 2D structures can be seen as 2D slabs sliced from the corresponding 3D framework perovskite along the (1 0 0) crystal plane. The corner-sharing inorganic $[MX_6]^{4-}$ octahedra interspaced by the large LA^+ cations forms the extended 2D unit layer. Each inorganic layer of a 2D layered perovskite has the same thickness, which is indicated by the n value. If n is infinite, the structure turns into a 3D perovskite of AMX_3. As n decreases to single digit, especially 1, the structure exhibits intrinsic quantum and dielectric confinement effects due to large dielectric contrast between the inorganic $[MX_6]^{4-}$ octahedral layers and organic cation; moreover, electrons and holes are strictly confined in the octahedral layers and strongly bound together via the Coulombic interaction. Therefore, these 2D perovskite can be regarded as overall assemblies of different isolated quantum wells (QWs) of 3D perovskites layer.[39]

3.10.3.2 Phase Transition

The framework structure of perovskite leads to inherent softness and dynamic disorder. Its intrinsic softness is shown by their Young's moduli being approximately ten times lower than those of Si or GaAs.[40] Its lattice structure changes with temperature because of the vibrational orientation of organic cations. Usually, perovskites have three main crystal phases, namely cubic, orthorhombic and tetragonal, as exhibited in Figure 3.174.[41]

For organic–inorganic hybrid perovskite methylammonium lead bromide $(MAPbBr_3)$[42] near 237K, the high-temperature cubic $Pm3m$ phases transform to a body-centered tetragonal $I4/mcm$ phase. Upon further cooling, all the tetragonal phases transform into a low-temperature orthorhombic $Pnma$ phase around 144.5K, the $[MX6]^{4-}$ octahedron experiences Jahn-Teller distortion, and the cation motion is restricted. The orthorhombic to tetragonal and tetragonal to cubic phase-transition temperatures for $MAPbI_3$ are 160±5 and 370±5K, respectively.[42,43] All-inorganic perovskite, by comparison, has a higher phase-transition temperature (370 and 420K for $CsPbBr_3$),[44] thereby resulting in higher thermal stability. Similar phase transitions also exist in Cl^-- and I^--based perovskites. These phase transitions are accompanied by the increase of dynamic disorder, especially the orientational motions of organic cations in organic-inorganic hybrid perovskite. See Table 3.18 for more phase-transition temperatures and structural parameters.

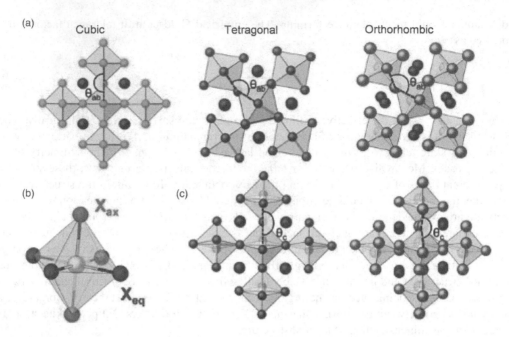

FIGURE 3.174 Typical cubic, tetragonal and orthorhombic crystal structures of AMX_3 perovskite with octahedral MX_6 geometry (top view) (a) and the corresponding side view pattern (c), where A=violet, M=gray, X=green, brown and cyan representatively. (b) Schematice of octahedral MX_6 with equatorial plane halide (X_{eq}, blue) and axial plane halide (X_{ax}, red). The phase of the perovskite adopted depends on the temperature of the lattice. (From Liu, W., Liu, Y., Wang, J., Wu, C., Liu, C., Xiao, L., Chen, Z., Wang, S. and Gong, Q., *Crystals*, 8, 216, 2018. With permission.)

TABLE 3.18

Crystal Phase and Groups of Common Perovskites

Perovskite	Halide	Phase	Group	Temperature	Reference
MAPbX$_3$	Br, I	Cubic	*Pm3m*	>237K for Br, >372K for I, >178K for Cl	[42,43]
		Tetragonal	*I4/mcm* or *I4/m* or *P4/mmm*	144.5–237K for Br, 162–372K for I, 172.9–178K for Cl	[42,43]
		Orthorhombic	*Pnma*	<144.5K for Br, <162K for I	[42,45]
	Cl	Orthorhombic	*P2221*	<172.9K	[42]
MASnX$_3$	I	Pseudocubic	*P4mm*	200–293K	[46]
		Tetragonal	*I4cm*	<200K	
FAPbX$_3$	I	Trigonal	*P3m1*	RT	[47]
		Trigonal	*P3*	150K	
FASnX$_3$	I	Orthorhombic	*Amm2*	340K	[46]
		Orthorhombic	*Imm2*	180K	
CsPbX$_3$	Br, I	Cubic	*Pm3m*	>420K for Br, >588K for I	[44]
		Orthorhombic	*Pnma*	370–420K for Br	[44]
CsSnX$_3$	Br, I	Orthorhombic	*Pnma*	443K	[48]
	Cl	Cubic	*Pm3m*	443K	

RT denotes room temperature (~293K).

3.10.3.3   Band Structures

3.10.3.3.1   Electronic Structure and Bandgap Engineering

The band-edge electronic states of lead halide perovskite are mainly constructed by $[MX_6]^{4-}$ octahedrals. For example, in the electronic band structure of MAPbI$_3$ (Figure 3.175(a)),[49] which is obtained by first principles density functional theory methods, the electronic states at the top of the valence bands are mainly determined by the occupied p-orbitals of I atoms and the s-orbitals of Pb atoms. Those at the bottom of the conduction band (CB) are of the unoccupied Pb atoms p-character and form the 1.5-eV bandgap of MAPbI$_3$. Meanwhile, the electronic states related to the cation define bonding and antibonding bands, which are arranged symmetrically around the bandgap and located several electron volts away from the band edges. As expected, the analysis of the electronic charge density indicates that the cation is ionized individually, and the PbI$_3$ octahedral network is negatively

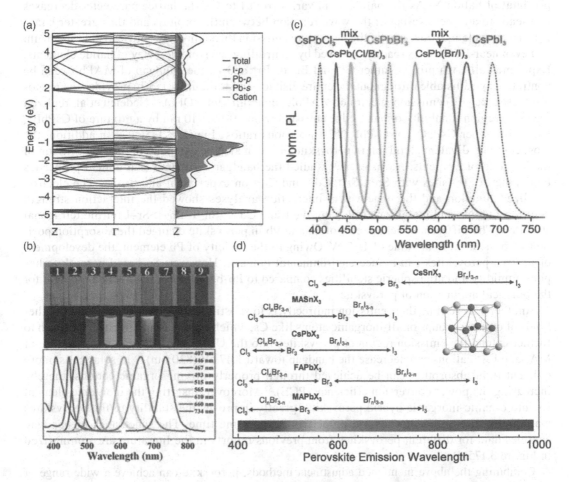

FIGURE 3.175 (a) Electronic band structure and partial density of the states of MAPbI$_3$ in the low-temperature orthorhombic phase by density functional theory. (From Filip, M., Eperon, G., Snaith, H. and Giustino, F., *Nat. Commun.*, 2014, 5, 5757. With permission.) (b) Optical images and photoluminescence (PL) emission spectra of MAPbX$_3$ quantum dots (QDs). (From Zhang, F., Zhong, H., Chen, C., Wu, X.-G., Hu, X., Huang, H., Han, J., Zou, D. and Dong, Y., *ACS Nano*, 9, 4533, 2015. With permission.) (c) The PL spectra of mixing CsPbBr$_3$ nanocrystals (NCs) with either CsPbI$_3$ or CsPbCl$_3$NCs in various ratios. (From Nedelcu, G., Protesescu, L., Yakunin, S., Bodnarchuk, M. I., Grotevent, M. J. and Kovalenko, M. V., *Nano Lett.*, 15, 5635, 2015. With permission.) (d) Diagrams of bandgap energies tuned by halide X for MAPbX$_3$, FAPbX$_3$, CsPbX$_3$ and CsSnX$_3$. (From Zhang, Q., Su, R., Du, W., Liu, X., Zhao, L., Ha, S. T. and Xiong, Q., *Small Methods*, 1, 1700163, 2017. With permission.)

charged. These findings manifest that the electronic structure of these metal-organic perovskites can be described by removing the cation from the structural model and using the positive background of MA^+ to compensate the negative charge of the PbI_3 network. An organic unit only affects the edge band structure through changing the bonding strength and angle between the halogen atoms and the Pb in the $[PbX_6]^{4-}$ octahedron. Advanced theoretical calculations have revealed that the density of states derived from the Pb p-orbital in the lower CB of the halide perovskites is much higher than that of GaAs, which lead to the much higher optical absorption of $MAPbI_3$.[50,51] Meanwhile, the strong coupling of s-p antibonding in $MAPbI_3$ perovskite results in small effective masses for both electrons and holes.[52] These results of theoretical calculations account for the $MAPbI_3$ perovskite's superior properties in comparison with other popular photovoltaic materials.

As mentioned before, the bandedge electronic states of lead halide perovskite are mainly derived from $[MX_6]^{4-}$ octahedrals. Therefore, the bandgap can generally be tuned by the p-orbital of halide X. As the halide atom varies from I to Cl, the lattice parameter decreases that leads to a larger overlap of the wave function between the orbitals and thus greater bandgap. The wavelength coverage of the perovskite emission throughout the entire visible spectrum and even near-IR bands is readily achieved by controlling the stoichiometry of halide elements. Experimentally, via mixing either Cl^- and Br^- or Br^- and I^-, the bandgap of $MAPbX_3$ can be continuously adjustable throughout the entire visible band.[13] Figure 3.175(b) provides the photoluminescence (PL) emission spectra of $MAPbX_3$ quantum dots (QDs).[53] Nedelcu et al. realized the large region tunability of band-edge emission from 400 to 710 nm by a mixture of $CsPbBr_3$ NCs with either $CsPbCl_3$ or $CsPbI_3$ NCs in various ratios (Figure 3.175(c)).[54] In addition, the most common divalent metal ion in perovskite Pb^{2+} can also be extended to the Ge or Sn ion of the V group of elements. Navas et al.[55] studied the bandgap structures and optical properties by doping in Pb^{2+} sites with Sn^{2+}, Sr^{2+}, Cd^{2+} and Ca^{2+} on experiment and theory. The electron localized function and the noncovalent interaction analyses showed the interaction strength between I atoms and metal ions following the trend Cd–I>Sn–I≈Pb–I>Sr–I for the tetragonal structure. The different stoichiometry of Sn and Pb in perovskite changed the absorption position with the minimum value of 1.15 eV. Owing to the toxicity of Pb element, the development of Pb-free perovskites can reduce environmental damage. However, Sn-based perovskite has poor humidity and atmospheric stabilities compared to Pb-based perovskite, it is a problem for the practical application of perovskite.

Another way to tune their emission indirectly is to substitute the organic cations with other kinds of organic groups or all-inorganic atoms like Cs, which serves as an effective approach to further extend the emission region of perovskites into the UV or near-IR region. A mixture of MA^+ and FA^+ cations can decrease the bandgap toward 1.47 eV (820 nm).[56] In solar cells, more efficient light absorption can be achieved through properly mixing organic cations, thereby increasing the power conversion efficiency (PCE).[57] Moreover, inserting the inorganic element Cs into organic-inorganic hybrid perovskites greatly improves the stability of the perovskites maintaining the excellent emission properties at the same time. The schematics of bandgap energies data for different perovskites from previous reports in the literature are summarized in Figure 3.175(d).

Combining the above mentioned adjustment methods, perovskite can achieve a wide range of bandgap and luminescence wavelength, which can be changed continuously from 405 to 1080 nm. This feature makes perovskite potential for applications in optoelectronic devices of different wavelengths.

3.10.3.3.2 Carrier Dynamics

3.10.3.3.2.1 Charge Carrier Recombination Dynamics in Perovskites Two kinds of quasiparticles can be generated upon photo-excitation in metal halide perovskites, free electron-hole pairs and free excitons, respectively. Among, the latter is electrically neutral. These two kinds of quasiparticles exhibit very different dynamics. Given the assumption that the number of electron is equal to that of holes, charge carrier recombination dynamics in perovskites can be expressed by the following rate equation[58]:

$$-\frac{dn}{dt} = k_1 n + k_2 n^2 + k_3 n^3 \tag{3.64}$$

where k_1, k_2 and k_3 are the rate constants of monomolecular, bimolecular and Auger recombination, respectively, and n is the photogenerated charge carrier density. The first-order process of monomolecular recombination is a geminate type excitonic recombination or a trap-related recombination, whereas the second term of bimolecular recombination is a non-geminate recombination of free charge carriers depending on both electron and hole densities. Auger recombination is a many-body effect in which an electron and a hole recombine along with transferring the energy or momentum to another electron or hole through collision. The timescale of monomolecular recombination depending on the energy, density and distribution of trap states is in the order of 10^6 s^{-1} for single and mixed halide lead perovskites,[4,59] which is mainly governed by the trap-assisted processes. Meanwhile, because of the self-doping by Sn^{4+} ions, the monomolecular recombination rate value of ~10^9 s^{-1} in lead-free $MASnI_3$ perovskites is three orders of magnitude greater than that of $MAPbI_3$.[60] On the other hand, due to the activation energy and depth of traps are different in diverse crystalline phase of perovskites, trap-assisted recombination rate in an orthorhombic phase is lower than that in cubic and tetragonal phases.[59]

3.10.3.3.2.2 Bimolecular Recombination in Perovskites For low-dimensional metal halide perovskites, the leading photogenerated species are excitons. Therefore, the carrier dynamics can be expressed as:

$$-\frac{dn}{dt} = k_1 n + k_2 n^2 - G \tag{3.65}$$

where the first term (k_1) indicates exciton trapping and exciton radiative recombination, and the second term (k_2) represents biexciton Auger recombination. Because the trap density is limited, the excitonic recombination can compare with the charge carrier trapping at low injected carrier density. Hence, perovskite multi-QWs (MQWs) exhibit near-constant high PLQY (≈60%) at carrier density below 10^{16} cm^{-3}.[61] The reduction in the dimension will also largely increase the second-order process rate of non-radiative Auger recombination, which is harmful for light emission under the condition of high-density carrier injection. For instance, on the occasion of the injected carrier density larger than 10^{16} cm^{-3}, the PLQY of perovskite MQWs significantly decreases with increasing carrier density.[61] Under high carrier injected density, the greatly increased Auger recombination rate of low-dimensional metal halide perovskites is detrimental for laser devices. Further reduction of the crystal size toward their exciton Bohr radius, the Auger recombination rates will increase greatly. The Auger coefficients in all-inorganic perovskites have been found to be approximately 5×10^{-29} cm^6/second, comparable to that of the organic-inorganic hybrid perovskites.[62,63] Biexciton Auger recombination in $CsPbX_3$ QDs usually occurs on a timescale of tens to hundreds of picoseconds, which requires a relatively higher pump fluence threshold for amplified spontaneous emission (ASE). Since the lifetime of trion Auger recombination is estimated to be four times longer than that of biexciton, it will result in a lower ASE threshold (1.2 $\mu J/cm^2$), thereby making it easier to realize ASE.[64,65]

3.10.3.3.2.3 Trapping and Self-Trapping Species Additionally, several distinct types of trapping or self-trapping species may also exist in halide perovskites, such as polarons, bound excitons, self-trapped excitons (STEs). Free excitons migrating through the crystal may interact with defects and localize at impurities to become bound excitons. Compared to the free excitons, these bound excitons are stable in energy, thereby resulting to a large Stokes shift of the outcome PL.[66] Polarons are formed through the carriers coupling to the local distortions of the lattice due to electron-phonon interactions. According to the coupling strength between electron and phonon, polarons are divided into small and large. The large polarons (Fröhlich polarons) extend over several lattice sites and are primarily formed from the deformation of the $PbBr_3^{3-}$ frameworks, irrespective of the cation type.[44] Meanwhile, the polaron formation time of $MAPbBr_3$ single crystals (0.3 ps) is less than half of that of $CsPbBr_3$ (0.7 ps).[67] The small polarons are confined to a volume of approximately one unit cell or less in space. Neukirch et al. demonstrated the strong coupling between those localized charged states and the local lattice distortions in $MAPbI_3$ with the MA^+ configurations assisting the creation of small polarons.[68]

Theoretical research reveals that the local perturbations of specific crystal lattice sites in 2D hybrid perovskites introduce the localized self-trapped species in the perovskite inorganic framework and lead to the formation of small polarons.[69] STEs are similar to the small polarons, which means that the neutral exciton generates a lattice distortion. By introducing transient lattice defects, the excitons can also be localized even without lattice defects. Because STEs exist only when excited, they can be treated as excited-state defects; with STEs decaying to the ground state, the lattice distortion disappears gradually. STEs are commonly appeared in quasi-2D perovskites, and it is now generally believed that the formation of STEs is the main mechanism underlying the large Stokes shift and broad-spectrum emission in the low dimensional materials.

3.10.3.3.3 Absorption and PL

The absorption and PL of perovskites can be tuned in the UV, visible and near-IR regions by bandgap engineering mentioned earlier as well as the external factors such as temperature and pressure. The effect of temperature is to cause phase transition and change exciton-phonon (EP) interactions through thermal expansion (TE). In the $APbI_3$ perovskite, as the temperature is increased, PL properties vary with a sequence of structural changes by changing the spin-orbit coupling (SOC) through octahedral tilt. In the higher temperature stable phase of cubic phase, the minimal octahedral tilt renders strong SOC. As a result, the bandgap decreases and both the absorption and PL spectra redshift accordingly. Further, the enhanced EP interactions at higher temperatures make the absorption and PL spectra broaden. Meanwhile, Du et al.[27] found that the joint contribution of TE and EP make the PL peak energy of $CsPbBr_3$ sphere little temperature-dependence PL peak below 150K and redshift in higher temperatures (upper panel in Figure 3.176(a)). PL linewidth broadening with varied temperatures was assigned to the participation of two phonon modes in the EP coupling fitting in the lower panel of Figure 3.176(a). Similar to the temperature, the influence of pressure is to change the SOC through the compression of the Pb–X bond, contraction of the Pb–X–Pb angle or increase in octahedral tilt, and amorphization of the perovskite crystal structure caused by pressure. In Figure 3.176(b), it is obvious that the special bandgap exhibited an initial redshift below 2.2 GPa followed by persistent blueshift during compression. The sudden reverse of the absorption and PL near 2.2 GPa indicated that the $FAPbBr_3$ crystal structure underwent a structural transition. With increasing the pressure further, the initial absorption edge and PL peak gradually fade until completely disappeared above 4.1 GPa. Then along comes the second absorption edge at higher energies and disappeared again at 8.0 GPa. The diminished and even disappeared absorption and PL could be ascribed to pressure-induced amorphization.[70]

Crystal size is another parameter that can tune the optical properties of perovskites. Quantum confinement effects play when the size of perovskites decreases below the exciton Bohr radius, which can be evidenced by the appearance of series spectral changes such as sharp excitonic peaks, narrow PL linewidth and blueshifted absorption and PL spectra (Figure 3.176(c)–(e)).[71–73] As the quantum confinement affects in $CsPbBr_3$ perovskites characterized in Figure 3.176(c), the absorption onset gradually blueshifts with a decrease in crystal size of 6.2–3.7 nm, which is associated with the blueshift of emission peaks from 498 to 467 nm.[71] The experimental and theoretical correlation of bandgap and size of $MAPbX_3$ are additionally provided in Figure 3.176(e).[73] Likewise, quantum confinement effects are also observed in perovskite NPLs and NWs. The thickness-dependent

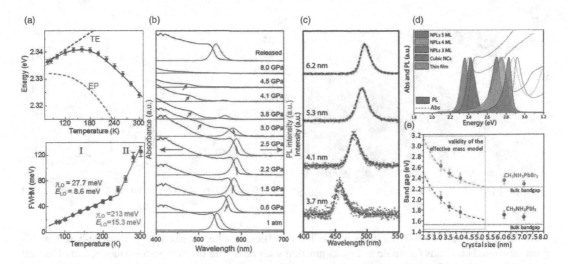

FIGURE 3.176 Effects of temperature, pressure, size and thickness on the optical properties of perovskites. (a) PL peak energy (upper) and linewidth (lower) of $CsPbBr_3$ sphere varied with temperature originated from the joint contribution of thermal expansion (TE) and EP. (From Du, W., Zhang, S., Wu, Z., Shang, Q., Mi, Y., Chen, J., Qin, C., Qiu, X., Zhang, Q. and Liu, X., *Nanoscale*, 11, 3145, 2019. With permission.) (b) Variation tendency of absorptions (cyan line) and PL spectra (red line) for $FAPbBr_3$ under pressure. Red and blue arrows indicate the evolution of the initial and second absorption and PL spectra as a function of pressure. (From Wang, L., Wang, K. and Zou, B., *J. Phys. Chem. Lett.*, 7, 2556, 2016. With permission.) (c) Size-dependent absorption and PL spectra of $CsPbBr_3$ NC perovskites and (d) $CsPbBr_3$ nanoplatelet (NPL). (From Dong, Y., Qiao, T., Kim, D., Parobek, D., Rossi, D. and Son, D. H., *Nano Lett.*, 18, 3716, 2018; Akkerman, Q. A., Motti, S. G., Srimath Kandada, A. R., Mosconi, E., D'Innocenzo, V., Bertoni, G., Marras, S., Kamino, B. A., Miranda, L. and De Angelis, F., *J. Am. Chem. Soc.*, 138, 1010, 2016. With permission.) (e) Theoretical and experimental bandgap values for $MAPbX_3$ varied with NC size. (From Malgras, V., Tominaka, S., Ryan, J. W., Henzie, J., Takei, T., Ohara, K. and Yamauchi, Y., *J. Am. Chem. Soc.*, 138, 13874, 2016. With permission.)

absorption and PL spectra of Br-based NPLs can be compared with those results from colloidal cube-shaped $CsPbBr_3$ NCs with a dimension of approximately 8.5 nm and from polycrystalline thin film. This comparison reveals that the cubes and the polycrystalline thin film have similar emission peaks, which suggests that the 8.5-nm cube-shaped NCs present solely weak quantum confinement effects. In contrast, the NPLs with a thickness of 3 nm or below present a blueshift of the optical band edge induced by the strong quantum confinement effects.

3.10.4 Applications of Perovskites

3.10.4.1 Solar Cell

In the last decade, perovskite metal halide has made an unprecedented advance in the field of solar cells, particularly $MAPbI_3$ as the main semiconductor of interest. The typical semiconductor $MAPbI_3$ forms crystalline films with the characters of long-range charge transport, large absorption coefficients, ultralow trap state densities and efficient charge collection, thereby yielding solar cells with the performance comparable to industry standard silicon, with a verified record PCE of 25.4%. Miyasaka's group first used hybrid perovskite as a light absorber in solar cells in 2009, and its PCE was less than 4%.[1] In this pioneering work, hybrid perovskites of $MAPbI_3$ were initially exploited as an alternative sensitizer for dye molecules in the conventional dye-sensitized solar cell configuration based on the liquid electrolyte. The PCE of perovskite-sensitized solar cells with identical configuration was promoted to 6% by Park et al.[74] However, the low PCE and the poor stability in these liquid perovskite-sensitized solar cells did not draw much attention. Until 2012, concurrent

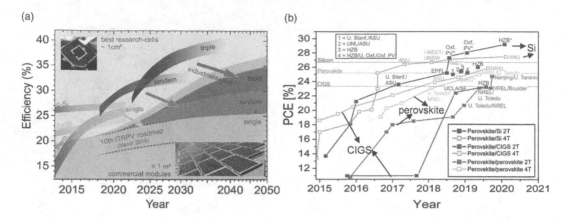

FIGURE 3.177 (a) General roadmap of the highly efficient solar modules of perovskite-based multi-junctions based on trends for single and multi-junction solar cells with the best performance reported as well as predictions. (b) PCE evolution from published papers or press releases (marked with *) for perovskite-based tandem solar cells in two-terminal (2T) and four-terminal (4T) configuration with silicon (Si), $Cu(In,Ga)Se_2$ (CIGS), or perovskite bottom cells. (From Jošt, M., Kegelmann, L., Korte, L. and Albrecht, S., *Adv. Energy Mater.*, 10, 1904102, 2020. With permission.)

breakthrough works[75,76] on solid-state perovskite solar cells with 10–11% PCEs sparked the research explosion on perovskite solar cells. Thereafter, hybrid halide perovskites have attracted widespread attention from researchers who have previously engaged in the research of different optoelectronic materials, and these research efforts include both the development of perovskite material/device and the essential understanding of material properties and device principles.

Figure 3.177(a) schematically visualizes the recent developments as well as the potential for perovskite single junctions comparing with Si and copper indium gallium selenide (CIGS) single junctions and presenting the results of the laboratory scale on the left side of the figure. Although the PCE of perovskite single junctions improves at the fastest speed over time, not all single junctions are expected to conquer the practical PCE limit of approximately 27% in the next few years.[77] This has driven the rise of the tandem technology that has already proved to surpass the practical limit of single-junction solar cell by reducing losses induced by thermalization and unabsorbed photons.[78,79] In addition, the perovskite-based triple-junction solar cells as proof-of-concept have already been reported firstly by Ref. [80]. As suggested in Figure 3.177(a), these triple junctions have even higher PCE potential and may make advanced module PCEs possible in the future.

In the last decade, the rapid progress in the PCE of lead halide perovskite photovoltaic devices (presented in Figure 3.177(b) from an initial value of 3.8–29.2%) has also promoted the developments of other types of perovskite-based photovoltaics. Figure 3.177(b) shows the PCE evolution of perovskite/silicon, perovskite/CIGS, and all-perovskite (perovskite/perovskite) tandem solar cells, in which the solid and open symbol arrows indicate the two- and four-terminal configurations, respectively.[81]

3.10.4.2   LED

Many of fundamental photo-physical properties of perovskite that promote solar cell PCE, such as low trap density, large absorption coefficient and long carrier diffusion length, also enhance light emission applications.

The works on light emission[82–84] and lasing[85] based on hybrid perovskites were published firstly in the 1990s. For instance, bulk crystals of the layered perovskite compound $(C_6H_5(CH_2)_2NH_3)_2$ PbI_4 showed high intense electroluminescence (EL) performed at a voltage of 24 V under liquid nitrogen atmosphere.[82] Thereafter, there were only a few works reported on light emission from perovskites, and the field seemed to hibernate for more than ten years. Until 2014, room temperature (RT) light emission based on low-temperature solution-processed 3D $MAPbX_3$ perovskites was demonstrated.[13,86] It promises highly compatible with unconventional substrates and monolithic integration with silicon-based electronics through the low-temperature processing rather than the expensive high-temperature and high-vacuum-based processing commonly exploited by solid-state light-emitting devices. It poses a very attractive alternative for cost-effective large-area optoelectronic applications. With the breakthrough in the performance of perovskite photovoltaic devices, $MAPbX_3$ were also shown excellent performance on ASE with the thresholds as low as 12 $\mu J/cm^2$. Additionally, typical solar cell architectures that with a mesoporous TiO_2 scaffold were also demonstrated to achieve ASE under very low optical pump fluence.[13] The LEDs were also firstly demonstrated based on the solution-processed halide-substituted $MAPbX_3$ family.[86] Through tuning the halide atom in the perovskite, the LEDs displayed near IR, green and red EL emission, with highest external quantum efficiencies (EQEs) of 3.4%. Apart from the devices based on the $MAPbX_3$ family, all-inorganic LEDs were realized from $CsPbBr_3$ thin films.[87] In addition to observing light emission in perovskite thin films, bright light emission has also been achieved in perovskite nanostructures.[88–92] The superiority of the tunability of the optical properties by substituting or mixing cation and anion plus the extended range provided by controlling size and morphology amount to a wide palette of possibilities for LEDs.

In perovskite LEDs (PeLEDs), the emitter layer may be consisted of the perovskites in 3D, layered or nanostructured forms sandwiched between electron- and hole-transport layers and injecting contacts (Figure 3.178(a)).[93] The performance of PeLEDs has rapidly improved from under 1% to over 20% of EQE in only half a decade (see the left panel of Figure 3.178(b)); the maximum brightness is approaching the QDs based light-emitting diode (QLED) and organic LED (OLED) ranges (plotted in the right panel of Figure 3.178(b)).[94] Although PeLEDs have not yet reached the level of industrial applications, remarkable achievements have been obtained by leveraging the developments of OLEDs/QLEDs over the last decade. Further breakthroughs in PeLED performance could be achieved solely via combining rational terminal materials design, defect-free systems and maximized EQEs. To yield high brightness, color quality, purity and white light emission, the material should be improved from several aspects, including materials design and exciton engineering, optimized charge-injection layers and emission tunability.

Furthermore, it has also achieved outstanding performance of LEDs based on solution-processed 2D perovskites with high EQEs, high color purity and wide spectral tunability profiting from the fundamental understanding of photo-physics and synthetic advances. The earliest LEDs using $n=1$ layered perovskite compound were reported in the 1990s and could function solely at liquid nitrogen temperatures.[82] By converting as-deposited $(PEA)_2PbBr_4$ from polycrystalline thin films f into single-crystal MPLs through solvent vapor annealing, the PLQY has been promoted from 10 to 26%.[95] Furthermore, color-pure violet LEDs were successfully demonstrated at RT with the EL peak centered at 410 nm and a narrow bandwidth of 14 nm due to natural quantum confinement. 2D perovskites with different values of n can be assembled to form an energy cascade structure that funnels photo-excitations to the lowest bandgap of light emitter and makes efficient radiative recombination possible even at low excitation densities.[10,96,97] The high-performance design with energy cascade has enabled EQEs of LEDs up to 11.7% in the near-IR region (~760 nm).[10]

The high performance of the PeLED devices for commercialization has been limited by several factors. For example, the high defect densities and incompetent morphologies of the perovskite layers seriously affect the carrier recombination, thereby leading to low EQE and high leakage current. Several parameters such as compositional and interfacial engineering, surface passivation, and the plasmonic effect may drastically affect the efficiency of PeLEDs.

3.10.4.3 Lasers

Lead halide perovskites are also an ideal candidate for high-performance lasing devices because of their combined merits of high defect tolerance, slow Auger recombination and direct bandgap

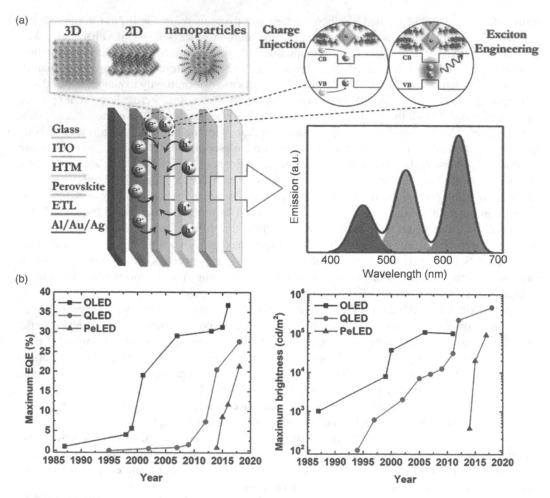

FIGURE 3.178 (a) Schematic of a perovskite light-emitting diode (PeLED) device concept. (From Veldhuis, S. A., Boix, P. P., Yantara, N., Li, M., Sum, T. C., Mathews, N. and Mhaisalkar, S. G., *Adv. Mater.*, 28, 6804, 2016. With permission.) (b) Evolution of representative external quantum efficiency (EQE) and representative maximum brightness for OLEDs, QLEDs and PeLEDs. (From Zhao, L. and Rand, B. P., *Inf. Disp.*, 34, 18, 2018. With permission.)

following large absorption coefficients. Halide perovskites can efficiently build up the population inversion and support higher material gain in a laser benefiting from its strong absorbance. Along with the low-cost facile processing of single-crystal perovskite as well as broad emission color tunability make lead halide perovskites a promising novel candidate material for nanophotonics. Using the thin film or nanostructure of perovskites as active materials, numerous valuable high-performance coherent light emission applications have been demonstrated (Figure 3.179 and Table 3.19).[98] For different lasing types, they have film and NCs edge-emitting ASE,[13,99–101] VCSELs,[102,103] whispering gallery mode (WGM) and Fabry-Perot (FP) mode microlasers,[14,22,92,102–112] distributed feedback (DFB) lasers[113–116] as well as newly burgeoning plasmonic nanolasers,[106,117] polariton lasers,[107,113,118] and bound in continuum (BIC) state lasers.[119] Meanwhile, the continuous-wave (CW) lasing, single-mode laser and two-photon lasing have been realized along with different lasing types.[120–122]

In 2014, Xing et al. firstly achieved the low-threshold RT ASEs with high stability in perovskite thin films.[13] Thereafter, the tunable ASE/lasing emission wavelength covering the entire visible

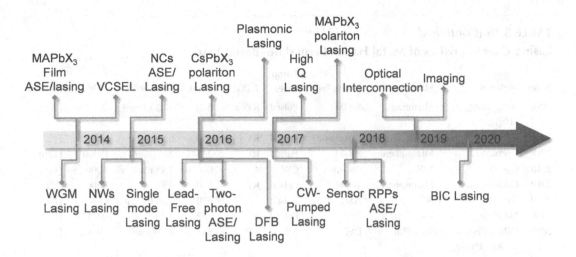

FIGURE 3.179 Home-made evolution trend of halide perovskite in coherent light emission applications over the past several years, illustrating breakthrough works on film and NCs ASE/lasing,[13,99,101] vertical-cavity surface-emitting lasers (VCSELs),[102,103] whispering gallery mode (WGM) and FP mode microlasers,[14,22,92,102–112] distributed feedback (DFB) lasing,[113–116] plasmonic nanolasers,[106,117,123] polariton lasers,[107,113,118] and bound in continuum (BIC) state lasers[119] as well as the CW lasing, single-mode laser and two-photon lasing.[120–122]

TABLE 3.19
Lasing Characteristics of Metal Halide Perovskite-Based Lasers

Years	Materials	Structure	Lasing Type	Pump Laser	T (K)	λ (nm)	Threshold	Q	Reference
2014	$MAPbX_3$	Thin film	ASE	Pulsed	RT	390–790	12 μJ/cm^2	NA	[13]
2014	$MAPbI_{3-x}Cl_x$	Thin film	VCSEL	Pulsed	RT	775	0.2 μJ/cm^2	~484	[102]
2014	$MAPbI_{3-x}X$	NP	WGM	Pulsed	RT	786	37 μJ/cm^2	900	[92]
2015	$CsPbX_3$	NC	ASE	Pulsed	RT	440–700	5 μJ/cm^2	NA	[99]
2015	$CsPbBr_3$	QDs	WGM	Pulsed	RT	525.4	11.6 μJ/cm^2	1637	[110]
2015	$MAPbX_3$	NW	FP	Pulsed	RT	500–800	220 nJ/cm^2	3600	[15]
2015	$MAPbX_3$	NW	FP	Pulsed	RT	551–777	11 μJ/cm^2	405	[22]
2015	$MAPbCl_xBr_{3-x}$	MD	WGM	Pulsed	RT	525–557	3.6 μJ/cm^2	430	[122]
2016	$MAPbI_3$	Thin film	DFB	Pulsed	RT	770–793	0.32 μJ/cm^2	400	[113]
2016	$CsPbX_3$	NP	WGM	Pulsed	RT	400–700	~2 μJ/cm^2	~4857	[105]
2016	$CsPbBr_3/CsPbCl_3$	NW/NP	FP/WGM	Pulsed	RT	425/530	~5 μJ/cm^2	1009	[14]
2016	$MAPbI_3$	MPL	WGM	Pulsed	RT	785	9 μJ/cm^2	1210	[106]
2016	$FAPbX_3$	NW	FP	Pulsed	RT	490–824	6 μJ/cm^2	2300	[111]
2016	$MAPbBr_3$	NW+MD	WGM	Pulsed	RT	550	21 μJ/cm^2	1460	[108]
2016	$MAPbI_3$	NW	Plasmonic	Pulsed	RT	785	13.5 μJ/cm^2	151	[117]
2017	$MAPbI_3$	Thin film	VCSEL	Pulsed	RT	778	7.6 μJ/cm^2	1100	[103]
2017	$MAPbI_3$	Thin film	DFB	CW	102	785	17 kW/cm^2	3140	[113]
2017	$CsPbCl_3$	NP	Polariton	Pulsed	RT	427	12 μJ/cm^2	300	[107]
2017	$CsPbBr_3$	NW	Polariton	CW	78	540	6 kW/cm^2	2300	[104]
2017	$MAPbBr_3$	MPL patterns	WGM	Pulsed	RT	550	2.3 μJ/cm^2	6000	[112]
2017	$CsPbX_3$	Microcubes	FP	Pulsed	RT	429–536	210 μJ/cm^2	1150	[124]
2018	$MAPbI_3$	Thin film	DFB	CW	RT	807	13 W/cm^2	1153	[115]
2018	$(BA)_2(MA)_{n-1}Pb_nI_{3n+1}$	Microflakes	WGM/FP	Pulsed	<230	620–690	20 μJ/cm^2	1347	[125]

(Continued)

TABLE 3.19 (*Continued*)

Lasing Characteristics of Metal Halide Perovskite–Based Lasers

Years	Materials	Structure	Lasing Type	Pump Laser	T (K)	λ (nm)	Threshold	Q	Reference
2018	$(BA)_2(MA)_{n-1}Pb_nI_{3n+1}$	Microring	WGM	Pulsed	RT	540	12.2 $\mu J/cm^2$	2600	[120]
2018	$CsPbBr_3$	NW	Plasmonic	Pulsed	RT	530	6.5 $\mu J/cm^2$	757	[123]
2018	$CsPbBr_3$	Microsphere	WGM	Pulsed	RT	533–543	203 $\mu J/cm^2$	15000	[126]
2018	$CsPbX_3$	NW	Polariton	CW	4	420–650	1.45 nW/cm^2	6000	[118]
2019	$CsPbBr_3$	Microsphere	WGM	Pulsed	RT	540	400	3600	[21]
2020	$CsPbBr_3$	NC	ASE	Pulsed	173	530	1.16 mJ/cm^2	NA	[100]
2020	$MAPbBr_3$	Thin film	BIC	Pulsed	RT	552	4.2 $\mu J/cm^2$	690	[119]
2020	$(NMA/PEA)_2(FA)_{n-1}Pb_nBr_{3n+1}$	Thin film	DFB	CW	RT	555	45 W/cm^2	694	[127]

Note: T means the temperature; Q is the abbreviation of quality factor; λ represents the emission wavelength; NA denotes the not available information; RT denotes room temperature (~293K).

range and near-IR emission have been dramatically developed with the metal halide perovskite micro/nanostructures in the shape of MPLs, NWs and NCs etc. served as gain materials. Zhu et al. reported high Q of 3600 FP mode lasing in perovskite NWs, in which the operation threshold as low as 220 nJ/cm^2.[15] Zhang et al. demonstrated a near-IR WGM nanolaser based on perovskite NPs.[92] Following closely, Yakunin et al. first realized the ASE and lasing from ~10 nm colloidal NCs of $CsPbX_3$ with low threshold.[99] In 2017, by integrating with DFB, the CW lasing under operating time over 1 hour in $MAPbI_3$ perovskite thin films was realized at $T \approx 100K$ by Giebink et al.[113] On account of the available strategy for designing small solid-state lasers, perovskites show great potentials for realizing electrically driven micro/nanolasers.

With the smooth structural facets serving as mirrors and the regular-shaped crystals themselves serving as the gain medium, single crystalline perovskite nanostructures can form optical resonators, for that the NPs naturally form WGM (see Figure 3.180(a)) and NWs active FP cavities (Figure 3.180(b)).

WGM microcavity has the superiorities of smaller model volume, high Q and small size for integration. The light is confined in the resonator with a spherical, polygonal or circular shape by total internal reflection along the resonator-surrounding interface, thereby providing efficient optical feedback. In 2014, the first WGM perovskite microlaser based on organic-inorganic perovskite $MAPbI_{3-a}X_a$ (X=Cl, Br, I) NPs was reported by Zhang et al. with a low threshold in the range 37–128 $\mu J/cm^2$ and a Q factor of 650–900, approximately.[92] Subsequently to improve the stability of gain materials, all-inorganic perovskite lasing was explored by the same group. In 2016, high coherent WGM lasing from all-inorganic $CsPbX_3$ were demonstrated with the wavelength tuning from 400 to 700 nm and a low threshold of ~2.0 $\mu J/cm^2$.[105] To reduce the light leakage on their vertical direction of WGM microcavities, some other kinds of WGM microcavities have been explored by the researcher in the lasing field. Using $CsPbBr_3$ perovskite microspheres as a gain medium, Tang et al. realized a two-photon single-mode laser displaying ultrahigh Q factor of 1.5×10^4 of WGM cavity.[126] In addition, utilizing the direct lithographic patterning of $MAPbX_3$ perovskite MPL on conventional substrates Liu et al. realized RT high Q (1210) WGM lasing with variable WGM cavity sizes. The versatile scalable top-down strategy offers another way for large-area reproducible light emission applications.

With the ultracompact physical size, efficient waveguiding and highly localized coherent output, semiconductor NWs are other promising candidates for fully integrated nanophotonic and optoelectronic devices. Compared to the bulk perovskites, perovskite NW laser is thus anticipated to show

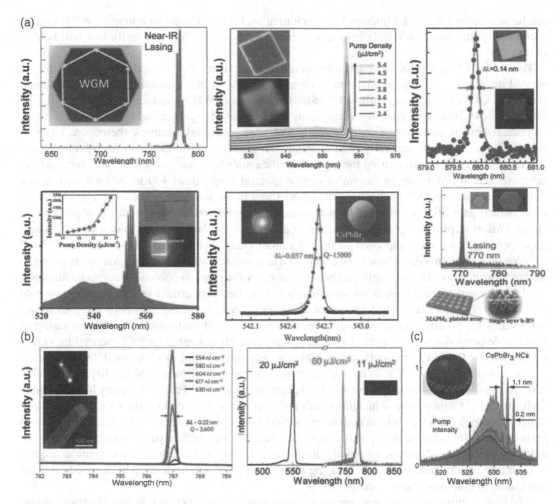

FIGURE 3.180 Perovskite crystals serving as the gain medium; nanowires and nanoplatelets naturally form whispering-gallery-mode (WGM) (a) and active FP cavities (b). Perovskite QD amplified spontaneous emission and lasing coupled with another optical cavity (c). (From Zhang, Q., Ha, S. T., Liu, X., Sum, T. C. and Xiong, Q., *Nano Lett.*, 14, 5995, 2014; Liao, Q., Hu, K., Zhang, H., Wang, X., Yao, J. and Fu, H., *Adv. Mater.*, 27, 3405, 2015; Zhang, Q., Su, R., Liu, X., Xing, J., Sum, T. C. and Xiong, Q., *Adv. Funct. Mater.*, 26, 6238, 2016; Wang, K., Sun, W., Li, J., Gu, Z., Xiao, S. and Song, Q., *ACS Photonics*, 3, 1125, 2016; Tang, B., Sun, L., Zheng, W., Dong, H., Zhao, B., Si, Q., Wang, X., Jiang, X., Pan, A. and Zhang, L., *Adv. Opt. Mater.*, 6, 1800391, 2018; Liu, X., Niu, L., Wu, C., Cong, C., Wang, H., Zeng, Q., He, H., Fu, Q., Fu, W. and Yu, T., *Adv. Sci.*, 3, 1600137, 2016; Zhu, H., Fu, Y., Meng, F., Wu, X., Gong, Z., Ding, Q., Gustafsson, M. V., Trinh, M. T., Jin, S. and Zhu, X. Y., *Nat. Mater.*, 14, 636, 2015; Xing, J., Liu, X. F., Zhang, Q., Ha, S. T., Yuan, Y. W., Shen, C., Sum, T. C. and Xiong, Q., *Nano Lett.*, 15, 4571, 2015; Yakunin, S., Protesescu, L., Krieg, F., Bodnarchuk, M. I., Nedelcu, G., Humer, M., De Luca, G., Fiebig, M., Heiss, W. and Kovalenko, M. V., *Nat. Commun.*, 6, 8056, 2015. With permission.)

more intriguing features. Zhu et al. firstly demonstrated the MAPbX$_3$ NW lasers, which exhibited ultralow lasing threshold (220 nJ/cm^2), a high Q of 3600 and broad tunability wavelength region from 500 to 800 nm.[15] To conquer their intrinsic instability of MAPbX$_3$ NW lasers, single-crystal FAPbX$_3$ NWs grown in low-temperature solution were used to act as waveguide active medium by Fu et al. Recently FAPbX$_3$ NWs show better thermal stability compared to MAPbX$_3$.[111] Under over 10^8 shots, the durable near-IR lasing (~820 nm) can still be maintained at RT. At the same time, Eaton et al. also reported ultra-stable NW lasing based on all-inorganic CsPbBr$_3$ perovskite, which can be maintained for over 1 hour and 10^9 excitation cycles under ambient atmosphere.[14] Recently, exciton-polariton was also widely investigated in 1D perovskite NW microcavity that will be discussed next.

Perovskite QDs, $CsPbX_3$ in particular, could further usher the advantages of metal halide perovskite into the realm of quantum confinement. Due to the quantum confinement effect, perovskite QDs exhibit narrow full-width-at-half maximum (FWHM) and remarkably high PLQYs (slightly above 70%) with emission range from UV to near-IR region. Single QD cannot form an individual optical resonator to support an optical waveguide and feedback thereupon. Used as gain medium, they have generally been coupled with another optical cavity to form laser devices (Figure 3.180(c)). Through coating the QDs onto silica microspheres, Yakunin et al.[99] successfully demonstrated WGM lasing in the entire visible spectral range from 440 to 700 nm with a low threshold of 5 $\mu J/cm^2$. Although perovskite QDs exhibit many favorable properties, their extremely short optical gain lifetimes affected by non-radiative Auger recombination limit the application in lasing. Park et al. provided practical strategies of a smooth core/shell interface to suppress the Auger decay process that notably improved the optical gain lifetime and thus reduced the ASE threshold.[128]

Very recently, perovskites have been considered as a great candidate for polariton lasers owing to the large exciton oscillator strength and binding energy. Compared to photonic laser, polariton laser can be generated without population inversion, which makes them attractive for ultralow-threshold coherent light source. In the last few years, there have been significant efforts to use lead halide perovskites for polariton lasing. In 2017, the first experimental realization of polariton lasing in lead halide perovskites was reported with an epitaxy-free all-inorganic $CsPbCl_3$ perovskite planar microcavity at RT.[107] Exciton-polariton was also widely investigated in 1D perovskite NW microcavity. Liu et al. as well as Evans et al. demonstrated exciton-polaritons in $MAPbBr_3$ and $CsPbBr_3$ NW, respectively, via spatially resolved PL spectroscopy.[104,129,130] Polariton lasing has also been realized in a 1D strong perovskite lattice with artificial periodic potentials, the so-called polariton lattices,[131] which open the route to realize topological lasers in perovskites with robust immunity to perturbations. Introducing the optical structure/lattice, such as broken parity-time symmetry and topology-protected BIC states, into cavity design provides not only alternative strategies to acquire ultrahigh Q factors and low-threshold lasing but also an approach to develop high-speed classical and quantum communication systems.[119]

The high stability under more severe operation circumstances, such as CW and electrical pumping, is still the primary obstacle toward practical applications. Besides, challenges, including the imperfect interface between perovskite and charge transport layer, carrier leakage and charge imbalance during injection need further efforts to conquer. To solve these difficulties, it must put strict requirements for the device design, which leads to little progress in electrically pumped perovskite lasers.

REFERENCES

1. Kojima, A., Teshima, K., Shirai, Y. and Miyasaka, T., *J. Am. Chem. Soc.*, 131, 6050, 2009.
2. Dong, R., Fang, Y., Chae, J., Dai, J., Xiao, Z., Dong, Q., Yuan, Y., Centrone, A., Zeng, X. C. and Huang, J., *Adv. Mater.*, 27, 1912, 2015.
3. Stranks, S. D., Eperon, G. E., Grancini, G., Menelaou, C., Alcocer, M. J. P., Leijtens, T., Herz, L. M., Petrozza, A. and Snaith, H. J., *Science*, 342, 341, 2013.
4. Wehrenfennig, C., Eperon, G. E., Johnston, M. B., Snaith, H. J. and Herz, L. M., *Adv. Mater.*, 26, 1584, 2014.
5. Murali, B., Saidaminov, M. I., Abdelhady, A. L., Peng, W., Liu, J., Pan, J., Bakr, O. M. and Mohammed, O. F., *J. Mater. Chem. C*, 4, 2545, 2016.
6. Li, Y., Shi, Z., Lei, L., Zhang, F., Ma, Z., Wu, D., Xu, T., Tian, Y., Zhang, Y., Du, G., Shan, C. and Li, X., *Chem. Mater.*, 30, 6744, 2018.
7. Zhou, H., Chen, Q., Li, G., Luo, S., Song, T.-B., Duan, H.-S., Hong, Z., You, J., Liu, Y. and Yang, Y., *Science*, 345, 542, 2014.

8. Jiang, Q., Zhang, L., Wang, H., Yang, X., Meng, J., Liu, H., Yin, Z., Wu, J., Zhang, X. and You, J., *Nat. Energy*, 2, 16177, 2016.
9. Cao, Y., Wang, N., Tian, H., Guo, J., Wei, Y., Chen, H., Miao, Y., Zou, W., Pan, K., He, Y., Cao, H., Ke, Y., Xu, M., Wang, Y., Yang, M., Du, K., Fu, Z., Kong, D., Dai, D., Jin, Y., Li, G., Li, H., Peng, Q., Wang, J. and Huang, W., *Nature*, 562, 249, 2018.
10. Wang, N., Cheng, L., Ge, R., Zhang, S., Miao, Y., Zou, W., Yi, C., Sun, Y., Cao, Y. and Yang, R., *Nat. Photonics*, 10, 699, 2016.
11. Ling, Y., Tian, Y., Wang, X., Wang, J. C., Knox, J. M., Perez-Orive, F., Du, Y., Tan, L., Hanson, K. and Ma, B., *Adv. Mater.*, 28, 8983, 2016.
12. Cho, H., Jeong, S.-H., Park, M.-H., Kim, Y.-H., Wolf, C., Lee, C.-L., Heo, J. H., Sadhanala, A., Myoung, N. and Yoo, S., *Science*, 350, 1222, 2015.
13. Xing, G., Mathews, N., Lim, S. S., Yantara, N., Liu, X., Sabba, D., Grätzel, M., Mhaisalkar, S. and Sum, T. C., *Nat. Mater.*, 13, 476, 2014.
14. Eaton, S. W., Lai, M., Gibson, N. A., Wong, A. B., Dou, L., Ma, J., Wang, L.-W., Leone, S. R. and Yang, P., *Proc. Natl. Acad. Sci. U.S.A.*, 113, 1993, 2016.
15. Zhu, H., Fu, Y., Meng, F., Wu, X., Gong, Z., Ding, Q., Gustafsson, M. V., Trinh, M. T., Jin, S. and Zhu, X. Y., *Nat. Mater.*, 14, 636, 2015.
16. Zhang, Q., Su, R., Du, W., Liu, X., Zhao, L., Ha, S. T. and Xiong, Q., *Small Methods*, 1, 1700163, 2017.
17. Pan, W., Wu, H., Luo, J., Deng, Z., Ge, C., Chen, C., Jiang, X., Yin, W.-J., Niu, G. and Zhu, L., *Nat. Photonics*, 11, 726, 2017.
18. Saidaminov, M. I., Haque, M. A., Savoie, M., Abdelhady, A. L., Cho, N., Dursun, I., Buttner, U., Alarousu, E., Wu, T. and Bakr, O. M., *Adv. Mater.*, 28, 8144, 2016.
19. Deng, H., Yang, X., Dong, D., Li, B., Yang, D., Yuan, S., Qiao, K., Cheng, Y.-B., Tang, J. and Song, H., *Nano Lett.*, 15, 7963, 2015.
20. Lian, Z., Yan, Q., Gao, T., Ding, J., Lv, Q., Ning, C., Li, Q. and Sun, J.-L., *J. Am. Chem. Soc.*, 138, 9409, 2016.
21. Saidaminov, M. I., Abdelhady, A. L., Murali, B., Alarousu, E., Burlakov, V. M., Peng, W., Dursun, I., Wang, L., He, Y. and Maculan, G., *Nat. Commun.*, 6, 7586, 2015.
22. Hang, H., Liu, X., Dong, J., Yu, H., Zhou, C., Zhang, B., Xu, Y. and Jie, W., *Cryst. Growth Des.*, 17, 6426, 2017.
23. Dong, Q., Fang, Y., Shao, Y., Mulligan, P., Qiu, J., Cao, L. and Huang, J., *Science*, 347, 967, 2015.
24. Guo, Y., Yin, X., Liu, J. and Que, W., *J. Mater. Chem. A*, 7, 19008, 2019.
25. Im, J.-H., Jang, I.-H., Pellet, N., Grätzel, M. and Park, N.-G., *Nat. Nanotechnol.*, 9, 927, 2014.
26. Fu, Y., Rea, M. T., Chen, J., Morrow, D. J., Hautzinger, M. P., Zhao, Y., Pan, D., Manger, L. H., Wright, J. C., Goldsmith, R. H. and Jin, S., *Chem. Mater.*, 29, 8385, 2017.
27. Du, W., Zhang, S., Wu, Z., Shang, Q., Mi, Y., Chen, J., Qin, C., Qiu, X., Zhang, Q. and Liu, X., *Nanoscale*, 11, 3145, 2019.
28. Zhong, Y., Liao, K., Du, W., Zhu, J., Shang, Q., Zhou, F., Wu, X., Sui, X., Shi, J. and Yue, S., *ACS Nano*, 14, 15605, 2020.
29. Gao, Y., Zhao, L., Shang, Q., Zhong, Y., Liu, Z., Chen, J., Zhang, Z., Shi, J., Du, W. and Zhang, Y., *Adv. Mater.*, 30, 1801805, 2018.
30. Liu, Z., Li, Y., Guan, X., Mi, Y., Al-Hussain, A., Ha, S. T., Chiu, M.-H., Ma, C., Amer, M. R. and Li, L.-J., *J. Phys. Chem. Lett.*, 10, 2363, 2019.
31. Leyden, M. R., Ono, L. K., Raga, S. R., Kato, Y., Wang, S. and Qi, Y., *J. Mater. Chem. A*, 2, 18742, 2014.
32. Tavakoli, M. M., Gu, L., Gao, Y., Reckmeier, C., He, J., Rogach, A. L., Yao, Y. and Fan, Z., *Sci. Rep.*, 5, 14083, 2015.
33. Wang, B. and Chen, T., *Adv. Sci.*, 3, 1500262, 2016.
34. Ng, A., Ren, Z., Shen, Q., Cheung, S. H., Gokkaya, H. C., So, S. K., Djurišić, A. B., Wan, Y., Wu, X. and Surya, C., *ACS Appl. Mater. Interfaces*, 8, 32805, 2016.
35. Peng, Y., Jing, G. and Cui, T., *J. Mater. Chem. A*, 3, 12436, 2015.
36. Goldschmidt, V. M., *Ber. Dtsch. Chem. Ges.*, 60, 1263, 1927.
37. Li, Z., Yang, M., Park, J.-S., Wei, S.-H., Berry, J. J. and Zhu, K., *Chem. Mater.*, 28, 284, 2016.
38. Cohen, B.-E., Wierzbowska, M. and Etgar, L., *Sustain. Energy Fuels*, 1, 1935, 2017.
39. Calabrese, J., Jones, N. L., Harlow, R. L., Herron, N., Thorn, D. L. and Wang, Y., *J. Am. Chem. Soc.*, 113, 2328, 1991.
40. Sun, S., Isikgor, F. H., Deng, Z., Wei, F., Kieslich, G., Bristowe, P. D., Ouyang, J. and Cheetham, A. K., *ChemSusChem*, 10, 3740, 2017.

41. Liu, W., Liu, Y., Wang, J., Wu, C., Liu, C., Xiao, L., Chen, Z., Wang, S. and Gong, Q., *Crystals*, 8, 216, 2018.
42. Poglitsch, A. and Weber, D., *J. Chem. Phys.*, 87, 6373, 1987.
43. Fan, Z., Sun, K. and Wang, J., *J. Mater. Chem. A*, 3, 18809, 2015.
44. Miyata, K., Atallah, T. L. and Zhu, X. Y., *Sci. Adv.*, 3, e1701469, 2017.
45. Wang, L., Williams, N. E., Malachosky, E. W., Otto, J. P., Hayes, D., Wood, R. E., Guyot-Sionnest, P. and Engel, G. S., *ACS Nano*, 11, 2689, 2017.
46. Stoumpos, C. C., Malliakas, C. D. and Kanatzidis, M. G., *Inorg. Chem.*, 52, 9019, 2013.
47. Levchuk, I., Osvet, A., Tang, X., Brandl, M., Perea, J. D., Hoegl, F., Matt, G. J., Hock, R., Batentschuk, M. and Brabec, C. J., *Nano Lett.*, 17, 3993, 2017.
48. Jellicoe, T. C., Richter, J. M., Glass, H. F. J., Tabachnyk, M., Brady, R., Dutton, S. N. E., Rao, A., Friend, R. H., Credgington, D. and Greenham, N. C., *J. Am. Chem. Soc.*, 138, 2941, 2016.
49. Filip, M., Eperon, G., Snaith, H. and Giustino, F., *Nat. Commun.*, 5, 5757, 2014.
50. Yin, W.-J., Yang, J.-H., Kang, J., Yan, Y. and Wei, S.-H., *J. Mater. Chem. A*, 3, 8926, 2015.
51. Motta, C., El-Mellouhi, F., Kais, S., Tabet, N., Alharbi, F. and Sanvito, S., *Nat. Commun.*, 6, 7026, 2015.
52. Yin, W. J., Shi, T. and Yan, Y., *Adv. Mater.*, 26, 4653, 2014.
53. Zhang, F., Zhong, H., Chen, C., Wu, X.-G., Hu, X., Huang, H., Han, J., Zou, B. and Dong, Y., *ACS Nano*, 9, 4533, 2015.
54. Nedelcu, G., Protesescu, L., Yakunin, S., Bodnarchuk, M. I., Grotevent, M. J. and Kovalenko, M. V., *Nano Lett.*, 15, 5635, 2015.
55. Navas, J., Sánchez-Coronilla, A., Gallardo, J. J., Hernández, N. C., Piñero, J. C., Alcántara, R., Fernández-Lorenzo, C., Desireé, M., Aguilar, T. and Martín-Calleja, J., *Nanoscale*, 7, 6216, 2015.
56. Koh, T. M., Fu, K., Fang, Y., Chen, S., Sum, T. C., Mathews, N., Mhaisalkar, S. G., Boix, P. P. and Baikie, T., *J. Phys. Chem. C*, 118, 16458, 2014.
57. Lee, J. W., Seol, D. J., Cho, A. N. and Park, N. G., *Adv. Mater.*, 26, 4991, 2014.
58. Herz, L. M., *Annu. Rev. Phys. Chem.*, 67, 65, 2016.
59. Milot, R. L., Eperon, G. E., Snaith, H. J., Johnston, M. B. and Herz, L. M., *Adv. Funct. Mater.*, 25, 6218, 2015.
60. Noel, N. K., Stranks, S. D., Abate, A., Wehrenfennig, C., Guarnera, S., Haghighirad, A.-A., Sadhanala, A., Eperon, G. E., Pathak, S. K. and Johnston, M. B., *Energy Environ. Sci.*, 7, 3061, 2014.
61. Xing, G., Wu, B., Wu, X., Li, M., Du, B., Wei, Q., Guo, J., Yeow, E. K. L., Sum, T. C. and Huang, W., *Nat. Commun.*, 8, 14558, 2017.
62. Eperon, G. E., Jedlicka, E. and Ginger, D. S., *J. Phys. Chem. Lett.*, 9, 104, 2018.
63. Dastidar, S., Li, S., Smolin, S. Y., Baxter, J. B. and Fafarman, A. T., *ACS Energy Lett.*, 2, 2239, 2017.
64. Castañeda, J. A., Nagamine, G., Yassitepe, E., Bonato, L. G., Voznyy, O., Hoogland, S., Nogueira, A. F., Sargent, E. H., Cruz, C. H. B. and Padilha, L. A., *ACS Nano*, 10, 8603, 2016.
65. Wang, L.-W., Califano, M., Zunger, A. and Franceschetti, A., *Phys. Rev. Lett.*, 91, 056404, 2003.
66. Chen, Z., Guo, Y., Wertz, E. and Shi, J., *Adv. Mater.*, 31, 1803514, 2019.
67. Miyata, K., Meggiolaro, D., Trinh, M. T., Joshi, P. P., Mosconi, E., Jones, S. C., De Angelis, F. and Zhu, X. Y., *Sci. Adv.*, 3, e1701217, 2017.
68. Neukirch, A. J., Nie, W., Blancon, J.-C., Appavoo, K., Tsai, H., Sfeir, M. Y., Katan, C., Pedesseau, L., Even, J. and Crochet, J. J., *Nano Lett.*, 16, 3809, 2016.
69. Yin, J., Li, H., Cortecchia, D., Soci, C. and Bredas, J.-L., *ACS Energy Lett.*, 2, 417, 2017.
70. Wang, L., Wang, K. and Zou, B., *J. Phys. Chem. Lett.*, 7, 2556, 2016.
71. Dong, Y., Qiao, T., Kim, D., Parobek, D., Rossi, D. and Son, D. H., *Nano Lett.*, 18, 3716, 2018.
72. Akkerman, Q. A., Motti, S. G., Srimath Kandada, A. R., Mosconi, E., D'Innocenzo, V., Bertoni, G., Marras, S., Kamino, B. A., Miranda, L. and De Angelis, F., *J. Am. Chem. Soc.*, 138, 1010, 2016.
73. Malgras, V., Tominaka, S., Ryan, J. W., Henzie, J., Takei, T., Ohara, K. and Yamauchi, Y., *J. Am. Chem. Soc.*, 138, 13874, 2016.
74. Im, J.-H., Lee, C.-R., Lee, J.-W., Park, S.-W. and Park, N.-G., *Nanoscale*, 3, 4088, 2011.
75. Lee, M. M., Teuscher, J., Miyasaka, T., Murakami, T. N. and Snaith, H. J., *Science*, 338, 643, 2012.
76. Heo, J. H., Im, S. H., Noh, J. H., Mandal, T. N., Lim, C.-S., Chang, J. A., Lee, Y. H., Kim, H.-J., Sarkar, A. and Nazeeruddin, M. K., *Nat. Photonics*, 7, 486, 2013.
77. Zhengshan, J. Y., Carpenter, J. V. and Holman, Z. C., *Nat. Energy*, 3, 747, 2018.
78. De Vos, A., *J. Phys. D Appl. Phys.*, 13, 839, 1980.
79. Marti, A. and Araújo, G. L., *Sol. Energ. Mat. Sol. C*, 43, 203, 1996.
80. Sahli, F., Werner, J., Kamino, B. A., Bräuninger, M., Monnard, R., Paviet-Salomon, B., Barraud, L., Ding, L., Leon, J. J. D. and Sacchetto, D., *Nat. Mater.*, 17, 820, 2018.

81. Jošt, M., Kegelmann, L., Korte, L. and Albrecht, S., *Adv. Energy Mater.*, 10, 1904102, 2020.
82. Era, M., Morimoto, S., Tsutsui, T. and Saito, S., *Appl. Phys. Lett.*, 65, 676, 1994.
83. Hong, X., Ishihara, T. and Nurmikko, A. V., *Solid State Commun.*, 84, 657, 1992.
84. Hattori, T., Taira, T., Era, M., Tsutsui, T. and Saito, S., *Chem. Phys. Lett.*, 254, 103, 1996.
85. Kondo, T., Azuma, T., Yuasa, T. and Ito, R., *Solid State Commun.*, 105, 253, 1998.
86. Tan, Z.-K., Moghaddam, R. S., Lai, M. L., Docampo, P., Higler, R., Deschler, F., Price, M., Sadhanala, A., Pazos, L. M. and Credgington, D., *Nat. Nanotechnol.*, 9, 687, 2014.
87. Yantara, N., Bhaumik, S., Yan, F., Sabba, D., Dewi, H. A., Mathews, N., Boix, P. P., Demir, H. V. and Mhaisalkar, S., *J. Phys. Chem. Lett.*, 6, 4360, 2015.
88. Schmidt, L. C., Pertegás, A., González-Carrero, S., Malinkiewicz, O., Agouram, S., Mínguez Espallargas, G., Bolink, H. J., Galian, R. E. and Pérez-Prieto, J., *J. Am. Chem. Soc.*, 136, 850, 2014.
89. Protesescu, L., Yakunin, S., Bodnarchuk, M. I., Krieg, F., Caputo, R., Hendon, C. H., Yang, R. X., Walsh, A. and Kovalenko, M. V., *Nano Lett.*, 15, 3692, 2015.
90. Zhang, D., Eaton, S. W., Yu, Y., Dou, L. and Yang, P., *J. Am. Chem. Soc.*, 137, 9230, 2015.
91. Hassan, Y., Song, Y., Pensack, R. D., Abdelrahman, A. I., Kobayashi, Y., Winnik, M. A. and Scholes, G. D., *Adv. Mater.*, 28, 566, 2016.
92. Zhang, Q., Ha, S. T., Liu, X., Sum, T. C. and Xiong, Q., *Nano Lett.*, 14, 5995, 2014.
93. Veldhuis, S. A., Boix, P. P., Yantara, N., Li, M., Sum, T. C., Mathews, N. and Mhaisalkar, S. G., *Adv. Mater.*, 28, 6804, 2016.
94. Zhao, L. and Rand, B. P., *Inf. Disp.*, 34, 18, 2018.
95. Liang, D., Peng, Y., Fu, Y., Shearer, M. J., Zhang, J., Zhai, J., Zhang, Y., Hamers, R. J., Andrew, T. L. and Jin, S., *ACS Nano*, 10, 6897, 2016.
96. Byun, J., Cho, H., Wolf, C., Jang, M., Sadhanala, A., Friend, R. H., Yang, H. and Lee, T. W., *Adv. Mater.*, 28, 7515, 2016.
97. Yuan, M., Quan, L. N., Comin, R., Walters, G., Sabatini, R., Voznyy, O., Hoogland, S., Zhao, Y., Beauregard, E. M. and Kanjanaboos, P., *Nat. Nanotechnol.*, 11, 872, 2016.
98. Zhang, Q., Shang, Q., Su, R., Do, T. T. H. and Xiong, Q., *Nano Lett.*, 21, 1903, 2021.
99. Yakunin, S., Protesescu, L., Krieg, F., Bodnarchuk, M. I., Nedelcu, G., Humer, M., De Luca, G., Fiebig, M., Heiss, W. and Kovalenko, M. V., *Nat. Commun.*, 6, 8056, 2015.
100. Li, S., Lei, D., Ren, W., Guo, X., Wu, S., Zhu, Y., Rogach, A. L., Chhowalla, M. and Jen, A. K. Y., *Nat. Commun.*, 11, 1192, 2020.
101. Veldhuis, S. A., Tay, Y. K. E., Bruno, A., Dintakurti, S. S. H., Bhaumik, S., Muduli, S. K., Li, M., Mathews, N., Sum, T. C. and Mhaisalkar, S. G., *Nano Lett.*, 17, 7424, 2017.
102. Deschler, F., Price, M., Pathak, S., Klintberg, L. E., Jarausch, D.-D., Higler, R., Hüttner, S., Leijtens, T., Stranks, S. D. and Snaith, H. J., *J. Phys. Chem. Lett.*, 5, 1421, 2014.
103. Chen, S., Zhang, C., Lee, J., Han, J. and Nurmikko, A., *Adv. Mater.*, 29, 1604781, 2017.
104. Evans, T. J. S., Schlaus, A., Fu, Y., Zhong, X., Atallah, T. L., Spencer, M. S., Brus, L. E., Jin, S. and Zhu, X. Y., *Adv. Opt. Mater.*, 6, 1700982, 2018.
105. Zhang, Q., Su, R., Liu, X., Xing, J., Sum, T. C. and Xiong, Q., *Adv. Funct. Mater.*, 26, 6238, 2016.
106. Liu, X., Niu, L., Wu, C., Cong, C., Wang, H., Zeng, Q., He, H., Fu, Q., Fu, W. and Yu, T., *Adv. Sci.*, 3, 1600137, 2016.
107. Su, R., Diederichs, C., Wang, J., Liew, T. C. H., Zhao, J., Liu, S., Xu, W., Chen, Z. and Xiong, Q., *Nano Lett.*, 17, 3982, 2017.
108. Wang, K., Sun, W., Li, J., Gu, Z., Xiao, S. and Song, Q., *ACS Photonics*, 3, 1125, 2016.
109. Xing, J., Liu, X. F., Zhang, Q., Ha, S. T., Yuan, Y. W., Shen, C., Sum, T. C. and Xiong, Q., *Nano Lett.*, 15, 4571, 2015.
110. Wang, Y., Li, X., Song, J., Xiao, L., Zeng, H. and Sun, H., *Adv. Mater.*, 27, 7101, 2015.
111. Fu, Y., Zhu, H., Schrader, A. W., Liang, D., Ding, Q., Joshi, P., Hwang, L., Zhu, X. Y. and Jin, S., *Nano Lett.*, 16, 1000, 2016.
112. Zhang, N., Sun, W., Rodrigues, S. P., Wang, K., Gu, Z., Wang, S., Cai, W., Xiao, S. and Song, Q., *Adv. Mater.*, 29, 1606205, 2017.
113. Jia, Y., Kerner, R. A., Grede, A. J., Rand, B. P. and Giebink, N. C., *Nat. Photonics*, 11, 784, 2017.
114. Shang, Q., Li, M., Zhao, L., Chen, D., Zhang, S., Chen, S., Gao, P., Shen, C., Xing, J. and Xing, G., *Nano Lett.*, 20, 6636, 2020.
115. Li, Z., Moon, J., Gharajeh, A., Haroldson, R., Hawkins, R., Hu, W., Zakhidov, A. and Gu, Q., *ACS Nano*, 12, 10968, 2018.
116. Saliba, M., Wood, S. M., Patel, J. B., Nayak, P. K., Huang, J., Alexander-Webber, J. A., Wenger, B., Stranks, S. D., Hörantner, M. T. and Wang, J. T. W., *Adv. Mater.*, 28, 923, 2016.

117. Yu, H., Ren, K., Wu, Q., Wang, J., Lin, J., Wang, Z., Xu, J., Oulton, R. F., Qu, S. and Jin, P., *Nanoscale*, 8, 19536, 2016.
118. Jiang, L., Liu, R., Su, R., Yu, Y., Xu, H., Wei, Y., Zhou, Z.-K. and Wang, X., *Nanoscale*, 10, 13565, 2018.
119. Huang, C., Zhang, C., Xiao, S., Wang, Y., Fan, Y., Liu, Y., Zhang, N., Qu, G., Ji, H. and Han, J., *Science*, 367, 1018, 2020.
120. Zhang, H., Liao, Q., Wu, Y., Zhang, Z., Gao, Q., Liu, P., Li, M., Yao, J. and Fu, H., *Adv. Mater.*, 30, 1706186, 2018.
121. Gu, Z., Wang, K., Sun, W., Li, J., Liu, S., Song, Q. and Xiao, S., *Adv. Opt. Mater.*, 4, 472, 2016.
122. Liao, Q., Hu, K., Zhang, H., Wang, X., Yao, J. and Fu, H., *Adv. Mater.*, 27, 3405, 2015.
123. Wu, Z., Chen, J., Mi, Y., Sui, X., Zhang, S., Du, W., Wang, R., Shi, J., Wu, X., Qiu, X., Qin, Z., Zhang, Q. and Liu, X., *Adv. Opt. Mater.*, 6, 1800674, 2018.
124. Hu, Z., Liu, Z., Bian, Y., Liu, D., Tang, X., Hu, W., Zang, Z., Zhou, M., Sun, L., Tang, J., Li, Y., Du, J. and Leng, Y., *Adv. Opt. Mater.*, 5, 1700419, 2017.
125. Liang, Y., Shang, Q., Wei, Q., Zhao, L., Liu, Z., Shi, J., Zhong, Y., Chen, J., Gao, Y. and Li, M., *Adv. Mater.*, 31, 1903030, 2019.
126. Tang, B., Sun, L., Zheng, W., Dong, H., Zhao, B., Si, Q., Wang, X., Jiang, X., Pan, A. and Zhang, L., *Adv. Opt. Mater.*, 6, 1800391, 2018.
127. Qin, C., Sandanayaka, A. S. D., Zhao, C., Matsushima, T., Zhang, D., Fujihara, T. and Adachi, C., *Nature*, 585, 53, 2020.
128. Park, Y.-S., Bae, W. K., Baker, T., Lim, J. and Klimov, V. I., *Nano Lett.*, 15, 7319, 2015.
129. Du, W., Zhang, S., Zhang, Q. and Liu, X., *Adv. Mater.*, 31, 1804894, 2019.
130. Zhang, S., Shang, Q., Du, W., Shi, J., Wu, Z., Mi, Y., Chen, J., Liu, F., Li, Y. and Liu, M., *Adv. Opt. Mater.*, 6, 1701032, 2018.
131. Su, R., Ghosh, S., Wang, J., Liu, S., Diederichs, C., Liew, T. C. H. and Xiong, Q., *Nat. Phys.*, 16, 301, 2020.

4 Energy Transfer Processes in Phosphors

Jiahua Zhang and Hao Wu

CONTENTS

4.1 INTRODUCTION

Energy transfer (ET) is a common process in the phosphors containing luminescent centers. It usually takes place between two luminescent centers. The quantitative theories of ET have been given by Förster [1] and Dexter [2]. The luminescent centers include rare earth ions, transition metal ions as well as color centers. The luminescent centers doped in a phosphor may either be identical or different. ET between the identical centers may lead to energy diffusion among these identical centers, largely increasing the possibility that energy is captured by nonradiative centers, thus resulting in the well-known phenomena of concentration quenching. ET from one type of center to another type of center produces sensitized luminescence. Sensitized luminescence refers to ET that enables a luminescent center (activator) having no appreciable absorption band covering a given excitation wavelength to emit radiation when another type of center (sensitizer) is effectively excited at the given wavelength. This is the typical result of ET from the sensitizer to the activator. The sensitizer is also called *donor*, which is denoted hereafter by *D*. The activator is also called *acceptor* and denoted by *A*.

As the luminescent centers *D* and *A* are codoped in a phosphor, ET may allow both *D* and *A* to emit, generating two color emissions upon a single-color excitation. Such a doubly doped system, or even triply doped system, exhibits attractive application for full color white light generation upon blue and/or UV light-emitting diode (LED) excitation.

DOI: 10.1201/9781003098690-4

4.2 SPECTROSCOPIC EVIDENCE FOR ET

ET from luminescent D to A necessarily requires spectral overlap between D emission band and A absorption band. ET enables A to emit when only D is excited. As a result, the photoluminescence (PL) excitation (PLE) spectrum of A emission contains the characteristic PLE spectrum of D emission in the D–A codoped phosphors. The abovementioned behavior is a strong spectroscopic evidence for D–A ET. The effects of ET are reflected by intensity decrease and lifetime shortening of D luminescence followed by intensity increase of A luminescence with the increase of A concentration. Accordingly, the emission intensity ratio of A to D increases on increasing A concentration. The ratio A/D can be conveniently used to evaluate the efficacy of ET.

4.3 DETERMINATION OF EFFICIENCY OF ET

The D centers have different transfer efficiencies for each other because the spatially random distribution of A centers results in a distribution of transfer rates. As a result, the efficiency of ET is an average efficiency of whole D centers. If the distribution function of transfer rate is $f(w)$, the radiative rate of D is γ, and nonradiative rate of D in the absence of ET is w_0, the efficiency of ET is

$$\eta_{ET} = \int_{0}^{\infty} \frac{w}{\gamma + w_0 + w} f(w)\,dw \tag{4.1}$$

The luminescence efficiency of D is also an average expressed as

$$\eta_D = \int_{0}^{\infty} \frac{\gamma}{\gamma + w_0 + w} f(w)\,dw \tag{4.2}$$

with consideration that γ and w_0 are independent on transfer rate, Eqs. (4.1) and (4.2) give

$$\eta_{ET} = 1 - \frac{\eta_D}{\eta_{D0}} \tag{4.3}$$

thus,

$$\eta_{ET} = 1 - \frac{I_D}{I_{D0}} \tag{4.4}$$

where I_D is the steady-state intensity of D luminescence under continuous wave (CW) light excitation. If the number of excitation light photons absorbed by D per time is P, $I_D = \eta_D P$, and thus, $\eta_D/\eta_{D0} = I_D/I_{D0}$. η_{D0} and I_{D0} are responsible for the absence of ET, i.e., $w = 0$. It is clear that ET reduces the intensity of D luminescence, and the decrement, $I_{D0} - I_D$, is the part transferred to A. The luminescence intensity of D can be obtained in PL spectra as D is excited by CW light excitation.

Meanwhile, D–A transfer can speed up the decay of D in its excited state and, thus, shorten the decay time of D. The time evolution function, i.e., the function with normalized initial intensity, of D luminescence after a pulse excitation is expressed as

$$I_D(t) = \int_{0}^{\infty} e^{-(\gamma + w_0 + w)t} f(w)\,dw \tag{4.5}$$

From Eqs. (4.2) and (4.5),

$$\int_0^\infty I_D(t)\,dt = \frac{\eta_D}{\gamma} \tag{4.6}$$

Eqs. (4.3) and (4.6) give

$$\eta_{ET} = 1 - \frac{\int_0^\infty I_D(t)\,dt}{\int_0^\infty I_{D0}(t)\,dt} \tag{4.7}$$

Here we introduce τ_0 and τ, defined by

$$\tau_0 = \int_0^\infty I_{D0}(t)\,dt \tag{4.8}$$

$$\tau = \int_0^\infty I_D(t)\,dt \tag{4.9}$$

Eq. (4.7) can be rewritten as

$$\eta_{ET} = 1 - \frac{\tau}{\tau_0} \tag{4.10}$$

Note that if $I_{D0}(t)$ is a single exponential decay function, τ_0 is just the decay time. Using Eqs. (4.4) and (4.10), the efficiency of ET can be calculated based on the decrement in D luminescence intensity or in D luminescence lifetime.

The transfer efficiency can also be calculated based on the emission intensity ratio of A and D. Let η_A be the quantum efficiency of A emission and use Eq. (4.3), the intensity ratio of A emission to D emission satisfies

$$\frac{I_A}{I_D} = \frac{\eta_A \eta_{ET}}{\eta_D} = \frac{\eta_A}{\eta_{D0}} \frac{\eta_{ET}}{1 - \eta_{ET}} \tag{4.11}$$

where the quantum efficiency of an emitting center can be calculated by the ratio of its emitting level lifetime defined as the integration of fluorescence decay curve with a normalized initial intensity as in Eq. (4.9) to its radiative lifetime τ_r, written as

$$\eta = 1 - \frac{\tau}{\tau_r} \tag{4.12}$$

where τ is the integration of fluorescence decay curve with a normalized initial intensity as in Eq. (4.9).

4.4 LIFETIMES

4.4.1 Excited State Lifetime

The excited state lifetime refers to the average time the molecule stays in its excited state after removal of excitation. Suppose D is excited with flashlight at $t = 0$. When no A is present, the excited state population of D decays exponentially after the excitation.

$$N_0(t) = N_0(0)\exp\left(-\frac{t}{\tau_0}\right) \tag{4.13}$$

where $N_0(t)$ and τ_0 are the excited state population of D at time t and the excited state lifetime, respectively, in the absence of A. The inverse of the lifetime is the total decay rate of excited D.

When A is present, the decay rate of excited D is increased by ET, which is strongly dependent on the D–A distance. If D and A ions distribute randomly in space, there is a transfer rate distribution. This means a lifetime distribution, resulting in multiexponential or nonexponential decay of D excited state population $N(t)$ in the presence of A. In this case, the average excited state lifetime ($\langle\tau\rangle$) can be calculated by

$$\langle\tau\rangle = \frac{\int_0^\infty N(t)\,dt}{N(0)} \tag{4.14}$$

In case of exponential decay for the absence of A as described in Eq. (4.13), one obtains

$$\langle\tau\rangle = \tau_0 \tag{4.15}$$

In isolated luminescent centers, the radiative transition is a single molecular process; i.e., the radiative rate is a constant. If all the centers have the same radiative rate, for instance, in case of the identical centers, the intensity is proportion to the emitting state population. This means the intensity decay function is identical to the emitting state population decay function. In this case, the right side of Eq. (4.14) is equal to the integral of the intensity decay function, expressed as,

$$\langle\tau\rangle = \int_0^\infty I(t)\,dt = \tau \tag{4.16}$$

In ET system, the multiexponential or nonexponential decay of D luminescence is generally resulted from different transfer rates. The radiative rate is the same for all the donors, indicating that D luminescence intensity is proportional to its excited state population. Therefore, Eq. (4.14) is suitable for calculating the average lifetime of the excited donors in ET system.

If a nonexponential decay of luminescence is resulted from different radiative rates, $I(t)$ no longer follows the same function as $N(t)$, and to achieve average excited state lifetime, using Eq. (4.16) is invalid. Luminescent materials with different radiative rates may be caused by codoping different types of luminescent centers or the same center on different lattice sites or mixing different phosphors etc.

4.4.2 FLUORESCENCE LIFETIME

In case of exponential decay of fluorescence intensity, the determination of fluorescence lifetime is simple. In case of complex decay, one may find an average fluorescence lifetime ($\bar{\tau}$) defined by [3]

$$\bar{\tau} = \frac{\int_0^\infty t\,I(t)\,dt}{\int_0^\infty I(t)\,dt} \tag{4.17}$$

The denominator is the number of total photons emitted, and the numerator is the sum of the delay times for each emitted photon after flash excitation. Hence, $\bar{\tau}$ means the average delay time for emitting a photon after excitation. Here, it should be noted that the $\langle\tau\rangle$ defined by Eq. (4.14) has different physical meaning from $\bar{\tau}$ defined by Eq. (4.17). $\langle\tau\rangle$ is the average excited state lifetime.

One may wonder why the average excited state lifetime is not identically equal to the average fluorescence lifetime. This is because the excited state at different delay times may have different emission efficiencies due to D centers having different ET rates. Only in case of exponential decay of D emission, i.e., D centers have the same transfer rate, one observes

$$\bar{\tau} = \langle \tau \rangle = \tau_0 \qquad (4.18)$$

4.5 THEORY OF ENERGY TRANSFER

D–A ET takes place generally by multipolar interaction or exchange interaction. Both the mechanisms rely on the separation between D and A. The former is a long-range interaction through coulomb field, and the latter is a short-range interaction through the overlapping of electronic wavefunctions of D and A. Quantitative theories of resonant ET by electric multipole-multipole interaction or exchange interaction have been given by Förster and Dexter.

4.5.1 Electric Multipolar Interaction

In electric multipolar interaction, the D–A transfer rate is proportional to an inverse power of distance r, written as

$$W_{ET} = \frac{\alpha}{r^m} \qquad (4.19)$$

where α is a rate constant for ET, $m = 6$, 8, and 10 corresponds to electric dipole-dipole interaction, electric dipole-quadrupole interaction, and electric quadrupole-quadrupole interaction, respectively.

In case of no diffusion among donors, or the average D–D transfer rate is much less than the average D–A transfer rate, the decay function of D luminescence is given by [3]

$$I(t) = \exp\left[-\frac{t}{\tau_0} - \frac{4\pi}{3}\Gamma\left(1 - \frac{3}{m}\right)x\alpha^{3/m}t^{3/m} \right] \qquad (4.20)$$

where x is the concentration of A. From Eq. (4.16), $\ln\{\ln[I_0(t)/I(t)]\}$ is proportional to $\ln(t)$ with a slop of $3/m$, $\ln[I(t)/I_0(t)]$ is proportional to $t^{3/m}$ with a slop of $-4\pi\,\Gamma\left((1-3/m)x\alpha^{3/m}/3\right)$, where $I_0(t) = \exp(-t/\tau_0)$. The parameters m and α can be obtained from the slops by fitting the measured decay curve. It is important to note that the fitting should be focused on the tail of the decay because Eq. (4.20) is obtained by assuming the nearest distance between a donor and an acceptor to be 0, leading to an infinite initial transfer rate.

Transfer critical distance (R_c) is an important parameter. This distance is defined as the distance between a donor and an acceptor for which the transfer rate is equal to the decay rate of the excited D in the absence of ET, then

$$\frac{\alpha}{R_c^m} = \tau_0^{-1} \qquad (4.21)$$

if m and α are obtained by fitting optical data using Eq. (4.20), R_c can be calculated by Eq. (4.21). The value of R_c can also be calculated by using the quantitative theories for D–A multipolar ET given by Förster and Dexter. For instance, the transfer rate for dipole-dipole interaction is given by Dexter equation [2]

$$W_{ET} = \frac{3\hbar^4 c^4 \sigma_A}{4\pi n^4 \tau_0 r^6} \int \frac{f_D(E)F_A(E)}{E^4}\,dE \qquad (4.22)$$

where n is the refractive index of the material, σ_A is the absorption cross-section of A, τ_0 is the lifetime of D in the absence of ET. $f_D(E)$ and $F_A(E)$ represent the shape of D emission and A absorption spectra, respectively, which are normalized in spectral area. The integrals are the energy overlap of the two spectra. When transfer rate is equal to the intrinsic decay rate of D, i.e., $W_{ET} = 1/\tau_0$, Eq. (4.22) can be solved for R_c.

There is another simple method for determination of R_c from the critical concentration (x_c) of acceptors for D–A system. The critical concentration is defined as the concentration at which the efficiency of ET is equal to the efficiency of D luminescence; i.e., the efficiency of D luminescence reaches one-half of its efficiency in the absence of A. The value of x_c can be obtained with steady-state luminescence measurement as the intensity of D luminescence is decreased to one-half of its intensity in the absence of A, expressed as

$$I_D\left(x_c\right) = \frac{I_{D0}}{2} \tag{4.23}$$

At acceptor concentration of x_c, one may think there is on average one A ion per volume $V/x_c n$, where V is the volume of the crystallographic unit cell, n is the number of lattice sites in the unit cell that can be occupied by A. The radius of a sphere centered at D with this volume can be calculated. Then the D–A critical distance R_c is proximately equal to the radius derived from

$$R_c = \left(\frac{3V}{4\pi x_c n}\right)^{1/3} \tag{4.24}$$

If D and A are identical centers, ET between them may result in concentration quenching. If the critical concentration of concentration quenching of the luminescence can be determined experimentally, the radius of a sphere with a volume occupied on average by one center is achievable using Eq. (4.24). When a space is fully filled by the spheres, the critical distance between two centers is proximately twice the radius of the sphere [4],

$$R_c = 2\left(\frac{3V}{4\pi x_c n}\right)^{1/3} \tag{4.25}$$

4.5.2 Exchange Interaction

Exchange interaction is a short-range interaction because the overlapping of the electron clouds of D and A is required. The rate of exchange interaction is given by [2]

$$W_{ET} = \left(\frac{2\pi}{\hbar}\right) K^2 \exp\left(-\frac{2r}{L}\right) \int f_D(E) F_A(E) dE \tag{4.26}$$

where K is a constant with dimension of energy, L is a constant called "the effective average Bohr radius", the integrals are the spectral overlap of D emission and A absorption. It is convenient to rewrite Eq. (4.26) by introducing R_c in the form,

$$W_{ET} = \tau_0^{-1} \exp\left[\frac{2(R_c - r)}{L}\right] \tag{4.27}$$

The decay of D luminescence is given by [3]

$$I(t) = \exp\left[-\frac{t}{\tau_0} - \gamma^{-3}\frac{4\pi}{3}R_c^3 g\left(e^\gamma\frac{t}{\tau_0}\right)x\right] \qquad (4.28)$$

where $\gamma = 2R_c/L$, g is a function of time. The exponential decay of transfer rate indicates that the exchange interaction is very sensitive to distance. The exchange transfer prefers the nearest or the second nearest D–A pairs. Such a behavior of exchange interaction is similar to the Perrin model, in which the transfer rate is assumed constant if A exists within a critical distance and is zero outside the range. This model is thought to be the most applicable to the exchange interaction. In Perrin model, the steady-state luminescence of D as a function of concentration of A is simply written as [3]

$$\frac{I_D}{I_{D0}} = \exp\left(-\frac{x}{x_c}\right) \qquad (4.29)$$

We have discussed the resonant ET. If the spectral overlap of D and A is small, the D–A ET by both electronic multipolar or exchange mechanisms requires phonon emission or absorption for compensating the energy difference between D and A transitions, called phonon-assisted ET. In this case, the transfer rate is strongly influenced by the energy separation, phonon energy, and temperature; the reader is referred to [5] for details.

4.5.3 DIFFUSION LIMITED ENERGY TRANSFER

In the previous discussion, the ET between donors is not considered. The ET between identical centers is called energy migration. This migration speeds up D–A transfer rate. Yokota and Tanimoto [6] have obtained a general solution for the donor decay function including both diffusion within the D system and D–A ET via dipole-dipole coupling. Their expression is

$$I(t) = \exp\left[-\frac{t}{\tau_0} - \frac{4}{3}\pi^{3/2}x(\alpha t)^{1/2}\left(\frac{1+10.87y+15.50y^2}{1+8.743y}\right)^{3/4}\right] \qquad (4.30)$$

where $y = D\alpha^{-1/3}t^{2/3}$, D is the diffusion constant between donors, depending on donor concentration. If the migration rate is much faster than D–A transfer rate, for instance, in case of high D concentration, the D–A transfer rate can keep its initial rate over the time, resulting in a single exponential decay of D luminescence [7]. This is called rapid diffusion limited ET.

4.6   ET IN PHOSPHORS

4.6.1 $Ba_2Lu_5B_5O_{17}$:Ce^{3+}, Tb^{3+} GREEN PHOSPHOR

Tb^{3+} is a green-emitting center due to the $^5D_4$–$^7F_5$ emission at 543 nm within the 4f configuration. However, Tb^{3+} has a small absorption cross section for excitation light due to parity forbidden transitions, resulting in low quantum conversion efficiency. $Ba_2Lu_5B_5O_{17}$ (BLBO):Ce^{3+} is an efficient blue phosphor. Adding Tb^{3+} in BLBO:Ce^{3+} can construct a Ce^{3+}–Tb^{3+} ET system and produce strong green emission of Tb^{3+}.

The PLE and PL spectra of Ce^{3+} or Tb^{3+} singly doped and codoped BLBO phosphors are presented in Figure 4.1. The PLE spectrum of BLBO:1% Ce^{3+} contains three distinctive bands in the UV region within 250 and 400 nm with the main band peaked at 348 nm. These PLE bands

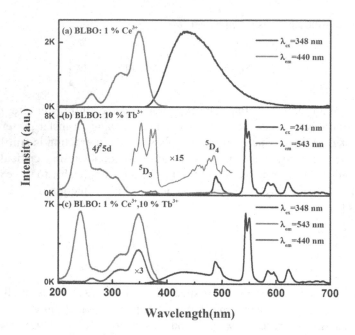

FIGURE 4.1   PLE and PL spectra of BLBO:1% Ce³⁺ (a), BLBO:10% Tb³⁺ (b), and BLBO:1% Ce³⁺, 10% Tb³⁺ (c). (Xiao Y., Hao Z.D., Zhang L.L. et al., J. Mater. Chem. C 6, 5984, 2018. With permission from Ref. [8].)

are assigned to the 4f–5d transitions of Ce³⁺. Upon 348-nm excitation, the PL spectrum exhibits a broad blue band peaked at 440 nm, which is assigned to the transition from the lowest 5d state to the 4f ground state of Ce³⁺. The PL spectrum of Tb³⁺ singly doped BLBO:10% Tb³⁺ shows a well-known group of emission lines with the strongest line located at 543 nm. These lines originated from the $^5D_4$–^7F_J (J = 6, 5, 4, and 3) transitions of Tb³⁺, peaked at 488, 543, 585, and 625 nm, respectively. The PLE spectrum of the Tb³⁺ singly doped sample includes transitions from the ground state $^7F_6$ to the $^5D_4$, $^5D_3$, and the 4f⁷5d states. Clearly, a big spectral overlap between Ce³⁺ PL bands and Tb³⁺ PLE bands corresponding to $^7F_6$–$^5D_4$ absorption is observed, indicating the possibility of ET from Ce³⁺ to Tb³⁺ according to the Dexter's theory. As shown in Figure 4.1(c), the ET is confirmed by the luminescence properties of Ce³⁺ and Tb³⁺-codoped BLBO. The PLE spectrum of Tb³⁺ emission at 543 nm shows the characteristic 4f–5d transition band of Ce³⁺ in the range of 250–400 nm for the BLBO:1% Ce³⁺, 10% Tb³⁺ sample. Moreover, upon Ce³⁺ excitation at 348 nm, not only a Ce³⁺ emission but also a strong Tb³⁺ emission appears in the PL spectrum. These results indicate the existence of effective ET from Ce³⁺ to Tb³⁺ in the BLBO:Ce³⁺, Tb³⁺ phosphor.

Considering the Ce³⁺ to Tb³⁺ ET, a set of BLBO:1% Ce³⁺, x Tb³⁺ (x = 0%, 2%, 10%, 15%, 20%, and 30%) samples was synthesized to investigate the effect of ET on the PL properties. The PL spectra for different Tb³⁺ concentrations, x, are shown in Figure 4.2(a). It is clear that with an increase in x, the intensity of the Ce³⁺ blue emission continuously decreases followed by the enhancement of the Tb³⁺ green emission; this reflects the enhanced ET. Correspondingly, blue to green emission color tuning is observed, as illustrated in the images of the phosphors under irradiation by a 365-nm mercury lamp, as shown in Figure 4.2(b). The color tuning is also displayed in the Commission Internationale de L'Eclairage (CIE) chromaticity diagram, showing a shift of the color points from blue (0.1685, 0.1483) for the absence of Tb³⁺ to green (0.3531, 0.4911) for 30% Tb³⁺. It may also be seen in Figure 4.2(a) that the Tb³⁺ emission intensity decreases with an increase in x from 20% to 30%. This decrease is attributed to the Tb³⁺–Tb³⁺ internal concentration quenching based on the experimental observation of a notable shortening of Tb³⁺ fluorescence lifetime as x is increased from 20% to 30%, as shown in Figure 4.2(c).

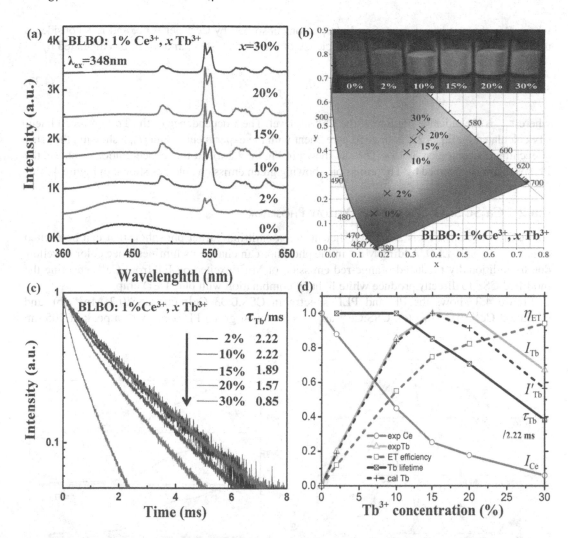

FIGURE 4.2 (a) PL spectra of the BLBO:1% Ce^{3+}, x Tb^{3+} phosphors under 348-nm excitation. (b) Emission color tuning as a function of Tb^{3+} concentration illustrated in the CIE 1931 chromaticity diagram together with the digital images of the phosphors obtained under 365-nm UV lamp irradiation. (c) Decay curves of Tb^{3+} emission in the phosphors after pulse excitation at 348 nm when monitored at 543 nm, herein, the lifetimes are defined as the area under the decay curves. (d) The x dependence of Ce^{3+}, Tb^{3+} emission intensities, Tb^{3+}-$^5D_4$ lifetimes, energy transfer efficiencies, and calculated Tb^{3+} emission intensities. The maximum of the emission intensities or the emission lifetimes is normalized. (Xiao Y., Hao Z.D., Zhang L.L. et al., J. Mater. Chem. C 6, 5984, 2018. With permission from Ref. [8].)

The x dependencies of Ce^{3+} emission intensity, denoted by I_{Ce}, Tb^{3+} emission intensity, denoted by I_{Tb}, and Tb^{3+} fluorescence lifetime, denoted by τ_{Tb}, are plotted in Figure 4.2(d), where each set of data is normalized to its maximum. The decrease of Tb^{3+} emission intensity with the increase of x is apparent. According to Eq. (4.4), the efficiency of ET can be calculated by

$$\eta_{ET} = 1 - \frac{I_{Ce}}{I_{Ce0}} \tag{4.31}$$

The intensity of Tb^{3+} emission excited by ET from Ce^{3+} can be written as

$$I_{Tb} = \eta_{Tb}\eta_{ET}P \tag{4.32}$$

where P is the number of excitation light photons absorbed by Ce^{3+} per time, and it is considered unchanged with x, η_{Tb} is the quantum efficiency of the $Tb^{3+5}D_4$ state, and it can be calculated by

$$\eta_{Tb} = \frac{\tau_{Tb}}{\tau_{Tbr}} \qquad (4.33)$$

where τ_{Tbr} is the radiative lifetime of the $Tb^{3+5}D_4$ level. The x dependence of the Tb^{3+} emission intensity calculated by Eq. (4.32) is in good agreement with the experimental data (I'_{Tb}), showing the same quenching concentration of 15% Tb^{3+}, as shown in Figure 4.2(d). At this concentration or above, the PL spectrum is governed by Tb^{3+} emission, showing green emission color, as shown in Figure 4.2(b).

4.6.2 $Ca_3Sc_2Si_3O_{12}$:Ce^{3+}, Mn^{2+} YELLOW PHOSPHOR

$Ca_3Sc_2Si_3O_{12}$ (CSS):Ce^{3+} is a highly efficient green-emitting garnet phosphor, that can be excited efficiently by blue LED. Adding Mn^{2+} in the phosphor can change its luminescence color to yellow due to additionally produced orange-red emission of Mn^{2+} excited by ET from Ce^{3+}, enabling the modified CSS to directly produce white light by combination with blue LED chip.

Figure 4.3 shows the PL and PLE spectra in CSS:0.03 Ce^{3+} (a), CSS:0.2 Mn^{2+} (b), and CSS:0.03 Ce^{3+}, 0.2 Mn^{2+} (c). CSS:Ce^{3+} appears a typical green PL band with a peak at 505 nm

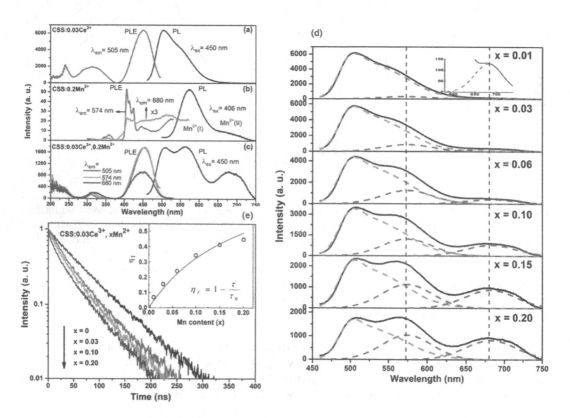

FIGURE 4.3   PLE and PL spectra for (a) CSS:0.03 Ce^{3+}, (b) CSS:0.2 Mn^{2+}, and (c) CSS:0.03 Ce^{3+}, 0.2 Mn^{2+}. (d) PL spectra (solid) of CSS:0.03 Ce^{3+}, x Mn^{2+} with x = 0.01, 0.03, 0.06, 0.10, 0.15, and 0.20, respectively, under 450-nm excitation. The individual emission of Ce^{3+}, Mn^{2+}(I), and Mn^{2+}(II) are also presented (dashed). (e) Time decay curves of Ce^{3+} emission at 505 nm for different x after pulse excitation at 450 nm with an inset: dependence of the ET efficiency on content x. (Liu Y.F., Zhang X., Hao Z.D. et al., J. Mater. Chem. 21, 16379, 2011. With permission from Ref. [9].)

and a shoulder around 540 nm, originated from the transitions from 5d to $^2F_{5/2}$ and $^2F_{7/2}$ of Ce^{3+}, respectively. Meanwhile, the PLE spectrum of the green emission exhibits an intense excitation band around 450 nm, well matching the emitting wavelength of the blue InGaN LEDs. The PL spectrum of Mn^{2+} singly doped CSS exhibits two emission bands, one is an orange emission band around 574 nm (named Mn^{2+}(I)) and the other is a red emission band around 680 nm (named Mn^{2+}(II)). The appearance of two emission bands indicates two Mn^{2+} sites in CSS. In general, the emission of Mn^{2+} originates from the spin-forbidden $^4T_1(4G)$–$^6A_1(6S)$ transition. The excitation spectra show typical forbidden d–d transitions of Mn^{2+} at 406 nm, resulting in very weak PL intensities compared to Ce^{3+}.

In Ce^{3+}, Mn^{2+} codoped CSS (c), the PL spectrum upon Ce^{3+} excitation at 450 nm exhibits not only the Ce^{3+} emission band at 505 nm but also Mn^{2+} emission bands at 574 and 680 nm, which is much stronger than that in Mn^{2+} singly doped CSS. The PLE spectra of the two Mn^{2+} emissions are dominated by Ce^{3+} PLE bands. These results give strong evidence for the effective Ce^{3+} to Mn^{2+} ET. The occurrence of ET can be clearly understood as noticing the spectral overlap between the Ce^{3+} emission band in CSS:0.03 Ce^{3+} and the Mn^{2+} excitation band in CSS:0.2 Mn^{2+}. Furthermore, it is found that the relative intensity of Mn^{2+}(II) to Mn^{2+}(I) is enhanced in the codoped CSS comparison with the Mn^{2+} singly doped CSS. This is attributed to a higher transfer efficiency to Mn^{2+}(II) than Mn^{2+}(I) because the Mn^{2+}(II) has a broader PLE band than Mn^{2+}(I), leading to a larger spectral overlap with the Ce^{3+} emission band.

Color tunable luminescence was realized in CSS:0.03 Ce^{3+}, x Mn^{2+} with a fixed Ce^{3+} concentration and variable Mn^{2+} concentrations, as shown in Figure 4.3(d). Each PL spectrum is decomposed into a Ce^{3+} band, a Mn^{2+}(I) band, and a Mn^{2+}(II) band. It is apparent that with increasing Mn^{2+} content, the Ce^{3+} emission reduces followed by the enhancement of Mn^{2+}(I) and the Mn^{2+}(II) emissions due to ET. For $x > 0.1$, the Mn^{2+} emissions are enhanced greatly, which enriches the emission in long wavelength visible region and consequently changes the phosphor from green emitting ($x = 0$) to the yellow emitting ($x = 0.2$) gradually. Thus, the Ce^{3+}, Mn^{2+} codoped CSS is able to generate white light based on blue LED excitation.

The Ce^{3+} to Mn^{2+} ET also results in lifetime shortening of donor Ce^{3+} emission, showing the continuously accelerated decay of the Ce^{3+} emission with the increase of Mn^{2+} concentration (Figure 4.3(e)). The transfer efficiency calculated using the Ce^{3+} emission lifetimes are also presented.

4.6.3 CaO:Ce^{3+}, Mn^{2+} Orange Phosphor

The Ce^{3+}–Mn^{2+} ET was also used to develop orange phosphor based on CaO host. The Ce^{3+} singly doped CaO is an efficient yellow phosphor with an emission band at 550 nm and a blue PLE band at 450 nm (Figure 4.4(a)). The Mn^{2+} singly doped CaO emits in the orange at 600 nm but with a low emission intensity due to weak absorption of excitation light by Mn^{2+} itself, as shown in Figure 4.4(b). One can find that Mn^{2+} has a PLE band at 550 nm that just coincides with the emission band of Ce^{3+}, implying an efficient ET from Ce^{3+} to Mn^{2+} if they are codoped in CaO. Indeed, in Ce^{3+} and Mn^{2+}-codoped CaO, the PLE band for monitoring Mn^{2+} emission at 600 nm exhibits only the typical Ce^{3+} PLE band peaked at 460 nm rather than the 420 and 550 nm PLE bands of Mn^{2+} itself, indicating efficient ET from Ce^{3+} to Mn^{2+} (Figure 4.4(c)). As a result, the PL spectrum appears not only strong Ce^{3+} emission band of 550 nm but also strong Mn^{2+} emission band at 600 nm when only Ce^{3+} is excited at 450 nm.

Based on Ce^{3+} to Mn^{2+} ET, the CaO phosphors doped with a fixed Ce^{3+} concentration but various Mn^{2+} concentrations exhibit tunable emitting colors from yellow to orange, as shown in Figure 4.4(d). With the increase of Mn^{2+} concentration, the Ce^{3+} emission reduces continuously followed by enhancement of the Mn^{2+} emission that already dominates the emission as x only higher than 0.01, reflecting highly efficient Ce^{3+} to Mn^{2+} ET. The quantum efficiency of ET was calculated by using Ce^{3+} emission intensity. The calculated transfer efficiencies for different concentrations of

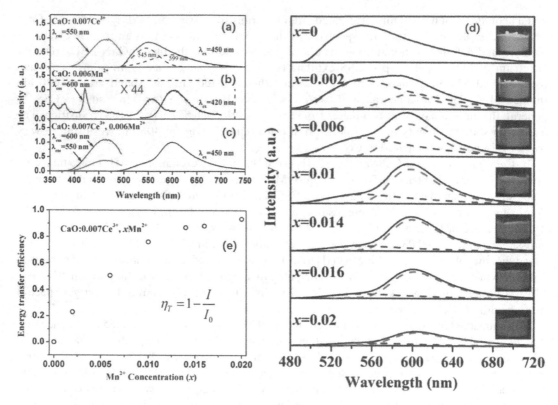

FIGURE 4.4 PLE and PL spectra for (a) CaO:0.007 Ce³⁺, (b) CaO:0.006 Mn²⁺, and (c) CaO:0.007 Ce³⁺, 0.006 Mn²⁺. The CaO:0.006 Mn²⁺ excitation and emission intensities are magnified 44 times. (d) PL spectra of CaO:0.007 Ce³⁺, x Mn²⁺ (x = 0, 0.002, 0.006, 0.01, 0.014, 0.016, and 0.02) excited under 450 nm at room temperature. The emission of Ce³⁺ (blue curves) and Mn²⁺ (red curves) are presented as dashed lines. The insets are the corresponding digital images. (e) Dependence of the energy transfer efficiency (η_{ET}) on Mn²⁺ doping concentration (x). (Feng L.Y., Hao Z.D., Zhang X. et al., Dalton Trans. 45, 1539, 2016. With permission from Ref. [10].)

Mn²⁺ are shown in Figure 4.4(e). The efficiency can reach as high as 94% for x = 0.02. Such a high transfer efficiency is attributed to the large spectral overlap between the Ce³⁺ emission band and the Mn²⁺ absorption band at 550 nm.

4.6.4 SUPER BROADBAND NEAR-INFRARED Ca₂LuZr₂Al₃O₁₂:Cr³⁺, Yb³⁺ GARNET PHOSPHOR

Super broadband near-infrared (NIR) phosphor converted light-emitting diodes (pc-LEDs) are future light sources in NIR spectroscopy applications such as food testing. The singly doped Ca₂LuZr₂Al₃O₁₂ (CLZA):Cr³⁺ phosphor shows a NIR emission band at 800 nm with a bandwidth of 150 nm, which is needed to be further broadened for applications. This problem is solved by doping Yb³⁺ in CLZA:Cr³⁺ to construct a Cr³⁺–Yb³⁺ ET system. Benefited from the superposition of Cr³⁺ emission and highly efficient Yb³⁺ emission excited by ET from Cr³⁺, a super broadband with the bandwidth of 320 nm and a simultaneously enhanced quantum efficiency of 77.2% are realized.

The PL and PLE spectra of Cr³⁺ singly doped and Ce³⁺, Yb³⁺ codoped CLZA are shown in Figure 4.5(a). The Cr³⁺ singly doped CLZA:8% Cr³⁺ shows a NIR band at 780 nm and three PLE bands as well as a PLE line. The emission band is originated from Cr³⁺·⁴T₂–⁴A₂ transition. The PLE peaks at 300, 460, 640, and 692 nm are assigned to the transition from Cr³⁺·⁴A₂ to ⁴T₁ (4P), ⁴T₁ (4F), ⁴T₂, and ²E transitions, respectively. The PL spectrum of the codoped CLZA appears an additional emission band peaked at 1032 nm, that is originated from Yb³⁺²F₅/₂–²F₇/₂ transition. The

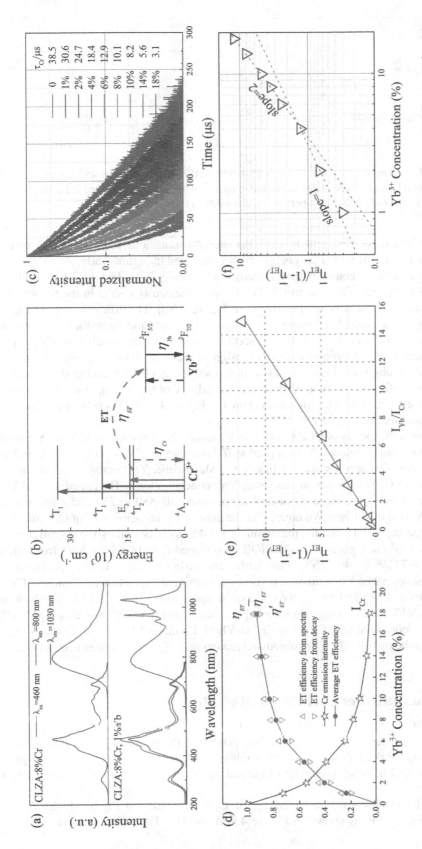

FIGURE 4.5 (a) PL and PLE spectra of CLZA:8% Cr³⁺ and CLZA:8% Cr³⁺, x Yb³⁺ (x = 1%). (b) Schematic energy level diagrams of Cr³⁺ and Yb³⁺ with involved ET process. (c) Decay curves of Cr³⁺ emission after pulse excitation with involved ET process. (d) The x dependence of Cr³⁺ emission intensity and ET efficiencies calculated based on Cr³⁺ emission intensities, Cr³⁺ emission lifetimes and the average of them. (e) The dependence of $\bar{\eta}_{ET}/(1-\bar{\eta}_{ET})$ on luminescence intensity ratio I_{Yb}/I_{Cr}. (f) Double-logarithmic plots of x dependence on $\bar{\eta}_{ET}/(1-\bar{\eta}_{ET})$. (He S., Zhang L.L., Wu H. et al., Adv. Optical Mater. 8, 1901684, 2020. With permission from Ref. [11].)

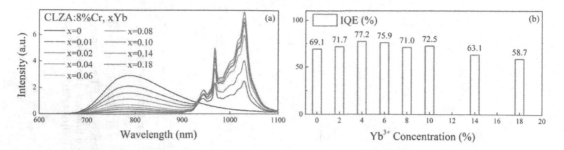

FIGURE 4.6 (a) PL spectra of CLZA:8% Cr^{3+}, x Yb^{3+} ($x = 0$–0.18) phosphors under 455-nm excitation by LD. (b) The change of IQE with x under 460-nm excitation by a xenon lamp. (He S., Zhang L.L., Wu H. et al., Adv. Optical Mater. 8, 1901684, 2020. With permission from Ref. [11].)

PLE spectrum of Yb^{3+} emission exhibits the Cr^{3+} characteristic transitions, indicating ET from Cr^{3+} to Yb^{3+}. The ET route with the schematic energy levels is depicted in Figure 4.5(b).

The decay pattern of Cr^{3+} emission shows lifetime shortening from 38.5 to 3.1 μs with the increase of x to 18% (Figure 4.5(c)) due to ET. The ET simultaneously results in the reduction of Cr^{3+} emission intensities on increasing x, as shown in Figure 4.5(d). The efficiency ($\bar{\eta}_{ET}$) of ET is obtained by averaging that calculated using Cr^{3+} emission lifetimes and intensities. The transfer efficiency increases with the increase of Yb^{3+} concentration and even approach 1 at high doping level of Yb^{3+} (Figure 4.5(d)). A proportional relationship between $\bar{\eta}_{ET}/(1-\bar{\eta}_{ET})$ and Yb^{3+} to Cr^{3+} intensity ratio I_{Yb}/I_{Cr} is observed (Figure 4.5(e)) as described in Eq. (4.11), meaning the validity of the calculated transfer efficiencies. Meanwhile, the x dependence of $\bar{\eta}_{ET}/(1-\bar{\eta}_{ET})$ shows a slope of 1 for small x and a slope close to 2 for high x, as shown in Figure 4.5(f). This behavior reflects the effect of dipole-dipole mechanism for ET.

Figure 4.6(a) shows PL spectra of CLZA:8% Cr, x Yb under 455-nm excitation. The Cr^{3+} singly doped sample shows a single emission band peaked at 780 nm with the FWHM of 150 nm. As Yb^{3+} is added, the Yb^{3+} emission appears due ET from Cr^{3+}. Meanwhile, Yb^{3+} emission grows up followed by the decrease of Cr^{3+} emission on increasing Yb^{3+} concentration x. The appearance of Yb^{3+} emission markedly expands the emission spectral range from 730–880 to 730–1050 nm, leading to a super broad FWHM of 320 nm. We notice that the sum of the emission area of Cr^{3+} and Yb^{3+} increases with increasing x and reach the maximum at $x = 4$%, as indicated by the overall internal quantum efficiency (IQE) in Figure 4.6(b). The IQE is enhanced from 69.1% for Yb^{3+} free sample to the maximum of 77.2% for 4% Yb^{3+}. With further increasing x from 4%, the IQE begins to decrease. The measurement of Yb^{3+} emission lifetimes determine the quantum efficiencies of Yb^{3+} emission in range from 88% to 83.5% for the codoped samples with x from 1% to 10%, which are much higher than 69.1% for Cr^{3+} singly doped sample. This means that Yb^{3+} acts as a highly efficient emitting center compared with Cr^{3+}. The Cr^{3+} to Yb^{3+} ET enables Yb^{3+} to emit instead of Cr^{3+}, therefore enhancing the overall IQE. The subsequent reduction of IQE is attributed to concentration quenching of Yb^{3+}.

4.6.5 Dynamical Analysis of ET in $Y_3Al_5O_{12}$:Ce^{3+}, Pr^{3+}

In $Y_3Al_5O_{12}$ (YAG):Ce^{3+}, Pr^{3+} the emission decay of Ce^{3+} is speeded up accompanied by nonexponential behavior with increasing Pr^{3+} concentration, exhibiting the typical effect of ET, as shown in Figure 4.7(a). Meanwhile, ET in form of cross relaxation (CR) between Pr^{3+} ions takes place, resulting in pronounced lifetime shortening of the red $^1D_2 \rightarrow {}^3H_4$ emission of Pr^{3+}, as shown in Figure 4.7(b).

The analysis of the measured decay curves of Ce^{3+} emission (in Figure 4.7(a)) is conducted based on Eq. (4.20), as shown in Figure 4.8(a) and (b). The parameters m and α are

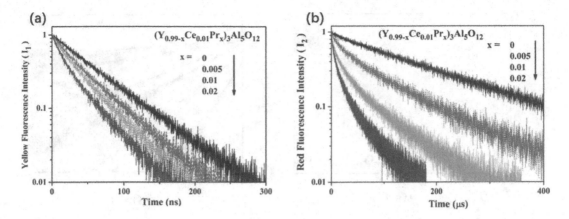

FIGURE 4.7 (a) Decay curves of Ce^{3+} yellow emission in $(Y_{0.99-x}Ce_{0.01}Pr_x)_3Al_5O_{12}$ for $x = 0$, 0.005, 0.01, and 0.02. (b) Decay curves of Pr^{3+} red emission in $(Y_{0.99-x}Ce_{0.01}Pr_x)_3Al_5O_{12}$ for $x = 0$, 0.005, 0.01, and 0.02. (Wang L., Zhang X., Hao Z.D. et al., J. Appl. Phys. 108(1), 093515, 2010. With permission from Ref. [12].)

determined to be 6 and 4.5×10^{-36} cm^6 s^{-1}, respectively. Using Eq. (4.21), R_c is calculated to be 0.81 nm.

Using the same method as mentioned earlier, the fitting of the decay curves of Pr^{3+} emission is performed, as shown in Figure 4.9(a) and (b). The parameters m, α, and R_c for $Pr^{3+}-Pr^{3+}$ ET are obtained. They are 6, 2.4×10^{-38} cm^6 s^{-1} and 1.3 nm, respectively. One can find the transfers for $Ce^{3+}-Pr^{3+}$, and for $Pr^{3+}-Pr^{3+}$ all belong to electric dipole-dipole interaction. The critical distance for $Pr^{3+}-Pr^{3+}$ transfer (1.3 nm) is longer than that for $Ce^{3+}-Pr^{3+}$ transfer (0.81 nm), indicating that $Pr^{3+}-Pr^{3+}$ transfer is a strong competing process with respective to $Pr^{3+}.^1D_2$ $\rightarrow {}^3H_4$ emission.

4.7 ROUND TRIP ET IN PHOSPHORS

In contrast to one-way ET described earlier, the round trip ET between centers A and B is described as after A transfers energy to B, B can transfer the energy back again to the same A not another A. The unique feature of round-trip ET exhibits efficient ET from B back to A because B in the nearest A–B pairs is preferentially excited by the first step ET from A to B. The back transfer from B to

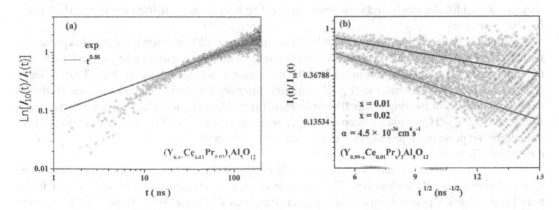

FIGURE 4.8 (a) ln–ln plot of $\ln[I_{10}(t)/I_1(t)]$ vs t for Ce^{3+} yellow emission in $(Y_{0.99-x}Ce_{0.01}Pr_x)_3Al_5O_{12}$ with $x = 0.02$. (b) Plotted $\ln[I_1(t)/I_{10}(t)]$ vs $t^{1/2}$ for Ce^{3+} yellow emission in $(Y_{0.99-x}Ce_{0.01}Pr_x)_3Al_5O_{12}$ with $x = 0.01$ and 0.02. (Wang L., Zhang X., Hao Z.D. et al., J. Appl. Phys. 108(1), 093515, 2010. With permission from Ref. [12].)

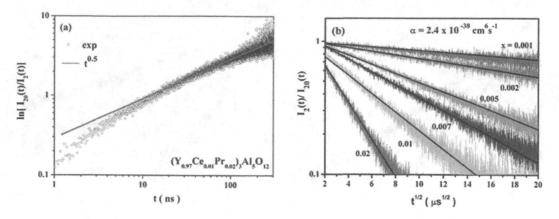

FIGURE 4.9 (a) ln–ln plot of $\ln[I_{20}(t)/I_2(t)]$ vs t for Pr^{3+} emission in $(Y_{0.99-x}Ce_{0.01}Pr_x)_3Al_5O_{12}$ with $x = 0.02$. (b) Plotted $\ln[I_2(t)/I_{20}(t)]$ vs $t^{1/2}$ for Pr^{3+} emission in $(Y_{0.99-x}Ce_{0.01}Pr_x)_3Al_5O_{12}$ with various x ($x = 0$–0.02). (Wang L., Zhang X., Hao Z.D. et al., J. Appl. Phys. 108(1), 093515, 2010. With permission from Ref. [12].)

A may mostly bring A to an excited state lower than its initial state. As a result, the round-trip ET provides an additional relaxation route from a high excited state of A down to its lower state with the help of B as an intermediate ion. Here we demonstrate two examples showing round trip ET.

4.7.1 ROUND TRIP ET IN $(Lu_{0.8}Sc_{0.2})_2O_3$:Tm^{3+}, Yb^{3+}

The round trip ET route is depicted in Figure 4.10(a). The round trip ET starts from Tm^{3+} $^3H_4$ state. After Tm^{3+} $^3H_4$ is populated, the Tm^{3+} can be de-excited to its ground state by transferring energy to Yb^{3+} to excite Yb^{3+} from ground state to $^2F_{5/2}$ excited state. Subsequently, the excited Yb^{3+} transfers energy back again to the same Tm^{3+} to bring it from the ground state to $^3H_5$ state. One can find that the round-trip ET in fact facilitates the population of $^3H_5$ from $^3H_4$. The evidence for that is shown in Figure 4.10(b).

The NIR emission spectra of Tm^{3+} singly doped (red dotted) or Tm^{3+}, Yb^{3+} codoped (black line) $(Lu_{0.8}Sc_{0.2})_2O_3$ upon Tm^{3+} $^3H_4$ excitation at 808 nm were depicted together for comparison because the round trip ET takes place only in the codoped samples. In Figure 4.10(b)–(e), the emission peaked at around 1030 nm appears only in the codoped samples is originated from Yb^{3+} $^2F_{5/2}$–$^2F_{7/2}$ transition. The lines in the range of 1300–1600 nm are responsible for Tm^{3+} $^3H_4$–$^3F_4$ transitions and they are scaled for the singly and codoped samples. The band peaked at 1620 nm is a part of Tm^{3+}: $^3F_4$–$^3H_6$ emissions.

Meanwhile, the emission spectra of codoped samples upon 980-nm excitation are also presented with a scaled Yb^{3+} emission for comparison with that upon 808-nm excitation. In Figure 4.10(b)–(e), the appearance of Yb^{3+} emission upon Tm^{3+} $^3H_4$ excitation is a direct evidence for ET from Tm^{3+}: $^3H_4$ to Yb^{3+} $^2F_{5/2}$. Besides that, the $^3F_4$–$^3H_6$ emission intensity for each codoped samples is always stronger than that for the singly doped sample upon Tm^{3+}: $^3H_4$ excitation and the extra intensity (red shaded in the $^3F_4$–$^3H_6$ emission) becomes larger with the increase of Yb^{3+} concentration, indicating additional population process of the $^3F_4$ state in the codoped samples through Tm^{3+}–Yb^{3+}–Tm^{3+} round trip ET, as sketched in Figure 4.10(a).

The extra $^3F_4$–$^3H_6$ emission is the contribution from Yb^{3+} to Tm^{3+} ET in the round trip process. The ratio of the extra $^3F_4$–$^3H_6$ emission to Yb^{3+} emission reflects the efficiency of Yb^{3+} to Tm^{3+} ET in the round-trip process. Upon 980-nm excitation, i.e., in the absence of round-trip ET, the $^3F_4$–$^3H_6$ emission is completely contributed from Yb^{3+} to Tm^{3+} ET, hence, the transfer efficiency is evaluated by the ratio of the $^3F_4$–$^3H_6$ emission to Yb^{3+} emission. One can find that the ratio for 808-nm excitation is much larger than that for 980-nm excitation. The behavior indicates that Yb^{3+} to Tm^{3+} ET in the round-trip

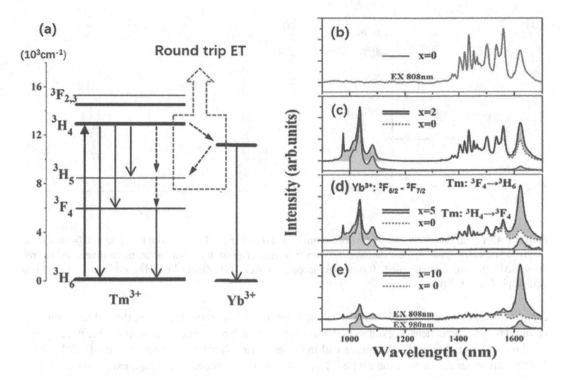

FIGURE 4.10 (a) Thulium-ytterbium energy level diagram with a Er–Yb–Er round trip ET route under 808-nm excitation. (b) (e) Emission spectra of samples $(Lu_{0.8}Sc_{0.2})_2O_3$:0.05% Tm, x% Yb with (b) x=0, (c) x=2, (d) x=5, (e) x=10 upon $^3H_4$ excitation at 808 nm or upon Yb^{3+} excitation at 980 nm. The Yb^{3+} emission intensity upon 980-nm excitation is scaled to that upon 808-nm excitation. (Liu W., Hao Z.D., Zhang L.L. et al., J. Alloys Compd. 696, 627, 2017. With permission from Ref. [13].)

process is more efficient than Yb^{3+} to Tm^{3+} ET upon direct excitation of Yb^{3+} by 980-nm light. The reason for that can be well explained as follows. In the round trip process, Yb^{3+} is excited by ET from Tm^{3+}, so that those Yb^{3+} ions with the nearest Tm^{3+} ion around are preferentially excited, whereas, in the direct excitation of Yb^{3+} by 980-nm light, the Yb^{3+} ions with the nearest Tm^{3+} or a distant Tm^{3+} around are excited with the same probability. At low doping level of Tm^{3+} (0.05%) as in this work, only a small fraction of Yb^{3+} ions have the nearest Tm^{3+} ion around. Therefore, ET from Yb^{3+} to Tm^{3+} upon 980-nm excitation is limited. In contrast, the preferential excitation of Yb^{3+} in the nearest Yb^{3+}–Tm^{3+} pairs in the round trip process leads to efficient ET from Yb^{3+} to Tm^{3+}.

4.7.2 ROUND TRIP ET IN Y_2O_3:Er^{3+}, Yb^{3+}

Figure 4.11(a) shows a round trip ET route in Er^{3+}–Yb^{3+} system. The ET starts from Er^{3+} green emitting $^4S_{3/2}$ state. After the $^4S_{3/2}$ is populated under 520-nm excitation, it is de-excited down to the $^4I_{13/2}$ state through ET to Yb^{3+} to bring it from $^2F_{7/2}$ ground state to $^2F_{5/2}$ excited state, generally called cross-relaxation process. Subsequently, the excited Yb^{3+} transfer energy back again to the same Er^{3+}, that is now in the $^4I_{13/2}$ excited state, to bring it further to red emitting $^4F_{9/2}$ state. It apparent that the round trip ET provides an additional population route of $^4F_{9/2}$ from the green $^4S_{3/2}$, so that the red to green intensity ratio is increased, as observed in PL spectra of Y_2O_3:0.002 Er^{3+}, x Yb^{3+} (Figure 4.11(b)) under 520-nm excitation of $Er^{3+2}H_{11/2}$, that can rapidly relax to the $^4S_{3/2}$ due to their proximity in energy.

One can observe that the red (660-nm) to green (560-nm) emission intensity ratio increases considerably with increasing x. It is noticed that the energy back transfer from Yb^{3+} to Er^{3+} involves two

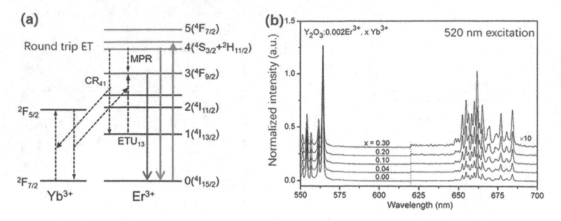

FIGURE 4.11 (a) Er³⁺–Yb³⁺ energy level diagram with a round trip ET route from Er³⁺ under 520-nm excitation (b) PL spectra upon $^2H_{11/2}$ excitation at 520 nm. The green emission intensities are normalized, and the red ones (620–700 nm) are magnified 10 times. Adapted from Ref. [14] (Zhang J.H., Hao Z.D., Li J. et al., Light Sci. Appl. 4, e239, 2015.)

excited ions, i.e., an excited Yb³⁺ and an excited Er³⁺ in its $^4I_{13/2}$ state. However, the red to green ratio is found to be independent on excitation density but only on Yb³⁺ concentration. It is, therefore, speculated that the round-trip ET takes place within the same Er³⁺–Yb³⁺ pair. Otherwise, the back ET from Yb³⁺ to any other Er³⁺ ions in the excited $^4I_{13/2}$ state will be enhanced under high excitation densities.

The round trip ET within the same Er³⁺–Yb³⁺ pair is supported by the time evolutions of PL, as shown in Figure 4.12. With increasing x, the appreciable shortening of the $^4S_{3/2}$ lifetime (Figure 4.12(a)) is due to the CR and the small shortening of the $^4F_{9/2}$ lifetime (Figure 4.12(b)) is

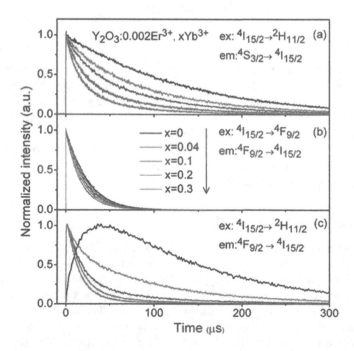

FIGURE 4.12 Time evolutions of the green emission after the green level is directly populated (a) and the red emission after the red level is directly populated (b) or the green level is directly populated (c). Adapted from Ref. [14] (Zhang J.H., Hao Z.D., Li J. et al., Light Sci. Appl. 4, e239, 2015.)

due to weak coupling of $Er^{3+4}F_{9/2} \rightarrow {}^4I_{15/2}$ with $Yb^{3+2}F_{5/2} \leftarrow {}^2F_{7/2}$ [6]. In the time evolution of the red emission after the ${}^4S_{3/2}$ is populated (Figure 4.12(c)), the codoped samples appear a fast buildup process unlike the Yb^{3+} free sample that appears a normal slow rising edge. Accordingly, the time evolution patterns in the presence of Yb^{3+} are the combined patterns of the fast buildup component and the slow buildup component. The decomposition of the time evolution patterns in Figure 4.12(c) demonstrates that the proportion of the fast buildup component in the total component is well consistent with the proportion of the increment in red to green ratio. It is concluded that the round-trip ET is a fast process. This behavior is attributed to the consequence of the round-trip ET taking place within the nearest Er^{3+}–Yb^{3+} pairs.

REFERENCES

1. Förster T., *Ann. Physik* 2, 55, 1948.
2. Dexter, D.L., *J. Chem. Phys.* 21, 836, 1953.
3. Inokuti M. and Hirayama F., *J. Chem. Phys.* 43, 1978, 1965.
4. Blasse G., *Phys. Lett.* 28A, 444, 1968.
5. Miyakawa T. and Dexter D.L., *Phys. Rev.* B1, 296, 1970.
6. Yokota M. and Tanimoto O., *J. Phys. Soc. Jpn.* 22, 779, 1967.
7. Huber D.L., *Phys. Rev. B* 20(6), 2307, 1979.
8. Xiao Y., Hao Z.D., Zhang L.L. et al., *J. Mater. Chem. C* 6, 5984, 2018.
9. Liu Y.F., Zhang X., Hao Z.D. et al., *J. Mater. Chem.* 21, 16379, 2011.
10. Feng L.Y., Hao Z.D., Zhang X. et al., *Dalton Trans.* 45, 1539, 2016.
11. He S., Zhang L.L., Wu H. et al., *Adv. Optical Mater.* 8, 1901684, 2020.
12. Wang L., Zhang X., Hao Z.D. et al., *J. Appl. Phys.* 108(1), 093515, 2010.
13. Liu W., Hao Z.D., Zhang L.L. et al., *J. Alloys Compd.* 696, 627, 2017.
14. Zhang J.H., Hao Z.D., Li J. et al., *Light Sci. Appl.*, 4, e239, 2015.

5 Upconversion Luminescence of Nanophosphors
Mechanisms and Properties

Langping Tu and Hong Zhang

CONTENTS

5.1 INTRODUCTION: PRINCIPLE OF UPCONVERSION LUMINESCENCE

As a nonlinear process, upconversion (UC) luminescence is highlighted by emitting high-energy (short wavelength) photon under low-energy (long wavelength) photon excitation. This phenomenon can be observed in numerous materials, from bulk to nanosized particles. In contrast with other competitors (such as second harmonic or coherent sum-frequency technology), which requires expensive ultra-fast pulse lasers (*e.g.*, femtosecond laser) to provide an extremely high-density excitation, Ln^{3+}-doped materials have been determined to be an ideal candidate for achieves photon

DOI: 10.1201/9781003098690-5

UC at a low-cost continuous-wave (CW) near-infrared (NIR) diode laser (*e.g.*, 808-nm, 980-nm or 1530-nm laser). This advantage can be attributed to: (i) the abundant ladder-like energy levels of Ln^{3+} (as shown in Figure 5.1, Dieke, G.H. and Crosswhite, H.M., *Appl. Opt.* 1963, 2, 675) and (ii) the relatively long lifetimes (microsecond to millisecond) of these energy levels. These features enable the Ln^{3+} ions in crystals to efficiently convert high-energy photons via successive photon absorption or energy transfer from each other.

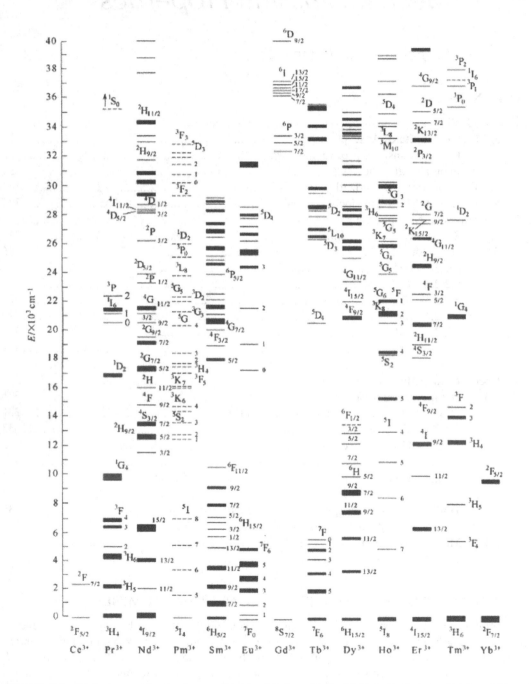

FIGURE 5.1 Dieke energy level scheme of lanthanide ions (Ce^{3+}→Yb^{3+}) in LaF$_3$ host. (Dieke, G.H. and Crosswhite, H.M., *Appl. Opt.*, 2, 675, 1963. With permission.)

To date, Ln^{3+}-doped materials are believed to have a large technical potential and wide applications in many fields, such as three-dimensional (3D) displays, bioimaging, anti-counterfeiting, solar cells, super-resolution nanoscopy and UC lasers. For example, for the bioimaging, UC materials allow the excitation wavelength to fall in the so-called biological window range (~650 nm–1300 nm), where human tissues have the smallest absorbance. Besides, the auto-fluorescence of the biological background can also be excluded due to the unique anti-Stokes emission mode.

Although the original idea of developing Ln^{3+}-doped UC materials can be traced back to Bloembergen (winner of the 1981 Nobel Prize in Physics) in 1959, the relevant UC phenomena were formally testified and implemented in the 1960s by Auzel and others. After decades of research, some parts of the mechanism are still not clear enough, which inspires people to continue to invest in the field.

Theoretically, UC phenomena are the statistical effects of numerous microscopic Ln^{3+}–Ln^{3+} interaction. It is, thus, necessary to recall the basic interaction mode of Ln^{3+} ions, that is, the energy transfer process. As shown in Figure 5.2, the excited state on one ion (named as donor) could "jump" to the other ion (named as acceptor) via the energy transfer process. In the case of a sufficiently large distance (R) between the two, the probability of energy transfer from the donor to the acceptor (P_{DA}) can be written as:

$$P_{DA}(R) = \frac{\sigma_A}{4\pi R^2 \tau_D} \int f_D(v) f_A(v) dv \tag{5.1}$$

where τ_D is the lifetime of the donor, σ_A is the absorption cross section of the acceptor, and $f_D(v)/f_A(v)$ are the emission/absorption spectra of the donor/acceptor, respectively (the integral represents the emission/absorption spectral overlap between the donor and the acceptor). In 1948, Förster, for the first time, treated the donor-acceptor energy transfer as a dipole-dipole Coulomb interaction (Förster, T., *Ann. Phys.* 1948, 2, 55). According to his calculation, the P_{DA} can be described as:

$$P_{DA}(R) = \frac{\left(R_0/R\right)^6}{\tau_D} \tag{5.2}$$

where τ_D is the actual lifetime of the donor excited state, including the influence of multiphonon nonradiative process, R_0 is the critical transfer distance for which energy transfer and spontaneous deactivation of the donor have equal probability, R is the real distance between acceptor and donor.

In 1953, Dexter modified Eq. (5.2) by introducing the higher multipole-multipole interactions (Dexter, D.L., *J. Chem. Phys.* 1953, 21, 836). In that case, the P_{DA} can be described as:

$$P_{DA} = \frac{\left(R_0/R\right)^s}{\tau_D} \tag{5.3}$$

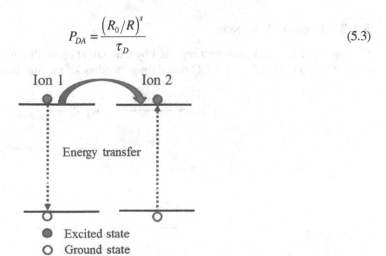

FIGURE 5.2   Schematic diagram Ln^{3+}–Ln^{3+} energy transfer process.

where S is a positive integer for three different interaction types:

$S=6$ for dipole-dipole interactions;
$S=8$ for dipole-quadrupole interactions;
$S=10$ for quadrupole-quadrupole interactions.

Based on these backgrounds, several UC mechanisms could be classified from the perspective of different Ln^{3+}–Ln^{3+} interaction modes.

5.1.1 Excited States Absorption

Excited states absorption (ESA) describes the simplest UC process, where the emission is caused by sequential absorption within the energy levels of a given ion (Figure 5.3). In principle, ESA could take place in many of the Ln^{3+} ions (*e.g.*, Er^{3+}, Tm^{3+}, Ho^{3+}, Pr^{3+}, Dy^{3+}). In detail, it could be divided into two types: (i) Ln^{3+} excited by one laser with fixed wavelength (Figure 5.3(a)), or (ii) Ln^{3+} excited by two lasers with different wavelengths (Figure 5.3(b)). However, it should be noticed that ESA only plays an important role when the doping concentration of Ln^{3+} is relatively low (typically, much less than 1%) to avoid interactions between Ln^{3+} ions (such as energy transfer or cross relaxation [CR]). Because of that, the materials based on ESA mechanism are restricted by low absorption and therefore low UC emission intensity.

5.1.2 Energy Transfer Upconversion

Energy transfer upconversion (ETU) which was also called as APTE (addition de photons par transferts d'énergie) in the early days is recognized as the dominant mechanism in most efficient UC systems. However, it was not really accepted until 1966. Auzel, who for the first time found the energy transfers, could take place between two excited Ln^{3+} ions at the initial step (Auzel, F., *C. R. Acad. Sci. (Paris)*, 1966, 262, 1016). As shown in Figure 5.4(a), UC can be achieved from the first ion (activator) in a higher excited state while the second ion (sensitizer) returns to its ground state. It should be noticed that sometimes excited state will jump to a lower energy level other than ground state, this special ETU process was also called as CR (Figure 5.4(b)). Compared with ESA, ETU allows a much higher doping concentration of Ln^{3+}, which offers much stronger absorption ability to the material. The UC efficiency of ETU is usually more than two orders of magnitude higher than that of ESA.

5.1.3 Photon Avalanche

Photon avalanche (PA) was first reported by Chivian *et al.* in Pr^{3+} (Chivian, J.S., et al., *Appl. Phys. Lett.*, 1979, 35, 124). The PA UC based on a "feedback" mechanism. A basic feature of PA is the

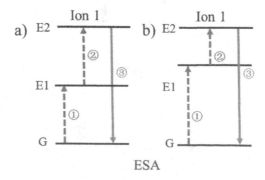

FIGURE 5.3 Schematic diagram of two types of ESA processes. (a) Single wavelength excitation. (b) Excitation of two different wavelengths.

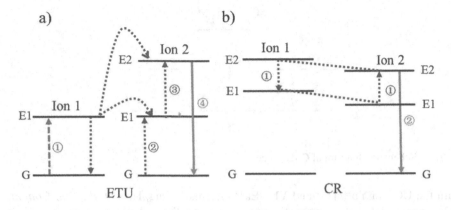

FIGURE 5.4 Schematic diagram of (a) ETU process, (b) cross relaxation process.

present of the threshold of excitation power. Below the threshold, very weak UC luminescence (UCL) can be produced, while the UC intensity increases by orders of magnitude above the excitation threshold. As shown in Figure 5.5, PA is a looping process that involves ESA for excitation light and an efficient CR that produces feedback. The level E1 of ion 2 is initially populated by nonresonant weak ground-state absorption. The looping process starts with the ESA process to elevate ion 2 (from level E1 to level E2). Then an efficient CR process of E2 (ion 2)+G (ion 1)→E1 (ion 2)+E1 (ion 1) occurs. Last, ion 1 transfers its energy to ion 2 to populate its level E1, forming a full loop. This looping results in the double of the population on the reservoir level (*i.e.*, E1 of ion 2). Subsequently, the feedback looping keeps going, increasing the population of the reservoir and UC emission levels dramatically. Robust UCL is, thus, produced. Unfortunately, the feedback loop for an efficient UC via PA mechanism only works well in very limited UC systems. Furthermore, the requirement of relatively high pump intensity also limits the PA UC in practical applications.

5.1.4 Cooperative Sensitization Upconversion

Cooperative sensitization upconversion (CSU) is also named as cooperative upconversion (CUC) in some literatures. It is a three-body interaction process. As shown in Figure 5.6, two excited adjacent sensitizer ions simultaneously transfer their energies to one activator ion, and the excited activator (ion 2) relaxes back to its ground state by emitting an UC photon. Since the emitters have no intermediate energy level matching the sensitizers (ions 1 and 3), CSU process performs a much lower probability (over four orders of magnitude lower than ETU) compared with the other mechanisms. In experiments, CUC requires very close proximity of interacting Ln^{3+} ions (less than about 0.5 nm), which could be used as a good signature of ions' clusters. Although CSU mechanism has been used

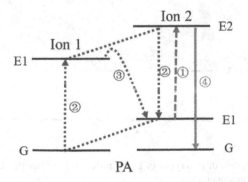

FIGURE 5.5 Schematic diagram of PA processes.

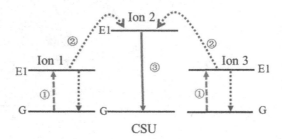

FIGURE 5.6 Schematic diagram of CSU process.

to explain the UCL in Yb^{3+}/Tb^{3+} and Yb^{3+}/Eu^{3+} systems (Liang, H.J., *et al.*, *Opt. Commun.* 2009, 282, 3028), more solid evidences are still necessary to confirm this mechanism.

5.1.5 ENERGY MIGRATION UPCONVERSION

Energy migration upconversion (EMU) is a special form of ETU. It should be noticed that the above-mentioned four UC mechanisms (*i.e.*, ESA, ETU, PA, CSU) all treat the complex UC emission as repeated interactions between adjacent Ln^{3+} ions. In fact, macroscopic UC emission belongs to a typical multibody interaction processes because one nanoparticle may contain thousands of Ln^{3+} ions or more. Besides, in most practical cases, only a small part of Ln^{3+} ions are photo-excited, and many Ln^{3+} still stay in the ground state. Moreover, the energy transfer between Ln^{3+} ions is usually (much) faster than other relaxation processes of the excited state. Therefore, from the microscopic point of view, the excited states will "swim" in the "ocean" of the unexcited Ln^{3+} ions before they contribute to UCL or relax back to the ground state. Different from the conventional ETU model, EMU highlights the "medium" role of a new type of Ln^{3+} ion, which is utilized to describe the "swimming" process of excited states. EMU also has two basic interaction modes. One example is shown in Figure 5.7(a) where

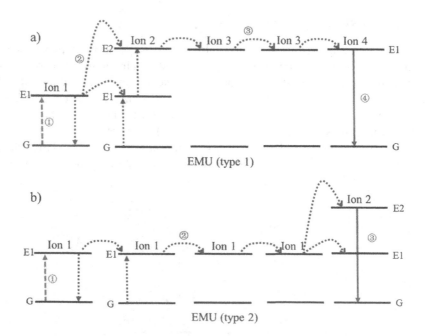

FIGURE 5.7 Schematic diagram of two types of EMU processes. (a) Energy migration after upconversion. (b) Energy migration before upconversion.

excited Ion 1 interacts with ion 2 in an ETU process. Ion 2 will transfer its energy to the neighboring ion 3, followed by a random hopping between the medium ions (ion 3) until being extracted by ion 4 to produce UC emission. The other form of the EMU process starts with migration of excited states and ends with an ETU process (Figure 5.7(b)). Although EMU process cannot be ignored in theory in bulk materials, the development of nanomaterials has injected new vitality into EMU research. Unlike bulk materials, the component/structure/size of nanomaterials can be precisely designed, which facilitates greatly the targeted study of energy migration effect. Recent advances have confirmed that with the help of multistep energy migration between medium ions. A long-range energy transport (reaches several to ten nanometers) between Ln^{3+} ions is possible (Wang, F., *et al.*, *Nat. Mater.*, 2011, 10, 968).

To sum up briefly, ETU (including EMU) is obviously the dominant mechanism of most efficient UC materials. Therefore, in the following introduction, we focus on the ETU process.

5.2 UPCONVERSION NANOPHOSPHORS

UC systems can be divided into three parts: host, sensitizer (donor) and activator (accepter). As shown in Figure 5.8, host provides the basic skeleton of the phosphor, as well as a proper crystal field environment for the sensitizer/activator ions, sensitizer and activator ions are doped in the host, and UC phenomena are evoked by sensitizer-activator interactions.

5.2.1 SENSITIZER

The role of sensitizer is to absorb light from pumping source. Therefore, a relatively large absorption cross section (*i.e.*, σ_{abs}) is essential (typically, at least one order of magnitude higher than relevant activator). Until now, there are three most commonly used sensitizers that are Yb^{3+} (working on ~980-nm excitation, σ_{abs}: ~10^{-20} cm^2), Nd^{3+} (working on ~800-nm excitation, σ_{abs}: ~10^{-19} cm^2) and Er^{3+} (working on ~1530-nm excitation, σ_{abs}: ~10^{-19} cm^2), respectively. As a comparison, the σ_{abs} of activator Er^{3+} ion at 980 nm is only ~10^{-21} cm^2. Noticing that influenced by crystal fields, the σ_{abs} values of Ln^{3+} may vary tens of percent in different materials. The sensitization process can be divided into two categories. In the first category, the sensitizer (*e.g.*, Yb^{3+}) directly interacts with the activator (*e.g.*, Er^{3+}, Tm^{3+}, Ho^{3+}), as shown in Figure 5.9. Besides, through Yb^{3+}–Yb^{3+} energy migration, sensitizer Yb^{3+} also plays the role of medium in EMU processes (*i.e.*, the type 2 in Figure 5.7). In the second category, sensitizer (Nd^{3+} or Er^{3+}) needs the cooperation of Yb^{3+} to complete the whole processes. Specifically, due to the complex energy levels scheme, it is difficult sometimes for sensitizer Nd^{3+} (or Er^{3+}) to transfer its energy to activator directly. However, the Nd^{3+} (or Er^{3+})→Yb^{3+} energy transfer are efficient. In that case, Yb^{3+} plays the role of a "bridge" for transferring the energy from Nd^{3+} (or Er^{3+}) to activator ions (such as Tm^{3+}, Ho^{3+}, Er^{3+}), as shown in Figure 5.10. This process is also called as co-sensitization. Benefited from the co-sensitization, the excitation band of some Ln^{3+} ions (*e.g.*, Tm^{3+} or Ho^{3+}) can be expanded to 800 nm and 1530 nm.

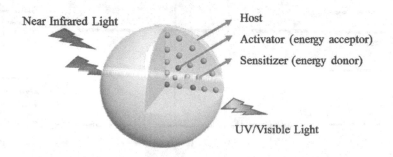

FIGURE 5.8 Typical structure of lanthanide-doped upconversion nanophosphors.

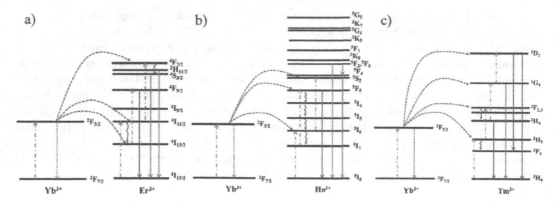

FIGURE 5.9 Schematic diagram of Yb^{3+} sensitized UC process with the activators: (a) Er^{3+}, (b) Ho^{3+} and (c) Tm^{3+}.

5.2.2 ACTIVATOR

The role of activators is to utilize their ladder-like energy levels to achieve UCL. Since each activator has its unique energy level structure, its UCL exhibits sharp emission peaks with differentiable spectroscopic fingerprints (as list in Table 5.1). It has been testified that most of the Ln^{3+} can produce UCL, including Pr^{3+}, Nd^{3+}, Sm^{3+}, Eu^{3+}, Gd^{3+}, Tb^{3+}, Dy^{3+}, Ho^{3+}, Er^{3+}, Tm^{3+}. Furthermore, it has been found that the emission peak positions of a given Ln^{3+} ion varies little with the host. Therefore, the output emission color can be adjusted by different activators.

In the following, the activator ions will be introduced one by one.

5.2.2.1 Er^{3+}

Er^{3+} was the first reported ion of UC phenomenon. And it seems that the previous numerous studies have not yet exhausted its UC properties. In the recent and expected future, it still seems to be the most studied Ln^{3+}.

Till now, Er^{3+} is the most efficient activator for green and red UCL. Due to the nearly perfect ladder-like energy levels, the role of Er^{3+} can be (i) a sensitizer (cooperate with Yb^{3+}, see also the Section 5.2.1), which provides the possibility of UC output of a broad range of activator ions (Tm^{3+},

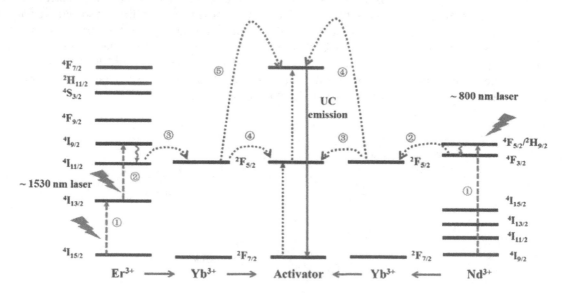

FIGURE 5.10 Schematic diagram of Er^{3+}/Yb^{3+} (left) and Nd^{3+}/Yb^{3+} (right) co-sensitization processes.

TABLE 5.1

The Major Emission Positions and the Corresponding Energy Transitions of Main Activators in UC Nanoparticles

Activators	Major Emissions (nm)	Energy Transitions
Pr^{3+}	489, 526, 548, 618, 652, 670, 732, 860	$^3P_0{\to}^3H_4$, $^1I_6{\to}^3H_5$, $^3P_0{\to}^3H_5$, $^3P_0{\to}^3H_6$, $^3P_0{\to}^3F_2$, $^3P_1{\to}^3F_3$, $^3P_0{\to}^3F_4$, $^1I_6{\to}^1G_4$
Nd^{3+}	430, 482, 525, 535, 580, 600, 664, 766	$^2P_{1/2}{\to}^4I_{9/2}$, $^2P_{1/2}{\to}^4I_{11/2}$, $^2P_{1/2}{\to}^4I_{13/2}$, $^4G_{7/2}{\to}^4I_{9/2}$, $^2P_{1/2}{\to}^4I_{15/2}$, $^4G_{7/2}{\to}^4I_{11/2}$, $^4G_{7/2}{\to}^4I_{13/2}$, $^4G_{7/2}{\to}^4I_{15/2}$
Sm^{3+}	555, 590	$^4G_{5/2}{\to}^6H_{5/2}$, $^4G_{5/2}{\to}^6H_{7/2}$
Eu^{3+}	590, 615, 690	$^5D_0{\to}^7F_1$, $^5D_0{\to}^7F_2$, $^5D_0{\to}^7F_4$
Gd^{3+}	204, 254, 278, 306, 312	$^6G_{7/2}{\to}^8S_{7/2}$, $^6D_{9/2}{\to}^8S_{7/2}$, $^6I_J{\to}^8S_{7/2}$, $^6P_{5/2}{\to}^8S_{7/2}$, $^6P_{7/2}{\to}^8S_{7/2}$
Tb^{3+}	490, 540, 580, 615	$^5D_4{\to}^7F_6$, $^5D_4{\to}^7F_5$, $^5D_4{\to}^7F_4$, $^5D_4{\to}^7F_3$
Dy^{3+}	570	$^4F_{9/2}{\to}^6H_{13/2}$
Ho^{3+}	542, 645, 658	$^5S_2{\to}^5I_8$, $^5F_5{\to}^5I_8$
Er^{3+}	411, 525, 545, 655	$^2H_{9/2}{\to}^4I_{15/2}$, $^2H_{11/2}{\to}^4I_{15/2}$, $^4S_{3/2}{\to}^4I_{15/2}$, $^4F_{9/2}{\to}^4I_{15/2}$
Tm^{3+}	294, 345, 368, 450, 470, 650, 690, 800	$^1I_6{\to}^3H_6$, $^1I_6{\to}^3F_4$, $^1D_2{\to}^3H_6$, $^1D_2{\to}^3F_4$, $^1G_4{\to}^3H_6$, $^1G_4{\to}^3F_4$, $^3F_3{\to}^3H_6$, $^3H_4{\to}^3H_6$

Ho^{3+}, Gd^{3+}, Eu^{3+} and Tb^{3+}) under 1530-nm irradiation, or (ii) an activator sensitized by Yb^{3+} (under 980-nm irradiation) or Nd^{3+}/Yb^{3+} combination (under 800-nm irradiation), producing robust green and red UC emission, or (iii) simultaneous sensitizer and activator, producing strong UCL via self-sensitization (*i.e.*, via Er^{3+}–Er^{3+} energy transfer). More interestingly, there are three possible excitation wavelengths in the self-sensitization: ~800 nm, ~980 nm and ~1530 nm, as shown in Figure 5.11.

5.2.2.2 Ho^{3+}

Ho^{3+} can also generate green and red UCL, with peaks close to that of Er^{3+} in the region. However, compared with Yb^{3+}/Er^{3+} combination, energy level mismatch between Yb^{3+} and Ho^{3+} is more

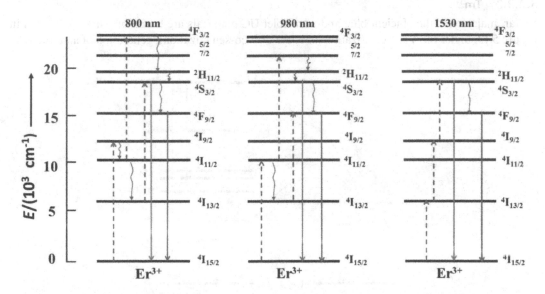

FIGURE 5.11 Schematic diagram of Er^{3+} self-sensitization, excited by 800 nm, 980 nm or 1530 nm, respectively, the dotted lines, solid curves and full arrows in the picture represent pump absorption (or Er^{3+}–Er^{3+} energy transfer), multiphonon relaxation and emission processes, respectively.

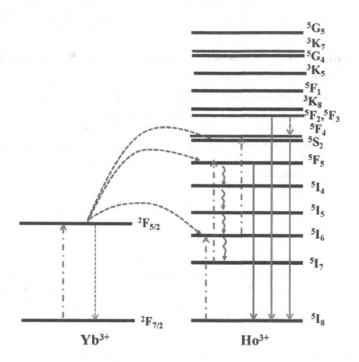

FIGURE 5.12 Schematic diagram of the UC processes in Yb^{3+}/Ho^{3+} combination.

serious (as shown in Figure 5.12), which leads to slightly lower UC efficiency. Besides, Ho^{3+} does exhibit UCL via ESA process, although it is extremely weak, whereas the UC properties of Ho^{3+} in most instances are ETU dominant (sensitized by Yb^{3+} or Nd^{3+}/Yb^{3+} combination). The candidates of excitation wavelengths for ESA are ~970 nm, 1150 nm, 646 nm or 754 nm (Zhou, J.J., *et al.*, *J. Mater. Chem. C.*, 2013, 48, 8023; Capobianco, J.A., *et al.*, *Chem. Mater.*, 2002, 14, 2915).

5.2.2.3 Tm^{3+}

So far, majority of the efficient blue and ultraviolet UC materials are based on Tm^{3+}. As shown in Figure 5.13, under the Yb^{3+} sensitization (or Nd^{3+}/Yb^{3+} co-sensitization), generation of an ultraviolet

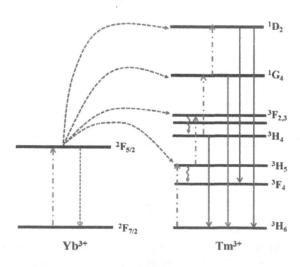

FIGURE 5.13 Schematic diagram of the UC processes in Yb^{3+}/Tm^{3+} combination.

photon needs energy transfer of three to five sensitizers to the activator. The high-energy emission of Tm^{3+} is expected to apply in some special fields, such as lithography, cancer cell killing and deep ultraviolet lasing.

5.2.2.4   Gd^{3+}

Due to the fact that its lowest excited state ($^6P_{7/2}$) is relatively high (32,000 cm^{-1}), Gd^{3+} is mainly used in downconversion studies. However, it is confirmed that deep-ultraviolet UCL can be produced by Gd^{3+} co-doped with Yb^{3+}/Ho^{3+} (Wang, L., et al., *J. Mater. Chem. C*, 2013, 1, 2485) or Yb^{3+}/Tm^{3+} (Wang, F., et al., *Nat. Mater.*, 2011, 10, 968). As shown in Figure 5.14, Gd^{3+} actually receives energy from the excited Ho^{3+} or Tm^{3+}, which is upconverted in advance via Yb^{3+} sensitization. Besides, in EMU mechanism studies, Gd^{3+} usually is employed as a "medium" ion, which transports the energy from upconverted Tm^{3+} to other activators (e.g., Tb^{3+}, Eu^{3+}, Sm^{3+}, Dy^{3+}). The detailed introduction can be seen in Section 5.1.5 (i.e., the type a of EMU, in Figure 5.7).

5.2.2.5   Nd^{3+}

Apart from the well-known role of a sensitizer around 800 nm (see also Section 5.2.1), Nd^{3+} can also act as an activator. As shown in Figure 5.15, UCL (from blue to red) can be produced by Nd^{3+}–Nd^{3+} energy transfer under 800-nm excitation. However, compared with the most widely used activators, i.e., Er^{3+}/Tm^{3+}/Ho^{3+}, its UC efficiency is orders of magnitude lower (Wei, W., et al., *J. Am. Chem. Soc.*, 2016, 138, 15130), which restricts its application.

5.2.2.6   Pr^{3+}

Yb^{3+}/Pr^{3+} combination is an alternative to provide white light UCL due to the RGB (red/green/blue) multiple UC bands of Pr^{3+}. The problem is that its UCL efficiency is orders of magnitude lower than that of Yb^{3+}/Er^{3+}/Tm^{3+} combination. Improving the ETU efficiency of Pr^{3+} remains a daunting challenge. On the other hand, Pr^{3+} can generate ultraviolet UCL under visible light excitation. although

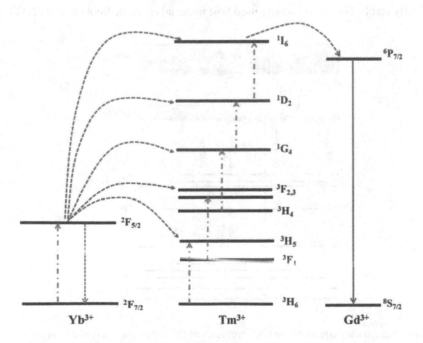

FIGURE 5.14   Schematic diagram of the UC processes of Gd^{3+}/Yb^{3+}/Tm^{3+}.

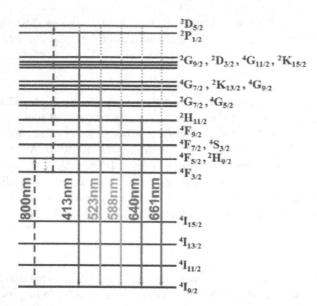

FIGURE 5.15 Schematic diagram of the UC processes of Nd^{3+}. (Wei, W., *et al.*, *J. Am. Chem. Soc.*, 138, 15130, 2016. With permission.)

the efficiency is low. As shown in Figure 5.16, the excitation wavelengths can be 473 nm or combination of 609 nm and 532 nm (Roman, K., *et al.*, *Phys. Rev. B.*, 2011, 84, 153413).

5.2.3 HOST

Selection of host materials is also critical for the UC properties. Basically, an ideal host material should be transparent in the excitation/emission spectral ranges, highly resistant to optical damage and chemically stable. So far, the widely used host materials include fluoride (*e.g.*, $NaYF_4$, $NaGdF_4$,

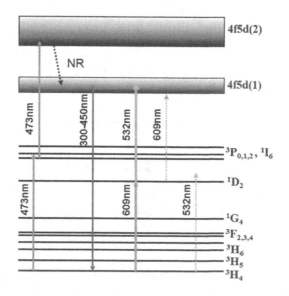

FIGURE 5.16 Schematic diagram of the ESA processes of Pr^{3+}. (Roman, K., *et al.*, *Phys. Rev. B.*, 84, 153413, 2011. With permission.)

$NaLuF_4$, LaF_3, CaF_2, $BaYF_5$), oxide (*e.g.*, Y_2O_3, Gd_2O_3, Lu_2O_3, YVO_4, ZrO_2, TiO_2, $LuPO_4$), sulfide (*e.g.*, Y_2O_2S, La_2O_2S), oxyfluoride (*e.g.*, YOF, GdOF) and halide (*e.g.*, GdOCl, $LaCl_3$). Among them, fluoride is relatively easy to be processed (or synthesized) into nanoscale. Thus, fluorides have become the most popular host in UC nanomaterials.

In most instances, host material affects the UC properties from two aspects: (i) lattice phonon-assisted nonradiative relaxation, and (ii) the local crystal field. Besides energy transfer, excited ions can relax to a lower energy level either by emitting a photon (radiative relaxation) or by transferring the energy to the surrounding lattice (nonradiative relaxation). Generally, the more the phonons needed to match the energy difference between the two electronic states, the lower the efficiency of the nonradiative process. Therefore, an ideal host for efficient UCL should have relatively low lattice phonon energy to circumvent nonradiative relaxation processes. Fluoride host materials, especially sodium rare-earth fluoride ($NaREF_4$), are so far the most effective host materials for UC, such as $NaYF_4$, $NaGaF_4$ and $NaLuF_4$ (cutoff phonon energy ~350 cm^{-1}). The cutoff phonon energies of Y_2O_3, YVO_4, Y_2O_2S and GdOCl hosts are ~550 cm^{-1}, ~890 cm^{-1}, ~520 cm^{-1} and ~500 cm^{-1}, respectively. The crystal phase is also important for an efficient UCL. The local site symmetry and crystal field strength around Ln^{3+} ions determine the intrinsic electronic transition rates. Taking $NaYF_4$ as an example, it is confirmed that Yb^{3+}/Er^{3+} in hexagonal phase $NaYF_4$ (β-$NaYF_4$) are able to generate 4.4 times higher UCL than their cubic counterparts (α-$NaYF_4$). In the cubic phase structure, the ion sites are highly symmetric, whereas in hexagonal phase structure, there are two types of low-symmetry cation sites where Na^+ and RE^{3+} distribute randomly. The low symmetry of crystal structure is beneficial to increasing the radiative transition rate of the excited states, which leads to more efficient UCL.

5.3 UPCONVERSION EFFICIENCY

5.3.1 DEFINITION AND MEASUREMENT

The UC efficiency is usually quantified by parameter UC quantum yield (QY). It is defined as below:

$$QY = \frac{\text{emitted UC photons}}{\text{absorbed photons}} \tag{5.4}$$

Obviously, UC QY is one of the most important parameters to characterize a phosphor. The first report on the UC QY measure of nanosized phosphor is published by Boyer and van Veggel in 2010 (Boyer, J.C., *et al.*, *Nanoscale*, 2010, 2, 1417). Integrating sphere is employed to collect and count all the absorbed near-infrared photons and emitted UC photons. According to the report, despite Yb^{3+}/Er^{3+} is recognized as one of the most efficient combinations, its UC efficiency is still relatively low. Under 150-W/cm^2 power density of 980-nm irradiation, the UC QY of the $NaYF_4$: 20% Yb, 2% Er bulk crystal is only 3±0.3% and below 0.3% in the nanosized particles. It should be noted, however, that the QY is not a very suitable parameter for characterizing nonlinear processes (such as UCL). Because in this case, its value depends on excitation power density. The absolute value of QY is difficult to obtain from measurement based on integrating sphere technique. QY value may also differ from setup to setup because the experimental conditions, such as excitation beam profile and collection configuration, are difficult to be exactly the same. However, QY values obtained on the same setup are very useful and reliable for comparison propose. So far, most of the reported QY values of UC nanophosphors fall in the range of 0.01–10%.

5.3.2 STRATEGIES TO ENHANCE THE UPCONVERSION EMISSION

Usually, the higher the power density, the higher the UC efficiency. The nonlinear relationship between UC intensity (*i.e.*, I) and excitation power density (p) can be described as $I \propto p^n$ (n=2,3,4…).

The n value represents the number of low energy excitation photons required to generate one high-energy UC photon. However, the value of "n" obtained in experiments are usually smaller than the expected integers, which may be attributed to the "saturation effect" of the intermediate energy levels of Ln^{3+} (Pollnau, M., et al., *Phys. Rev. B*, 2000, 61, 3337).

Obviously, the present UC QY values are unsatisfactory, especially considering the restricted excitation power density requested in practical application, *e.g.*, in clinics, where the QY values are probably lower than 0.1%. Therefore, exploring new methods to enhance the emission efficiency (or intensity) is one of the most exciting frontiers in UC research. It is difficult to comprehensively review the massive research activities and progresses here. Instead, we focus on several primary categories, including (i) core-shell structure design, (ii) organic dye sensitization, (iii) employing plasmonic fields and (iv) Ln^{3+} concentration quenching combating.

5.3.2.1 Core-Shell Structure Design

Because of the large surface-to-volume ratio, nanosized material features the significant "surface effect". The surface ions are exposed directly to the surface-related quenchers, such as surface defects, high-energy vibrations of solvent and/or ligand molecules. These surface quenchers could lead to a strong nonradiative relaxation of the Ln^{3+} ions and inevitably engender a low UC efficiency to the bare core nanostructure. Based on this understanding, a popular way to enhance greatly the UC efficiency is the construction of a hierarchical core-shell system, *i.e.*, to employ an inert crystalline shell of an undoped material (*e.g.*, $NaYF_4$, CaF_2, *etc.*) around each doped nanocrystal. It has been demonstrated that by providing nanoparticles with a protective shell with a thickness of a few nanometers, UC emission is significantly increased (10–1000 times). The remarkable effect of surface passivation can be attributed to two reasons: (i) excited states located on or near the surface can be quenched directly by neighboring quenching centers; (ii) via energy migration processes, even the energy contained in the center of nanophosphors can migrate a long distance to the surface and finally be quenched. For most UC nanophosphors, the second reason is more prominent, especially considering the concentration of "transport ion" (*e.g.*, Yb^{3+}) in the nanophosphors is usually relatively high.

5.3.2.2 Organic Dye Sensitization

The large surface-to-volume ratio can also contribute to the UC emission of nanophosphors. Although the sensitizers (*e.g.*, Nd^{3+} and Yb^{3+}) ions have been used to increase the absorption of UC systems, the absorption of Ln^{3+} sensitizers is still several orders of magnitude lower compared with the widely used linear emission materials (such as organic dyes or quantum dots). Starting from 2012, infrared organic dyes have become popular candidates for antenna ligand to improve the absorption of UC nanoparticles (Zou, W., et al., *Nat. Photonics*, 2012, 6, 560). According to the report, through the organic infrared dye (IR-806), sensitization the extinction coefficient of $NaYF_4$: Yb^{3+}, Er^{3+} UC nanoparticle has been significantly improved by about 5×10^6 times. Take advantage of that, the entire UC emission output of dye-sensitized nanoparticles was largely enhanced to ~3300 folds. It should be noticed that according to Eq. (5.4), this strategy only improves the UC emission intensity, rather than QY. On the other hand, organic dyes are susceptible to harmful photobleaching. Photostability concerns impede the expansion of this strategy.

5.3.2.3 Employing Plasmonic Fields

In the case of the electromagnetic interaction between metal and incident light of a specific wavelength, oscillating electrons at the interface of metallic structures will produce so-called localized surface plasmon resonances (LSPR). If the interaction distance is appropriate, LSPR can positively affect the QY of phosphors such as dyes, quantum dots, as well as UC nanoparticles. In detail, there are three strategies:

a. Depositing upconversion nanoparticles onto a layer of gold islands, dense metal nanoparticles (Au or Ag) or 3D plasmonic antennas.
b. Attaching or self-assembling metallic nanoparticles on the surface of UC nanoparticles.
c. Utilizing multilayer structures such as nanoparticle-silica-metal.

All these strategies are highly dependent on a precise tuning of the distance between nanoparticles and metallic structures.

5.3.2.4 Combating the Concentration Quenching Effect

Until about 2010, the commonly used doping concentration of Ln^{3+} ions in nanophosphors was relatively low, $i.e.$, in the range of 0.2–2% for activators ($e.g.$, Er^{3+}, Tm^{3+} or Ho^{3+}) and 20–40% for sensitizer ($e.g.$, Yb^{3+}). In that case, more than 50% grid point is "wasted" on the inert ions, such as Y^{3+} or Lu^{3+}. From this point of view, one of the direct ways to improve the UC intensity is to raise the dopant concentrations of active Ln^{3+} ions ($i.e.$, sensitizer or activator) inside the nanoparticle. However, this positive expectation is hindered by the so-called concentration quenching effect. According to the conventional understanding, further increasing the doping concentration of active Ln^{3+} will encourage the cascade energy transfer processes between Ln^{3+}, and the absorbed energy will be more likely to be captured by the potential quenching sites in the particle, rather than contributing to the UC emission. In recent years, the development in nanomaterials has discovered several strategies to counteract the "concentration quenching" effect and the optimal doping concentration of Ln^{3+} has been significantly increased, as well as the UC efficiency (enhancement factor in the range 10–100).

5.3.2.4.1 Excitation Energy Confining

The physical picture responsible for concentration quenching is the migrated excited states being captured by energy quenchers. Therefore, spatial confinement of excitation energy can help prevent the energy dissipation. There are two strategies in practice: (i) eliminating the "quenching site" in the material as much as possible. Theoretically, if a nanoparticle is completely "quencher free", or the influence of quenching site can be negligible, there will be no quenching of the migrated energy. A typical example is the construction of $NaErF_4$–$NaYF_4$ UC nanosystem (Johnson, N.J., $et\ al.$, $J.\ Am.\ Chem.\ Soc.$, 2017, 139, 3275). As a nanostructure, most of the quenching sites are located on the particle surface, which could be deactivated by an inert shell ($i.e.$, $NaYF_4$) coating. By blocking the energy dissipation path from luminescent core to the surface quencher, the optimal doping concentration of Er^{3+} ion increases from below 20% (in bare core structure) to 100%. (ii) Confining the excitation energy via controlling energy migration pattern. A typical example is the development of orthorhombic KYb_2F_7 nanocrystals (Wang, J., $et\ al.$, $Nat.\ Mater.$, 2014, 13, 157), in which the Yb^{3+} ions are distributed in arrays of tetrad clusters. The Yb^{3+}–Yb^{3+} distance within the cluster (3.41 Å–3.55 Å) is shorter than the intercluster distance (3.75 Å–3.83 Å). Since the energy migration possibility (P) is inversely proportional to the high orders of the ion-to-ion distance (R), $i.e.$, $P \propto 1/R^S$ (S=6, 8, 10), excitation energy is more likely to be confined to the sublattice domain. By cutting off the long-distance energy migration, the dissipation of excitation energy to quenching sites is minimized even in the absence of a protection shell. Consequently, the optimal doping concentration of sensitizer Yb^{3+} is promoted from ~20 to ~98% (the rest 2% is taken by activator Er^{3+}).

5.3.2.4.2 High-Density Excitation

A high excitation power density is employed to compensate for the negative effect of "concentration quenching". At an extremely high excitation power density ($e.g.$, above 10^6 W cm^{-2}), large percentage ($e.g.$, over 50%) of the sensitizers in the system will be excited. In that case, a new balance between the active Ln^{3+}, excitation irradiance and quenching sites will be established. Since quenching sites can only exert a limited negative effect, the bottleneck of effective UC emission will be moved to the quantity of activators, which may not enough to convert the absorbed energy of sensitizers in time. It was confirmed that under a fixed dopant concentration of sensitizer Yb^{3+} (20%), the quenching concentration of Tm^{3+} increases with the excitation density and reaches up to 8% under the high excitation irradiance of 2.5×10^6 W cm^{-2} (Zhao, J.B., *et al.*, *Nat. Nanotechnol.*, 2013, 8, 729), much higher than the 0.2–0.5% observed under low excitation conditions (below 100 W cm^{-2}). A similar result was also observed for Er^{3+}, which was promoted from 2 to 20% (Gargas, D.J., *et al.*, *Nat. Nanotechnol.*, 2014, 9, 300). However, it is crucial for many applications that request lower excitation power to compromise safe operation conditions. Because of that, this strategy is mostly used in special occasions, such as the optical imaging/detection on single nanoparticle level.

5.3.2.4.3 Dye Sensitization

In addition to high-density excitation, another alternative approach to promote excitation processes in UC is dye sensitization. This strategy is to certain extent similar to the high-density excitation approach, but with relatively low equipment requirements. As an example, using indocyanine green dyes to sensitize NaYF$_4$: Nd UC nanoparticles can increase the optimal doping concentration of the activator Nd^{3+} from 2 to 20%, and increase UCL by about ten times by excitation at 800 nm (Wei, W., *et al.*, *J. Am. Chem. Soc.*, 2016, 138, 15130). However, dye-sensitized nanoparticles generally suffer from photobleaching due to the limited stability of dye molecules.

Overall, the realization of the "most efficient" UC nanophosphors depends on practical conditions, such as material structure, excitation power density, *etc.* The "optimal results" in a particular environment may not be easily extended to more complex situations.

5.4 UPCONVERSION DYNAMICS

Luminescence dynamics is another fundamental aspect of UC phosphors. An in-depth understanding on UC dynamics is not only key to elucidating luminescence mechanism, but also highly needed in practical applications, such as optical multiplexing detection in time scale (Lu, Y.Q., *et al.*, *Nat. Photonics*, 2014, 8, 32).

As shown in Figure 5.17, UC dynamic curves usually exhibit both rise and decay components under nanosecond pulse excitation. The measured decay "lifetimes" are normally much longer than the intrinsic lifetime of the emissive state. This feature is related to the slow population processes from photo-excited sensitizers to the activators.

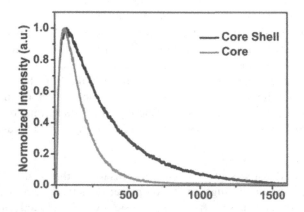

FIGURE 5.17 A typical example of the temporal behaviors of the green UC emission (~540 nm) of the core and core-shell NaYF$_4$: Yb^{3+}, Er^{3+} nanostructures (under the nanosecond pulsed 980-nm excitation.)

5.4.1 TRADITIONAL ANALYSIS APPROACH: SIMULTANEOUS RATE EQUATIONS

Traditional approaches in analyzing UC dynamics are usually based on the simultaneous rate equations (Eq. (5.5)), which describe all the population and depopulation paths of each energy level:

$$\frac{dN_i}{dt} = \sum \text{population rate} - \sum \text{depopulation rate}$$

$$= \sum_j \left(N_j A_{ji}^{ED} - N_i A_{ij}^{ED} \right) - \sum_j \left(N_j A_{ji}^{MD} - N_i A_{ij}^{MD} \right)$$

$$+ \left(N_{i+1} W_{i+1,i}^{NR} - N_i W_{i,i-1}^{NR} \right) + \sum_{ij,kl} \left(N_j N_l C_{ji,lk}^{ET} - N_i N_k C_{ij,kl}^{ET} \right) \tag{5.5}$$

where N_i is the population density of each energy level, A_{ij} are the Einstein coefficients for electric dipole (ED) and magnetic dipole (MD) radiative transitions from energy level i to j. $W_{i+1,i}^{NR}$ is the nonradiative multiphonon relaxation (NMPR) rate constant from energy level $i+1$ to i. $C_{ij,kl}^{ET}$ is the microscopic energy transfer parameter for the transfer of energy via the sensitizer $i{\rightarrow}j$ transition and the activator $k{\rightarrow}l$ transition. The A_{ij} values can be calculated by Judd-Ofelt theory (Judd, B.R., *Phys. Rev.*, 1962, 127, 750; Ofelt, G.S., *J. Chem. Phys.*, 1962, 37, 511), while the NMPR rate is treated with a modified energy gap law, and a related description of phonons is used to calculate phonon-assisted energy transfer constants. Finally, the intensity of any given UC emission is proportional to the product of the population of the emissive energy levels and the corresponding radiative transition rates.

Although this analysis approach (simultaneous rate equations) is the most widely used one, there are still some shortcomings that cannot be ignored. For example, (i) it remains a challenge to assign precise values to all the relevant parameters in these complex rate equations; (ii) the obtained result is the average over the interactions of sensitizers and activators at all possible distances. And there is no parameter representing the effect of energy migration process, which is especially important in some heterogeneous nanostructures. Because of these limitations, the calculated results out of the rate equations can only qualitatively explain the experimental results.

5.4.2 MODIFICATION OF RATE EQUATIONS

In the past few decades, some modified dynamic models have been proposed. For example, the models suggested by Zusman-Burshtein and Yokota-Tanimoto take the energy migration effect between sensitizers into consideration. The Zusman-Burshtein model is more suitable for the situation where the sensitizer-sensitizer interaction is much stronger than the sensitizer-activator one (Burshtein, A., *Zhurnal. Eksperimentalnoi. Teor. Fiz.*, 1972, 62, 1695), whereas the Yokota-Tanimoto model is more suitable for the relatively weak energy migration processes (Yokota, M. and Tanimoto, O., *J. Phys. Soc. Jpn.*, 1967, 22, 779). However, both of them are made for linear process, *i.e.*, downshift luminescence. Although Grant's model allows one to describe the nonlinear nature of ETU, which is, however, restricted by the approximation that the energy-migration rate is infinitely high (Grant, W.J.C., *Phys. Rev. B.*, 1971, 4, 648). The model presented by Zubenko *et al.* combines the approaches of Zusman and Grant, thus considering simultaneously the finite energy-migration rate in Zusman's equations and the nonlinear nature in Grant's model (Zubenko, D.A., *et al.*, *Phys. Rev. B*, 1997, 55, 8881). Accordingly, the dynamics of excited states population density $n(t)$ under short-pulsed light excitation conditions can be described by Eq. (5.6) (assuming dipole-dipole interaction):

$$n(t) = \frac{n(0)\exp(-t/\tau_D)}{1 + n(0)\tau_D \left(2\pi^2/3 \right)\sqrt{C_{DA}/\tau_0} \left\{ \sqrt{1+\tau_0/\tau_D}\,\mathrm{erf}\left(\sqrt{t\left(1/\tau_0 + 1/\tau_D\right)} \right) - \exp(-t/\tau_D)\mathrm{erf}\left(\sqrt{t/\tau_0} \right) \right\}} \tag{5.6}$$

where $n(0)$ is the initial excited state's population density; C_{DA} is the micro-parameter of donor-acceptor interaction; τ_0 is the time of excitation hopping to non-excited donors; τ_D is the excitation lifetime; and parameter $\mathrm{erf}(x) = 2/\sqrt{\pi} \int_0^x \exp(-t^2)dt$ represents the error function. However, the application of Zubenko's model should be done with caution since its solution is based on very complex parameters that are difficult to be obtained precisely.

5.4.3 Monte Carlo Simulation on UC Dynamics

UC dynamics can be seen as the final result of multiple interactions among an ensemble of sensitizers and activators. Apart from rate equations, Monte Carlo simulation has also been introduced to deal with the UC dynamics. This strategy can be traced back to 1990s (Jacob, L.P., *et al.*, *IEEE J. Quantum. Electron.*, 1997, 33, 845). The basic simulation model was set up as a 3D random walk of excited states in the sublattice (consisted by Ln^{3+} ions). UC photons are attributed to the "collisions" of walking excited states on the sites of activators. Compared with rate equation approach, Monte Carlo simulation has some unique advantages, such as (i) from microscopic point of view, interaction modes between the active Ln^{3+} are very limited, *i.e.*, sensitizer-to-sensitizer, sensitizer-to-activator and activator-to-activator. More importantly, most of the interaction parameters can be reasonably obtained from experiments or quantum mechanical calculations (Kushida, T., *J. Phys. Soc. Jpn.*, 1973, 34, 1318), which lays a solid foundation for the simulation; and (ii) it is particularly powerful in dealing with the situation of complicated nanosystems, such as energy migration at the core-shell interface of heterogeneous structures. In 2018, Zuo *et al.* reported an upgraded Monte Carlo simulation model by introducing the time evolution process (Zuo, J., *et al.*, *Angew. Chem. Int. Ed.*, 2018, 57, 3054). And the simulation results are in quasi-quantitative agreement with the experimental results. Furthermore, under the guidance of the simulation, precisely tailoring the excitation energy migration paths in core-shell structure was realized experimentally, which has been successfully applied to tune in a wide range the UC dynamic curves (either the rise or decay edge).

Despite these efforts, further development of Monte Carlo simulation strategy on UC is necessary, which may be focused on the following aspects: (i) more detailed UC processes should be taken into consideration, such as CR, anisotropic energy migration, *etc.*; (ii) more interaction parameters still need to be testified by experiment.

5.5 SUMMARY AND PERSPECTIVE

In this chapter, the fundamental physical concepts and basic UC mechanisms of UCL are introduced. The comprehension achieved so far on the microscopic interactions involved in UC is reviewed.

Despite the huge progress in synthesis and spectral control of UC nanomaterials, relatively low UC efficiency remains the main obstacle to the further development and applications of the UCL nanomaterials. To find roadmaps to further improving the UC efficiency, it is necessary to have a more comprehensive understanding of the microscopic interactions involved in every stages of UC from excitation to UCL generation. Recent advances in nanomaterial chemistry and computer simulation have provided favorable means for solving these challenges. Clarifying the microscopic effect of various factors, such as energy migration, internal impurity quenching *etc.* on UCL, is becoming an urgent issue which can lay a solid foundation for the smart design of the UC nanomaterials. Based on the progress achieved so far we can be cautiously optimistic that the UC mechanism can be better understood in the near future.

6 Organic/Polymer Luminescent Materials and Devices

Junbiao Peng

CONTENTS

6.1 INTRODUCTION

Luminescent materials can be divided into two categories: inorganic luminescent materials and organic luminescent materials. Compared with inorganic materials, organic light-emitting materials have a wider range of light-emitting colors due to their diverse compounds and flexible molecular design. In recent years, organic electroluminescent diodes have become research hotspots due to their advantages of self-luminescence, fast response, ultrathin, and bendable features.

DOI: 10.1201/9781003098690-6

According to the molecular weight, organic light-emitting materials can be divided into two categories: small molecules and polymers. Small-molecule luminescent materials have the advantages of easy adjustment of molecular structure, simpler synthesis process, easier purification, etc., and vacuum evaporation process is normally used to cast films. Owing to the rapid development of high-performance luminescent materials and devices, the display technology of small-molecule light-emitting diodes based on vacuum evaporation has become increasingly mature and has been applied in products such as mobile phones and TVs. Polymer light-emitting materials have the advantages of good thermal stability, simple device preparation process, easy realization of a large-area device, and become the first choice for solution processing technology.

6.2 SMALL-MOLECULE LUMINESCENT MATERIALS

Small-molecule luminescent materials generally refer to luminescent materials with a smaller molecular weight of about 500–2000. During use, they are generally doped into host materials with certain carrier transport properties to improve their concentration quenching effect. Light emission is realized on small molecules through energy transfer from host to guest.

6.2.1 FLUORESCENT MATERIALS

Fluorescent materials are mainly aromatic conjugated structures, and their photoelectric properties can be changed by introducing unsaturated or steric hindrance groups. Commonly used green light materials are coumarin derivative C545-T and quinacridone derivative QA, which have high fluorescence quantum efficiency and thermal stability, but their concentration quenching effect is serious; one way to suppress concentration quenching is to introduce substituents to increase steric hindrance. The design of pure red light emission materials is relatively difficult, because of their small energy gap, excitons are prone to non-radiative recombination, and the materials usually have a larger conjugate structure, which is prone to fluorescence quenching due to strong π-π stacking. Commonly used red light emission materials are 4-(dicyanomethylene)-2-methyl-6-[p-(dimethylamino)styryl]-4H-pyran (DCM) derivatives that are represented by 4-(dicyanomethylene)-22t-butyl-6-(1,1,7,7-tetramethyljulolidyl-9-enyl)-4H-pyran (DCTJB). Another type of red emission material is fused-ring hydrocarbon derivatives that have a large π-π conjugate plane and good rigidity. In contrast to red light materials, blue light materials have higher energy but are less stable. At present, blue light materials are mainly fused-ring materials, such as fluorene derivatives, diarylanthracene derivatives, and stilbene derivatives. The design of blue light materials should choose a suitable conjugate structure to avoid redshift of the emission spectrum.

6.2.2 PHOSPHORESCENT MATERIALS

At present, most phosphorescent materials are composed of heavy metals and organic frameworks, such as Ir, Pt, and Au. Among them, Ir complexes are the most studied phosphorescent materials. They are easy to modify, have high efficiency, have good thermal stability, and have a shortly excited state lifetime, which reduces the probability of triplet excitons returning to the ground state through non-radiative transition. Ir complexes can change their photophysical properties through the selection and optimization of the ligand structure to achieve different colors of light emission, such as typical green light material tris (2-phenylpyridine) iridium (Ir(ppy)$_3$), red light material tris(1-phenyl isoquinolinato-C2,N) iridium (III) (Ir(piq)$_3$), blue light material iridium bis(4,6-difluorophenyl)-pyridinato-N,C2'-picolinate (FIrpic).

At present, compared with fluorescent materials, phosphorescent materials have absolute advantages in green, orange, and red organic light-emitting diodes (OLEDs). The prepared devices show ultra-high-efficiency and ultra-long device life (the lifetime of devices made of green phosphorescent materials exceeds 100,000 hours), and these excellent properties have greatly promoted the application of OLEDs in the field of flat-panel displays and lighting. Phosphorescent devices emit light through both triplet excitons and singlet excitons, the synthesis of blue phosphorescent materials with high efficiency and stability is still a hot and difficult problem. In addition to color purity and efficiency, the issue of material stability is the main obstacle for blue phosphorescent materials to meet application requirements. And in addition to the development of blue phosphorescent materials with high color purity, efficiency, and stability, the host materials, electron transport, hole blocking, and hole transport materials that match the blue light materials are all facing relatively strict requirements.[1]

In recent years, with the in-depth research on organic light-emitting materials, in addition to traditional fluorescence and phosphorescence, new mechanisms of low-cost and high luminous efficiency all-organic luminescent materials have also been discovered, such as thermally activated delayed fluorescence (TADF) materials, thermal exciton materials, and free radical materials.

6.2.3 New Organic/Polymer Luminescent Materials

6.2.3.1 TADF Luminescent Materials

Due to the small energy level difference between the singlet state (S1) and the triplet state (T1) of the TADF material, the triplet excitons in the excited state are converted into singlet excitons through the process of intersystem crossover under the external thermal assistance. The delayed fluorescence produced can achieve 100% internal quantum efficiency, which is one of the current hotspots in organic electronics research. The maximum external quantum efficiency (EQE) of blue, green, red, and white OLED devices based on small-molecule TADF materials reached 36.7%,[2] 29.7%,[3] 29.2%,[4] and 25.1%, respectively.[5] In addition, TADF polymer materials and their devices have also made good progress. The EQE of its blue, green, red, and white light devices reached 23.5%,[6] 23.8%,[7] 19.4%,[8] and 10.4%,[9] respectively. TADF materials are considered to be a new generation of luminescent materials with good application prospects after fluorescent materials and phosphorescent materials.[10]

6.2.3.2 Thermal Exciton Luminescent Materials

Charge transfer (CT) and locally excited (LE) state molecules are designed and synthesized through hybridization principles to synthesize new hybridized local and CT (HLCT) luminescent materials with hybrid CT states. Compared with the reverse intersystem crossing between the TADF materials through the T1 to the S1 excited state, the HLCT materials use high-energy excited state (Tm–Sn, m, n > 1) to achieve the transition from Tm to S1 by the reverse intersystem crossing process (thermal exciton process), so HLCT materials are also called "thermal exciton" materials. Beginning in 2008, Yuguang Ma's research group took the lead in researching HLCT materials[11,12] and found that triphenylamine-substituted anthracene derivatives showed high electroluminescence efficiency (η_{ext} = 6.19%). The S1 is about 50%. In 2012, Li et al. used triphenylamine-phenanthrimidazole (TPA-PPI) as the emission layer material to obtain a deep blue electroluminescent device.[13] The maximum current efficiency of the device was 57 cd/A, the EQE exceeded 5%, and the S1 is about 28%. Therefore, using this new principle is expected to design and synthesize OLED materials with better light-emitting performance and stability.[14]

6.2.3.3 Free Radical Luminescent Materials

The luminescence mechanism of conventional organic/polymer luminescent materials is based on closed-shell molecules. In contrast, the luminescence of free radical luminescent materials is derived from the characteristics of organic radical molecules. Generally, 25% of singlet and 75% of triplet excitons are formed when charge is injected into closed-shell molecules, and two doublet excitons and one quartet exciton are formed when electrons are injected into open-shell (radical) molecules. Peng et al. believed that the formation probability of both doublet excitons was 25%, and the formation probability of quartet excitons was 50%, so the exciton utilization rate of the electro-luminescence of radical molecules was 50%.[15] Stable free radical molecules have a wide range of applications in optoelectronic devices. They can be used as light-emitting materials and as doping materials for electron transport layers to reduce driving voltage and increase electron injection rate. At present, in the experiments of Li Feng et al., radical OLED devices with emission wavelength in the deep-red region had made great progress.[16,17] This kind of luminescent material based on organic free radical molecules has attracted widespread attention.

6.3 POLYMER LUMINESCENT MATERIALS

As early as 1990, Friend et al. of the Cavendish Laboratory at the University of Cambridge in the United Kingdom reported for the first time a polymer electroluminescence device prepared using a conjugated polymer poly(*p*-phenylene vinylene) (PPV).[18] The EL phenomenon was observed at a low voltage of 2.2 eV, but the efficiency of the device is only 0.05%. After 30 years of develop-ment of polymer light-emitting devices, although the EQE and brightness of polymer light-emitting devices have been greatly improved, the EQE of OLED devices prepared with phosphorescent dyes has exceeded 20%, the actual production efficiency has also increased from the early 0.1 lm/W to 20 lm/W, but the device lifetime is still one of the difficult issues to be solved. The following describes four common types of polymer material systems.

6.3.1 INTRODUCTION OF POLYMER LUMINESCENT MATERIALS

Polymer electroluminescent materials mainly include polyfluorene (PFO), PPV, polyvinylcarbazole (PVK), and polythiophene (PAI). The following mainly introduces these types of polymer lumines-cent materials.

6.3.1.1 Polyfluorene (PFO) and its Derivatives

Polyfluorene-based materials are currently one of the most studied and promising polymer light-emitting materials. In 1989, Fukuka's research group synthesized polyfluorene for the first time using ferric chloride oxidative polymerization.[19] PFO has a good thermal stability and exhibits high fluorescence quantum efficiency in solution and film; its structure is shown in Figure 6.1.

Polyfluorene has good modifiability. By modifying the main chain of fluorene, a series of deriva-tives can be designed and synthesized. By introducing different substituent groups, such as aliphatic carbon chain and aromatic hydrocarbon, the luminescence performance and solubility of polyfluo-rene can be effectively adjusted. Due to the wide bandgap of polyfluorene, a narrow bandgap emitter can be linked to the main or side chain of polyfluorene by copolymerization, and a polyfluorene copolymer, the emission spectrum of which covers the entire visible light region, can be obtained by means of energy transfer. In addition, polyfluorene and its derivatives have higher fluorescence efficiency and wider energy gap, and after structural design, it is also widely used as a host material for light-emitting devices.

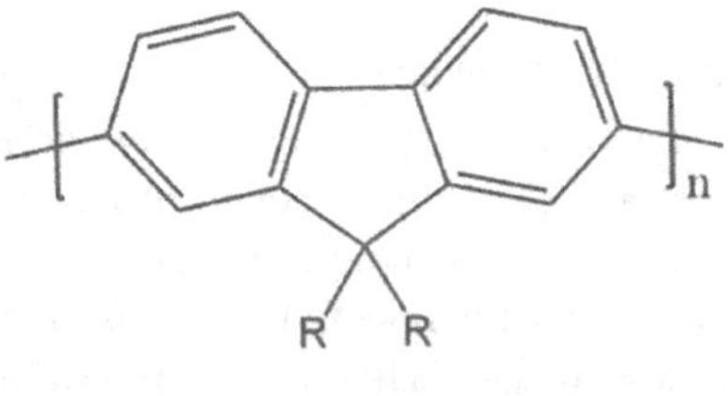

FIGURE 6.1   Schematic diagram of the structure of polymer PFO.

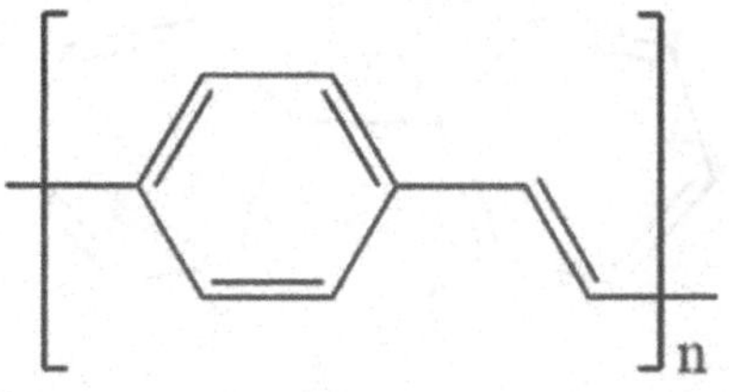

FIGURE 6.2 Schematic diagram of the structure of polymer PPV.

Wei Yang's research group[20] used polyfluorene to copolymerize with dibenzothiophene-S, S-dioxide isomer (FSO) monomers with strong electron affinity to form a class of blue light materials. The introduction of thiofluorene units not only improved the electron injection and transport capabilities but also improved the stability of the device and avoided the problem of green light emission of polyfluorene blue material even under heating conditions. The luminous efficiency of the prepared device is 3.7 cd/A, and the color coordinate is (0.16, 0.07). In 2006, Yong Cao's group synthesized a series of random narrow bandgap of fluorene-based copolymers that are easily soluble in organic solvents such as toluene and methylene chloride.[21]

6.3.1.2 Poly(p-phenylene vinylene) (PPV) and Its Derivatives

In the 1990s, the Richard Friend research group of the University of Cambridge applied conjugated polymer PPV to electroluminescent devices for the first time, opening up a new field of conjugated polymer applications. The structure of polymer PPV is shown in Figure 6.2.

PPV is a typical linear conjugated polymer material. Theoretical and experimental studies have shown that the introduction of chemical substituent groups, such as $-Cl$ and $-NO_2$, on the benzene ring of PPV can significantly change its optical and electrical properties. Researchers found that in addition to adjusting the energy gap of the material through molecular design to obtain different emission wavelengths, it is also possible to improve the processing properties of PPV by synthesizing precursor polymers and introducing flexible chains;[22] therefore, it is considered that PPV derivatives are polymer materials with photoelectricity, processability, and applicability.

Figure 6.3 is a schematic diagram of the structure of some PPV derivatives. Because PPV alone is poor in processability, flexible branches are generally introduced on the benzene ring to improve the solubility of the material; for example, the alkyl chain in the polymer poly(2-methoxy, 5-(2′-ethylhexyloxy)-1,4-phenylenevinylene) (MEH-PPV)[23] improves its solubility, making the polymer easy to dissolve in organic solvents, such as tetrahydrofuran and chloroform. The introduction of large steric hindrance groups on the side chain of PPV can distort the main chain structure and realize the adjustment of the emission wavelength. The silicon group introduced in the polymer poly(2-dimethyloctylsilyl-1, 4-phenylenevinylene) (DMOS-PPV)[24] increases the energy gap of the polymer, to change the light color to green. M. Helbig et al. designed and synthesized the polymerphenyl substituted poly(phenylenevinylene) (CN-PPV) containing the cyano structure that can

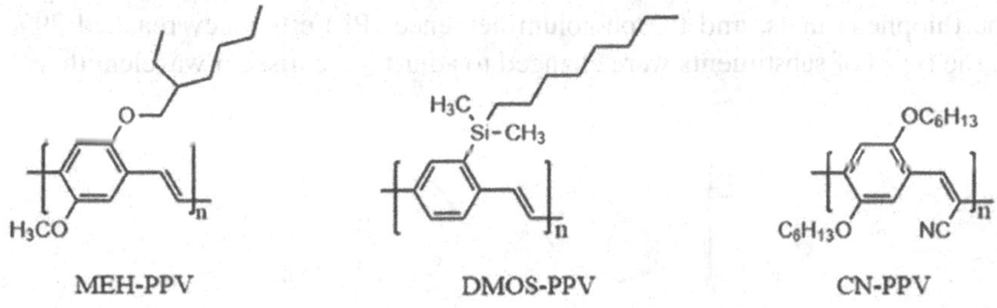

MEH-PPV DMOS-PPV CN-PPV

FIGURE 6.3 Derivatives of polymer PPV.

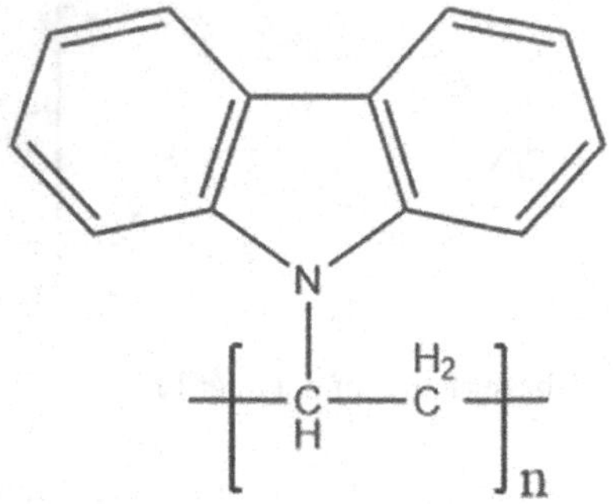

FIGURE 6.4   Schematic diagram of the structure of polymer PVK.

enhance the electron affinity of the polymer, which is conducive to the transportation of electrons in the device.[25]

6.3.1.3   Polyvinylcarbazole (PVK) and its Derivatives

PVK is a kind of nonconjugated polymer with carbazole unit in the side chain position. It has both hole transport and good electron blocking abilities. Therefore, it is often used as a hole transport layer in electroluminescent devices. At the same time, the PVK has also been widely used by researchers as the host material in phosphorescent doped systems because it has good film-forming properties and higher triplet energy levels. Its structural formula is shown in Figure 6.4.

Junji Kido et al. prepared a device with PVK as the luminescent material, the structure is ITO/PVK/TAZ/Alq/Mg:Ag, the device starts to emit light at a low voltage of 4 V, the emission peak is at 410 nm, and the luminous brightness reaches 700 cd/m^2.[26] A. Kukhta et al. reported a novel copolymer with moieties capable of charge transport and light emission on the basis of polyvinyl-carbazole and 1,8-naphthalimide; white light emission can be easily realized by tuning the content of naphthalimide moieties or a number of naphthalimide derivatives.[27]

6.3.1.4   Polythiophene (PAI) and its Derivatives

Polythiophene was first reported to be synthesized by Yamamoto in 1980.[28] It is a type of conjugated polymer material that has been widely studied at present. Its advantages are having good stability and easy structural modification, mainly by introducing substituents on its conjugated groups. The structure of polythiophene is shown in Figure 6.5.

With the deepening of research, the products of copolymerization of PAI and its derivatives with other monomer molecules have attracted much attention. Its copolymers not only have the advantages of their respective homopolymers but also can adjust their emission wavelength, stability, and luminous efficiency through molecular design, making it easier to meet application requirements. Richard E et al. synthesized a copolymer of alkyl-substituted and -unsubstituted thiophene through molecular design and studied the effect of the effective conjugation length of the main chain on the color of light.[29] Pei et al. increased the luminous efficiency by inserting substituted phenyl groups between the thiophene units, and the photoluminescence (PL) efficiency reached 29%.[30] At the same time, the types of substituents were changed to adjust the emission wavelength.

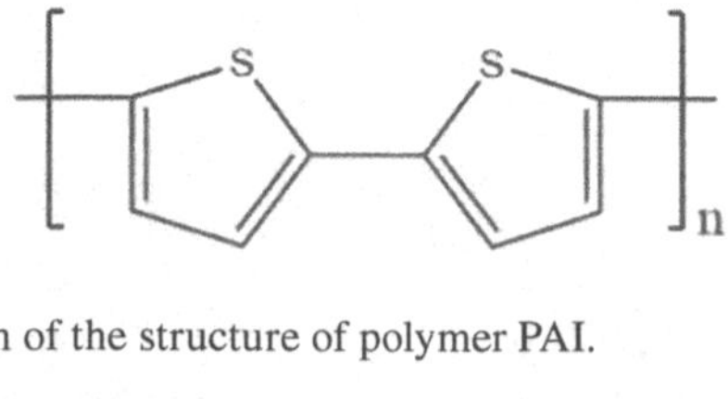

FIGURE 6.5   Schematic diagram of the structure of polymer PAI.

6.3.2 Interface Problems of Polymer Light-Emitting Devices

For small-molecule light-emitting devices, the preparation of the functional layer can achieve a clear layered structure by thermal evaporation. However, for solution-processed polymer organic light-emitting devices, the primary problem is how to achieve a clear layered structure. Since organic materials have relatively similar solubility, the solvent used in the preparation of the upper film may corrode and damage the lower one, causing serious interfacial miscibility problems. In order to solve the interface problems that may occur during solution processing of polymer light-emitting devices, researchers have been working on developing new interface materials and appropriate device construction strategies. Some methods to solve the problem of interfacial solubility will be briefly introduced as the following.

(1) To modify the chemical structure of conventional materials. Through chemical modification to make them have good solubility in strong polar solvents, and nonpolar solvents are their poor solvents.[31] In this way, the materials can be resistant to specific organic solvents. (2) With the help of chemical modification to introduce functional groups that can undergo cross-linking reactions under light or heating conditions,[32] and during the film preparation process, the film is irradiated or heated to form a cross-linked structure to resist corrosion by organic solvents. (3) To use ultrahigh molecular weight polymer materials combined with high-temperature annealing treatment,[33] the thermal annealing treatment of the film after spin coating can make the polymer chain segment move locally, reduce the free volume of the film, and achieve the maximum degree of entanglement between polymer chains, which may make the film have certain solvent resistance characteristics. (4) Choose orthogonal solvent system for the upper and lower film layers, so that the solvent used in the upper material cannot dissolve the lower one.

6.4 MECHANISM OF ORGANIC ELECTROLUMINESCENT DEVICES

6.4.1 Working Principle of Organic Electroluminescent Devices

OLED has a "sandwich" structure in which the light-emitting layer is sandwiched between two electrodes. Its structure is shown in Figure 6.6(a). Its working principle is that under a forward direct current drive, electrons and holes are injected from the cathode and the anode, respectively. Under electric field, they migrate to the opposite electrode, and the electron-hole pairs form excitons in the emission layer, resulting in radiative recombination and light emission. With the in-depth understanding of the mechanism of OLED devices, the device structure has been optimized, and a multilayer device structure is proposed. As shown in Figure 6.6(b), each functional layer embodies different functions, and the device performance has been significantly improved.

At present, the theory of organic semiconductors is still in the stage of continuous improvement. In order to get better understanding of its mechanism, the theory of inorganic semiconductors is generally still used to explain the electronic processes of organic semiconductors. The light-emitting process of OLED can be divided into the following four steps: (1) carrier injection;

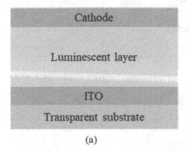
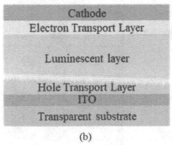

FIGURE 6.6 Typical OLED device structure. (a) Single-layer device. (b) and (c) Multilayer device.

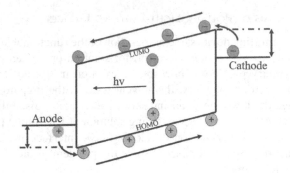

FIGURE 6.7   Basic working principle of organic light-emitting diode.

(2) carrier transport; (3) carrier recombination and formation of excitons; (4) exciton attenuation, recombination, and yielding light emission. The process is shown in Figure 6.7.

6.4.1.1   Carrier Injection

Generally, the intrinsic carrier concentration of organic semiconductors is low, and carriers in OLEDs are all obtained by injection, so effective carrier injection is the basis for ensuring device performance. Holes must overcome the energy barrier between the anode and the highest occupied molecular orbital (HOMO) of the light-emitting layer, while electrons must overcome the energy barrier between the cathode and the lowest unoccupied molecular orbital (LUMO) of the light-emitting layer. The smaller the barrier, the easier the injection of carriers. Therefore, the selection of electrodes and the barriers between the electrode and the organic interface have a great influence on the performance of OLEDs.[34]

There are currently two theoretical mechanisms for charge injection in organic semiconductors: tunnel penetration and thermal electron emission injection theories, as shown in Figure 6.8. Thermionic emission injection theory means that when the electrode and the semiconductor material form a Schottky barrier, if the width of the barrier is much smaller than the mean free path of electrons, the collision of electrons in the barrier region can be ignored. At this time, the height of the barrier plays a major role in transportation. And as long as the electron has enough energy to cross the apex of the barrier, it can enter the organic layer, so the thermionic emission current cannot be ignored when the temperature is high. The carrier injection in the tunneling mechanism requires a sufficiently high electric field strength to allow the carriers to overcome the barrier. When the wavelength of electrons or holes is not less than the width of the barrier, they can tunnel into the semiconductor through quantum mechanical effects instead of crossing barriers.

Although OLED is dual-carrier injection, the study of its current injection and limitation through single-carrier injection and transport mechanism have found that the current injected under a low

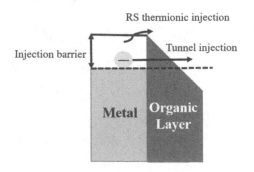

FIGURE 6.8   OLED charge injection mechanism.

electric field is provided by the thermionic emission of holes or electrons. Under the low-voltage field, the barrier width is large, and there is no tunnel injection; when the electrode contact is non-ohmic contact, the current is limited by the electrode injection current. However, heat injection under a high electric field cannot meet the current demand of the device, so heat injection is supplemented by tunnel injection. The increase of current makes the organic layer generate absolute charge concentration and form space charge limitation. In addition, the study of the electrical characteristics of single-layer OLEDs through numerical simulation found that when the injection barrier is less than 0.3 eV, the electrode injection current at room temperature can provide the device with the current required for space charge limitation.[35]

6.4.1.2 Carrier Transportation

The electrons of organic semiconductors are localized, and the electrons are basically restricted to the molecular range. Therefore, how the carriers are transported in the organic semiconductor transport layer and light-emitting layer after carrier injection is worthy of attention. The transport mode of carriers in organic semiconductors is generally divided into energy band model, jump model, and polaron model according to the relative relationship between electroacoustic coupling energy and electronic coupling energy between molecules.[36]

Energy band model: For some organic materials such as rubrene, the intermolecular electronic coupling energy is relatively large, and the electron delocalization is relatively strong. The carrier transport is considered inorganic semiconductors, showing a "band-like" model.[37] As the temperature rises, nuclear vibration strengthens, phonon scattering increases, and the mobility of such semiconductor carriers decreases.

Jump model: In addition to the abovementioned highly stacked organic materials, the distances between molecules in organic semiconductors are generally far apart, and the van der Waals force and electronic coupling energy are weak, so electrons are generally confined within the molecules and are highly localized. To transport electrons in organic semiconductors, they need to "jump" from one molecule to another. The Marcus theoretical simulation believes that carrier diffusion is a thermally activated adiabatic transition process, and carriers no longer diffuse at a temperature of 0K.[38] Based on the Fermi Golden Rule, Shuai et al. proposed that the quantum nuclear tunneling effect needs to be considered. They believe that quantum tunneling caused by nuclear vibration can reduce the activation energy of electronic transitions, and electrons can transition even at 0K.[39]

Polaron model: Polaron refers to the whole of the distortion potential field generated by electrons and their surroundings in the crystal lattice through interaction. Its size is determined by the region where the lattice polarization occurs when coupled with electrons (or holes). If the lattice constant is larger, it is a large polaron otherwise it is a small polaron. The small polaron assumes an energy band transmission model at low temperature, and the transfer between lattice points does not change the phonon state; the temperature rise needs to consider the influence of lattice vibration, and the effective mass of the electron increases. When the temperature is higher than the critical temperature, the small polaron is basically localized. Holstein's small polaron model of one-dimensional molecular crystals under local electroacoustic action can describe this type of transition.

6.4.1.3 Carrier Recombination and Formation of Excitons

Excitons are bound states formed by the interaction of electron-hole pairs by electrostatic Coulomb force. For OLED, the excitons are injected electrons and holes jumping and migrating in opposite directions, and they are formed under the action of Coulomb force after they meet in the light-emitting layer.

According to the different types of electron orbits in excitons, excitons can be divided into singlet (S) and triplet (T) excitons. In singlet excitons, the electron spin direction in the excited state is antisymmetric (S = 0), and the multiple state (2S + 1) is 1; in the triplet exciton, electron spin direction is symmetric (S = 1), and the multiple state (2S + 1) is 3. The electrons in the OLED are injected from

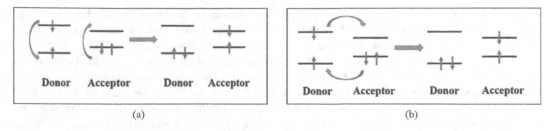

FIGURE 6.9 (a) Förster energy transfer. (b) Dexter energy transfer electronic structure diagram.

the outside. According to the calculation of the spin-statistics theory, the generation ratio of triplet and singlet excitons is 3:1.

Excitons are unstable when they are in an excited state, and there will be energy transfer before electron-hole recombination. In OLED, there are mainly Förster and Dexter energy transfers. Förster energy transfer is mainly the resonance energy transfer produced by the interaction of dipoles between molecules. The excited state donor molecule D transfers energy to excite the acceptor molecule A, and the molecule D returns to the ground state and thus the molecule A is excited, which can also be understood as the acceptor molecule absorbs the photon before the donor molecule emits the photon, as shown in Figure 6.9 (a). This process is mainly related to the distance between the donor and the acceptor, the emission transition dipole of the donor and the acceptor absorption transition dipole, so the main influencing factors are the donor fluorescence spectrum and the acceptor absorption spectrum overlap integral; molecular radiation rate; acceptor extinction coefficient; and donor-acceptor distance.

Dexter energy transfer, or electron exchange transfer, is a multilevel coupling between excitons and adjacent ground state molecules. The occurrence of the process requires that the electron clouds of the acceptor molecular orbital overlap each other, the emission spectrum of the donor and the absorption spectrum of the acceptor must overlap, and the spin properties before and after the electron exchange are unchanged.

6.4.1.4   Exciton Attenuation and Recombination Luminescence

Excitons are in an unstable excited state, so there will be a series of electronic processes in the excitons that will eventually bring the energy back to the ground state (as shown in Figure 6.10): (1) the singlet excitons radiate from S1 to S0 and emit fluorescence; (2) Sn to S1 and Tn to T1 internal conversion; (3) S1 to T1 intersystem crossing and its reverse process (reverse intersystem crossing); (4) T1 to S0 radiation transition, phosphorescence luminescence.

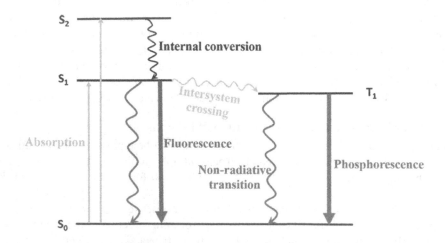

FIGURE 6.10   Typical intramolecular photophysical process.

For fluorescent materials, process (1) is spin-allowed fluorescent radiation, but for the carrier injection mechanism, process (4) is spin-forbidden; since singlet excitons only account for 25%, the luminous efficiency of fluorescent materials is low. The phosphorescence of the process (4) can be made possible by the heavy metal effect, and the efficiency can be improved. The molecular design can make the energy level difference between T1 and S1 small, realize the reverse intersystem crossing, and use triplet excitons to generate delayed fluorescence to make the utilization rate of excitons reaching 100%, so full utilization of triplet luminescence and reduction of non-radiative transition are important research directions of organic electroluminescent devices.

6.4.2 Organic Electroluminescent Devices

According to the light-emitting mechanism of organic light-emitting materials used in light-emitting devices, organic electroluminescent devices are mainly divided into three types: fluorescence, phosphorescence, and TADF, as shown in Figure 6.11.

Fluorescent electroluminescent device: According to the statistical law of electron spin, the ratio of singlet to triplet excitons among the excitons generated by electrical excitation is 1:3. The singlet exciton radiation transitions to emit fluorescence. In this process, since the spin orbit is the same, the transition from the singlet to the ground state is allowed by the spin, so the fluorescence lifetime is generally nanoseconds. Since the spin orbits are opposite, the transition from the T1 to the ground state is spin forbidden, so the phosphorescence lifetime is milliseconds. In fluorescence-based OLEDs, the device can effectively utilize 25% of singlet excitons, while triplet excitons are converted into heat in the form of non-radiative transitions. Therefore, the theoretical maximum internal quantum efficiency of fluorescent OLEDs is 25%.

Although fluorescent OLED has a long life and good stability, its low efficiency limits its development. Therefore, the research of fluorescent OLED mainly focuses on how to improve device efficiency, such as developing high-efficiency blue fluorescent materials, optimizing device structure, and improving light extraction rate. There are several ways to optimize the device structure: (1) lower the injection barrier, such as lowering the injection barrier through a gentle gradient energy level. (2) Choose a functional layer with high carrier mobility, such as introducing an electron transport layer and a hole transport layer to improve the carrier injection of the device, or use N-doping technology to improve the electron transport ability of the device to balance the carrier injection and reduce the turn-on voltage.[40] (3) Regulate the energy transfer between the host and guest by selecting the appropriate host and guest materials. For example, the emission spectrum of the host material and the absorption spectrum of the guest material have a large overlap to enhance the energy transfer between the host and guest. (4) Increasing the number of excitons, such as the use of T1-T1 annihilation and excimer complex mechanisms to use T1 excitons. Phosphorescent OLED and TADF devices can also refer to these optimization strategies.[41]

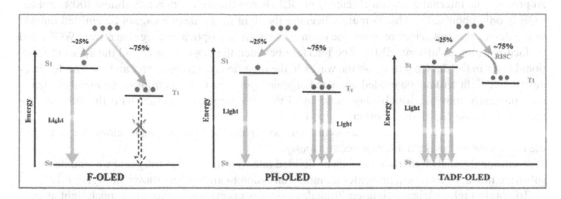

FIGURE 6.11 The dynamic light-emitting process of electro-excitons in the third-generation OLED.

Phosphorescent light-emitting device: Phosphorescent materials can convert singlet excitons into triplet excitons through intersystem crossing and realize the radiation transition from triplet to ground state through the strong spin-orbit coupling (SOC) effect, so phosphorescent OLEDs use both singlet and triplet excitons at the same time and can achieve 100% internal quantum efficiency. Therefore, the efficiency of phosphorescent OLED is much higher than that of fluorescent OLED, but due to the long lifetime of triplet excitons, under high current density, triplet excitons may rapidly accumulate and lead to quasiparticle interactions such as triplet-triplet annihilation and triplet-polaron annihilation. The quenching of such excitons will cause a rapid roll-off in efficiency.

The efficiency roll-off can be improved in the following ways: (1) reduce the excited state lifetime, through molecular design, and increase the microcavity effect, and so on. (2) Choose a matching host material. The HOMO energy level of the host material should be higher than that of the guest material to avoid energy transfer from the guest material to the host material; the host material with high carrier mobility can also make the carriers evenly distributed in the light-emitting layer, avoiding exciton aggregation and quenching. (3) Change the mobility of the transport layer to achieve carrier balance, such as doping the electron transport layer to increase its electron mobility,[42] or introducing an intermediate layer to reduce the concentration of holes injected into the light-emitting layer.[43]

TADF light-emitting OLED device: The highest theoretical value of its internal quantum efficiency can be 100%. TADF materials can convert triplet excitons generated by electrical excitation into singlet excitons through inverse intersystem transitions. Therefore, 25% of singlet excitons in TADF devices emit light through radiation transitions, with a lifetime of nanoseconds; 75% of triplet exciton first converts to the singlet exciton and then transitions to the ground state, with a lifetime of microseconds and delayed fluorescence. Therefore, choosing a small ΔE_{ST} material to enhance the intersystem transition is the key to realizing high-efficiency TADF devices.

In recent years, exciplexes are also expected to realize high-efficiency TADF devices.[44] Exciplexes are excited state complexes formed by donor and acceptor molecules. The HOMO is mainly distributed in the donor molecule, and the LUMO is mainly distributed in the acceptor molecule, achieving a small ΔE_{ST}. At the same time, the exciplex can be used to adjust the device structure and carrier transport characteristics. Due to the excellent carrier transport capabilities of the donor and acceptor molecules of the exciplex, the donor and acceptor can be used as the hole transport layer and the electron transport layer, respectively, which simplifies the device structure. When the exciplex is used as a host material, the carrier injection can be balanced by adjusting the ratio of donor molecules to acceptor molecules, instead of some more complicated methods, such as doping in the light-emitting layer or introducing an intermediate layer.[45,46]

6.4.3 LIGHT EXTRACTION

At present, the internal quantum efficiency of OLEDs can theoretically reach almost 100%, but the EQE is only about 20%. This is mainly because the light in the device cannot be emitted outside of the device due to various reasons; the main reasons are the optical waveguide effect (WGE) and surface plasmon polaritons (SPPs). The interface between the organic layer and the adjacent functional layer in the device will cause the waves in the organic layer to interfere, and the interference-destructive light will not be guided out of the organic layer. The free electrons in the metal electrode are constantly moving, which may cause local excess charges to interact with the surrounding charges to cause oscillating motion, which is plasma oscillation, which generates transverse and vertical electric fields; the transverse electric field will make the light propagate along the surface of the conductor and make it attenuate continuously.

Traditional OLED light radiation can be divided into four modes: ITO-organic waveguide mode, substrate mode, SPPs mode, and external mode (emission to air),[47,48] as shown in Figure 6.12.

To obtain high-efficiency light-emitting devices, it is necessary to extract as much light as possible, and internal and external light extraction technology can be used.

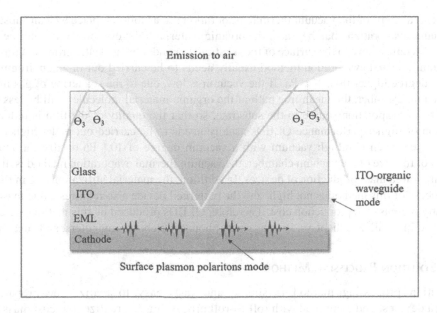

FIGURE 6.12 Schematic diagram of the structure of a multilayer OLED device and its different modes of light transmission.

The internal light extraction technology mainly starts from the internal structure of the device to reduce the optical WGE and the surface plasmon element effect. The more commonly used method is to introduce a low refractive index layer, scattering medium layer or transparent electrode patterning between the ITO and the substrate. However, due to the relatively complicated control of the internal structure, the internal light extraction technology needs to be further studied.

The external light extraction technology mainly reduces the total reflection of the interface by modifying the outer surface of the substrate and using the principles of scattering and concentrating light. The simplest solution is to roughen the surface of the substrate and enhance the scattering and exit interface area by increasing the roughness of the substrate surface. Microlens array has the function of converging and diverging light radiation. It can serve as a buffer layer with a matching refractive index and increase the light-emitting surface. Other methods include surface patterned films and surface scattering media.[49]

6.5   DISPLAY DEVICE MANUFACTURING PROCESS

At present, the methods for preparing organic light-emitting devices can be divided into two categories, vacuum thermal evaporation and solution processing methods, such as spin coating and printing. Small-molecule organic material devices are prepared by vacuum thermal evaporation method, and polymer materials are prepared by solution processing method. Below, vacuum thermal evaporation method and the typical spin coating and inkjet printing methods in the solution processing method are briefly introduced.

6.5.1   VACUUM THERMAL EVAPORATION METHOD

At present, the vacuum thermal evaporation method is most commonly used for preparing OLED display devices based on small-molecule materials, and it is also the current preparation method for OLED commercialization. The method has the characteristics of mature technology, formation of good quality films, easy doping, and stable performance of the OLED device, but it also has disadvantages such as difficulty in large-area preparation and expensive vacuum equipment.

The general steps of the vacuum thermal evaporation are as follows: place a clean substrate in a high-vacuum evaporation chamber, heat the organic material to be evaporated to sublime, and the material molecules move to the surface of the substrate to condense, nucleate, grow, and form a film. The vacuum thermal evaporation method usually needs to be carried out in an environment with a vacuum degree higher than 10^{-4} Pa. If the vacuum is low, due to the scattering of gas molecules in the vacuum chamber, the mean free path of the organic material molecules will be less than the distance from evaporation source to the substrate, so that the quality of the film is reduced. The preparation of higher performance OLEDs usually needs to be carried out under higher vacuum conditions, and even ultrahigh vacuum with a vacuum degree of 10^{-6} Pa or 10^{-7} Pa. Due to the limitation of the size of the vacuum chamber, the vacuum thermal evaporation method is difficult to realize large-area mass production of devices. In addition, the material utilization rate in the evaporation process of this method is not high, and the patterned device also requires a fine mask, which will greatly increase the production cost. Therefore, OLEDs produced by this method are currently mainly used in small-size displays, such as smart phones, notebook computers, and smart watches.

6.5.2 Solution Processing Method

The solution processing method is simple and fast, easy to realize the preparation of large-area devices, and combined with roll-to-roll processing can realize low-cost mass production.[50] However, compared with the vacuum method, the compactness of the thin film prepared by solution processing method, the lifetime, stability, and luminous efficiency of the device need to be improved. The molecular weight of the polymer is relatively large, and it is difficult to evaporate to a substrate by using the vacuum thermal evaporation method, and it is generally prepared by a solution processing method. In the preparation of multilayer devices by the solution method, problems such as mutual dissolution between different organic functional layers and interface work function matching between metal and organic layers are likely to occur. These problems can be solved by designing functional cross-linkable molecules or using different organic solvents.

6.5.2.1 Spin Coating Method

The spin coating method for preparing thin films has the characteristics of simple operation, low cost, and rapid film formation and has become one of the most commonly used methods for preparing thin film devices.

Spin coating generally includes three steps: precursor configuration, high-speed rotation, and solvent volatilization to form a film. Spin coating can be divided into static and dynamic spin coating. Static spin coating is to first dissolve the material in a solvent to form a solution, drop a certain amount of the solution in the middle of the substrate, and then rotate; dynamic spin coating is to drop the solution during the low-speed rotation of the substrate. In the high-speed rotation step, the spin coater drives the substrate to rotate at a high speed, so that the solution is spread out from the center of the substrate to the periphery under the action of centrifugal force, and a thin film is formed on it. Finally, the solvent and oxygen in the film are volatilized and removed by heating and drying on a heating table.

The thickness, related properties, and corresponding device performance of the film prepared by the spin coating method are not only affected by the material itself but also by the process parameters during the preparation process, such as spin coating time and rotation speed. The influence of process parameters on device performance is one of the factors that need special attention when preparing devices by spin coating. Although the spin coating method does not require a vacuum environment, most of the solution will be thrown off the substrate during this process, resulting in low material utilization. Moreover, this method is difficult to prepare patterned devices. In addition, due to the limitation of the process itself, the spin coating method cannot be used to manufacture large-scale color displays in industry.

6.5.2.2 Inkjet Printing

Inkjet printing has the characteristics of high material utilization and large-size and high-resolution displays. It is one of the most promising organic light-emitting display device preparation technologies for large-scale commercial production.[51] The main principle of inkjet printing is that the ink nozzle is driven by the printing signal, and ink droplets are directly ejected from the nozzle to a specific position on the surface of the substrate to realize the printing of the patterned device. In the inkjet printing process, the ink formula, the design of the print head and the printing system, the evaporation and drying process of the solvent, and the interface contact characteristics of the ink and the substrate will have a great impact on the performance of the prepared device. Among them, the ink formula is the key to determining the performance of the device, and its viscosity and surface tension also affect the printing accuracy and the resolution of the display. In addition, the volume of inkjet droplets is very small. Due to capillary action, the solvent volatilization and drying process are prone to "coffee ring" effect, which will directly affect the uniformity of the film.

6.6 CONCLUSION

Organic/polymer light-emitting materials and devices have been widely used in the field of display and lighting and have a broader development space in the field of flexible display, which will have a significant impact on the future information technology and industry. This chapter mainly introduces the organic/polymer electroluminescent materials, device structure, device physical mechanism, thin film fabrication technology, and so on. In materials, the typical structure of luminescent materials are mainly introduced, including fluorescent materials, phosphorescent materials, and all fully organic high-efficiency luminescent materials; in device structure, the multilayer structure and the function of each layer are mainly introduced; in device physics, the process of carrier injection, transportation, and recombination are mainly introduced; in film fabrication process, the vacuum evaporation method and solution processes are mainly introduced for display and lighting applications. For the realization of high-efficiency electroluminescent devices, the light out-coupling principle and method are also introduced.

REFERENCES

[1] Sooyong Lee, and Hwajeong Kim. 2021. Progress in Organic Semiconducting Materials with High Thermal Stability for Organic Light-Emitting Devices. *InfoMat* 3, no. 1: 61–81. https://doi.org/10.1002/inf2.12123.

[2] Ting-An Lin, and Tanmay Chatterjee. 2016. Sky-Blue Organic Light Emitting Diode with 37% External Quantum Efficiency Using Thermally Activated Delayed Fluorescence from Spiroacridine-Triazine Hybrid. *Advanced Materials* 28, no. 32: 6976–83. https://doi.org/10.1002/adma.201601675.

[3] Jin Won Sun, and Jeong-Hwan Lee. 2014. A Fluorescent Organic Light-Emitting Diode with 30% External Quantum Efficiency. *Advanced Materials* 26, no. 32: 5684–8. https://doi.org/10.1002/adma.201401407.

[4] Weixuan Zeng, and Hsin-Yu Lai. 2018. Achieving Nearly 30% External Quantum Efficiency for Orange-Red Organic Light Emitting Diodes by Employing Thermally Activated Delayed Fluorescence Emitters Composed of 1,8-Naphthalimide-Acridine Hybrids. *Advanced Materials* 30, no. 5: 1704961. https://doi.org/10.1002/adma.201704961.

[5] Jie Liang, and Chenglong Li. 2018. Novel Blue Bipolar Thermally Activated Delayed Fluorescence Material as Host Emitter for High-Efficiency Hybrid Warm-White OLEDs with Stable High Color-Rendering Index. *Advanced Functional Materials* 28, no. 17: 1707002. https://doi.org/10.1002/adfm.201707002.

[6] Fernando B. Dias, and Thomas J. Penfold. 2017. Photophysics of Thermally Activated Delayed Fluorescence Molecules. *Methods and Applications in Fluorescence* 5, no. 1: 12001. https://iopscience.iop.org/article/10.1088/2050-6120/aa537e/pdf.

[7] Xinxin Ban, and Aiyun Zhu. 2017. Design of Encapsulated Hosts and Guests for Highly Efficient Blue and Green Thermally Activated Delayed Fluorescence OLEDs Based on a Solution-Process. *Chemical Communications* 53, no. 86: 11834–7. https://pubs.rsc.org/en/content/articlepdf/2017/cc/c7cc06967g.

[8] Yanjie Wang, and Yunhui Zhu. 2017. Bright White Electroluminescence From a Single Polymer Containing a Thermally Activated Delayed Fluorescence Unit and a Solution-Processed Orange OLED Approaching 20% External Quantum Efficiency. *Journal of Materials Chemistry C* 5, no. 41: 10715–20. https://pubs.rsc.org/en/content/articlepdf/2017/tc/c7tc03769d.

[9] Chensen Li, and Roberto S. Nobuyasu. 2017. Solution-Processable Thermally Activated Delayed Fluorescence White OLEDs Based on Dual-Emission Polymers with Tunable Emission Colors and Aggregation-Enhanced Emission Properties. *Advanced Optical Materials* 5, no. 20: 1700435. https://doi.org/10.1002/adom.201700435.

[10] Jin-Ming Teng, and Yin-Feng Wang. 2020. Recent Progress of Narrowband TADF Emitters and their Applications in OLEDs. *Journal of Materials Chemistry C* 8, no. 33: 11340–53. https://pubs.rsc.org/no/content/articlepdf/2020/tc/d0tc02682d.

[11] Soo-Kang Kim, and Bing Yang. 2008. Exceedingly Efficient Deep-Blue Electroluminescence from New Anthracenes Obtained Using Rational Molecular Design. *Journal of Materials Chemistry* 18, no. 28: 3376. https://pubs.rsc.org/en/content/articlepdf/2008/jm/b805062g.

[12] Soo-Kang Kim, and Bing Yang. 2009. Synthesis and Electroluminescent Properties of Highly Efficient Anthracene Derivatives with Bulky Side Groups. *Organic Electronics* 10, no. 5: 822–33. https://doi.org/10.1016/j.orgel.2009.04.003.

[13] Weijun Li, and Dandan Liu. 2012. A Twisting Donor-Acceptor Molecule with an Intercrossed Excited State for Highly Efficient, Deep-Blue Electroluminescence. *Advanced Functional Materials* 22, no. 13: 2797–803. https://doi.org/10.1002/adfm.201200116.

[14] Shi Tang, and Weijun Li. 2012. Highly Efficient Deep-Blue Electroluminescence Based on the Triphenylamine-Cored and Peripheral Blue Emitters with Segregative HOMO–LUMO Characteristics. *Journal of Materials Chemistry* 22, no. 10: 4401–8. https://pubs.rsc.org/ko/content/articlepdf/2012/jm/c1jm14639d.

[15] Zhigang Shuai, and Qian Peng. 2017. Organic Light-Emitting Diodes: Theoretical Understanding of Highly Efficient Materials and Development of Computational Methodology. *National Science Review* 4, no. 2: 224–39. https://doi.org/10.1093/nsr/nww024.

[16] Qiming Peng, and Ablikim Obolda. 2015. Organic Light-Emitting Diodes Using a Neutral Π Radical as Emitter: The Emission from a Doublet. *Angewandte Chemie International Edition* 54, no. 24: 7091–5. https://doi.org/10.1002/ange.201500242.

[17] Ablikim Obolda, and Xin Ai. 2016. Up to 100% Formation Ratio of Doublet Exciton in Deep-Red Organic Light-Emitting Diodes Based on Neutral π-Radical. *ACS Applied Materials & Interfaces* 8, no. 51: 35472–8. https://pubs.acs.org/doi/pdf/10.1021/acsami.6b12338.

[18] J. H. Burroughes, and D. D. C. Bradley. 1990. Light-Emitting Diodes Based On Conjugated Polymers. *Nature* 347, no. 11: 539–41. https://www.nature.com/articles/347539a0.

[19] M. Fukuda, and K. Sawada. 1989. Fusible Conducting Poly(9-alkylfluorene) and Poly(9,9-Dialkylfluorene) and their Characteristics. *Japanese Journal of Applied Physics* 28, no. 8: 1433–5. https://iopscience.iop.org/article/10.1143/JJAP.28.L1433/pdf.

[20] Yuanyuan Li, and Hongbin Wu. 2009. Enhancement of Spectral Stability and Efficiency on Blue Light-Emitters via Introducing Dibenzothiophene-S,S-Dioxide Isomers into Polyfluorene Backbone. *Organic Electronics* 10, no. 5: 901–9. https://doi.org/10.1016/j.orgel.2009.04.021.

[21] Yangjun Xia, and Jie Luo. 2006. Novel Random Low-Band-Gap Fluorene-Based Copolymers for Deep Red/Near Infrared Light-Emitting Diodes and Bulk Heterojunction Photovoltaic Cells. *Macromolecular Chemistry and Physics* 207, no. 5: 511–20. https://doi.org/10.1002/macp.200500517.

[22] D. Braun, and A. J. Heeger. 1991. Improved Efficiency in Semiconducting Polymer Light-Emitting Diodes. *Journal of Electronic Materials* 20, no. 11: 945–8. https://doi.org/10.1007/BF02816037.

[23] R. Jakubiak, and L. J. Rothberg. 1999. Reduction of Photoluminescence Quantum Yield by Interchain Interactions in Conjugated Polymer Films. *Synthetic Metals* 101, no. 1: 230–3. https://doi.org/10.1016/S0379-6779(98)00215-X.

[24] S. Hoger, and J. J. McNamara. 1994. Novel Silicon-Substituted, Soluble Poly(Phenylenevinylene)s: Enlargement of the Semiconductor Bandgap. *Chemistry of Materials* 6, no. 2: 171–3. https://doi.org/10.1021/cm00038a012.

[25] H. H Horhold, and M. Helbig. 1987. Poly(Phenylenevinylene)s-Synthesis and Redoxchemistry of Electroactive Polymers. *Makromolekulare Chemie Macromolecular Symposia* 12, no. 1: 229–58. https://doi.org/10.1002/masy.19870120112.

[26] Junji Kido, and Kenichi Hongawa. 1993. Bright Blue Electroluminescence From poly(N-Vinylcarbazole). *Applied Physics Letters* 63, no. 19: 2627–9. https://doi.org/10.1063/1.110402.

[27] Alexander Kukhta, and Eduard Kolesnik. 2006. Spectral and Luminescent Properties and Electroluminescence of Polyvinylcarbazole with 1,8-Naphthalimide in the Side Chain. *Journal of Fluorescence* 16, no. 3: 375–8. http://dx.doi.org/10.1007/s10895-005-0064-6.

[28] T. Yamamoto, and K. Sanechika. 1980. Preparation of Thermostable and Electric-Conducting Poly(2,5-Thienylene). *Journal of Polymer Science: Polymer Letters Edition* 18, no. 1: 9–12. https://doi.org/10.1002/pol.1980.130180103.

[29] R. E. Gill, and G. G. Malliaras. 1994. Tuning of Photo- and Electroluminescence in Alkylated Polythiophenes with Well-Defined Regioregularity. *Advanced Materials* 6, no. 2: 132–5. https://doi.org/10.1002/adma.19940060206.

[30] Jian Pei, and Wang-Lin Yu. 2000. A Novel Series of Efficient Thiophene-Based Light-Emitting Conjugated Polymers and Application in Polymer Light-Emitting Diodes. *Macromolecules* 33, no. 7: 2462–71.

[31] X. Gong, and S. Wang. 2005. Multilayer Polymer Light-Emitting Diodes: White-Light Emission with High Efficiency. *Advanced Materials* 17, no. 17: 2053–8. https://doi.org/10.1002/adma.200590088.

[32] Wen-Yi Hung, and Chi-Yen Lin. 2012. A New Thermally Crosslinkable Hole Injection Material for OLEDs. *Organic Electronics* 13, no. 11: 2508–15. https://doi.org/10.1016/j.orgel.2012.06.023.

[33] Chandramouli Kulshreshtha, and Woo Sik Jeon. 2011. High Molecular Weight PVK as an Interlayer in Green Phosphorescent OLEDs. *Molecular Crystals and Liquid Crystals* 550, no. 1: 225–32. https://doi.org/10.1080/15421406.2011.599753.

[34] I. H. Campbell, and T. W. Hagler. 1996. Direct Measurement of Conjugated Polymer Electronic Excitation Energies Using Metal/Polymer/Metal Structures. *Physical Review Letters* 76, no. 11: 1900–3. https://doi.org/10.1103/PhysRevLett.76.1900.

[35] G. G. Malliaras, and J. C. Scott. 1999. Numerical Simulations of the Electrical Characteristics and the Efficiencies of Single-Layer Organic Light Emitting Diodes. *Journal of Applied Physics* 85, no. 10: 7426–32. https://doi.org/10.1063/1.369373.

[36] Wei-Qiao Deng, and William A. Goddard. 2004. Predictions of Hole Mobilities in Oligoacene Organic Semiconductors from Quantum Mechanical Calculations†. *The Journal of Physical Chemistry B* 108, no. 25: 8614–21. https://doi.org/10.1021/jp0495848.

[37] Ling Tang, and MengQiu Long. 2009. The Role of Acoustic Phonon Scattering in Charge Transport in Organic Semiconductors: A First-Principles Deformation-Potential Study. *Science in China Series B: Chemistry* 52, no. 10: 1646–52. https://link.springer.com/content/pdf/10.1007/s11426-009-0244-3.pdf.

[38] Rudolph A. Marcus. 1997. Electron Transfer Reactions in Chemistry Theory and Experiment. *Journal of Electroanalytical Chemistry* 438, no. 1: 251–9. https://doi.org/10.1016/S0022-0728(97)00091-0.

[39] Guangjun Nan, and Xiaodi Yang. 2009. Nuclear Tunneling Effects of Charge Transport in Rubrene, Tetracene, and Pentacene. *Physical Review B, Condensed Matter and Materials Physics* 79, no. 11: 115203. https://doi.org/10.1103/PhysRevB.79.115203.

[40] Jiun-Haw Lee, and Meng-Hsiu Wu. 2005. High Efficiency and Long Lifetime OLED Based on a Metal-Doped Electron Transport Layer. *Chemical Physics Letters* 416, no. 4: 234–7. https://doi.org/10.1016/j.cplett.2005.09.104.

[41] Xinyi Cai, and Shi-Jian Su. 2018. Marching Toward Highly Efficient, Pure-Blue, and Stable Thermally Activated Delayed Fluorescent Organic Light-Emitting Diodes. *Advanced Functional Materials* 28, no. 43: 1802558. https://doi.org/10.1002/adfm.201802558.

[42] Yunfei Li, and Yuying Hao. 2013. A Single-Heterojunction Electrophosphorescence Device with High Efficiency, Long Lifetime and Suppressive Roll-Off. *Synthetic Metals* 164: 12–6. https://doi.org/10.1016/j.synthmet.2012.11.024.

[43] Yu-Cheng Chen, and Ying-Chien Fang. 2012. The Investigation of Two Different Types of Multiple-Quantum-Well Structure on Fluorescent White Organic Light Emitting Devices. *ECS Journal of Solid State Science and Technology* 1, no. 2: R66–71. https://doi.org/10.1149/2.009202jss.

[44] Thanh Ba Nguyen, and Hajime Nakanotani. 2020. The Role of Reverse Intersystem Crossing Using a TADF-Type Acceptor Molecule on the Device Stability of Exciplex-Based Organic Light-Emitting Diodes. *Advanced Materials* 32, no. 9: 1906614. https://doi.org/10.1002/adma.201906614.

[45] Peng Xiao, and Junhua Huang. 2018. Recent Advances of Exciplex-Based White Organic Light-Emitting Diodes. *Applied Sciences* 8, no. 9: 1449. https://doi.org/10.3390/app8091449.

[46] Ziqi Wang, and Chao Wang. 2019. The Application of Charge Transfer Host Based Exciplex and Thermally Activated Delayed Fluorescence Materials in Organic Light-Emitting Diodes. *Organic Electronics* 66: 227–41. https://doi.org/10.1016/j.orgel.2018.12.039.

[47] Amin Salehi, and Xiangyu Fu. 2019. Recent Advances in OLED Optical Design. *Advanced Functional Materials* 29, no. 15: 1808803. https://doi.org/10.1002/adfm.201808803.

[48] Kanchan Saxena, and V. K. Jain. 2009. A Review on the Light Extraction Techniques in Organic Electroluminescent Devices. *Optical Materials* 32, no. 1: 221–33. https://doi.org/10.1016/j.optmat.2009.07.014.

[49] Tae-Wook Koh, and Jung-Min Choi. 2010. Optical Outcoupling Enhancement in Organic Light-Emitting Diodes: Highly Conductive Polymer as a Low-Index Layer on Microstructured ITO Electrodes. *Advanced Materials* 22, no. 16: 1849–53. https://doi.org/10.1002/adma.200903375.

[50] Benjamin Thomas Mogg, and Tim Claypole. 2016. Flexographic Printing of Ultra-Thin Semiconductor Polymer Layers. *Translational Materials Research* 3, no. 1: 15001. http://cronfa.swan.ac.uk/Record/cronfa26114.

[51] Charles A. Annis. 2019. Roadmapping Strategies for Rapidly Diversifying FPD Applications and Manufacturing Technologies. *SID International Symposium Digest of Technical Papers* 50, no. 1: 762–4. https://doi.org/10.1002/sdtp.13032.

Index

Note: Locators in *italics* represent figures and **bold** indicate tables in the text.

Printed in the United States
by Baker & Taylor Publisher Services

Printed in the United States
by Baker & Taylor Publisher Services